ESE 2024

Main Examination

UPSC ENGINEERING SERVICES EXAMINATION

Topicwise

Conventional Practice Questions

Civil Engineering

PAPER-II

MADE EASY
PUBLICATIONS

MADE EASY Publications Pvt. Ltd.

Corporate Office: 44-A/4, Kalu Sarai (Near Hauz Khas Metro Station), New Delhi-110016
E-mail: infomep@madeeasy.in
Contact: 9021300500
Visit us at: www.madeeasypublications.org

ESE Main Examination • Conventional Practice Questions :
Civil Engineering PAPER-II

First Edition: 2023

ESE 2024 Main Examination

Conventional Practice Questions

Civil Engineering

PAPER-II

CONTENTS

01 Fluid Mechanics, Open Channel Flow & Hydraulic Machines

1. Fluid Properties

Level-1

1.1 A 6 cm diameter soap bubble is to be enlarge by blowing air into it. Taking the surface tension of soap solution to be 0.039 N/m. Determine the work input required to inflate the bubble diameter to 6.9 cm.

[5 Marks]

Sol:

$$\text{Initial surface area} = 4\pi r_1^2$$

$$\text{Final surface area} = 4\pi r_2^2$$

$$\text{Change in surface area } (\Delta A) = 4\pi\left(r_2^2 - r_1^2\right)$$

$$\text{Work required} = \text{Change in surface energy}$$

$$= T \cdot \Delta A$$

$$W = \frac{0.039 \times 4\pi}{10^4}\left[\left(\frac{6.9}{2}\right)^2 - \left(\frac{6}{2}\right)^2\right] = 1.42 \times 10^{-4} \text{ N-m}$$

1.2 A Glass tube of 2 mm internal diameter is immersed in an oil of mass density 960 kg/m^3 to a depth of 10 mm. If a pressure of 172 N/m^2 is needed to form a bubble which is just released. Determine the surface tension of oil.

[10 Marks]

Sol:

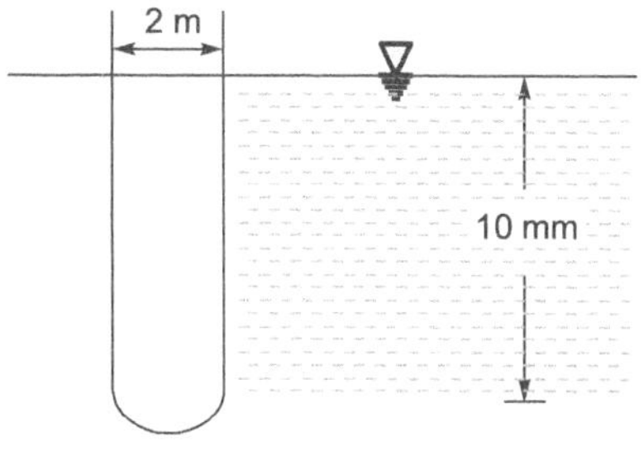

$$\text{External pressure on bubble, } \rho gh = 960 \times \frac{9.81 \times 10}{1000} = 94.176 \text{ N/m}^2$$

$$\text{Inside pressure of bubble} = 172 \text{ N/m}^2$$

$$\text{Difference in pressure, } \Delta P = 172 - 94.176 = 77.824 \text{ N/m}^2$$

$$\Delta P \text{ of bubble} = \frac{2T}{R} \quad \text{(Only one contact layer)}$$

$$77.824 = \frac{2 \times T}{1 \times 10^{-3}}$$

Solving, $T = 0.0389$ N/m

1.3 As the nutrients are carried to upper parts of a plant by tiny tubes due to action of capillarity. Determine how high will the water solution will rise in a tree of 0.002 mm diameter tube as a result of the capillary effect.

Assume the water solution at 20°C

Contact angle = 15°

[5 Marks]

Sol:

The rise in a capillary is given by

$$h = \frac{2T\cos\theta}{\rho r g}$$

where T = Surface tension = 0.073 N/m (at 20°C)

θ = Contract angle = 15°C (given)

r = Radius of capillary tube = $\frac{0.002}{2}$ = 0.001 mm

Putting, $$h = \frac{2\times 0.073\times \cos 15^\circ}{10^3\times 0.001\times 10^{-3}\times 9.81}$$

h = 14.38 m (Ans)

1.4 A spherical shell made of a material with a density of 1600 kg/m^3 is placed in water. If the inner and outer radii of shell are R_1 = 6 cm, R_2 = 8 cm determine the percentage of the shell's total volume that would be submerged.

[10 Marks]

Sol:

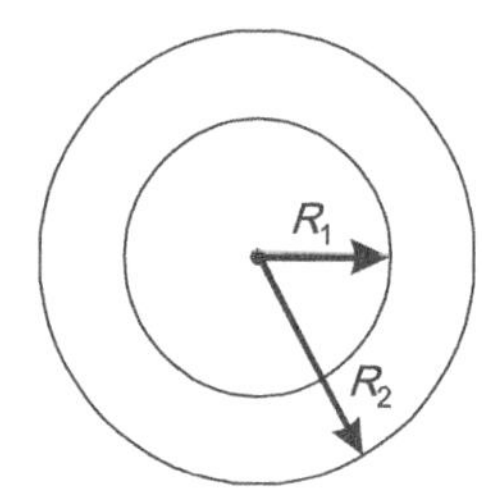

$$\text{Volume of shell} = \frac{4}{3}\pi\left(R_2^3 - R_1^3\right)$$

Given R_1 = 6 cm, R_2 = 8 cm

$$\text{Total weight of shell} = 1800\times g\times\frac{4}{3}\pi(0.08^3 - 0.06^3)$$

Also, the Buoyant force = Weight or displaced fluid

= Volume displaced × g × 1000

Now, equating,

$$1600\times g\times\frac{4}{3}\pi\left(0.08^3 - 0.06^3\right) = \text{Volume displaced}\times g\times 1000$$

$$\text{Volume displaced} = 1982.80\times 10^{-6}\text{ m}^3 = 1982.8\text{ cm}^3$$

$$\%\text{ submerged} = \frac{1982.8\text{ cm}^3}{\frac{4}{3}\pi\times(8)^3}\times 10 = 92.5\%$$

Level-2

1.5 Two immiscible Newtonian liquids flow steadily between two large parallel plates under the influence of an applied pressure gradient. The lower plate is fixed while the upper plate is pulled with a constant velocity of U = 10 m/sec. Thickness, h of each layer on fluid is 0.5 m. The velocity profile of each layer is given by

$$V_1 = 6 + ay - 3y^2 \qquad -0.5 \le y \le 0$$
$$V_2 = b + cy - 9y^2 \qquad 0 \le y \le +0.5$$

Where a, b and c are constant

(a) Determine the values of constants a, b and c.

(b) Develop an expression for viscosity ratio (μ_1/μ_2) = ?

(c) Determine the forces and their directions exerted by liquids on both plates if μ_1 = 10^{-3} Pa sec and each plate has a surface area of 4 m^2.

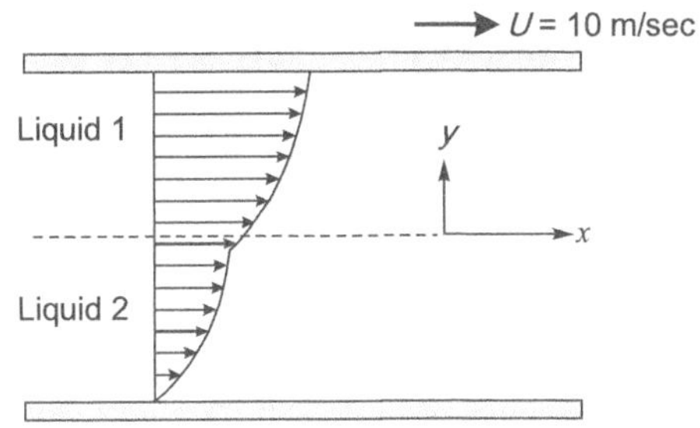

[15 Marks]

Sol:

(a)

For $-0.5 \le v \le 0$

$$V = 6 + ay - 3y^2$$

at $y = -0.5 \quad \Rightarrow y = 0$

$$6 + a(-0.5) - 3(-0.5)^2 = 0$$

Solving $\quad a = 10.5$

For $0 \le y \le +0.5$

$\Rightarrow \quad V = b + cy - 9y^2$

at $y = 0.5$ m, $v = 10$ m/sec (No slip condition)

$$y = 0.5 \text{ m}$$

$$b + c(0.5) - 9(0.5)^2 = 10$$

$$b + 0.5c = 12.25$$

Also at interface, $\quad V_1 = V_2$

$$6 + 10.5(0) - 3(0) = b + c(0) - 9(0)^2$$

$$b = 6$$

$$c = 12.5$$

(b)

$$V_1 = 6 + 10.5y - 3y^2$$

$$V_2 = 6 + 125y - 9y^2$$

Now at interface green stress is equal

$$\tau_1 = \tau_2$$

$$\mu_2 \frac{\partial V_1}{\partial y} = \mu_1 \frac{\partial V_2}{\partial y}$$

$$\frac{\mu_1}{\mu_2} = \left[\frac{10.5 - 6y}{12.5 - 18y}\right]_{at\, y=0} = \frac{10.5}{12.5} = 0.84$$

(c) Force on bottom plate $= \tau_0 \times$ Area

$$\tau_0 = \mu_2 \left|\frac{\partial V_1}{\partial y}\right|_{at\, y=-0.5\text{ m}} = \frac{10^{-3}}{0.84}[10.5 - 6y]_{y=-0.5\text{m}} = 16.07 \times 10^{-3} \text{ N/m}^2$$

$$\text{Force} = 16.07 \times 10^{-3} \times 4 = 64.28 \times 10^{-3} \text{ N}$$

On top plate,

$$\tau_0 = \mu_1 \left|\frac{\partial V_2}{\partial y}\right|_{at\, y=+0.5\text{ m}}$$

$$= 10^{-3}\,[12.5 - 18y]_{y=0.5\text{ m}}$$

$$= 10^{-3}\,[12.5 - 18 \times 0.5] = 3.5 \times 10^{-3} \text{ N/m}^2$$

Force $= 14 \times 10^{-3}$ N **(Ans)**

2. Fluid Pressure Measurements, Buoyancy and Rigid Body Motion

Level-1

2.1 A 10g of wood of square section 0.36 m × 0.36 m and specific gravity 0.8 floats in water one edge is depressed and released causing the dog to roll. Estimate the period of the roll.

[10 Marks]

Sol:

Since time period is given by,

$$T = 2\pi\sqrt{\frac{K^2}{g \times GM}}$$

where, GM = Metacentric height = $\frac{I}{V_{Dis}} - B_G$

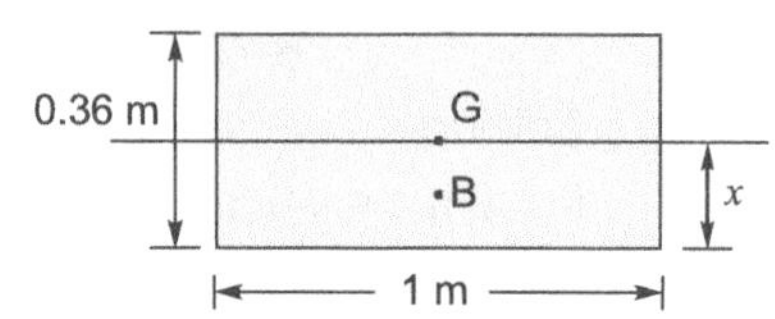

Considering 1 m length of log

For vertical equilibrium

$$0.8 \times \gamma_w \times 0.36 \times 0.36 \times 1 = \gamma_w \times 0.36\,(x) \times 1$$

Solving, $x = 0.288$ m

$$BG = \frac{0.36}{2} - \frac{0.288}{2} = 0.036 \text{ m}$$

$$I = \frac{1 \times 0.36^3}{12} \qquad V_{Dis} = 0.36 \times 1 \times 0.288$$

$$\text{Metacentric height, } GM = \frac{1 \times 0.36^3}{12 \times 0.36 \times 1 \times 0.288} - 0.36 = 1.5 \times 10^{-3}$$

$$k = \text{Radius of gyration} = \sqrt{\frac{I}{A}} = \sqrt{\frac{1 \times 0.36^3}{12 \times 1 \times 0.36}} = 0.104 \text{ m}$$

$$\text{Putting, } T = 2\pi\sqrt{\frac{0.104^2}{9.81 \times 1.5 \times 10^{-3}}} = 5.384 \text{ sec}$$

2.2 Water rises to level 4 m in the pipe attached to the tank ABCD shown in figure.

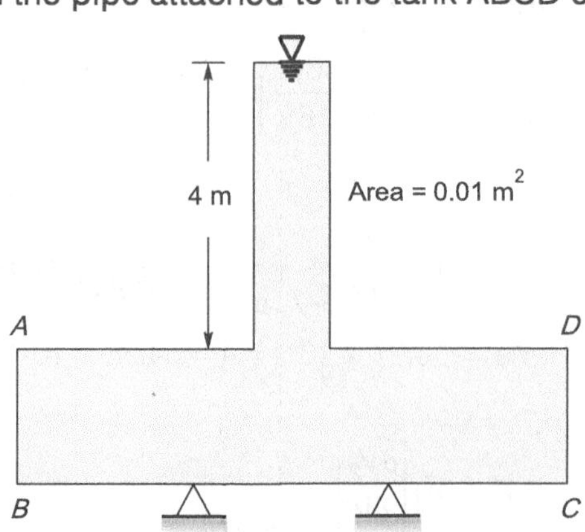

(i) Compute the total pressure on the bottom of the tank having area 6.5 m × 2.5 m.

(ii) Compare the total weight or water with the result in (ii).

[8 Marks]

Sol:

(i) Pressure intensity on bottom BC is uniform and is equal to

$$= 9.81 \times (4 + 2) = 58.86 \text{ kN/m}^2$$

$$\text{Total force acting} = 58.86 \times (6.5 \times 2.5)$$

$$= 956.475 \text{ kN}$$

(ii)

$$\text{Total weight of water} = 9.81\,(6.5 \times 2.5 \times 2 + 4 \times 0.01)$$

$$= 319.217 \text{ kN}$$

Thus the total weight of water is much less than force due to total pressure, this is known as Pascal's paradox.

2.3 **Two chambers with the same fluid at their base are separated by a 30 cm diameter miston whose weight is 25 N, as shown in figure. Calculate the gauge pressure in chambers A and B.**

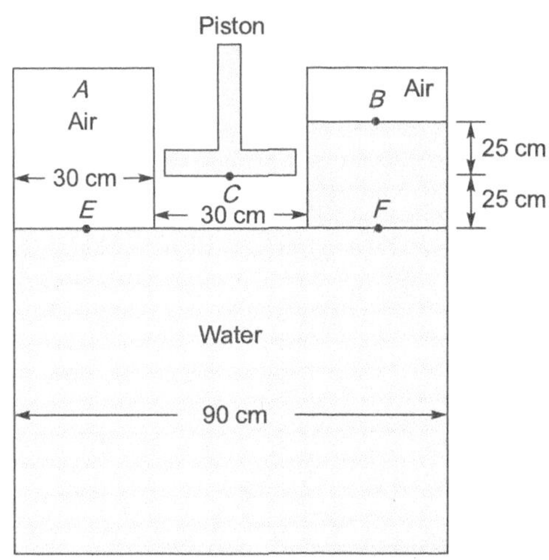

[10 Marks]

Sol:

For left limb /Arm :

Equating pressure at the level of E

$$P_A = \frac{25\,N}{\frac{\pi}{4} \times \left(\frac{30}{100}\right)^2} + \gamma_w \times 0.25$$

Solving, $P_A = 353.86 \text{ N/m}^2 + 9810 \times 0.25$

$$P_A = 2806.36 \text{ N/m}^2$$

For right arm

Equating pressure at level of C

$$P_B + \gamma_w \times 0.25 = \frac{25}{\frac{\pi}{4} \times (0.3)^2}$$

Solving, $P_B = 353.86 \text{ N/m}^2 - 9810 \times 0.25 = -2098.64 \text{ N/m}^2$ (Vacuum)

2.4 **An inclined manometer gives a reading of 5 cm along the tube inclined at 15° to the horizontal. The measuring liquid has a specific gravity of 1.45. Find the pressure applied on the basin side taking into account the correction applied due to the change in level of the liquid in the basin which has a diameter of 5 cm and that of inclined tube 5 mm.**

[6 Marks]

Sol:

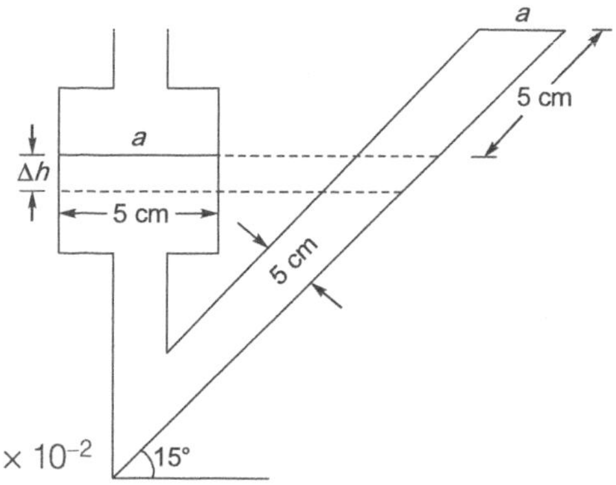

Effective vertical reading = 5 sin 15° + Δh

By conservation of volume

$$\frac{\pi}{4} \times \Delta h \times (50)^2 = \frac{\pi}{4} \times 5^2 \times 5$$

Solving Δh = 0.05 cm

Effective reading = 5 sin 15° + 0.05

= 1.344 cm

Hence, Pressure applied, $P = \rho_m g h_{eff}$

$P = 1.45 \times 10^3 \times 9.81 \times 1.344 \times 10^{-2}$

$= 191.17 \text{ N/m}^2$ **(Ans)**

2.5 **The flow of water from a reservoir is controlled by a 2 m wide L-shape gate hinged at point *A*, as shown in figure. If it is desired that the gate open when the water height is 3.3 m. Determine the mass of the required weight *W*.**

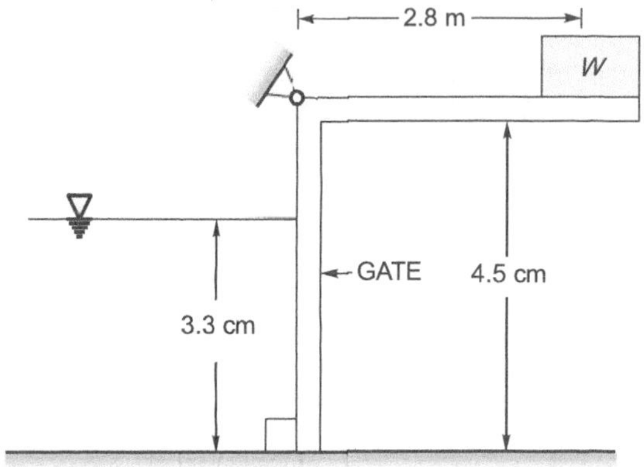

[10 Marks]

Sol:

Considering hydrostatic distribution of pressure,

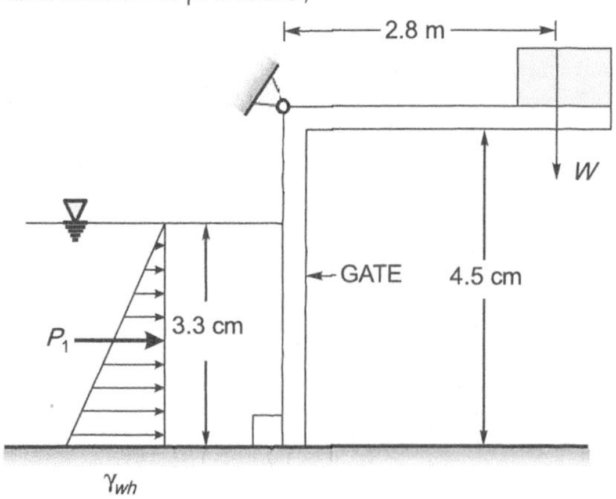

$$P_1 = \frac{1}{2} \times \gamma_w \times h \times h \times B = \frac{1}{2} \times 9.81 \times 3.3 \times 3.3 \times 2 = 106.83 \text{ kN}$$

$$\text{Location of } P_1 = \frac{3.3}{3} = 1.1 \text{ m} \quad \text{(from bottom)}$$

Distance from hinge = 4.5 – 1.1 = 3.4 m

Now, equating the moment, about hinge

$$\Rightarrow \qquad \Sigma M_{hinge} = 0$$

$$106.83 \times 3.4 = W \times 2.8$$

Solving,
$$W = 129.72 \text{ kN}$$

$$m = \frac{129.72 \times 10^3}{9.81}$$

$$m = 13223.6 \text{ kg}$$

2.6 A metallic cube 30 cm side and weighing 500 N is lowered into a tank containing a two fluid layer of water and mercury. Top edge of the cube is at water surface. Determine the position of block at water-mercury interface when it has attained equilibrium.

[12 marks]

Sol:

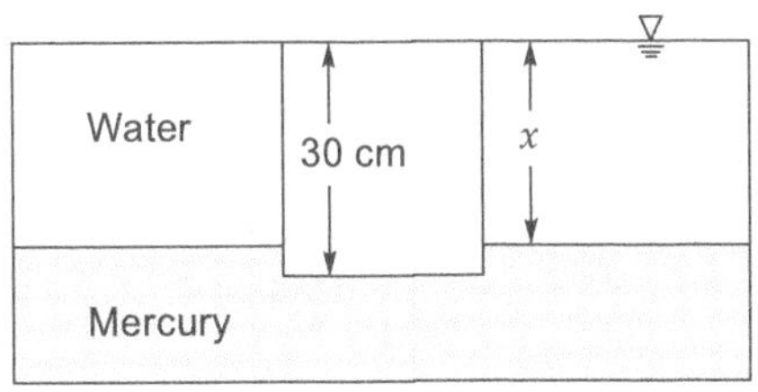

Let the top edge of the cube be at a distance x from the water-mercury interface.

Thus, the force of buoyancy is given by

F_B = Volume of cube in water × $\rho_w g$ + Volume of cube in mercury × ρ_{Hg} × g

$$F_B = \left(\frac{30}{100} \times \frac{30}{100} \times x\right) \times 1000 \times 9.81 + \frac{30}{100} \times \frac{30}{100} \times \left(\frac{30}{100} - x\right) \times 13600 \times 9.81$$

$$\Rightarrow \qquad F_B = 882.9x + 12007.44\,(0.3 - x)$$

Given, Weight of cube, $W = 500$ N

$$\therefore \qquad W = F_B$$

$$\Rightarrow \qquad 500 = 882.9x + 12007.44\,(0.3 - x)$$

$$\Rightarrow \qquad 500 = 882.9x + 3602.232 - 12007.44\,x$$

$$\Rightarrow \qquad 11124.54x = 3102.232$$

$$\Rightarrow \qquad x = 0.2789 \text{ m}$$

$$= 27.89 \text{ cm} \qquad \text{[From the top edge of the block]}$$

2.7 A square plate of side 1 m is immersed in water with it's diagonal vertical. The upper corner is 0.5 m below the free surface, as shown in figure below. Find the hydrostatic force on the plate and depth of centre of pressure from free surface of water.

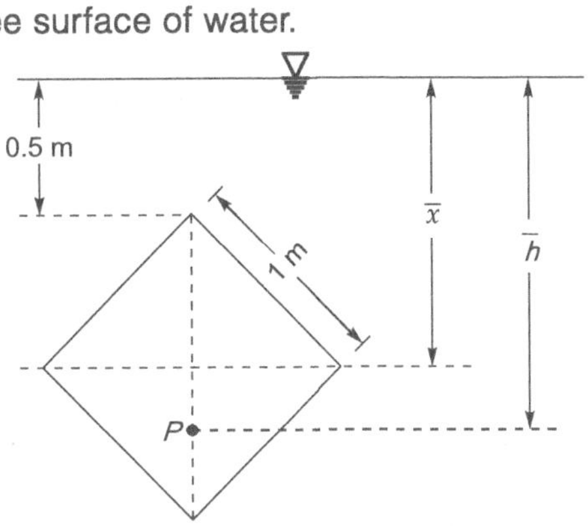

[12 marks]

Sol:

Given side of square plate, $L = 1$ m

$\therefore$ Area, $A = L^2 = 1$ m^2

$$\bar{x} = \frac{L}{\sqrt{2}} + 0.5 = 1.207 \text{ m}$$

Hydrostatic force on plate, $F_x = \rho g A \bar{x}$

$$= 1000 \times 9.81 \times 1 \times 1.207 = 11840.67 \text{ N}$$

Depth of centre of pressure, $\bar{h} = \bar{x} + \frac{I_{CG}\sin^2\theta}{A\bar{x}}$

where, $\theta = 90°$

and $I_{CG} = \frac{L^4}{12} = \frac{1}{12} = 0.083 \text{ m}^4$

$\therefore$ $\bar{h} = 1.207 + \frac{0.083}{1 \times 1.207} = 1.276$ m from free surface of water

Level-2

2.8 **An open tank of oil 5 m long contains 2 m of oil (specific gravity 0.8). If the tank accelerated upto 30° include planes at 3.6 m/sec², what is the angle of the oil surface makes with the horizontal? Also find the pressure intensities at the bottom of the vessel at the front and the rear ends? [Assume container horizontal]**

[15 Marks]

Sol:

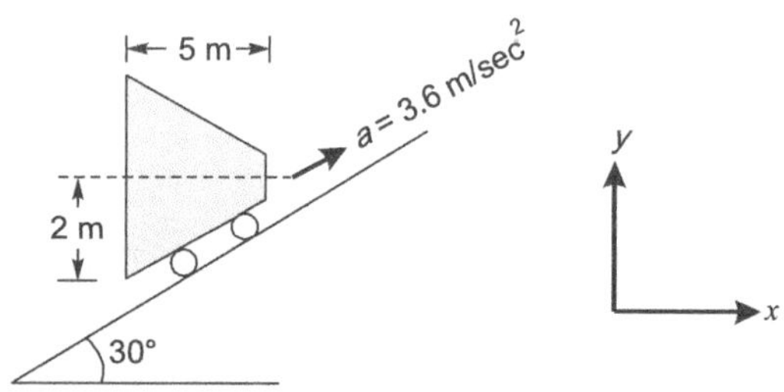

Acceleration in horizontal direction,

$$a_x = 3.6 \cos 30° = 3.12 \text{ m/sec}^2$$

Acceleration in vertical direction,

$$a_y = 3.6 \sin 30° = 1.8 \text{ m/sec}^2$$

Now using, $\tan\theta = \frac{a_x}{a_y + g} = \frac{3.12}{1.8 + 9.81} = 0.2687 \text{ m/sec}^2$

$$\theta = 15°2' \text{ with the horizontal}$$

Depth at the rear (deep) end is

$$h_1 = 2 + \left(\frac{5}{2}\tan 15°2'\right) = 2.67 \text{ m}$$

Pressure intensity at the bottom of tank at rear end is

$$P_B = \rho(a_v + g)h_1 = 1000\,(1.8 \times 9.81) \times 0.8 \times 2.67 = 24.8 \text{ kN/m}^2$$

Similarly the depth of oil in front (shallow) end is

$$h_2 = 2 - \left(\frac{5}{2}\tan 15°2'\right) = 1.33 \text{ m}$$

Pressure intensity at the bottom of the tank of the front end is

$$P_D = \rho(a_v + g)h_2 = 800\,(1.8 + 9.81) \times 1.33 = 12.35 \text{ kN/m}^2$$

2.9 The cylindrical tank in figure has a hemispherical end cap ABC. Compute the total horizontal and vertical forces exerted on ABC by oil and water.

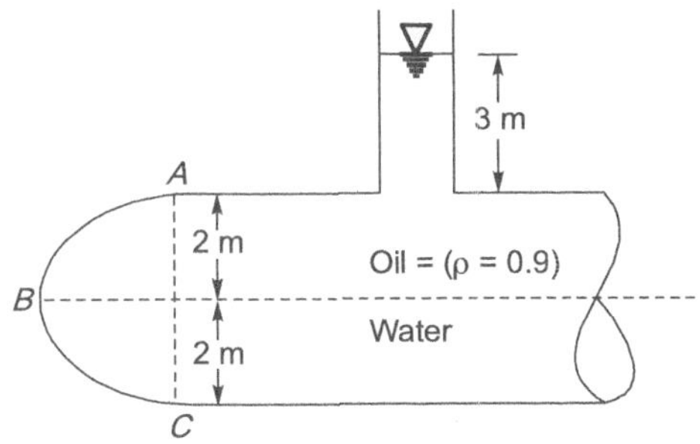

[15 Marks]

Sol:

$$\text{For horizontal force} = F_{H_1} + F_{H_2}$$

$$F_{H_1} = \text{Force on } AB = \rho g\bar{h}A_V$$

$$\bar{h} = \text{Location of C.G.} = \left(3 + 2 - \frac{4r}{3\pi}\right)$$

$$= \left(5 - \frac{4\times 2}{3\pi}\right) = 4.15 \text{ m} \quad \text{(From top)}$$

$$AV = \frac{1}{2}\times(\pi r^2) \quad \text{(Hemispheral leads to produce semi-circle)}$$

$$= \frac{1}{2}\times\left(\pi\times 2^2\right) = 6.28 \text{ m}^2$$

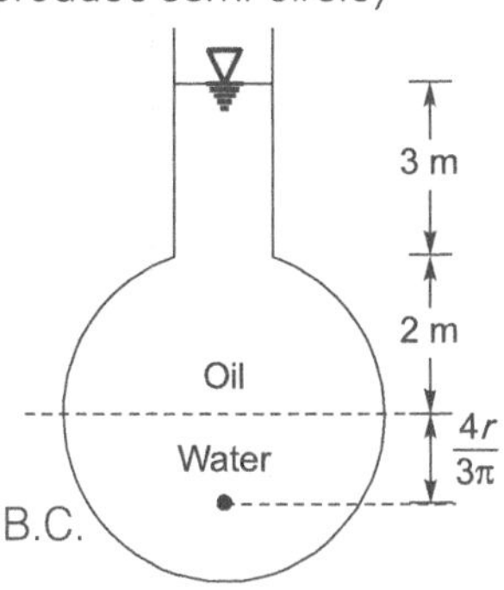

Putting, $F_{H_1} = (0.9\times 9.81)\times 4.15\times 6.28$

$$= 230.1 \text{ kN} (\leftarrow)$$

$$F_{H_2} = \text{Horizontal force on } BC$$

$$= \left(\rho g\bar{h}\right)\times A_V$$

$$= \text{(Pressure of C.G.)} \times \text{Vertical projection of B.C.}$$

$$= \left(0.9\times 9.81\times 5 + 1\times 9.81\times\frac{4r}{3\pi}\right)\times 6.28$$

$$= 329.69 \text{ kN} (\leftarrow)$$

$$\text{Total horizontal force} = 230.1 + 329.69 = 559.79 \text{ kN}$$

For vertical force

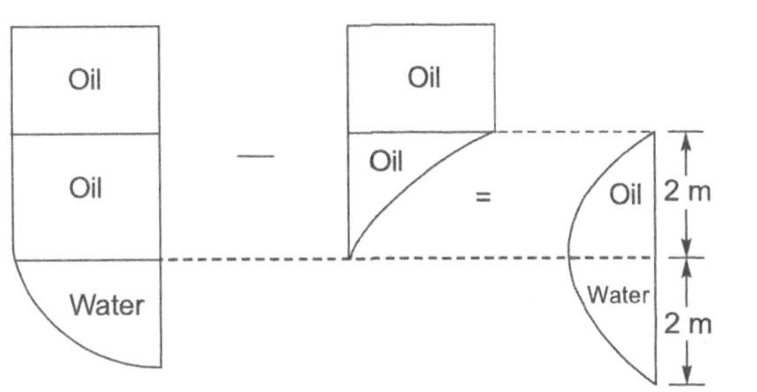

$F_{V_1}(\downarrow)$ $F_{V_2}(\uparrow)$ Net force = F_V

$$\text{Net vertical force, } F_V = \frac{1}{4}\left(\frac{4}{3}\pi r^3\right)\left(\gamma_w + 0.9\gamma_w\right) = \frac{1}{3}\times\pi\times 2^3\times\gamma_w\times 1.9 = 156.07 \text{ kN}$$

2.10 Figure below shows a conical vessel having its outlet at A to which a U-tube manometer is connected. The reading of the manometer as given in the figure shows when the vessel is empty. Find the reading of the manometer when the vessel is completely filled with water.

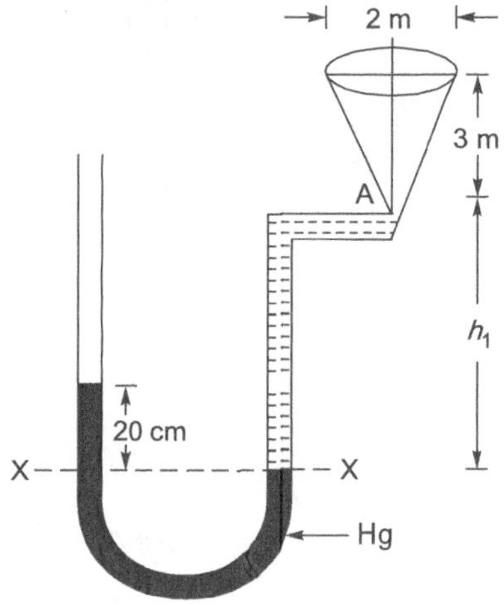

[20 marks]

Sol:

Vessel is empty.

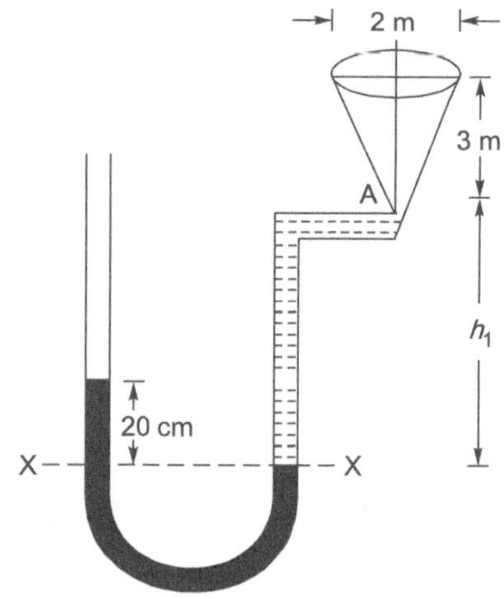

Difference of mercury level

h_2 = 20 cm

Let h_1 = Height of water above X-X

Sp. gr. of mercury, $S_2 = 13.6$

Sp. gr. of water, $S_1 = 1.0$

Density of mercury, $\rho_2 = 13.6 \times 1000$ kg/m^3

Density of water, $\rho_1 = 1000$ kg/m^3

Equating the pressure above datum line X-X, we have

$$\rho_2 \times g \times h_2 = \rho_1 \times g \times h_1$$

$$\Rightarrow \quad 13.6 \times 1000 \times 9.81 \times 0.2 = 1000 \times 9.81 \times h_1$$

$$\Rightarrow \quad h_1 = 2.72 \text{ m of water}$$

Vessel is full of water: When vessel is full of water, the pressure in the right limb will increase and mercury level in the right limb will go down. Let the distance through which mercury goes down in the right limb be, y cm as shown in figure. The mercury will rise in the left by a distance of y cm. Now the datum line is Z-Z. Equating the pressure above the datum line Z-Z.

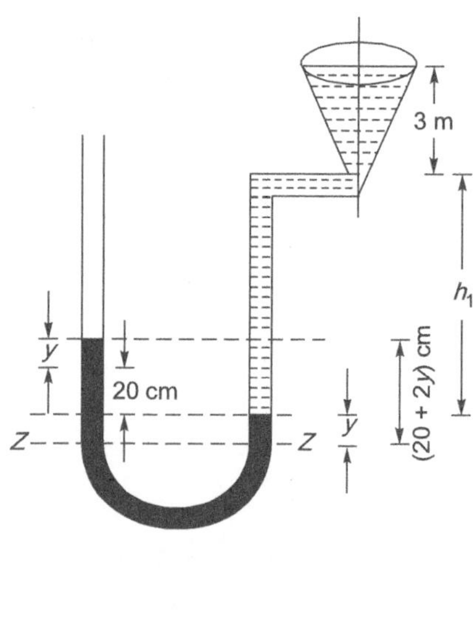

Pressure in left limb = Pressure in right limb

$$\Rightarrow 13.6 \times 1000 \times 9.81 \times \left(\frac{20+2y}{100}\right) = 1000 \times 9.81 \times \left(3 + h_1 + \frac{y}{100}\right)$$

$$\Rightarrow \quad 13.6 \times \left(\frac{20+2y}{100}\right) = \left(3 + 2.72 + \frac{y}{100}\right)$$

$$\Rightarrow \quad 272 + 27.2y = 5.72 \times 100 + y$$

$$\Rightarrow \quad 26.2y = 300$$

$$\Rightarrow \quad y = 11.45 \text{ cm}$$

The difference of mercury levels in two limbs

$$= (20 + 2y) \text{ cm of mercury}$$

$$= 20 + 2 \times 11.45 = 20 + 22.90 = 42.90 \text{ cm of mercury}$$

$$\therefore \quad \text{Reading of manometer} = 42.90 \text{ cm}$$

3. Fluid Kinematics

Level-1

3.1 If $\phi = \frac{-A}{2\pi}\log r$, where A is a positive constant, determine Y and plot the typical equipotential lines and stream-lines. Identify the flow pattern.

[12 Marks]

Sol:

$$\phi = -\frac{A}{2\pi}\log r$$

Above velocity potential function shows radial flow like source or sink flow.
For such flows, the radial and tangential components in terms of ϕ are given by

$$u_r = \frac{\partial \phi}{\partial r} \text{ and } u_\theta = \frac{1}{r}\frac{\partial \phi}{\partial \theta}$$

$$\therefore \quad u_r = \frac{\partial}{\partial r}\left[-\frac{A}{2\pi}\log r\right]$$

$$\Rightarrow \quad u_r = -\frac{A}{2\pi}\cdot\frac{1}{r} \quad \text{...(i)}$$

Also

$$u_\theta = \frac{1}{r}\frac{\partial}{\partial \theta}\left[-\frac{A}{2\pi}\log r\right] = 0$$

Also the radial and tangential component in terms of ψ are given by

$$u_r = \frac{1}{r}\frac{\partial \psi}{\partial \theta} \text{ and } u_\theta = -\frac{\partial \psi}{\partial r}$$

$$\therefore \quad \frac{1}{r}\frac{\partial \psi}{\partial \theta} = -\frac{A}{2\pi r} \quad \text{[From (i)]}$$

$$\Rightarrow \quad \frac{\partial \psi}{\partial \theta} = -\frac{A}{2\pi}$$

$$\Rightarrow \quad \partial \psi = -\frac{A}{2\pi}d\theta$$

$$\Rightarrow \quad \psi = -\frac{A}{2\pi}\theta + C_1$$

Let $\psi = 0°$, when $\theta = 0°$,

then $0 = 0 + C_1$ $\quad \therefore C_1 = 0$

Hence the equation of stream function becomes

$$\psi = -\frac{A}{2\pi}\theta$$

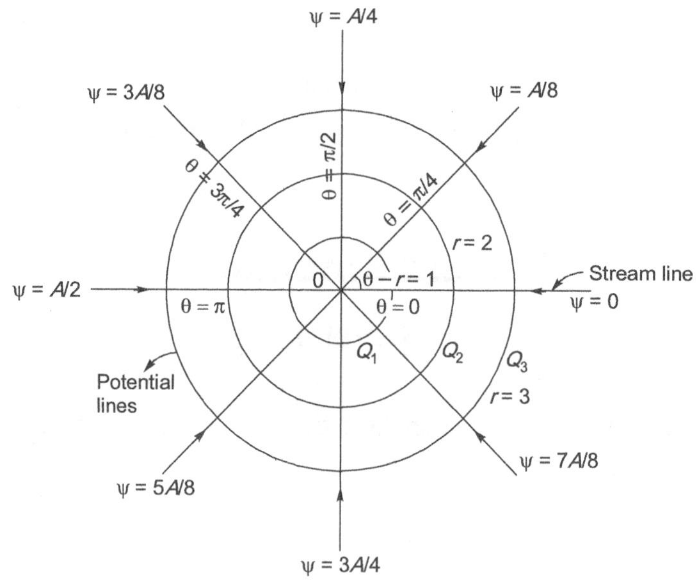

Fig. Flow Net

The flow net shown above represents typical sink flow.

3.2 The expression for stream function is described by $y = x^3 - 3xy^2$. Indicate whether the flow is rotational or irrotational. Determine the value or velocity potential ϕ, if it exists.

[10 Marks]

Sol:

Given data : $y = x^3 - 3xy^2$

Checking of rotation ability
$$\nabla^2\psi = \frac{\partial^2\psi}{\partial x^2} + \frac{\partial^2\psi}{\partial y^2} = \frac{\partial^2}{\partial x^2}(x^3 - 3xy^2) + \frac{\partial^2}{\partial y^2}(x^3 - 3xy^2)$$
$$= \frac{\partial}{\partial x}(3x^2 - 3y^2) + \frac{\partial}{\partial y}(0 - 6xy) = (6x - 0) + (0 - 6x) = 0$$

$\therefore$ Velocity potential ϕ exists

Using Cauchy – Reiman equation
$$\frac{\partial\phi}{\partial x} = \frac{\partial\psi}{\partial y}, \quad \frac{\partial\phi}{\partial y} = -\frac{\partial\psi}{\partial x}$$

$\therefore$
$$\frac{\partial\phi}{\partial x} = \frac{\partial}{y}(x^3 - 3xy^2) = 0 - 6xy$$

Integrating both sides
$$\phi = -\frac{6x^2}{2}\cdot y + f(y) + c$$
$$\phi = -3x^2 y + f(y) + c \quad \text{... (i)}$$

also,
$$\frac{\partial\phi}{\partial y} = -\frac{\partial\psi}{\partial x} = -\frac{\partial}{\partial x}(x^3 - 3xy^2)$$
$$= -(3x^2 - 3y^2) = 3y^2 - 3x^2$$

$\Rightarrow$
$$\frac{\partial\phi}{\partial y} = 3y^2 - 3x^2$$

Integrating both sides
$$\phi = \frac{3y^3}{3} - 3x^2 y + f(x) + c$$

$$= y^3 - 3x^2y + f(x) + c \quad \text{... (ii)}$$

from equation (i) and (ii) $\phi = y^3 - 3x^2y + c$

where c is a constant of integration.

Level-2

3.3 If velocity field is given by $U = x^2 - y^2 + x$ and $V = -(2xy + y)$, determine the velocity potential function and stream function.

[15 Marks]

Sol:

For given flow $U = x^2 - y^2 + x$

$$V = -(2xy + y)$$

Checking

(i) $$\frac{\partial U}{\partial x} + \frac{\partial V}{\partial y} = \frac{\partial(x^2 - y^2 + x)}{\partial x} + \frac{\partial(-2xy - y)}{\partial y}$$

$$= 2x + 1 - 2x - 1 = 0 \Rightarrow \text{Flow is possible}$$

(ii) $$\omega_Z = \frac{1}{2}\left(\frac{\partial V}{\partial x} - \frac{\partial U}{\partial y}\right) = \frac{1}{2}\left(\frac{\partial(-2xy - y)}{\partial x} - \frac{\partial(x^2 - y^2 + x)}{\partial y}\right)$$

$$= \frac{1}{2}[-2y + 2y] = 0 \Rightarrow \text{Flow is irrotational}$$

As flow is irrotational, velocity potential function exists

$$U = -\frac{\partial \phi}{\partial x}$$

$\Rightarrow$ $$\partial \phi = -Udx = -(x^2 - y^2 + x)dx$$

$\Rightarrow$ $$\int d\phi = -\int (x^2 - y^2 + x)dx$$

$\Rightarrow$ $$\phi = -\frac{x^3}{3} + xy^2 - \frac{x^2}{2} + f(y) + C \quad \text{...(i)}$$

Also $$V = -\frac{\partial \phi}{\partial y}$$

$\Rightarrow$ $$\partial \phi = -Vdy = +(2xy + y)\,dy$$

$\Rightarrow$ $$\int \partial \phi = \int [2xy + y]dy$$

$\Rightarrow$ $$\phi = xy^2 + \frac{y^2}{2} + f(x) + C \quad \text{...(ii)}$$

From eq. (i) and (ii) $$\phi = -\frac{x^3}{3} + xy^2 + \frac{y^2}{2} - \frac{x^2}{2} + C$$ where C is constant

$$U = -\frac{\partial \psi}{\partial y}$$

$\Rightarrow$ $$\partial \psi = -Udy = -(x^2 - y^2 + x)dy$$

$\Rightarrow$ $$\int \partial \psi = -\int (x^2 - y^2 + x)dy$$

$\Rightarrow$ $$\psi = -x^2y + \frac{y^3}{3} - xy + f(x) + C_1 \quad \text{...(iii)}$$

Also $$V = \frac{\partial \psi}{\partial x}$$

$\Rightarrow$ $$\partial \psi = V dx = -(2xy + y)\, dx$$

$\Rightarrow$ $$\psi = -x^2y - xy + f(y) + C_1 \quad \text{...(iv)}$$

From eq. (iii) and (iv)

$$\psi = -x^2y + \frac{y^3}{3} - xy + C_1 \quad \text{(Where } C_1 \text{ is constant)}$$

3.4 **If the velocity components of two-dimensional flow are**

$$u_x = \frac{y^3}{3} + 2x - x^2y \text{ and } u_y = xy^2 - 2y - \frac{x^3}{3}$$

Obtain :

(i) Stream function – ψ

(ii) Discharge passing between two points (2, 3) and (3, 4).

[12 Marks]

Sol:

Checking if flow is possible by **applying continuity equation**,

$$\frac{\partial u_x}{\partial x} + \frac{\partial u_y}{\partial y} = \frac{\partial}{\partial x}\left(\frac{y^3}{3} + 2x - x^2y\right) + \frac{\partial}{\partial y}\left(xy^2 - 2y - \frac{x^3}{3}\right)$$

$$= (0 + 2 - 2xy) + (2xy - 2 - 0) = 0$$

$\Rightarrow$ Flow is there

we know, $$-\frac{\partial \psi}{\partial y} = u_x = \frac{y^3}{3} + 2x - x^2y$$

$$\frac{\partial \psi}{\partial y} = x^2y - \frac{y^3}{3} - 2x$$

Integrating both sides $$\psi = \frac{x^2y^2}{2} - \frac{y^4}{4 \times 3} - 2xy + f(x) + c \quad \text{... (i)}$$

also, $$\frac{\partial \psi}{\partial y} = u_y = xy^2 - 2y - \frac{x^3}{3}$$

Integrating both sides

$$\psi = \frac{x^2y^2}{2} - 2y \cdot x - \frac{x^4}{4 \times 3} + f(y) + c \quad \text{... (ii)}$$

from equation (i) and (ii)

stream function, $$\psi = \frac{x^2y^2}{2} - \frac{y^4}{12} - \frac{x^4}{12} - 2xy + c$$

$$\psi_{(2,3)} = \frac{2^2 \times 3^2}{2} - \frac{3^4}{12} - \frac{2^4}{12} - 2 \times 2 \times 3 + c = -2.0833 + c$$

$$\psi_{(3,4)} = \frac{3^2 \times 4^2}{2} - \frac{4^4}{12} - \frac{3^4}{12} - 2 \times 3 \times 4 + c = 19.9167 + c$$

Discharge between (3, 4) and (2, 3) per unit width

$$Q = \psi_{(3,4)} - \psi_{(2,3)} = 22 \text{ units}$$

3.5 In a three dimensional flow, the components of velocity are
$u = xy$, $v = 4yz^3$ and $w = -(yz + z^4)$
Test whether the continuity equation for incompressible fluid flow is satisfied.
Determine the acceleration vector at point (1,1,1).

[12 Marks]

Sol:

The continuity equation for a 3-*D* steady incompressible flow is given by

$$\frac{\partial u}{\partial x} + \frac{\partial v}{\partial y} + \frac{\partial w}{\partial z} = 0$$

Here,

$$u = xy \Rightarrow \frac{\partial u}{\partial x} = y; \frac{\partial u}{\partial y} = x; \frac{\partial u}{\partial z} = 0$$

$$v = 4yz^3 \Rightarrow \frac{\partial v}{\partial x} = 0; \frac{\partial v}{\partial y} = 4z^3; \frac{\partial v}{\partial z} = 12yz^2$$

$$w = -(yz + z^4) \Rightarrow \frac{\partial w}{\partial x} = 0; \frac{\partial w}{\partial y} = -z; \frac{\partial w}{\partial z} = -(y + 4z^3)$$

$$\therefore \quad \frac{\partial u}{\partial x} + \frac{\partial v}{\partial y} + \frac{\partial w}{\partial y} = y + 4z^3 - (y + 4z^3)$$

$$\Rightarrow \quad \frac{\partial u}{\partial x} + \frac{\partial v}{\partial y} + \frac{\partial w}{\partial y} = 0 \quad \text{(Hence proved)}$$

For steady flow, $a_x = u\frac{\partial u}{\partial x} + v\frac{\partial u}{\partial y} + w\frac{\partial u}{\partial z}$

$$\Rightarrow \quad a_x = (xy \times y) + (4yz^3 \times x) + [-(yz + z^4) \times 0]$$

$$\Rightarrow \quad a_x = xy^2 + 4xyz^3 \quad \text{...(i)}$$

also, $a_y = u\frac{\partial v}{\partial x} + v\frac{\partial v}{\partial y} + w\frac{\partial v}{\partial z}$

$$\Rightarrow \quad a_y = (xy \times 0) + (4yz^3 \times 4z^3) - (yz + z^4)(12\,yz^2)$$

$$\Rightarrow \quad a_y = 16\,yz^6 - 12y^2z^3 - 12yz^6$$

$$\Rightarrow \quad a_y = 4yz^6 - 12y^2z^3 \quad \text{...(ii)}$$

and $a_z = u\frac{\partial w}{\partial x} + v\frac{\partial w}{\partial y} + w\frac{\partial w}{\partial z}$

$$\Rightarrow \quad a_z = (xy \times 0) + 4yz^3 \times (-z) + (yz + z^4)(y + 4z^3)$$

$$\Rightarrow \quad a_z = -4yz^4 + y^2z + 4yz^4 + yz^4 + 4z^7$$

$$\Rightarrow \quad a_z = y^2z + yz^4 + 4z^7 \quad \text{...(iii)}$$

We know that acceleration vector $\vec{a}$ is given by

$$\vec{a} = a_x\hat{i} + a_y\hat{j} + a_z\hat{k}$$

$$\Rightarrow \quad \vec{a} = (xy^2 + 4xyz^3)\hat{i} + (4yz^6 - 12y^2z^3)\hat{j} + (y^2z + yz^4 + 4z^7)\hat{k}$$

$\therefore$ Acceleration vector $\vec{a}$ at point (1, 1, 1) will be

$$\vec{a} = (1 \times 1^2 + 4 \times 1 \times 1 \times 1^3)\hat{i} + (4 \times 1 \times 1^6 - 12 \times 1^2 \times 1^3)\hat{j} + (1^2 \times 1 + 1 \times 1^4 + 4 \times 1^7)\hat{k}$$

$$\Rightarrow \quad \vec{a} = 5\hat{i} - 8\hat{j} + 6\hat{k} \text{ (Ans.)}$$

3.6 Find the convective acceleration at the middle of a pipe which converges uniformly from 0.4 m diameter to 0.2 m diameter over 2 m length. The rate of flow is 20 lit/s. If the rate of flow changes uniformly from 20 lit/s to 40 lit/s in 30 seconds, find the total acceleration at the middle of the pipe at 15th second.

[20 marks]

Sol:

Given:

$$D_1 = 0.4 \text{ m}; D_2 = 0.2 \text{ m}, L = 2 \text{ m}, Q = 20 \text{ lit/s} = 0.02 \text{ m}^3/\text{s}$$

(i) convective acceleration at middle when Q = 20 lit/s

In this case, the rate of flow is constant and equal to 0.02 m^3/s. The velocity of flow is in x-direction only. Hence this is one-dimensional flow and velocity components in y and z directions are zero i.e., $v = 0$, $w = 0$.

$$\therefore \quad \text{Convective acceleration} = u\frac{\partial u}{\partial x} \quad ...(1)$$

Let us find the value of u and $\frac{\partial u}{\partial x}$ at a distance x from inlet.

The diameter (D_x) at a distance x from inlet or at section X-X is given by,

$$D_x = 0.4 - \frac{0.4 - 0.2}{2} \times x = (0.4 - 0.1x) \text{ m}$$

The area of cross-section (A_x) at section X-X is given by,

$$A_x = \frac{\pi}{4}D_x^2 = \frac{\pi}{4}(0.4 - 0.1x)^2$$

Velocity (u) at the section X-X in terms of Q (i.e., in terms of rate of flow)

$$u = \frac{Q}{\text{Area}} = \frac{Q}{A_x} = \frac{Q}{\frac{\pi}{4}D_x^2} = \frac{4Q}{\pi(0.4 - 0.1x)^2}$$

$$= \frac{1.273Q}{(0.4 - 0.1x)^2} = 1.273\, Q(0.4 - 0.1x)^{-2} \text{ m/s} \quad ...(2)$$

To find $\frac{\partial u}{\partial x}$, we must differentiate equation (2) with respect to x.

$$\frac{\partial u}{\partial x} = \frac{\partial}{dx}\left[1.273Q[0.4 - 0.1x]^{-2}\right]$$

$$= 1.273 \times (-2)Q[0.4 - 0.1x]^{-3} \times (-0.1)$$

$$\frac{\partial u}{\partial x} = 0.2546Q\,[0.4 - 0.1x]^{-3}$$

So, convective acceleration at x = 1 m

$$\frac{u\partial u}{\partial x} = 1.273Q[0.4 - 0.1x]^{-2} \times 0.2546Q[0.4 - 0.1x]^{-3}$$

$$= 0.05335 \text{ m/sec}^2$$

Case II: Total acceleration at middle of pipe at 15th second.

Here Q changes from 0.02 m^3/s to 0.04 m^3/s in 30 seconds and we need to find the total acceleration at x = 1 m and t = 15 seconds.

Total acceleration = Convective acceleration + Local acceleration (at t = 15 seconds)

The rate of flow at t = 15 seconds is given by

$$Q = Q_1 + \frac{Q_2 - Q_1}{30} \times 15 \text{ where } Q_2 = 0.04 \text{ m}^3\text{/s and } Q_1 = 0.02 \text{ m}^3\text{/s}$$

$$= 0.02 + \frac{(0.04 - 0.02)}{30} \times 15 = 0.03 \text{ m}^3\text{/s}$$

The velocity (u) and gradient $\left(\frac{\partial u}{\partial x}\right)$ in terms of Q are given by equations (2) and (3) respectively.

$$\therefore \text{Convective acceleration} = u \cdot \frac{\partial u}{\partial x}$$

$$= [1.273Q(0.4 - 0.1x)^{-2}] \times [0.2546Q(0.4 - 0.1x)^{-3}]$$

$$= 1.273 \times 0.2546Q^2 \times (0.4 - 0.1 \times 1)^{-5}$$

$\therefore$ Convection acceleration (when $Q = 0.03$ m³/s and x = 1 m) = 0.12 m/s² ...(4)

$$\text{Local acceleration} = \frac{\partial u}{\partial t} = \frac{\partial}{\partial t}[1.273Q(0.4 - 0.1x)^{-2}]$$

[$\because$ u from equation (2) is $u = 1.273Q(0.4 - 0.1x)^{-2}$]

$$= 1.273 \times (0.4 - 0.1x)^{-2} \times \frac{\partial Q}{\partial t}$$

$$= 1.273 \times (0.3)^{-2} \times \frac{0.02}{30} = 0.00943 \text{ m/s}^2 \quad ...(5)$$

$$\left[\because \frac{\partial Q}{\partial t} = \frac{Q_2 - Q_1}{t} = \frac{0.04 - 0.02}{30} = \frac{0.02}{30}\right]$$

Hence adding equations (4) and (5), we get total acceleration.

$\therefore$ Total acceleration = Convective acceleration + Local acceleration

$$= 0.12 + 0.00943 = 0.1294 \text{ m/s}^2$$

4. Fluid Dynamics

Level-1

4.1 A cylindrical tank of water rotates in solid body rotation, counterclockwise about its vertical axis at angular speed $\vec{\omega}$ = 135 rpm. Calculate the vorticity of fluid particles in the tank.

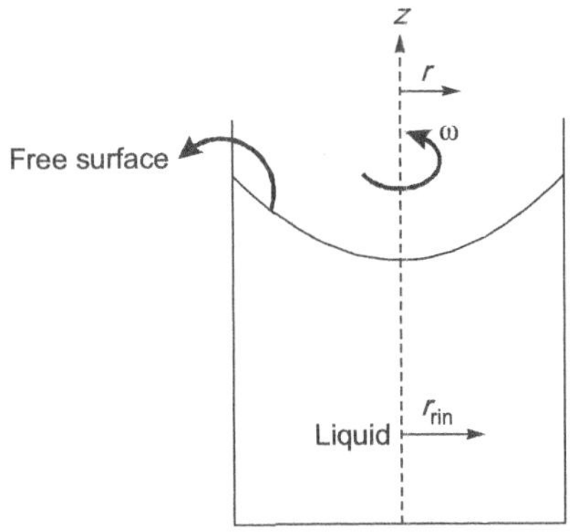

[6 Marks]

Sol:

For any distance r from the axis the circulation will be given as,

$$\Gamma = \int V.dl \qquad \text{where } dl = rd\theta;\ V = r\omega$$

Putting
$$\Gamma = \int r\omega \cdot rd\theta = r^2\omega \int_0^{2\pi} d\theta = r^2\omega[2\pi - 0]$$

Now vorticity is given by $\Omega = \dfrac{\Gamma}{\text{Area enclosed}}$

$$\Omega = \frac{\Gamma}{\pi r^2} = \frac{r^2\omega(2\pi)}{\pi r^2} = 2\omega$$

$\Omega = 2\omega$ (same as the result in linear problem)

$$= 2\times\left[135\times\frac{2\pi}{60}\right] = 28.26 \text{ rad/sec}$$

4.2 **A two dimensional flow pattern is as shown in figure. Water is flowing out an opening with a velocity of 20 m/sec and is reflected against a construction as shown. Determine the force 'F' that is needed to keep the construction in its position.**
The width of the opening from which water comes is equal to 0.10 m. The weight of the shaded mass of water can be neglected.
Is the velocity at point 'B' more, less or equal to the velocity at point 'A'? Explain with basic equation. 'A' and 'B' are not situated at the boundary (friction is neglected).

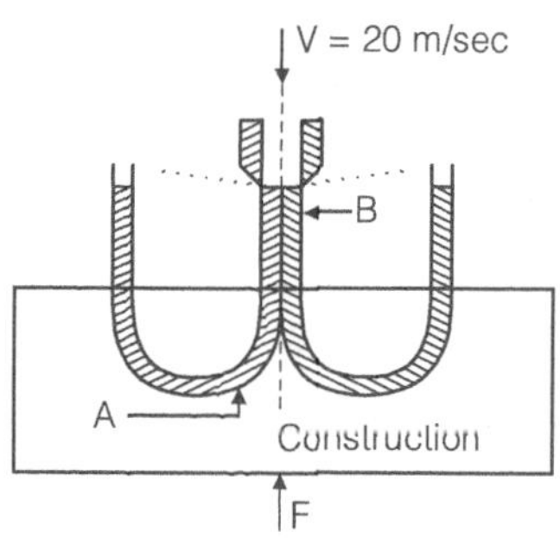

[8 Marks]

Sol:

Given data:

Velocity, $V = 20$ m/sec

Let velocity at two outlets be V and V′

Assume, $V = V' = 20$ m/sec

Width of opening = 0.10 m

Consider per unit length, i.e. L = 1.0 m

∴ Area, $A = 0.1 \times 1 = 0.1 \text{ m}^2$

Now, $Q = A.V = 0.1 \times 20$
$= 2 \text{ m}^3/\text{sec}$

∴ Force, $F = \rho Q[V_2 - V_1]$
$= \rho Q[V - (-V)]$

(since V is in downward and V′ is in upward direction)

∴ $F = 1000 \times 2[20 + 20]$
$= 80{,}000$ kg

Since the velocity is uniform throughout therefore velocity at point B will be equal to that at point A.

4.3 A sharp crested rectangular weir 1.0 m long with end contractions suppressed and a 90° notch weir are placed in the same weir box, with the vertex of the 90° V-notch weir 15 cm below the rectangular weir crest. Determine the head on the V-notch weir when the rectangular weir discharges its maximum amount above the discharge of the V-notch weir. Assume 'C_d (coefficient of discharge) to be the same for both the weirs.

[10 Marks]

Sol:

Since the rectangular weir is sharp crested with end contractions suppressed therefore total head on the rectangular weir will be same as that of the v-notch.

i.e. $H_2 = H_1 - 0.15$

Here, H_2 and H_1 are the heads over rectangular weir and v-notch.

Now discharge through rectangular weir,

$$Q_R = \frac{2}{3} C_d \sqrt{2g} \cdot L\ H_2^{3/2}$$

and discharge through v-notch weir

$$Q_V = \frac{8}{15} C_d \sqrt{2g} \cdot (H_1)^{5/2} \cdot \tan\left(\frac{90°}{2}\right)$$

$\because$ $$Q_R = Q_V$$

$\Rightarrow$ $$\frac{2}{3} C_d \cdot \sqrt{2g} \cdot L\,(H_2)^{3/2}$$

$$\frac{8}{15} C_d \sqrt{2g} \cdot (H_1)^{5/2} \cdot \tan 45°$$

$\Rightarrow$ $$(H_2)^{3/2} = 0.8(H_1)^{5/2} (\because L = 1 \text{ m})$$

$\Rightarrow$ $$H_2 = (0.8)^{\frac{2}{3}} (H_1)^{\frac{5}{2} \times \frac{2}{3}}$$

$\Rightarrow$ $$H_2 = 0.8617\,(H_1)^{5/3}$$

$\because$ $$H_2 = H_1 - 0.15$$

$\Rightarrow$ $$0.8617\,(H_1)^{5/3} = H_1 - 0.15$$

$\Rightarrow$ $$H_1 \sim 0.2180 \text{ m}$$

Thus head over v-notch = 0.2180 m

4.4 In a free cylindrical vortex of water, the tangential velocity at radial distance of 0.1 m from the axis of rotation is found to be 10 m/sec and the intensity of pressure is 200 kN/m^2. Find the pressure intensity at a radial distance of 0.2 m from the axis.

[10 marks]

Sol:

Given : Free cylindrical vortex $\Rightarrow V \alpha \frac{1}{R};\ V = \frac{C}{R}$

at $R = 0.1$ m, $V = 10$ m/s $$10 = \frac{C}{0.1}$$

$$C = 1$$

$\Rightarrow$ $$VR = 1$$

Fundamental equation, $$dP = \frac{\rho \cdot v^2}{r} \cdot dr - \rho g \cdot dz = \frac{\rho \cdot (1/r)^2}{r} \cdot dr - \rho g dz$$

$$dP = \frac{\rho}{r^3} \cdot dr - \rho g dz$$

Integrating both sides
$$\int_{P_1}^{P_2} dP = \int_1^2 \frac{\rho}{r^3} \cdot dr - \int_1^2 \rho g dz$$

$$P_2 - P_1 = -\frac{\rho}{2}\left[\frac{1}{r_2^2} - \frac{1}{r_1^2}\right] - \rho g(z_2 - z_1)$$

Here,
$$P_1 = 200 \times 10^3 \text{ N/m}^2, r_1 = 0.1 \text{ m}$$
$$r_2 = 0.2 \text{ m}, P_2 \text{ is to be found}$$
$$\rho = \rho_\omega = 10^3 \text{ kg/m}^3$$

and
$$z_2 - z_1 = 0$$

$\Rightarrow$
$$P_2 - 200 \times 10^3 = -\frac{10^3}{2}\left[\frac{1}{0.2^2} - \frac{1}{0.1^2}\right]$$

$\Rightarrow$
$$P_2 = 237.5 \text{ kN/m}^2$$

4.5 **(a) Determine the torque about point *O* required to prevent the notation of sprinkler as shown in figure.**

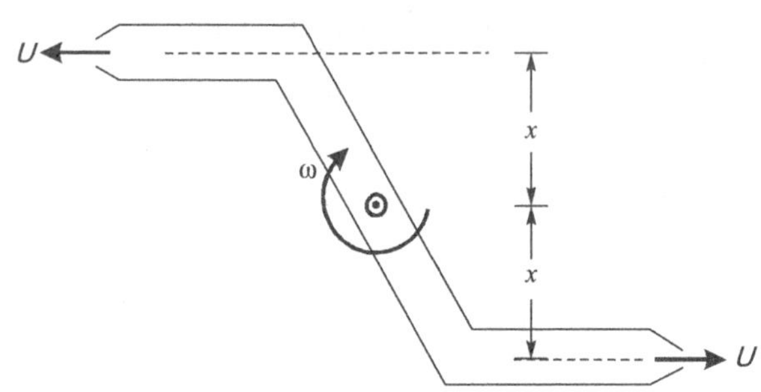

(b) A lawn sprinkler shown in figure discharges 0.314 *l*/sec through both the nozzles. Assuming no friction, determine its speed of rotation.

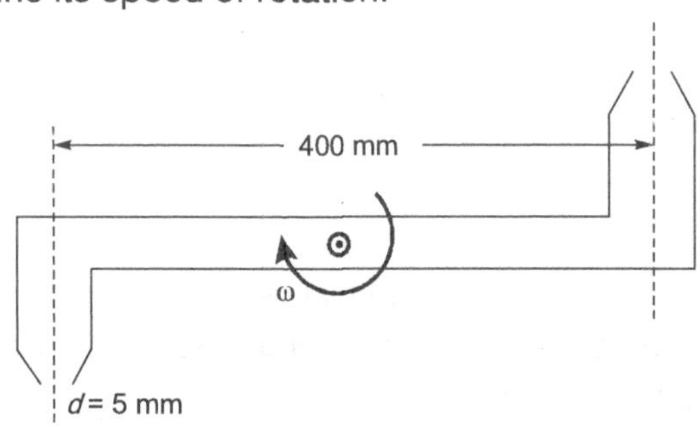

[12 Marks]

Sol:

(a) Applying momentum equation,

$$T = \dot{m}[V_1 r_1 + V_2 r_2]$$

where,
$$V_1 = V_2 = \text{Relative velocity}$$
$$= U - x \times \omega$$

Since no rotation, $\omega = 0$.
$$V_1 = V_2 = U$$
$$T = \dot{m}[Ux + Ux]$$
$$= \rho\theta \times 2Ux$$
$$= 2A\rho U^2 x \qquad \textbf{(Ans)}$$

(b) Given : No friction

⇒ Means torque requirement is zero.

$$T = \dot{m}(V_1 r_1 + V_2 r_2) = 0$$

$$V_1 = v - r_1\omega$$

$$V_2 = v - r_2\omega$$

$$V_2 = \frac{Q/2}{A} = \frac{0.314\times10^{-3}}{2\times\frac{\pi}{4}\times\left(5\times10^{-3}\right)^2} = \frac{16\text{ m/sec}}{2} = 8\text{ m/sec}$$

$$r_1 = r_2 = \frac{400}{2} = 200\text{ mm}$$

$$[2\times(8-0.2\times\omega)\times0.2] = 0$$

Solving,

$$8 = 0.2\omega$$

$$\omega = 40\text{ rad/sec} = 381.97\text{ rpm}$$

4.6 **A 5 mm diameter nozzle is discharging a vertically upward jet of water with a velocity of 10 m/s at the nozzle. What is the maximum height to which this jet will reach? If a flat plate weighing 1.5 N is balanced on the jet, what will be its height from the nozzle?**

[10 Marks]

Sol:

Diameter of nozzle, d = 5 mm = 0.005 m

Velocity of water at nozzle, V_1 = 10 m/s

Let h be maximum height to which jet will rise.

At this velocity of water, $V_2 = 0$

Using energy equation at these two sections;

$$\frac{P_1}{\rho g}+\frac{V_1^2}{2g}+0 = \frac{P_2}{\rho g}+\frac{V_2^2}{2g}+h$$

∵ Pressure is atmosphere ⇒ $p_1 - p_2$

$$\Rightarrow \quad \frac{V_1^2}{2g} = h$$

$$\Rightarrow \quad h = \frac{(10)^2}{2g} = 5.1\text{m}$$

Let 'a' be the area of jet at which flat plate is supported.

Thrust on the plate = Weight of plate

Discharge, $Q = A_1 \times V_1$ (at the nozzle)

$$= \frac{\pi}{4}\times d_1^2\times V_1 = \frac{\pi}{4}\times(0.005)^2\times10 = 1.963\times10^{-4}\text{ m}^3\text{/sec}$$

At the plate location, $Q = aV_2$ (where velocity of striking the plate = V_2)

$$\Rightarrow \quad V_2 = \frac{Q}{a}$$

$$\therefore \quad \text{Thrust on plate} = \rho_w a V_2^2$$

$$\Rightarrow \quad 1.5 = 1000\times a\times\frac{Q^2}{a^2}$$

$$\Rightarrow \quad a = \frac{1000Q^2}{1.50} = \frac{1000 \times (1.963 \times 10^{-4})^2}{1.5} = 2.57 \times 10^{-5}\ \text{m}^2$$

$$\Rightarrow \quad d_2 = \sqrt{\frac{4}{\pi}a} = \sqrt{\frac{4}{\pi}\left(2.57 \times 10^{-5} \times 10^6\ \text{mm}^2\right)} = 5.72\ \text{mm}$$

Let z be height from nozzles of the plate.

∴ Using Energy equation,

$$\Rightarrow \quad \frac{P_1}{\rho g} + \frac{V_1^2}{2g} + 0 = \frac{P_2}{\rho g} + \frac{V_2^2}{2g} + z$$

$$\Rightarrow \quad z = \frac{V_1^2 - V_2^2}{2g}$$

$$V_2 = \frac{4Q}{\pi d_2^2} = \frac{4 \times 1.963 \times 10^{-4}}{\pi \times (5.72 \times 10^{-3})^2} = 7.638\ \text{m/s}$$

$$\therefore \quad Z = \frac{(10)^2 - (7.638)^2}{2g} = 2.123\ \text{m from the nozzle.}$$

Level-2

4.7 **A semi-spherical tank of radius R is completely filled with water. Now a hole of cross-sectional area A_h and discharge coefficient C_d at the bottom of tank is fully opened. Develop an expression for the time needed to empty the tank completely?**

[10 Marks]

Sol:

Assume a thin cylindrical strip of radius r and depth dh

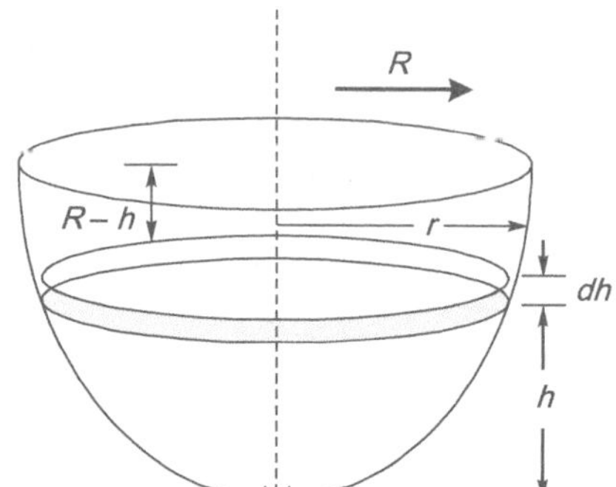

In time dt the total volume outflow $= q \times dt$

$= A \times V \times dt$

Equating $\quad A_h \times V \times dt = \pi r^2 dh$

Instant velocity $(V) = \sqrt{2gh}$

$$C_d \times A_h \times \sqrt{2gh}\, dt = \pi r^2 dh$$

Also from diagram, $\quad r^2 = R^2 - (R-h)^2$

Putting $\quad C_d \times A_h \times \sqrt{2g} \times dt = \dfrac{\pi\left(R^2 - (R-h)^2\right)dh}{\sqrt{h}}$

$$C_d A_h \sqrt{2g} \times dt = \frac{\pi\left(R^2 - R^2 - h^2 + 2Rh\right)dh}{\sqrt{h}}$$

$$C_d A_h \sqrt{2g} \times dt = \frac{\pi\left(2Rh - h^2\right)dh}{\sqrt{h}}$$

Integrating both side and solving,

$$C_d A_h \sqrt{2g}\int_0^t dt = \pi\int_R^0 \left(2Rh^{5/2} - h^{\frac{3}{2}}\right)dh$$

$$t = \frac{\pi}{C_d A_h \sqrt{2g}}\left[\frac{2}{5}R^{5/2} - \frac{4}{5}R^{5/2}\right]$$

$$t = \frac{14}{15}\frac{\pi}{C_d}\frac{R^{5/2}}{\sqrt{2g}}$$

4.8 A closed cylindrical tank 2 m high 1 m in diameter contains 1.5 m of water. When the angular velocity about its longitudinal axis (vertical) is constant at 20 rad/sec. How much of the bottom of the tank is uncovered? If the pressure at the centre of bottom of tank is to be zero. What horizontal speed (in rpm) is required?

[15 Marks]

Sol:

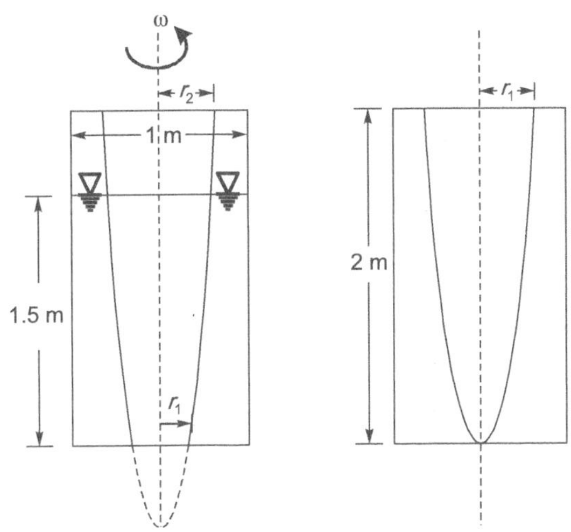

Case I Case II

For case I : Equating velocity of air

$$\frac{1}{2}\pi r_2^2(2+x_1) - \frac{1}{2}\pi r_1^2 \times x_1 = \frac{\pi}{2}\times 1^2 \times 0.5$$

$$r_2^2(2+x_1) - r_2^2 \times x_1 = \frac{1}{2}\times(1)^2\times 0.5 \quad \text{...(i)}$$

Also from equation of parabola

$$x_1 = \frac{\omega^2 r_1^2}{2g} \quad \text{...(ii)}$$

$$(2+x_1) = \frac{\omega^2 r_2^2}{2g} \quad \text{...(iii)}$$

Hence,
$$(2+x_1)-(x_1) = \frac{\omega^2}{2g}\left(r_2^2 - r_1^2\right)$$

From (i)

$$x_1 \times \left(r_2^2 - r_1^2\right) + 2r_2^2 = 0.28$$

$$x_1 \times \frac{2g\times 2}{\omega^2} + 2r_2^2 = 0.28$$

Given ω = 20 rad/sec

$$\frac{4g}{\omega^2}\times x_1 - 2\times\frac{(2+x_1)\times 2g}{\omega^2} = 0.25$$

Solving for x_1

$$x_1(4g + 2\times 2g) + 8g = 0.25\,\omega^2$$

$$x = 0.274 \text{ m}$$
$$r_1 = 0.116 \text{ m}$$

Case II :

$$z = \frac{\omega^2 r_1^2}{2g}$$

$\Rightarrow$
$$z = 2 = \frac{\omega^2 r_1^2}{2g} \quad ...(i)$$

Equating volume of air,

$$\frac{1}{2} \times \pi r_1^2 \times 2 = \frac{\pi}{4} \times 1^2 \times 0.5$$
$$r_1 = 0.353 \text{ m}$$

From (i),
$$\omega = 17.74 \text{ rad/sec}$$

4.9 (i) A 90° elbow in a horizontal pipe is used to direct water flow upwards at a rate of 40 kg/sec. The diameter of the entire elbow is 10 cm. The elbow discharges water into the atmosphere, and thus the pressure at the exit is local atmospheric pressure. The elevation difference between the centres of the exit and inlet of elbow is 50 cm. The weight of the elbow and the water in it is considered to be negligible. Determine

(a) The gauge pressure at the center of inlet of elbow.

(b) The anchoring force needed.

Take momentum flux correction factor to be 1.03 at both inlet and outlet.

(ii) Repeat the problem when an identical elbow is attached to existing elbow so that fluid makes a U-turn?

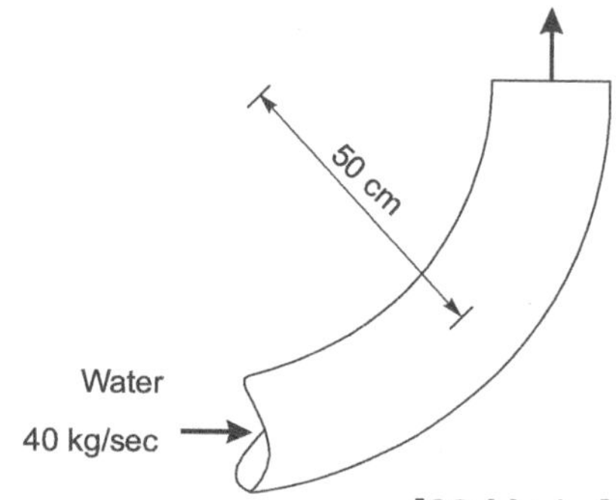

[20 Marks]

Sol:

$$\dot{m} = \rho Q = 40 \text{ kg/sec}$$
$$10^3 \times \frac{\pi}{4} \times (0.1)^2 \times V = 40$$
$$V = 5.1 \text{ m/sec}$$

Applying Bernoulli's equation between entry point (1) and exit point (2)

$$\frac{P_1}{\rho g} + \frac{V_1^2}{2g} + Z_1 = \frac{P_2}{\rho g} + \frac{V_2^2}{2g} + Z_2 + H_L \; (\to 0)$$

$V_1 = V_2$, $P_2 = P_{atm} = 0$

$$\frac{P_1}{\rho g} = Z_2 - Z_1 = 0.5 \text{ m}$$
$$P_1 = 0.5 \times 10^3 \times 9.81$$
$$= 4.905 \times 10^3 \text{ N/m}^2 \text{ (Gauge)}$$

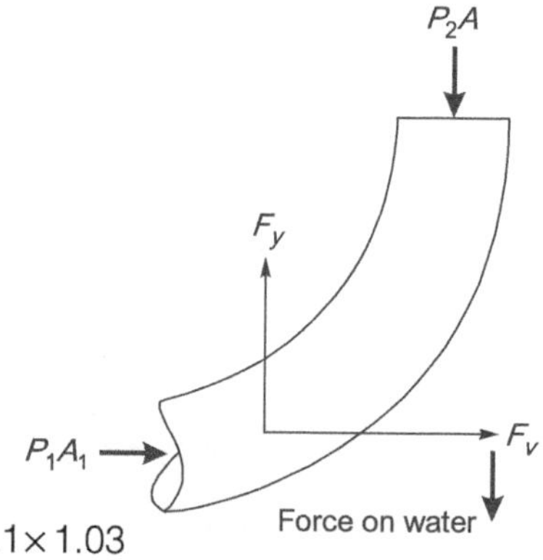

(b) For Anchoring force in X-direction

For anchoring force in x-direction,

$$0 - 1.03 (\dot{m}u) = P_1A + F_x$$
$$F_x = -4.905 \times 10^3 \times \frac{\pi}{4} \times (0.1)^2 - 40 \times 5.1 \times 1.03$$
$$= -38.5 - 40 \times 5.1 \times 1.03 = 248.5 \text{ N } (\rightarrow)$$

$$\text{Force on elbow} = 248.6 \text{ N} (\leftarrow)$$

In y-direction

$$\dot{m}v - 0 = F_y - P_2A$$
$$F_y = 40 \times 5.1 \times 1.03 + 0 \quad (P_2 = 0)$$
$$= 210.12 \text{ N} \quad \text{(On water)}$$
$$F_y = -210.12 \text{ N} (\downarrow) \quad \text{(On elbow)}$$

4.10 For the given condition, a venturimeter is being used as a suction device to introduce petrol into an air stream. By using continuity and Bernoulli's equation, estimate the volume flow rate (Q) of air required to such the petrol in the air stream.
Given: $\rho_{fuel} = 800$ kg/m^3; $h = 150$ mm and $\rho_{air} = 1.2$ kg/m^3
Assume air to be an ideal fluid.

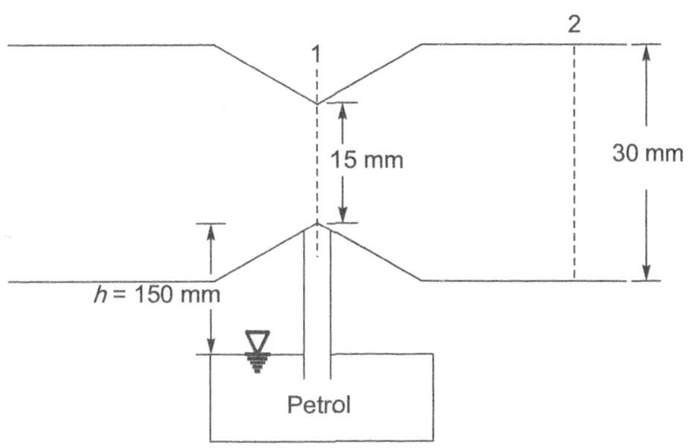

[12 Marks]

Sol:

Assume the rise of petrol is h above its free surface up to throat

Now,
$$\Delta P = \rho_{Petrol}\, gh \quad \text{(below atm. pressure)}$$
$$P_a - P_1 = \rho_{Petrol}\, gh$$

By continuity,
$$V_1A_1 = V_2A_2$$

Since
$$A_2 = 4A_1$$
$$V_1 = 4V_2$$

Now Apply Bernoulli's equation between (1) and (2) (Horizontal $Z_1 = Z_2$)

$$\frac{P_1}{\gamma} + \frac{V_1^2}{2g} = \frac{P_2}{\gamma} + \frac{V_2^2}{2g}$$
$$\frac{V_1^2 - V_2^2}{2} = \frac{P_2 - P_1}{\rho} \quad (P_2 = P_a)$$
$$\frac{15V_2^2}{2} = \frac{P_a - P_1}{\rho}$$
$$\frac{15V_2^2}{2} = \frac{\rho_{liq}\, gh}{1.2}$$
$$\frac{15V_2^2}{2} = \frac{800 \times 9.81 \times 0.15}{1.2}$$

Solving
$$V_2 = 11.44 \text{ m/sec}$$

$$Q_2 = \frac{\pi}{4} \times (0.03)^2 \times 11.44$$
$$= 8.1 \times 10^{-3} \text{ m}^3\text{/sec} = 8.1\ l\text{/sec}$$

4.11 A flow nozzle is used for measuring the discharge in the laboratory. The approach flow pipe has a 100 mm diameter pipe and the throat diameter of the flow nozzle is 60 mm. The coefficient of velocity is 0.90. The differential mercury manometer shows 13.5 cm of Hg. Obtain the discharge in litres per second.

[12 Marks]

Sol:

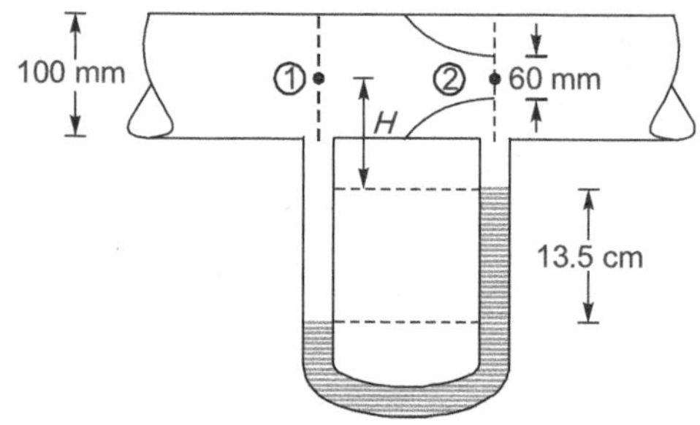

The brief schematic of the flow nozzle is given below:

Let the flow nozzle be horizontally laid and water is flowing through the arrangement. Let H be the distance of upper meniscus of mercury from point (1).

Applying pressure equation between (1) and (2), we get

$$\frac{p_1}{w} + H + \frac{13.5}{100} - \frac{13.5}{100} \times 13.6 - H = \frac{p_2}{w}$$

$$\Rightarrow \quad \frac{p_1}{w} - \frac{p_2}{w} = \frac{13.5}{100}(13.6 - 1)$$

$$\Rightarrow \quad \frac{p_1}{w} - \frac{p_2}{w} = 1.701 \text{ m of water} \quad \text{... (i)}$$

Now applying Bernoulli's equation between (1) and (2), we get

$$\frac{P_1}{w} + \frac{V_1^2}{2g} + Z_1 = \frac{p_2}{w} + \frac{V_2^2}{2g} + Z_2 \qquad [Z_1 = Z_2]$$

$$\Rightarrow \quad \frac{p_1}{w} - \frac{p_2}{w} = \frac{V_2^2 - V_1^2}{2g}$$

But from (i) $\quad \frac{p_1}{w} - \frac{p_2}{w} = 1.701 \text{ m} = 33.374 \text{ m}^2/\text{s}^2$

$$\therefore \quad V_2^2 - V_1^2 = 2 \times 9.81 \times 1.701 = 33.37$$

Also $\quad Q = A_1V_1 = A_2V_2$

$$\therefore \quad Q^2 \times \left[\frac{A_1^2 - A_2^2}{A_1^2 A_2^2}\right] = 33.37$$

$$\Rightarrow \quad Q = \frac{A_1A_2\sqrt{33.37}}{\sqrt{A_1^2 - A_2^2}}$$

$$\Rightarrow \quad Q = \frac{\frac{\pi}{4} \times (0.1)^2 \times \frac{\pi}{4} \times (0.06)^2 \times \sqrt{33.37}}{\frac{\pi}{4} \times \sqrt{(0.1)^4 - (0.06)^4}}$$

$$\Rightarrow \quad Q = 0.0175 \text{ m}^3/\text{s} = 17.5 \text{ litres per second}$$

Note: The coefficient of velocity (0.90) is redundant here.

4.12 A pump is installed in a pipeline of diameter 0.2 m. Close to the pump, gauges are installed on both sides. The gauge installed on the suction side indicates a pressure of 0.25 m vacuum of mercury and the gauge installed on the delivery side indicates a pressure of 40 m of water. A venturimeter 0.2 m × 0.1 m is also installed to measure the discharge of water flowing through the pipeline. The differential manometer installed between the main and the throat shows a deflection of 0.3 m of mercury (13.6). Taking coefficient of discharge for venturimeter as 1.0, compute the pump power in kW.

[15 Marks]

Sol:

Given data: Inlet pipe diameter, $d_1 = 0.2$ m,

Throat diameter, $d_2 = 0.1$ m,

Differential manometer reading, $x = 0.3$ m of Hg,

Specific gravity of Hg = 13.6,

Coefficient of discharge, $C_d = 1.0$

$\Rightarrow$ Venturi head will be given as

$$h = 0.3\,(13.6 - 1)$$

$\Rightarrow$ $h = 3.78$ m of water

Since $C_d = 1$, therefore there are no losses which means the theoretical discharge will also be the actual discharge

$$\Rightarrow \quad Q = C_d \frac{A_1 A_2 \sqrt{2gh}}{\sqrt{A_1^2 - A_2^2}} = \frac{1.0 \times \frac{\pi}{4} \times (0.2)^2 \times \frac{\pi}{4} \times (0.1)^2 \times \sqrt{2 \times 9.81 \times 3.78}}{\sqrt{\left[\frac{\pi}{4} \times (0.2)^2\right]^2 - \left[\frac{\pi}{4} \times (0.1)^2\right]^2}}$$

$$= \frac{\frac{\pi}{4} \times (0.2)^2 \times \frac{\pi}{4} \times (0.1)^2 \times \sqrt{74.1636}}{\frac{\pi}{4}\sqrt{(0.2)^4 - (0.1)^4}}$$

$$\Rightarrow \quad Q = 0.0699 \text{ m}^3/\text{s}$$

Applying Bernoulli's equation between 1 and 2,

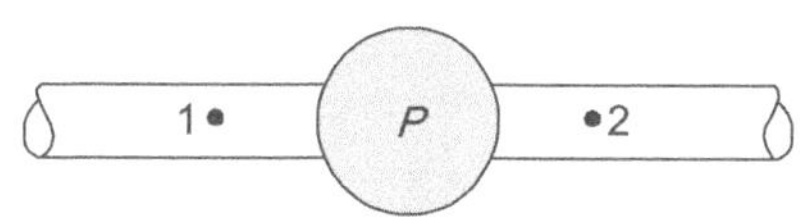

$$\frac{p_1}{w} + \frac{V_1^2}{2g} + Z_1 + H_p = \frac{p_2}{w} + \frac{V_2^2}{2g} + Z_2$$

H_p is added as some work is done by the pump.

Assuming the points 1 and 2 are at the same level

i.e. $Z_1 = Z_2$.

Also, the diameter of both the pipes is same, hence the velocities V_1 and V_2 are same.

$$\therefore \quad \frac{p_1}{w} + H_p = \frac{p_2}{w}$$

$$-\frac{13.6 \times 1000 \times 9.81 \times 0.25}{1000 \times 9.81} + H_p = 40$$

$$\Rightarrow \quad H_p = 40 + 3.4 = 43.4 \text{ m}$$

Now pump power is given by

$$P = wQH_p = \frac{1000 \times 9.81 \times 0.0699 \times 43.4}{1000} = 29.76 \text{ kW.}$$

4.13 A horizontal stream of air flows past a vertical circular smoke stack at a velocity of 30 m/s. The static pressure in the air stream for upstream of the smoke stack is 101 kN/m^2. Calculate the stagnation pressure and pressure at a point 90° from the flow directions on the surface of the stack if the local velocity at that point is 60 m/s. Assume mass density of air as 1.2 kg/m^3. On the surface of the cylinder, velocity of air stream in any direction θ is given as $U_\theta = -2U\sin\theta$, where U = Velocity of horizontal air stream.

[20 marks]

Sol:

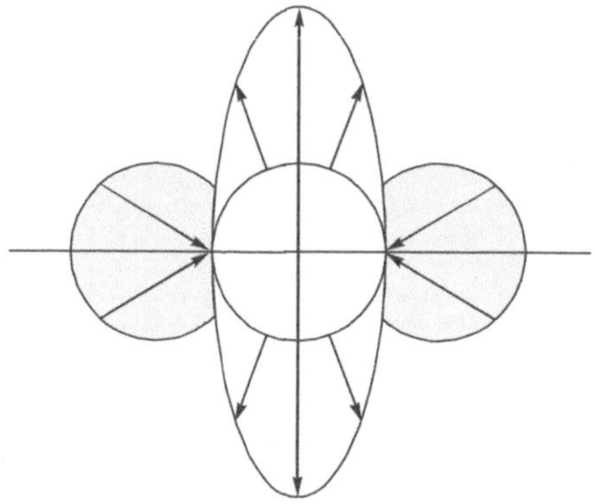

$U = 30$ m/s, $P_0 = 101$ kN/m^2

On the surface of cylinder,

$$U_\theta = -2U\sin\theta$$

$$U_r = 0$$

$\therefore$ At

$$\theta = 90°$$

$$U_\theta = -2U \times 1$$

$$U_\theta = -2 \times 30 = -60 \text{ m/s}$$

Applying Bernoulli's equation,

$$\frac{P_0}{\rho g} + \frac{U^2}{2g} = \frac{P}{\rho g} + \frac{V^2}{2g}$$

$$\Rightarrow \quad \frac{P_0}{\rho g} + \frac{U^2}{2g} = \frac{P}{\rho g} + \frac{[-2U\sin\theta]^2}{2g}$$

$$\rightarrow \quad \frac{P - P_0}{\rho} = \frac{U^2}{2}\left(1 - 4\sin^2\theta\right)$$

$$C_P = \frac{P - P_0}{\frac{1}{2}\rho U^2} = \left(1 - 4\sin^2\theta\right)$$

At stagnation point, $\theta = 0°$ and $180°$

$$\therefore \quad C_P = 1$$

$$\therefore \quad P - P_0 = \frac{1}{2}\rho U^2$$

$$\Rightarrow \quad P = 101 \times 10^3 + \frac{1}{2} \times 1.2 \times 30^2 = 101540 \text{ N/m}^2$$

$$= 101.54 \text{ kN/m}^2$$

At 90° from flow direction, $\theta = 90°$

$$\therefore \quad C_P = 1 - 4\sin^2 90° = 1 - 4 = -3$$

$$\Rightarrow \quad \frac{P - P_0}{\frac{1}{2}\rho U^2} = -3$$

$$\Rightarrow \qquad P = P_0 - \frac{3}{2}\times \rho U^2 = 101\times 10^3 - \frac{3}{2}\times 1.2\times 30^2$$

$$= 99380 \text{ N/m}^2 = 99.38 \text{ kN/m}^2$$

4.14 A convergent-divergent mouthpiece is fitted to the side of a tank. The discharge through the mouthpiece under a constant head of 1.5 m is 5 litres/second. The head loss in the divergent portion is 0.1 times the kinetic head at outlet. Find the throat and exit diameters, if separation pressure is 2.5 m and atmospheric pressure head is 10.3 m of water.

[20 marks]

Sol:

Given: Constant head, $H = 1.5$ m

Discharge, $Q = 5$ litres/sec $= 0.005$ m^3/s

Head loss in divergent portion,

$h_L = 0.1 \times$ Kinetic head at outlet

H_c (or H_{sep}) $= 2.5$ m (absolute)

$H_a = 10.3$ m of water

Let dia at throat is d_c and dia. at outlet is d_1.

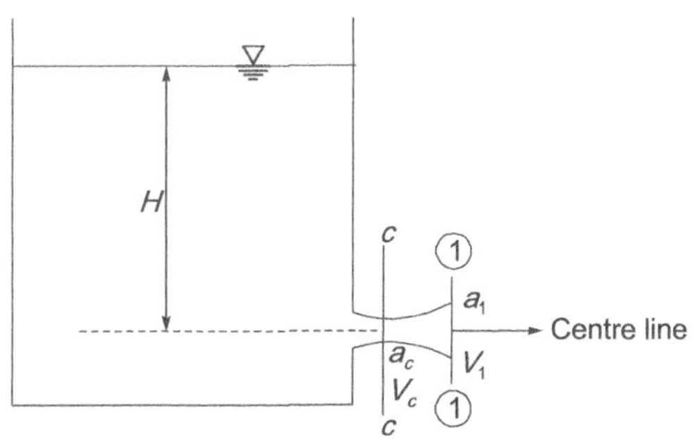

Applying Bernoulli's equation to the free water surface and throat section.

$$\frac{p}{\rho g} + \frac{V^2}{2g} + Z = \frac{p_c}{\rho g} + \frac{V_c^2}{2g} + Z_c$$

Taking the centre line of mouthpiece as datum

$$H_a + 0 + H = H_c + \frac{V_c^2}{2g}$$

$$\therefore \qquad \frac{V_c^2}{2g} = H_a + H - H_c = 10.3 + 1.5 - 2.5 = 9.3 \text{ m of water}$$

$$\therefore \qquad V_c = \sqrt{2\times 9.81\times 9.3} = 13.508 \text{ m/s}$$

Now, $$Q = a_c \times V_c$$

$$\Rightarrow \qquad 0.005 = \frac{\pi}{4} d_c^2 \times 13.508$$

$$\Rightarrow \qquad d_c = 0.0217 \text{ m} = 2.17 \text{ cm} = \text{Throat diameter}$$

Now, applying Bernoulli's equation to the free water surface and outlet of mouthpiece

$$\frac{p}{\rho g}+\frac{V^2}{2g}+Z = \frac{p_1}{\rho g}+\frac{V_1^2}{2g}+Z_1+h_L$$

$$H_a+0+H = H_a+\frac{V_1^2}{2g}+0+0.1\times\frac{V_1^2}{2g} \qquad \left\{\because \frac{p_1}{\rho g}=H_a\right\}$$

$$\Rightarrow \quad H = \frac{V_1^2}{2g}+0.1\times\frac{V_1^2}{2g}=\frac{1.1V_1^2}{2g}$$

$$\Rightarrow \quad V_1 = \sqrt{\frac{2gH}{1.1}}=\sqrt{\frac{2\times 9.81\times 1.5}{1.1}} = 5.1725 \text{ m/s}$$

Now,

$$Q = A_1V_1$$

$$\Rightarrow \quad 0.005 = \frac{\pi}{4}d_1^2\times V_1$$

$$\Rightarrow \quad d_1 = \sqrt{\frac{4\times 0.005}{\pi\times 5.1725}} = 0.035 \text{ m} = 3.5 \text{ cm} = \text{Exit diameter}$$

5. Weir and Notches

Level-1

5.1 Find the discharge of water through a stepped notch having the following dimensions ($L \times H$). Top portion 1 m × 0.4 m, middle portion 0.9 m × 0.3 m and the bottom portion 0.5 m × 0.22 m. The water level is at top of the notch. Assume $C_d = 0.6$.

[10 Marks]

Sol:

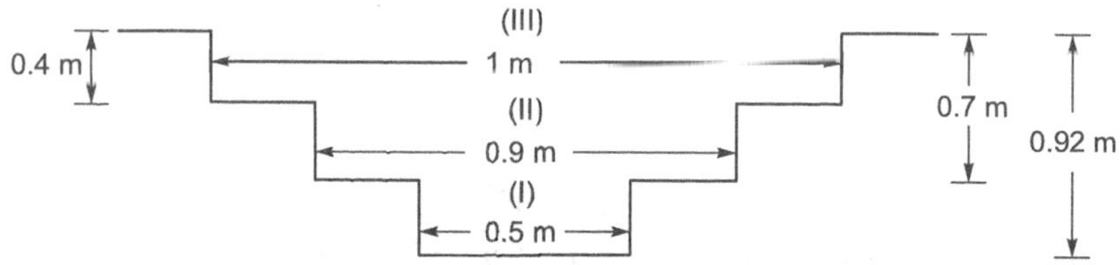

$$\text{Total discharge} = Q_1+Q_2+Q_3$$

$$Q_1 = \frac{2}{3}C_d\sqrt{2g}\times\left(0.92^{3/2}-0.7^{3/2}\right)\times 0.5$$

$$= 0.2642 \text{ m}^3\text{/sec} \qquad (C_d = 0.6)$$

$$Q_2 = \frac{2}{3}C_d\sqrt{2g}\times\left(0.7^{3/2}-0.4^{3/2}\right)\times 0.9 = 0.533 \text{ m}^3\text{/sec}$$

$$Q_3 = \frac{2}{3}C_d\sqrt{2g}\times\left(0.4^{3/2}\right)\times 1 = 0.451 \text{ m}^3\text{/sec}$$

$$Q = Q_1+Q_2+Q_3 = 0.2642+0.533+0.451 = 1.2482 \text{ m}^3\text{/sec}$$

5.2 In a laboratory experiment with a right angled V-notch, the discharge is calculated to be 0.05 m³/min, taking discharge coefficient of 0.63. However, the actual discharge is known to be 0.0555 m³/min. Assuming that this discrepancy is due to an error in measuring head above the sill. Calculate the error in millimeters.

[10 Marks]

Sol:

Given data : $Q_{act} = 0.0555\ m^3/min$, $Q_{th} = 0.05\ m^3/min$, $C_d = 0.63$

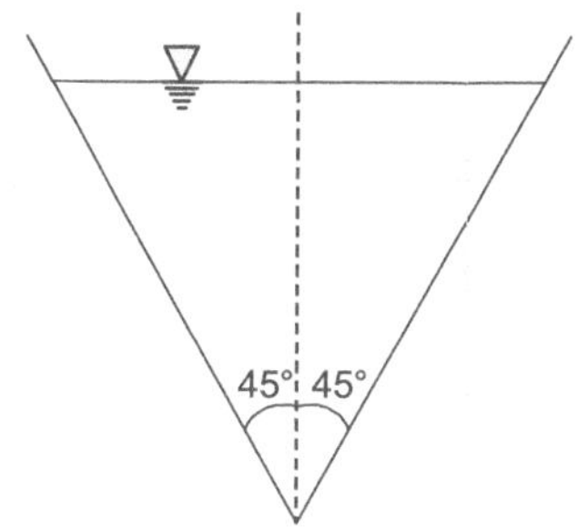

$$\frac{dQ}{Q} = \frac{5}{2}\frac{dH}{H} \quad \left(\text{From } Q \propto H^{5/2}\right)$$

$$\frac{dH}{H} = \frac{2}{5}\times\frac{0.0555-0.05}{0.055}$$

$$dH = \frac{2}{5}\times\frac{0.0055}{0.055}\times H = ??$$

Also,

$$Q = \frac{8}{15}C_d\sqrt{2g}\times H^{5/2}\tan\frac{\theta}{2}$$

$$\frac{0.05}{60} = \frac{8}{15}\times 0.63\times\sqrt{2\times 9.81}\times H^{5/2}$$

Solving, $H = 0.052\ m$

$$dH = \frac{2}{5}\times\frac{0.0055}{0.055}\times 0.052 = 1.89\ mm$$

5.3 The catchment area of a tank is 40 km^2. The maximum rainfall over the catchment area is 4 cm/hr and only 40% of the rain reaches to the tank. The waste water weir has to carry piers of 0.6 m wide and 2 m centre to centre apart for supporting a footway. Calculate the number of piers. If depth of water over the weir is not move then 0.6 m. [Take $C_d = 0.6$]

[10 Marks]

Sol:

Applying the equation of discharge for rectangular weir,

$$Q = \frac{2}{3}\times C_d\times\sqrt{2g}\times l_{eff}(H)^{3/2}$$

$$Q = 40\times 10^6\times\frac{0.04}{3600}\times 0.4 = 177.77\ m^3/sec$$

$$177.77 = \frac{2}{3}\times 0.6\times\sqrt{2\times 9.81}\times l_{eff}\times(0.6)^{3/2}$$

Solving, $l_{eff} = 215.89\ m$

Effective length of one spillway section $= L - 0.01\ nH$

L = Clear span $= 2 - 0.6 = 1.4\ m = 1.4 - 0.1\times 2\times 0.6 = 1.28\ m$

$$\text{Number of piers} = \frac{215.8}{1.28} = 168 \text{ piers}$$

5.4 The flow rate of water flowing in a 5 m-wide channel is to be measured with a sharp-crested triangular weir 0.5 m above the channel bottom with a notch angle of 80°. If the flow depth upstream from the weir is 1.5 m, determine the flow rate of water through the channel. Take $C_d = 0.6$.

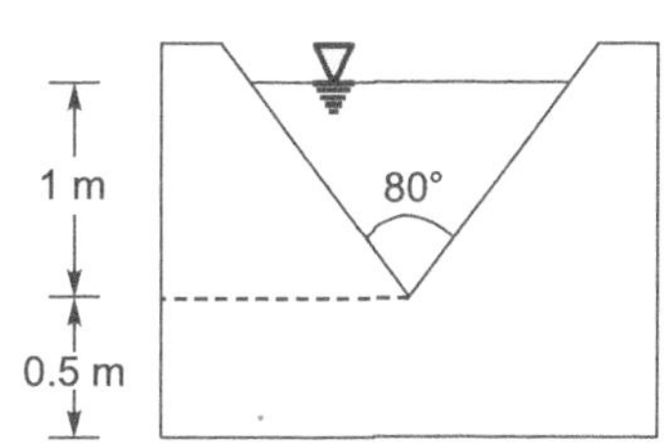

[6 Marks]

Sol:

For sharp crested triangular weir the discharge is given by,

$$Q = \frac{8}{15} C_d \sqrt{2g} \times H^{5/2} \tan\frac{\theta}{2}$$

where, H = Head over the weir/notch = 1 m

θ = 80° (given)

Putting $$Q = \frac{8}{15} \times 0.6 \times \sqrt{2 \times 9.81} \times (1)^{5/2} \tan 40°$$

$$= 1.19 \text{ m}^3/\text{sec} \qquad \text{(Ans)}$$

Level-2

5.5 Determination of discharge over a broad crested weir 30 m long with rounded entrance. The measured upstream level over the crest being 65 cm. The width of channel of approach is 40 m and its depth below the crest of weir is 60 cm. Apply the correction for velocity of approach after 1st approximation. [Take C_d = 0.95]

[10 Marks]

Sol:

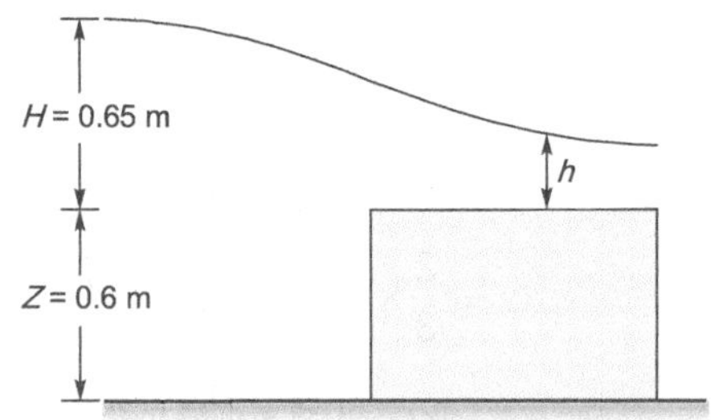

$$Q = 1.7\, GL \times H^{3/2} = 1.7 \times 0.95 \times 30 \times (0.65)^{3/2} = 25.39 \text{ m}^3/\text{sec}$$

$$V = \frac{Q}{A} = \frac{25.39}{(40)(0.65+0.6)} = 0.51 \text{ m/sec}$$

$$Q = 1.7 C_d \times L \times \left[(H + h_a)^{3/2} - h_a^{\ 3/2}\right]$$

where, h_a = Velocity head = $\frac{V_a^2}{2g} = \frac{(0.51)^2}{2 \times 9.81} = 0.0131$

$$Q = 1.7 \times 0.95 \times 30 \times \left[(0.65 + 0.0131)^{3/2} - (0.0131)^{3/2}\right] = 29.099 \text{ m}^3/\text{sec}$$

5.6 A cylindrical tank 2.5 m diameter and 12 m long is installed with its axis horizontal. It holds water up to a depth of 2 m. There is a circular opening of 7.5 cm in diameter at the bottom, which is kept plugged. If the plug is removed to drain the water completely, estimate how long does it take to completely drain the water. C_d for the opening is 0.6

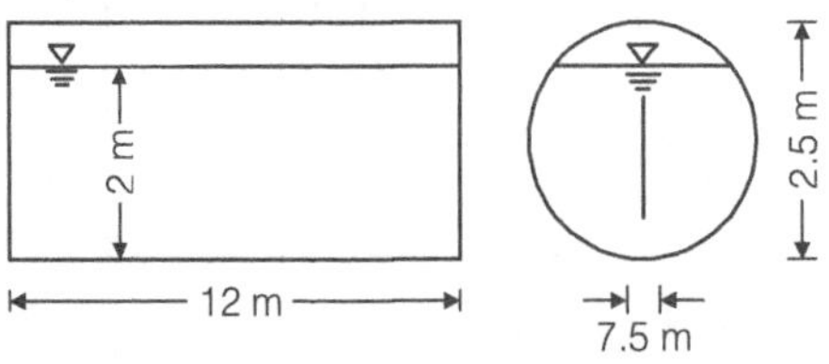

[10 Marks]

Sol:

As we know

Time required to completely drain the water out of the tank is given by

$$T = \frac{4L}{3C_d \times a \times \sqrt{2g}}\left[(2R)^{3/2} - (2R - H_1)^{3/2}\right]$$

Where, R = Radius of tank

L = Length of tank

H_1 = Height of liquid

a = Area of opening

Given,

Diameter of tank = 2.5 m

∴ Radius of tank R = 1.25 m

Length of tank L = 12 m

H_1 = 2 m

C_d = 0.6

$$a = \frac{\pi}{4} \times (7.5 \times 10^{-2})^2 = 4.418 \times 10^{-3}\ m^2$$

∴ $$T = \frac{4 \times 12}{3 \times 0.6 \times 4.418 \times 10^{3} \times \sqrt{2 \times 9.81}}$$

$$= \left[(2.5)^{3/2} - (2.5 - 2)^{3/2}\right] = 4904.68 \text{ seconds}$$

$$= 81.74 \text{ minutes}$$

6. Flow Through Pipes

Level-1

6.1 Calculate the velocity of propagation of pressure wave in steel pipe $E = 2.07 \times 10^5$ MPa of 2.5 diameter carrying Kerosene ($R_D = 0.8$) and $K = 1.43 \times 10^3$ MPa. The pipe thickness is 2 cm.

[6 Marks]

Sol:

$$\text{Velocity of pressure wave } (c) = \sqrt{\frac{\frac{k}{\rho}}{1 + \frac{DK}{Et}}}$$

$$c = \sqrt{\frac{\frac{1.43 \times 10^3 \times 10^6}{10^3 \times 0.8}}{1 + \frac{250 \times 1.43 \times 10^3}{2.07 \times 10^5 \times 2}}}$$

Solving, $c = 0.979 \times 10^3$ m/sec

6.2 A 15 cm diameter pipe is joined to a 30 cm diameter pipe by a reducing flange. For water flowing at a rate of 0.115 m^3/s, the head loss when water flows from the smaller to the larger diameter pipe is

[8 Marks]

Sol:

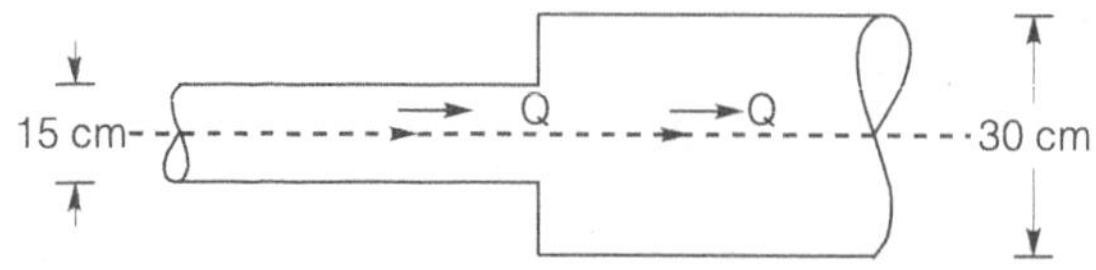

Total head loss, $h_L = \frac{(v_1 - v_2)^2}{2g}$

From continuity equation

$$A_1 v_1 = A_2 v_2$$

$$\Rightarrow \quad \frac{\pi}{4} d_1^2 . v_1 = \frac{\pi}{4} d_2^2 . v_2$$

$$\Rightarrow \quad d_1^2 v_1 = d_2^2 . v_2$$

$$\Rightarrow \quad (15)^2 v_1 = (30)^2 v_2$$

$$\Rightarrow \quad v_1 = 4v_2$$

As the pipes are connected in series therefore discharge in both the Pipe must be same

$\Rightarrow Q_1 = Q_2 = Q = 0.115$ m³/sec

Also $Q = A_1 v_1$

$$\Rightarrow \quad 0.115 = \frac{\pi}{4}(0.15)^2 . v_1$$

$$\Rightarrow \quad v_1 = 6.508 \text{ m/sec}$$

$$\therefore \quad v_2 = \frac{v_1}{4} = 1.627 \text{ m/s}$$

From equation (i)

$$h_L = \frac{(v_1 - v_2)^2}{2g} = \frac{(6.508 - 1.627)^2}{2 \times 9.81} = 1.214 \text{ m}$$

6.3 A straight pipe AB, of length 10 m, tapers from a diameter of 40 cm at A to 20 cm at B. The centre line of the pipe is so located that the end B is 2 m above the level of A. Liquid of specific gravity 0.9 flows through the pipe at 150 litres/second. Pressure gauges connected at A and B show the readings of 60 kPa and 40 kPa, respectively. Determine the direction of flow.

[10 Marks]

Sol:

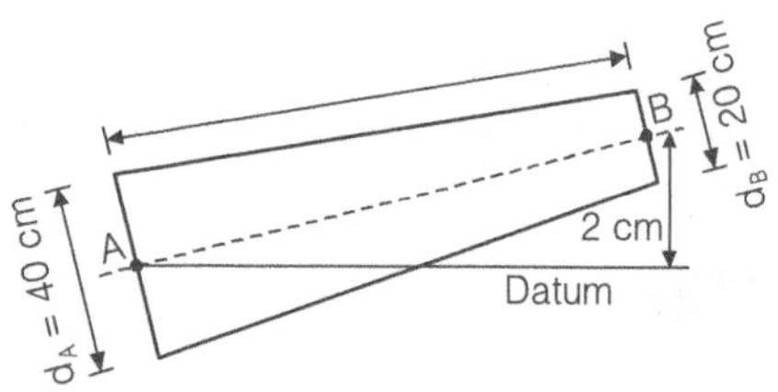

At A: Diameter,

$$d_A = 40 \text{ cm} = 0.4 \text{ m}$$

$\therefore$ Area at A

$$A_A = \frac{\pi}{4} \times 0.4^2 = 0.1256 \text{ m}^2$$

At B: Diameter,

$$d_B = 20 \text{ cm} = 0.2 \text{ m}$$

$\therefore$ Area of B,

$$A_B = \frac{\pi}{4} \times 0.2^2 \approx 0.0314 \text{ m}^2$$

Pressure at A = 60 kPa

$$p_A = 60 \text{ kN/m}^2$$

Pressure at B

$$p_B = 40 \text{ kPa} = 40 \text{ kN/m}^2$$

$$Q = 150 \text{ litres/sec} = 0.15 \text{ m}^3\text{/sec}$$

$$\therefore \quad V_A = \frac{Q}{A_A} = \frac{0.15}{0.1256} = 1.19 \text{ m/s}$$

$$V_B = \frac{Q}{A_B} = \frac{0.15}{0.0314} = 4.78 \text{ m/s}$$

∴ Total head at A

$$h_A = \frac{p_A}{\gamma_w} + \frac{V_A^2}{2g} + Z_A = \frac{60}{0.9 \times 9.81} + \frac{(1.19)^2}{2 \times 9.81} + 0 = 6.8 + 0.072$$

$$= 6.872 \text{ m}$$

Total head at B,

$$h_B = \frac{p_B}{\gamma_w} + \frac{V_B^2}{2g} + Z_B = \frac{40}{0.9 \times 9.81} + \frac{4.78^2}{2.981} + 2 = 4.53 + 1.162 + 2 = 7.698 \text{ m}$$

Since head at B in greater then head at A, the liquid flows from B to A

6.4 Power is to be transmitted hydraulically along a distance of 8 km through a number of 10 cm diameter pipes, laid in parallel. The pressure at the discharge end is maintained constant at 6650 kPa. Determine the minimum number of pipes required to ensure an efficiency of at least 90% when the power delivered at discharge end is 150 kW. Take friction coefficient f = 0.0075 for all pipes and neglect losses other than pipe friction.

[10 Marks]

Sol:

$$\text{Power delivered} = \dot{m}g(H - h_f) = 6650 \text{ KPa}$$

$$\text{Head form, } (H - h_f) = \frac{6650 \times 10^3}{10^3 \times 9.81} = 677.88 \text{ m}$$

$$150 \times 10^3 = \dot{m}g \times 677.88$$

Solving, $Q = 22.56$ *l*/sec

Given efficiency = 90%

$$0.9 = \frac{H - h_f}{H}$$

$$0.9 = \frac{677.88}{H}$$

⇒ $H = 753.2$ m

$h_f = 75.32$ m

From Darcy Weisback equation,

$$h_f = \frac{8(Q(n)^2)}{\pi^2 g} \times \frac{fL}{D^6}$$

$$75.36 = \frac{8 \times (0.02256)^2}{n^2 \times \pi^2 \times g} \times \frac{0.0075 \times 4 \times 8 \times 10^3}{(0.1)^5}$$

Solving, $n = 3.66$

Provide 4 number of pipes.

6.5 Two pipes A and B are connected in parallel between two points M and N as shown in the figure. Pipe A is of 80 mm diameter, 900 m long and its friction factor is 0.015. Pipe B is of 100 mm diameter, 700 m long and its friction is 0.018. A total discharge of 0.030 m³/s is entering the parallel pipes throughout the division at M. Calculate the discharge in the two pipes A and B.

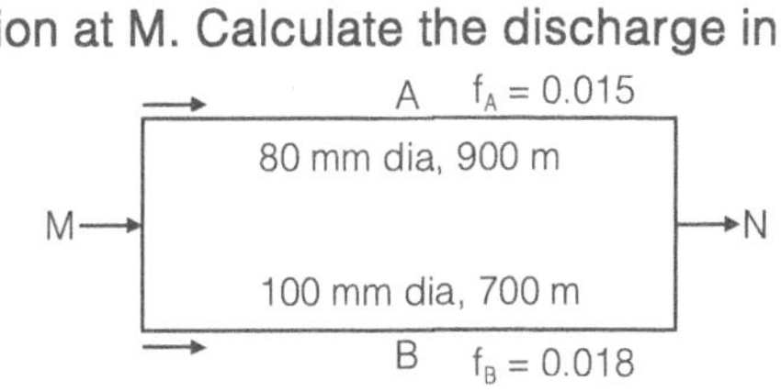

[12 Marks]

Sol:

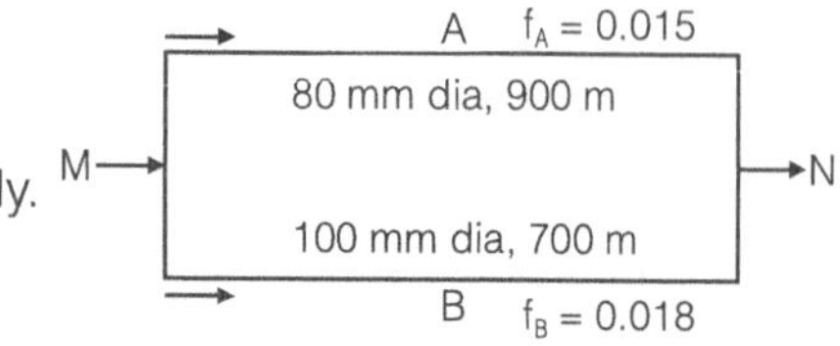

Total discharge, $Q = 0.03\ m^3/s$

Let Q_A and Q_B be the discharges in pipe A and pipe B respectively.

Length of pipe A, $L_A = 900\ m$

Diameter of pipe A, $d_A = 80\ mm = 0.08\ m$

Friction factor of pipe A, $f_A = 0.015$

Friction factor of pipe B, $f_B = 0.018$

Diameter of pipe B, $d_B = 100\ mm = 0.1\ m$

Length of pipe B, $L_B = 700\ m$

$$Q = Q_A + Q_B \quad ...(i)$$

Also,. head loss across pipe A = Head loss across pipe B

$$\Rightarrow \frac{f_A L_A V_A^2}{2gd_A} = \frac{f_B L_B V_B^2}{2gd_B}$$

$$\Rightarrow \frac{0.015 \times 900 \times V_A^2}{0.080} = \frac{0.018 \times 700 \times V_B^2}{0.1}$$

$$\Rightarrow V_A = 0.864\ V_B$$

$$\therefore Q_A = \frac{\pi}{4} \times d_A^2 \times V_A = \frac{\pi}{4} \times (0.08)^2 \times 0.864 V_B = 4.34 \times 10^{-3}\ V_B$$

$$Q_B = \frac{\pi}{4} \times d_B^2 \times V_B = \frac{\pi}{4} \times (0.1)^2 \times V_B = 7.85 \times 10^{-3}\ V_B$$

Putting the value of Q_A and Q_B in eq. (i) we get

$$0.03 = 4.34 \times 10^{-3}\ V_B + 7.85 \times 10^{-3}\ V_B$$

$$\Rightarrow V_B = 2.461\ m/s$$

$$\therefore V_A = 2.126\ m/s$$

Hence,

$$Q_A = \frac{\pi}{4} \times (0.08)^2 \times 2.126 = 0.01\ m^3/s$$

$$Q_B = \frac{\pi}{4} \times (0.1)^2 \times 2.461 = 0.02\ m^3/s$$

Level-2

6.6 Three reservoirs *A*, *B* and *C* are connected by a pipe system shown in figure. Find the discharge into or from the reservoirs *B* and *C*. If the rate of flow from reservoir *A* is 60 *l*/sec. Find the height of water level in the reservoir *C*. Z_c [Take, friction coefficient, $f = 0.006$]

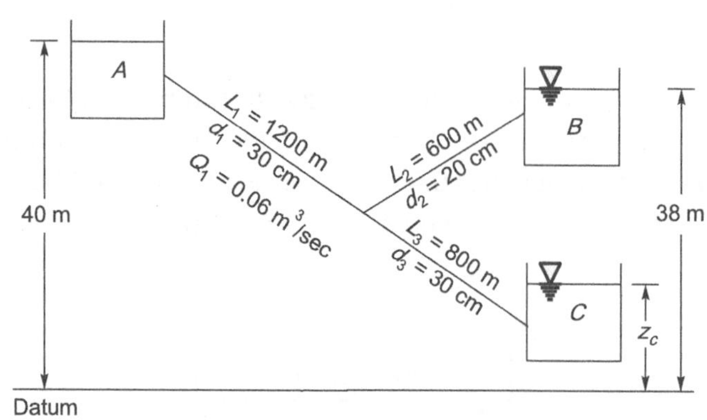

[20 Marks]

Sol:

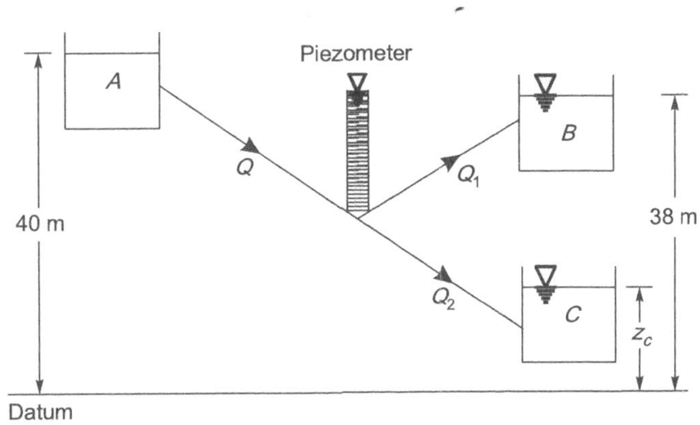

Now, applying Bernoulli's equation between junction and reservoir A

$$\frac{P_A}{\rho g} + \frac{V_A^2}{2g} + Z_A = \frac{P_J}{\rho g} + \frac{V_J^2}{2g} + Z_J + h_{fL}$$

$$40 - h_{fL} = \left(\frac{P_J}{\rho g} + Z_B\right)$$

Taking $$h_{fL} = \frac{fLV^2}{2gd}$$

$$= \frac{0.006 \times 1200 \times 0.85^2}{2 \times 9.81 \times 0.3} = 3.54 \text{ m}$$

$$V = \frac{0.06}{\frac{\pi}{4} \times 0.3^2} = 0.85 \text{ m/sec}$$

Hence elevation of Piezometer = 40 – 3.54 = 36.47 m

Since $Z_B > Z_J = 0$

Hence, flow is from *B* to *J*.

Apply Bernoulli's equation between *B* and *J*

$$h_{fL} = \frac{0.006 \times 4 \times 600 \times V_2^2}{2 \times 9.81 \times 0.2} = 38 - 36.47$$

Solving $V_2 = 0.645$ m/sec

$Q_2 = 20.2$ *l*/sec

$Q_3 = 60 + 20.26 = 80.26$ *l*/sec

Apply Bernoulli's equation between junction and *C*

$$h_{fL} = 36.47 - Z_c$$

where $$V_3 = \frac{80.26 \times 10^{-3}}{\frac{\pi}{4} \times 0.3^2} = 1.14 \text{ m/sec}$$

$$36.47 - Z_c = \frac{0.006 \times 4 \times 800 \times 1.14^2}{2 \times 9.81 \times 0.3}$$

Solving $Z_c = 32.233$ m **(Ans)**

6.7 Between reservoir 1 and 2, a pump *P* and a turbine *T* are provided. The pipe line characteristics are shown in figure. Away with Piezometric heads at four places. Determine

(i) The discharge and power supplied by pump.

(ii) Power removed by turbine.

(iii) Elevation of reservoir 2.

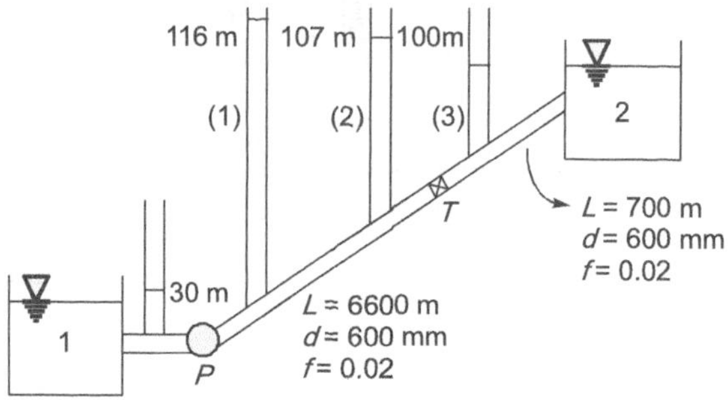

[15 Marks]

Sol:

Neglecting all minor losses. Applying Bernoulli's equation between Piezometer (1) and (2)

$$h_{fL} = \frac{fLV^2}{2gd} \quad \text{(Darcy-Weisbach)}$$

$$116 - 107 = \frac{0.02 \times 6600 \times V^2}{2 \times 9.81 \times 0.6}$$

Solving, $V = 0.896$ m/sec

Discharge, $Q = A \times V$

$$= \frac{\pi}{4} \times 0.6^2 \times 0.896 = 0.253 \text{ m}^3\text{/sec}$$

Apply Bernoulli's equation between Piezometer (2) and (3),

$$\text{Head loss} = 107 - 100.9 = 6.1 \text{ m}$$

$$\text{Hence power removed by turbine} = \dot{m}gh_f$$

$$= \rho Q \times g \times h_f = 10^3 \times 0.253 \times 9.81 \times 6.1 = 15.13 \text{ kW}$$

$$\text{Head added by pump} = 116 - 30 = 86$$

$$\text{Power supplied} = \dot{m}gh$$

$$= 10^3 \times 0.253 \times 9.81 \times 86 = 213.44 \text{ kW}$$

Now apply Bernoulli's equation between Piezometric (3) and (2)

$$100.9 - Z_2 = \frac{fLV^2}{2gd} = \frac{0.02 \times 700 \times 0.896^2}{2 \times 9.87 \times 0.6}$$

Solving, $Z_2 = 99.945$ m

6.8 A 700 mm diameter water main for domestic water supply runs horizontally for 1600 m and then branches into two 450 mm diameter pipes of 3000 m length each. In the first branch, the entire discharge is withdrawn at the uniform rate along its length, while in the second branch half of the discharge entering the pipe is withdrawn at a uniform rate along its length. Assume $f = 0.025$ for all the pipes, the flow entering 700 mm diameter pipe to be 0.3 m³/sec and pressure at the outlet of two branches to be atmospheric and calculate the head loss between entrance and exit.

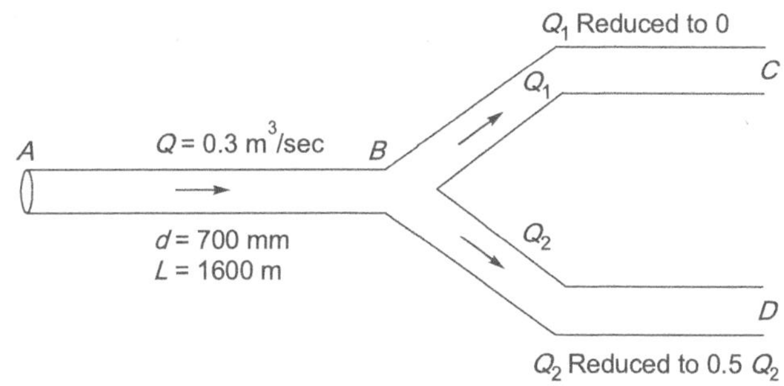

[20 Marks]

Sol:

$$\text{Head loss in } AB = \frac{8Q^2}{\pi^2 g} \times \frac{fL}{D^5} = \frac{8 \times 0.3^2}{\pi^2 \times 9.81} \times \frac{0.025 \times 1600}{(0.7)^6} = 1.77 \text{ m}$$

$$\text{Head loss in } BC = \frac{1}{3}\left[\frac{8Q_1^2}{\pi^2 g} \cdot \frac{fL}{D^5}\right]$$

Head loss in BD = ??

Using
$$dh_f = \frac{8Q_x}{\pi^2 g} \times \frac{f \times dx}{d^5}$$

$$Q_x = Q - q \times x$$

$$h_f = \int dh_f = \frac{8}{\pi 2g} \times \frac{f}{d^5} \int_{x=0}^{L} (Q - qx)^2\, dx$$

$$= \frac{8}{\pi^2 g} \times \frac{f}{d^5} \times \frac{1}{-q} \times \frac{1}{3}\left[(Q - qx)^3\right]_0^L$$

$$= -\frac{1}{3q}\left[\frac{8}{\pi^2 g} \times \frac{f}{d^5}\right]\left[(Q - qL)^3 - Q^3\right]$$

Since
$$qL = \frac{Q}{2} \qquad \text{(Given)}$$

$$= -\frac{1}{3q}\left[\frac{8}{\pi^2 g} \times \frac{f}{d^5}\right]\left[\frac{Q^3}{8} - Q^3\right] = -\frac{1}{3q}\left[\frac{8}{\pi^2 g} \times \frac{f}{d^5}\right]\left[\frac{7}{8}Q^3\right]$$

$$= \frac{2L}{3 \times Q}\left(\frac{8}{\pi^2 g} \times \frac{f}{d^5}\right)\left(\frac{7}{8}Q^3\right) = \frac{14}{3}\left[\frac{Q_2^2}{\pi^2 g} \times \frac{fL}{D_2^5}\right]$$

Since head loss in BC = Head loss in BD

$$\frac{8}{3} \times \frac{fL_1 Q_1^2}{\pi^2 g D_1^5} = \frac{14}{3} \times \frac{fL_2 Q_2^2}{\pi^2 g D_2^5}$$

$$4Q_1^2 = 7Q_2^2$$

$\Rightarrow$
$$Q_1 = 1.33Q_2$$

Also
$$Q = Q_1 + Q_2$$

$$0.3 = Q_2 + 1.33Q_2$$

$\Rightarrow$
$$Q_2 = \frac{0.3}{2.33} = 0.129 \text{ m}^3/\text{sec}$$

$$Q_1 = 0.3 - 0.129 = 0.171 \text{ m}^3/\text{sec}$$

$$\text{Head loss in } BC = \frac{8\times 0.025\times 3000\times (0.171)^2}{3\times (3.142)^2\times 9.81\times (0.45)^5} = 3.274 \text{ m}$$

$$\text{Total head loss} = 1.77 + 3.274 = 5.044 \text{ m}$$

6.9 Two reservoirs are connected by a 40 cm diameter and 750 m long pipe. The level difference of the water surface levels of the two reservoirs is 10 m. The pipeline has to cross a ridge at a distance of 250 m from the upper reservoir. Find the maximum height of the pipeline at the ridge point above the upper reservoir water level at which the pipeline can be laid and compute the discharge in the pipeline. The local barometric pressure is 10 m of water, the vapour pressure at the maximum local temperature is 2.5 m of water and a positive pressure of 1.5 m is desirable above this vapour pressure. Assume coefficient of friction of the pipe 'f' = 0.025.

[20 Marks]

Sol:

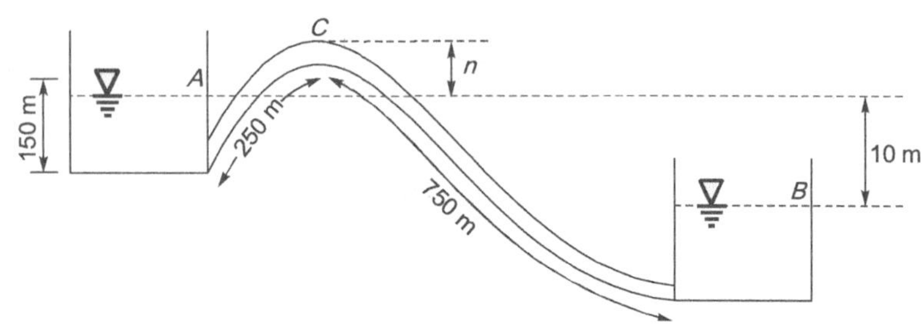

Let the maximum height of the pipeline at ridge point above the upper reservoir water level be 'h' and let discharge be 'Q' in the pipeline.

Condition given : $\frac{P_C}{\rho g}$ i.e. pressure head at $C \geq \frac{P_v}{\rho g} + 1.5$ m

$$\frac{P_C}{\rho g} \geq 2.5 + 1.5$$

$$\frac{P_C}{\rho g} \geq 4 \text{ m}$$

Total head at A, $$H_A = \frac{P_A}{\rho g} + z_A = 10 + 10 = 20 \text{ m}$$

Total head at B, $$H_B = \frac{P_B}{\rho g} + z_B = 10 + 10 = 10 \text{ m}$$

Total head at C, $$H_C = \frac{P_C}{\rho g} + z_C = 4 + (10 + h) = 14 + h$$

Head loss between A and C,

$$H_A - H_C = h_{f_{AC}}$$

$$20 - (14 + h) = \frac{8\cdot f\cdot L_{AC}\cdot Q^2}{\pi^2 g d^5}$$

$$6 - h = \frac{8\times 0.025\times 250\cdot Q^2}{\pi^2\times 9.81\times (0.4)^5}$$

$$6 - h = 50.4314\, Q^2 \qquad \text{... (i)}$$

Head loss between C and B

$$H_C - H_B = h_{f_{CB}}$$

$$(14+h)-10 = \frac{8\cdot f\cdot L_{CB}\cdot Q^2}{\pi^2 g d^5}$$

$$4+h = \frac{8\times 0.025\times 500}{\pi^2\times 9.81\times(0.4)^5}\times Q^2$$

$$4+h = 100.8629\,Q^2 \quad \text{... (ii)}$$

Adding equation (i) and (ii)

$$10 = 151.2943\,Q^2$$

$\Rightarrow$ $$Q = 0.25709\ \text{m}^3/\text{s}$$

from equation (ii)

$\therefore$ $$h = 2.6667\ \text{m}$$

6.10 **Water flows from a reservoir through a pipe of 0.15 m diameter and 180 m long to a point 13.5 m below the open surface of the reservoir. Here it branches into two pipes each of 0.1 m diameter one of which is 48 m long discharging to atmosphere at a point 18 m below reservoir level, and the other 60 m long discharging to atmosphere 24 m below reservoir level. Assuming a constant friction coefficient of 0.032, calculate the discharge from each pipe. Neglect any losses at the junction.**

[20 Marks]

Sol:

The figure below shows the layout of the reservoir and pipes.

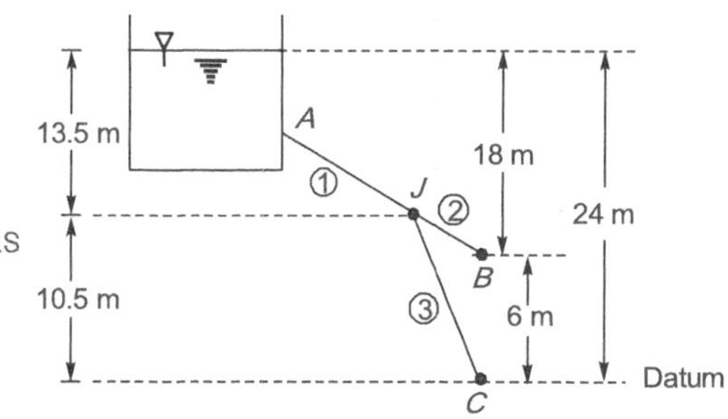

Given data:

$d_1 = 0.15$ m $\quad d_2 = 0.1$ m $\quad d_3 = 0.1$ m

$L_1 = 180$ m $\quad L_2 = 48$ m $\quad L_3 = 60$ m

Friction factor, $f = 0.032\times 4 = 0.128$ for all pipes

As per Darcy-Weisbach equation head loss is expressed as

$$h_f = rQ^n = \left(\frac{fL}{12.1d^5}\right)Q^2$$

If open to atm. then head loss is expressed as

$$h_f = \left(\frac{fL}{12.1d^5}\right)Q^2 + \frac{V^2}{2g}$$

$\therefore$ $$r_1 = \frac{fL_1}{12.1(d_1)^5} = \frac{0.128\times 180}{12.1\times(0.15)^5} = 25075$$

$$r_2 = \frac{fL_2}{12.1(d_2)^5}+\frac{1}{12.1(d_2)^4} = \frac{0.128\times 48}{12.1\times(0.1)^5}+\frac{1}{12.1\times(0.1)^4} = 51603.306$$

$$r_3 = \frac{fL_3}{12.1(d_3)^5}+\frac{1}{12.1(d_3)^4} = \frac{0.128\times 60}{12.1\times(0.1)^5}+\frac{1}{12.1\times(0.1)^4} = 64297.521$$

Taking flow towards junction J as positive and away from junction J as negative. Assuming piezometric head at J as $H_j = 10.5$ m

1st trial

Pipe	r	Estimated h_f	$Q=\left(\frac{h_f}{r}\right)^{1/2}$	$\left\lvert\frac{Q}{h_f}\right\rvert$
1	25075	24 − 10.5 = 13.5	0.0232	0.00172
2	51603.306	6 − 10.5 = − 4.5	− 0.00934	0.00207
3	64297.521	0 − 10.5 = − 10.5	− 0.0128	0.00122
		Total	$\Delta Q = 0.0011$	0.00501

$$\therefore \quad \Delta H_j = \frac{n \times \Delta Q}{\Sigma \left|\frac{Q}{h_f}\right|} = \frac{2 \times 0.0011}{0.00501} = +0.44 \text{ m}$$

H_j for 2nd trial = 10.5 + 0.44 = 10.94 m

2nd trial

Pipe	r	Estimated h_f	$Q = \left(\frac{h_f}{r}\right)^{1/2}$	$\left\|\frac{Q}{h_f}\right\|$
1	25075	24 – 10.94 = 13.06	0.0228	0.00174
2	51603.306	6 – 10.94 = – 4.94	– 0.0098	0.001984
3	64297.521	0 – 10.94 = – 10.94	– 0.013	0.00119
		Total	ΔQ = 0	0.004914

$$\therefore \quad \Delta H_j = \frac{n \times \Delta Q}{\Sigma \left|\frac{Q}{h_f}\right|} = \frac{2 \times 0}{0.004914} = 0$$

Hence, the refined value of H_j = 10.94 m

Applying Bernoulli's equation between A and J, we get

$$H_A = H_J + r_1 Q_1^2$$

$$\Rightarrow \quad 24 = 10.94 + 25075 Q_1^2$$

$$\Rightarrow \quad Q_1^2 = 5.2084 \times 10^{-4}$$

$$\Rightarrow \quad Q_1 = 0.0228 \text{ m}^3\text{/s}$$

Applying Bernoulli's equation between J and B, we get

$$H_J = H_B + r_2 Q_2^2$$

$$\Rightarrow \quad 10.94 = 6 + 51603.306\, Q_2^2$$

$$\Rightarrow \quad Q_2^2 = 9.573 \times 10^{-5}$$

$$\Rightarrow \quad Q_2 = 0.00978 \text{ m}^3\text{/s}$$

By continuity equation, we get

$$Q_3 = Q_1 - Q_2 = 0.0228 - 0.00978 = 0.01302 \text{ m}^3\text{/s}$$

$$Q_1 = 0.0228 \text{ m}^3\text{/s (towards junction)}$$

$$Q_2 = 0.00978 \text{ m}^3\text{/s (away from junction)}$$

$$Q_3 = 0.01302 \text{ m}^3\text{/s (away from junction)}$$

Alternate Method

Applying Bernauli's between point A and J (considering Jasdalum and gauge pressures)

$$\frac{P_A}{\rho g} + \frac{V_A^2}{2} + Z_A = \frac{P_J}{\rho g} + \frac{V_J^2}{2g} + Z_J + h_f$$

$$13.5 = \frac{P_J}{\rho g} + 0 + 7.828 V_1^2 \quad \text{...(i) } [V_A = 0]$$

Similarly, applying Bernauli's between J and B

$$\frac{P_J}{\rho g} + \frac{V_J^2}{2g} + Z_J = \frac{P_B}{\rho g} + \frac{V_2^2}{2g} + Z_B + h_f$$

$$\frac{P_J}{\rho g} = 3.182\,V_2^2 - 4.5 \qquad \text{...(ii)}$$

and applying Bernauli's between J and C

$$\frac{P_J}{\rho g} + \frac{V_J^2}{2g} + Z_J = \frac{P_C}{\rho g} + \frac{V_3^2}{2g} + Z_C + h_f$$

$$\frac{P_J}{\rho g} = 3.965\,V_3^2 - 10.5 \qquad \text{...(iii)}$$

From eq. (i) and (ii)

$$18 = 3182\,V_2^2 + 7.828\,V_1^2 \qquad \text{...(iv)}$$

Similarly from eq. (i) and (iii)

$$24 = 3.965\,V_3^2 + 7.828\,V_1^2 \qquad \text{...(v)}$$

Applying continuity equation at junction J

$$A_1V_1 = A_2V_2 + A_3V_3$$

$$\frac{\pi}{4} \times 0.15^2 V_1 = \frac{\pi}{4} \times 0.1^2 \times V_2 + \frac{\pi}{4} \times 0.1^2 \times V_3$$

$$2.25V_1 = V_2 + V_3 \qquad \text{...(vi)}$$

Consider $V_2 = mV_1$

Solving eq. (iv), (v) and (vi)

$$m = 0.964$$

$$V_1 = 1.292 \text{ m/s} \Rightarrow Q_1 = 0.0228 \text{ m}^3\text{/s}$$

$$V_2 = 1.245 \text{ m/s} \Rightarrow Q_2 = 0.00978 \text{ m}^3\text{/s}$$

$$V_3 = 1.662 \text{ m/s} \Rightarrow Q_3 = 0.0180 \text{ m}^3\text{/s}$$

(direction of flow at shown in fig.)

6.11 **A horizontal pipe line 40 m long is connected to a water tank at one end and discharges freely into the atmosphere at the other end. For the first 25 m of its length from the tank, the pipe is 150 mm in diameter and its diameter is suddenly enlarged to 300 mm thereafter. The height of water level in the tank is 8 m above the centre of the pipe. Considering all losses of head which occur, determine the rate of flow. Take coefficient of friction, f = 0.01 for both sections of the pipe.**

[20 marks]

Sol:

Given:

Total length of pipe, $L = 40$ m

Length of 1st pipe, $L_1 = 25$ m

Diameter of 1st pipe, $d_1 = 150$ mm $= 0.15$ m

Length of 2nd pipe, $L_2 = 40 - 25 = 15$ m

Diameter of 2nd pipe, $d_2 = 300$ mm $= 0.3$ m

Height of water, $H = 8$ m

Co-efficient of friction, $f = 0.01$

Applying Bernoulli's theorem to the free surface of water in the tank and outlet of pipe as shown in figure and taking reference line passing through the centre of pipe,

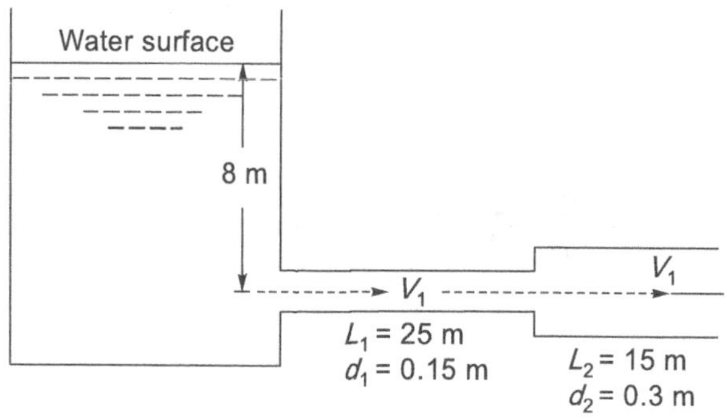

$$0 + 0 + 8 = \frac{p_2}{\rho g} + \frac{V_2^2}{2g} + 0 + \text{all losses}$$

$$\Rightarrow \quad 8.0 = 0 + \frac{V_2^2}{2g} + h_i + h_{f_1} + h_e + h_{f_2} \quad \text{...(i)}$$

where

$$h_i = \text{loss of head at entrance} = 0.5\frac{V_1^2}{2g}$$

$$h_{f_1} = \text{head lost due to friction in pipe 1} = \frac{4 \times f \times L_1 \times V_1^2}{d_1 \times 2g}$$

$$h_e = \text{head lost due to sudden enlargement} = \frac{(V_1 - V_2)^2}{2g}$$

$$h_{f_2} = \text{head lost due to friction in pipe 2} = \frac{4 \times f \times L_2 \times V_2^2}{d_2 \times 2g}$$

But from continuity equation, we have

$$A_1 V_1 = A_2 V_2$$

$$\Rightarrow \quad V_1 = \frac{A_2 V_2}{A_1} = \frac{\frac{\pi}{4} d_2^2 \times V_2}{\frac{\pi}{4} d_1^2} = \left(\frac{d_2}{d_1}\right)^2 \times V_2 = \left(\frac{0.3}{0.15}\right)^2 \times V_2 = 4V_2$$

Substituting the value of V_1 in different head losses, we have

$$h_i = \frac{0.5V_1^2}{2g} = \frac{0.5 \times (4V_2)^2}{2g} = \frac{8V_2^2}{2g}$$

$$h_{f_1} = \frac{4 \times 0.01 \times 25 \times (4V_2)^2}{0.15 \times 2 \times g} = \frac{4 \times 0.01 \times 25 \times 16}{0.15} \times \frac{V_2^2}{2g} = 106.67\frac{V_2^2}{2g}$$

$$h_e = \frac{(V_1 - V_2)^2}{2g} = \frac{(4V_2 - V_2)^2}{2g} = \frac{9V_2^2}{2g}$$

$$h_{f_2} = \frac{4 \times 0.01 \times 15 \times V_2^2}{0.3 \times 2g} = \frac{4 \times 0.01 \times 15}{0.3} \times \frac{V_2^2}{2g} = 2.0\frac{V_2^2}{2g}$$

Substituting the values of these losses in equation (i), we get

$$8.0 = \frac{V_2^2}{2g} + \frac{8V_2^2}{2g} + 106.67\frac{V_2^2}{2g} + \frac{9V_2^2}{2g} + 2\frac{V_2^2}{2g}$$

$$\Rightarrow \quad 8.0 = \frac{V_2^2}{2g}[1 + 8 + 106.67 + 9 + 2] = 126.67\frac{V_2^2}{2g}$$

$$\Rightarrow \quad V_2 = \sqrt{\frac{8.0 \times 2 \times g}{126.67}} = \sqrt{\frac{8.0 \times 2 \times 9.81}{126.67}} = \sqrt{1.2391} = 1.113 \text{ m/s}$$

$$\therefore \quad \text{Rate of flow, } Q = A_2 \times V_2 = \frac{\pi}{4}(0.3)^2 \times 1.113 = 0.07867 \text{ m}^3\text{/s} = 78.67 \text{ litres/s}$$

6.12 Two reservoirs with 20 m difference in their water levels are connected by a 300 mm diameter pipe line of 2000 m length. Calculate the discharge. If a parallel pipeline of 300 mm diameter is attached to the last 1000 m length of existing pipe, determine the percentage change in discharge. Take only wall friction into account. Assume friction factor, $f = 0.04$ in Darcy Weisbach's formula.

[20 marks]

Sol:

Given data: $h_L = 20$ m; $d = 300$ mm;

$L = 2000$ m, $f = 0.04$

The difference in levels of the two reservoir will be the total head loss. This head loss may be calculated by Darcy Weisbach's formula as :

$$h_L = \frac{fLQ^2}{12.1d^5}$$

$$\Rightarrow \quad 20 = \frac{0.04 \times 2000 \times Q^2}{12.1 \times (0.3)^5}$$

$$\Rightarrow \quad Q = 0.0857 \text{ m}^3\text{/s}$$

Now, when a pipe with same diameter is attached parallel to the existing pipeline to the last 1000 m, the arrangement will be as shown below.

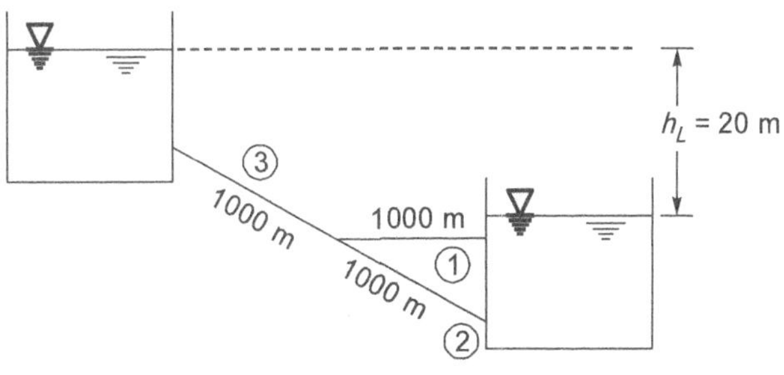

Let the discharge in the pipe 1, 2 and 3 be Q_1, Q_2 and Q_3 respectively.

$$\therefore \quad Q_3 = Q_1 + Q_2 \quad \text{[From continuity equation]}$$

$$h_{L_1} = h_{L_2} \quad \text{[Condition for parallel pipe connection]}$$

$$\frac{fL_1Q_1^2}{12.1d_1^5} = \frac{fL_2Q_2^2}{12.1d_2^5} \quad \text{[where } L_1 = L_2 = 1000 \text{ m and, } d_1 = d_2 = 300 \text{ mm]}$$

$$\Rightarrow \quad Q_1 = Q_2$$

$$\therefore \quad Q_3 = Q_1 + Q_1 = 2Q_1$$

$$\Rightarrow \quad Q_1 = \frac{Q_3}{2}$$

Also, $$Q_1 = Q_2 = \frac{Q_3}{2}$$

$\therefore$ Total head loss will be

$$h_L = h_{L_3} + h_{L_1}$$

$$\Rightarrow \quad 20 = \frac{fL_3Q_3^2}{12.1d^5} + \frac{fL_1Q_1^2}{12.1d_1^5}$$

$$\Rightarrow \quad 20 = \frac{0.04 \times 1000 \times Q_3^2}{12.1 \times (0.3)^5} + \frac{0.04 \times 1000 \times Q_3^2}{4 \times 12.1 \times (0.3)^5}$$

$$\Rightarrow \quad 20 = \frac{0.04 \times 1000 \times Q_3^2}{12.1 \times (0.3)^5}\left[1 + \frac{1}{4}\right]$$

$$\Rightarrow \quad Q_3 = 0.1084 \text{ m}^3/\text{s}$$

It is evident from the value of Q_3 that the discharge has increased. Thus pipes are laid in parallel to increase the discharge.

$$\text{Percentage increase in discharge} = \left(\frac{Q_3 - Q}{Q}\right) \times 100 = \frac{0.1084 - 0.0857}{0.0857} \times 100$$

$$= 26.49\% \simeq 26.5\%$$

6.13 **Determine the optimum ratio between the diameter of the pipe before expansion and the diameter of the pipe after expansion so that pressure rise may be maximum for sudden expansion in pipe flow. What will be the corresponding pressure rise?**

[20 marks]

Sol:

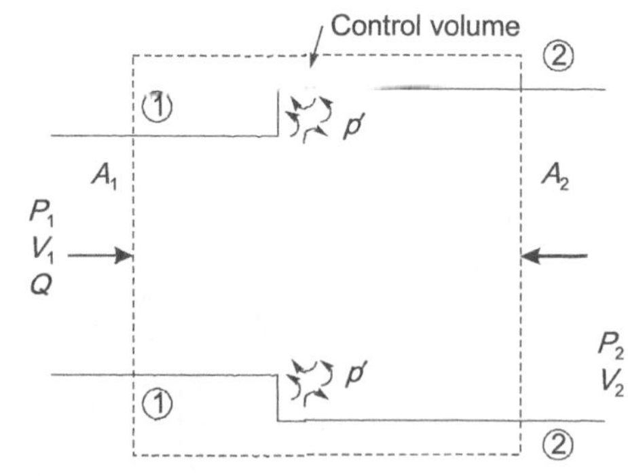

Applying momentum theorem in the control volume

$$P_1A_1 + p'(A_2 - A_1) - P_2A_2 = \rho Q(V_2 - V_1)$$

From experiment it is found that $p' = P_1$

$$\Rightarrow \quad (P_2 - P_1)A_2 = \rho Q(V_1 - V_2)$$

From the continuity equation

$$Q = A_2V_2$$

$$P_2 - P_1 = \frac{\rho V_2}{A_2} \times A_2(V_1 - V_2)$$

$$\Rightarrow \qquad \Delta P = \rho V_2 (V_1 - V_2)$$

$$\Rightarrow \qquad \Delta P = \rho \times \frac{Q}{\frac{\pi}{4} \times d_2^2}\left(\frac{Q}{\frac{\pi}{4} \times d_1^2} - \frac{Q}{\frac{\pi}{4} \times d_2^2}\right)$$

$$\Delta P = \frac{\rho Q^2}{\left(\frac{\pi}{4}\right)^2}\left(\frac{1}{d_2^2 \times d_1^2} - \frac{1}{d_2^4}\right)$$

For maximum pressure rise,

$$\frac{d\Delta P}{dd_2} = 0$$

$$\Rightarrow \qquad \left[\frac{-2d_2^{-3}}{d_1^2} + 4d_2^{-5}\right] = 0$$

$$\Rightarrow \qquad \frac{d_1}{d_2} = \frac{1}{\sqrt{2}}$$

Corresponding pressure rise,

$$\Delta P = \frac{\rho Q}{\frac{\pi}{4} \times d_2^2}\left[\frac{Q}{\frac{\pi}{4} \times d_1^2} - \frac{Q}{\frac{\pi}{4} \times d_2^2}\right]$$

$$\Delta P = \rho \times \frac{V_1}{2}\left[V_1 - \frac{V_1}{2}\right]$$

$$\Delta P = \frac{\rho V_1^2}{4}$$

7. Boundary Layer

Level-1

7.1 A wind tunnel has a cross-section at its inlet of 1 m by 1.0 m and a length of 10 m. Wind at uniform velocity of 15 m/sec enters the tunnel at 20°C. Determine the cross-sectional dimensions at the end of the test section which will yield zero pressure gradient along its length. Assume velocity distribution in turbulent boundary layer to follow the law.

$$\frac{u}{V} = \left(\frac{y}{\delta}\right)^{1/5}$$

[10 Marks]

Sol:

Because of the growth of boundary layer, the centre line or free stream velocity increases and hence there will be negative pressure gradient in the tunnel. To make the free stream velocity at both the sections the same $\left(\text{so that } \frac{\partial p}{\partial x} \simeq 0\right)$ each side needs to be displaced backwards by δ^*, the displacement thickness. Hence tunnel dimension at exit will be $(1 + 2\delta^*)$.

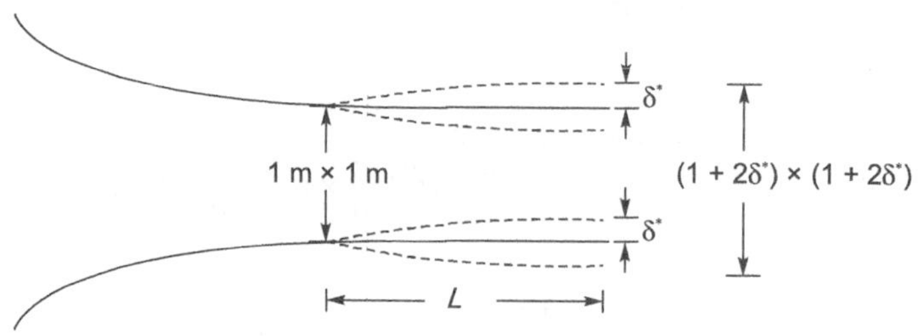

$$R = \frac{Vx}{V} = \frac{15\times10}{1.53\times10^{-5}} = 9.804\times10^{6}$$

Assuming the boundary layer to be turbulent from beginning,

$$\frac{\delta}{x} = \frac{0.377}{(R_{ex})^{1/6}} = \frac{0.377}{(9.804\times10^{6})^{1/5}} = 0.01507 \text{ m}$$

$$\delta = 0.01507\times10 = 0.1507 \text{ m}$$

Since the velocity distribution in turbulent layer is given by

$$\frac{U}{U_\infty} = \left(\frac{y}{\delta}\right)^{1/5}$$

$$\delta^* = \int_0^\delta\left(1-\frac{U}{V_\infty}\right)dy = \int_0^\delta 1-\left(\frac{y}{\delta}\right)^{1/5} dy$$

Solving, $\delta^* = \frac{\delta}{6} = \frac{0.1507}{6} = 0.0251 \text{ m}$

$$1 + 2\delta^* = 1 + 2\times0.0251 = 1.0502 \text{ m}$$

Cross-section of tunnel at exit = 1.0502 m × 1.0502 m

7.2 **Define boundary layer thickness, displacement thickness, momentum thickness.**

[6 Marks]

Sol:

Boundary layer thickness: It is the distance from the boundary surface in which velocity reaches 99% of the free stream velocity.

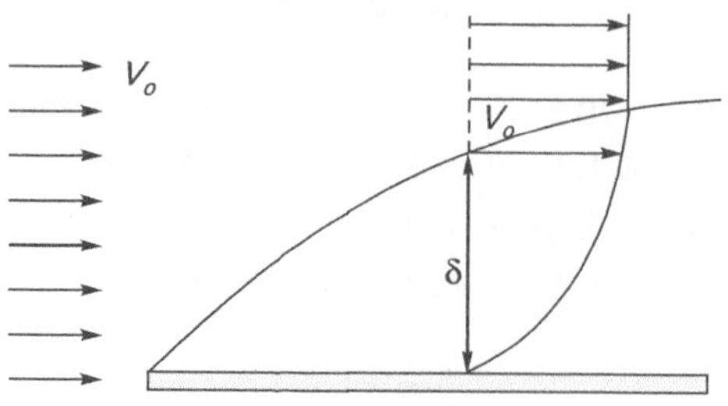

V_0 = free stream velocity
δ = boundary layer thickness

Displacement thickness: Displacement thickness is defined as the distance by which the boundary should be shifted in order to compensate for reduction in mess flow rate on account of boundary layer formation.

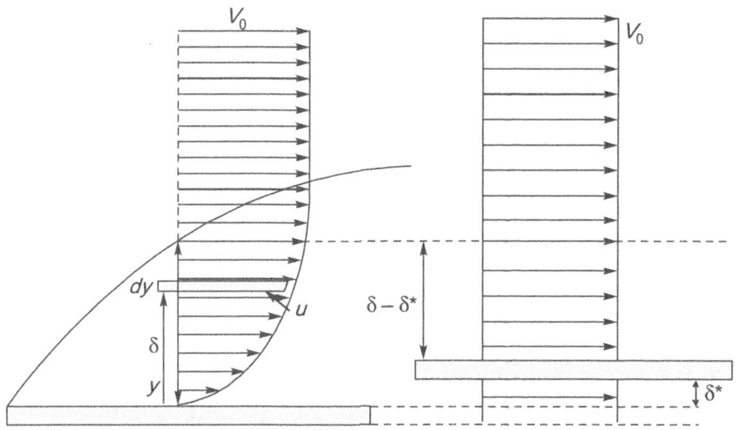

Considering unit width inside

$$\int_0^\delta \rho u\,dy = \rho(\delta-\delta^*)\times 1\times V_o$$

$$\Rightarrow \quad -\int_0^\delta \rho u\,dy+\int_0^\delta \rho V_o\,dy = \rho V_o\delta^*$$

$$\int_0^\delta \rho(V_0-u)\,dy = \rho V_o\delta^*$$

$$\delta^* = \int_0^\delta\left(1-\frac{u}{V_0}\right)dy$$

Momentum thickness: Momentum thickness is defined as the distance measured perpendicular to the boundary of the solid body, by which the boundary should be displaced to compensate for the reduction in momentum of the flowing fluid on account of boundary layer formation. It is denoted by 'θ'.

$$\theta = \int_0^\delta \frac{u}{V_0}\left(1-\frac{u}{V_0}\right)dy$$

7.3 **For the velocity profile $\frac{u}{U_\infty}=\left(\frac{y}{\delta}\right)^{1/7}$, calculate the ratio of displacement thickness and momentum thickness.**

[15 Marks]

Sol:

We know, the displacement thickness is given by

$$\delta^* = \int_0^\delta\left(1-\frac{u}{V_\infty}\right)dy$$

Given,

$$\frac{u}{U_\infty} = \left(\frac{y}{\delta}\right)^{1/7}$$

$$\therefore \quad \delta^* = \int_0^\delta\left[1-\left(\frac{y}{\delta}\right)^{1/7}\right]dy = \frac{\delta}{8} \qquad \text{...(i)}$$

The momentum thickness is given by

$$\theta = \int_0^\delta \frac{u}{U_\infty}\left(1-\frac{u}{V_\infty}\right)dy = \int_0^\delta \left(\frac{y}{\delta}\right)^{1/7}\left[1-\left(\frac{y}{\delta}\right)^{1/7}\right]dy$$

$\therefore$
$$\theta = \frac{7}{72}\delta \qquad \text{...(ii)}$$

Hence,
$$\frac{\delta^*}{\theta} = \frac{\delta/8}{7\delta/72} = \frac{\text{Displacement thickness}}{\text{Momentum thickness}} = \frac{9}{7}$$

$\Rightarrow$
$$\frac{\delta^*}{\theta} = 1.2857$$

7.4 **2.0 m wide and 6.0 m long plate moves at a velocity of 1.75 m/s parallel to its length through a stationary air of density 1.22 kg/m^3 and viscosity of 1.8 × 10^{-5} Ns/m^2.**
Determine the drag force on one side of a plate assuming
(i) Laminar flow boundary layer
(ii) Turbulent flow boundary layer

[5 Marks]

Sol:

Given : free – stream velocity, U_∞ = 1.75 m/s; Density, ρ = 1.22 kg/m^3; Dynamic viscosity, μ = 1.8 × 10^{-5} Ns/m^2

(i) Assuming laminar flow boundary layer

Using exact solution,

Average skin friction coefficient,

$$\bar{C}_{f\cdot x} = 2\times C_{f\cdot x} = 2\times\frac{0.664}{\sqrt{R_{ex}}} = \frac{1.328}{\sqrt{R_{ex}}}$$

Here,
$$R_{ex} = \frac{\rho\cdot U_\infty\cdot x}{\mu} = \frac{1.22\times1.75\times6}{1.8\times10^{-5}}$$

$$R_{ex} = 7.12\times10^5$$

Drag force,
$$F_D = \left(\bar{C}_{f\cdot x}\times\frac{1}{2}\rho\cdot U_\infty^2\right)\times A$$

$$= \frac{1.328}{\sqrt{7.12\times10^5}}\times\frac{1}{2}\times1.22\times1.75^2\times(2\times6)$$

$$= 0.0353\text{ N} = 3.53\times10^{-2}\text{ N}$$

(ii) Assuming turbulent flow boundary layer

$$R_{ex} = \frac{\rho\cdot U_\infty\cdot x}{\mu} = \frac{1.22\times1.75\times6}{1.8\times10^{-5}} = 7.12\times10^5$$

Average skin friction coefficient,
$$\bar{C}_{f\cdot x} = \frac{0.073}{(R_{ex})^{1/5}} = \frac{0.073}{(7.12\times10^5)^{1/5}} = 4.93\times10^{-3}$$

Drag force,
$$F_D = \bar{C}_{f\cdot x}\times\frac{1}{2}\rho\cdot U_\infty^2\times A$$

$$F_D = 4.93\times10^{-3}\times\frac{1}{2}\times1.22\times1.75^2\times(2\times6)$$

$$= 0.1105\text{ N} = 11.05\times10^{-2}\text{ N}$$

Level-2

7.5 Repeat the flat plate momentum analysis by replacing the equation

$$u(x,y) = U\left(\frac{2y}{\delta} - \frac{y^2}{\delta^2}\right) \qquad 0 \le y \le \delta\,(x)$$

By using a trigonometric profile approximation

$$\frac{u}{U} \simeq \sin\left(\frac{\pi y}{2\delta}\right)$$

Compute momentum interval estimates of C_f, $\frac{\theta}{x}, \frac{\delta^*}{x}$.

[20 Marks]

Sol:

From given new profile $\frac{U}{U_\infty} = \sin\left(\frac{\pi}{2}\frac{y}{\delta}\right)$

(i) δ^* = Displacement thickness

$$= \int_{y=0}^{y=\delta}\left(1-\frac{U}{U_\infty}\right)dy$$

$$= \int_0^\delta\left(1-\sin\left(\frac{\pi}{2}\cdot\frac{y}{\delta}\right)\right)dy = \frac{2\delta}{\pi}\left[\frac{\pi}{2}-1\right]$$

$$\frac{\delta^*}{\delta} = \frac{2}{\pi}\left[\frac{\pi}{2}-1\right] \qquad \text{...(i)}$$

where,

$$\tau_{ox} = \mu\frac{\partial u}{\partial y} = \mu\frac{\partial}{\partial y}\sin\left(\frac{\pi}{2}\times\frac{y}{\delta}\right)U_\infty$$

$$= \mu\cos\left[\frac{\pi}{2}\cdot\frac{y}{\delta}\right]\cdot\frac{\pi}{2}\cdot\frac{1}{\delta}\cdot U_\infty = \frac{\mu\pi}{2\delta}\cdot U_\infty\cos\left(\frac{\pi}{2}\cdot\frac{y}{\delta}\right)$$

At boundary condition $y = 0$

$$\tau_{ox} = \frac{\mu\pi}{2\delta}\cdot U_\infty$$

Now using,

$$\theta = \int\frac{U}{U_\infty}\left(1-\frac{V}{V\infty}\right)dy$$

$$0 = \int_{y=0}^{y=\delta}\sin\left(\frac{\pi y}{2\pi}\right)\left[1-\sin\left(\frac{\pi}{2}\cdot\frac{y}{\delta}\right)\right]dy$$

use $\frac{\pi}{2}\cdot\frac{y}{\delta} = t$ and solve

$$= \frac{2\delta}{\pi}\left[\int_0^{\pi/2}\sin t\,dt - \int_0^{\pi/2}\sin^2 t\,dt\right]$$

$$= \frac{2\delta}{\pi}\left[-[\cos t]_0^{\pi/2} - \frac{1}{2}\int_0^{\pi/2}(1-\cos 2t)\,dt\right]$$

Solving, $$\theta = \frac{2\delta}{\pi}\left(1-\frac{\pi}{4}\right) \quad \text{...(ii)}$$

Now using Von-Korman momentum integral equation

$$\frac{\tau_{ox}}{\rho V_\infty^2} = \frac{\partial\theta}{\partial x}$$

$$\frac{U_\infty \dfrac{\mu\pi}{2\delta}}{\rho V_\infty^2} = \frac{\partial}{\partial x}\frac{2\delta}{\pi}\left[1-\frac{\pi}{4}\right]$$

Solving, $$V_\infty \frac{11.464\mu}{\rho U_\infty^2} = \delta \cdot \frac{\partial\delta}{\partial x}$$

$$U_\infty \frac{11.464\,\mu}{\rho U_\infty^2}\int \partial x = \int \delta\,\partial\delta$$

Solving, $$U_\infty \frac{11.464\,\mu}{\rho U_\infty^2} x = \frac{\delta^*}{2}$$

$$\frac{\delta}{x} = \frac{4.788}{\sqrt{l_{ex}}} \quad \text{...(iii)}$$

From (i), $$\delta^* = \frac{2\delta}{\pi}\left(\frac{\pi}{2}-1\right)$$

Putting value of δ from (3)

$$\frac{\delta^*}{x} = \frac{1.738}{\sqrt{R_{ex}}} \quad \text{(Ans)}$$

From (2)

$$\theta = \frac{2\delta}{\pi}\left(1-\frac{\pi}{4}\right) = 0.137\delta$$

$$\frac{\theta}{x} = \frac{0.655}{\sqrt{R_{ex}}} \quad \text{(Ans)}$$

Now, $$C_{fx} = \frac{\Gamma_{ox}}{\frac{1}{2}\rho U_\infty^2} = \frac{\mu\dfrac{\pi}{2\delta}\cdot U_\infty}{\frac{1}{2}\rho U_\infty^2}$$

$$C_{fx} = \frac{\mu \times \pi}{\rho U_\infty S} = \frac{(x)u\pi\sqrt{R_{ex}}}{\rho U_\infty (x) 4.788}$$

Solving, $$\frac{C_{fx}}{x} = \frac{0.655}{\sqrt{R_{ex}}} \quad \text{(Ans)}$$

7.6 Air flows at 10 m/s past a smooth rectangular flat plate 0.3 m wide and 3 m long. Assuming that the turbulence level in the incoming stream is low and that transition occurs at $R_e = 5 \times 10^5$, calculate the ratio of total drag when the flow is parallel to the length of the plate to the total drag when the flow is parallel to width of the plate.

Take density of air as 1.24 kg/m^3 and kinematic viscosity of air as 0.15 stoke.

[20 marks]

Sol:

Given :

$$U = 10 \text{ m/s};\ b = 0.3 \text{ m};\ L = 3 \text{ m}$$

Reynolds number for laminar Boundary Layer

$$= 5 \times 10^5$$

Given, $\rho = 1.24 \text{ kg/m}^3$ and $\nu = 0.15$ stoke.

$\therefore$ $\rho = 1.24 \text{ kg/m}^3$ and $\nu = 0.15 \text{ stoke} = 0.15 \times 10^{-4} \text{ m}^2/\text{s}$

(i) **Drag when flow is parallel to the length of the plate**

Let x = the distance from leading edge upto which laminar boundary exists.

$$\therefore \quad 5 \times 10^5 = \frac{\rho \times U \times x}{\mu} = \frac{U \times x}{\nu} = \frac{10 \times x}{0.15 \times 10^{-4}}$$

$$\Rightarrow \quad x = \frac{5 \times 10^5 \times 0.15 \times 10^{-4}}{10} = 0.75 \text{ m}$$

Now the drag force on the plate on one side

= Drag due to laminar boundary layer + Drag due to turbulent boundary layer ...(i)

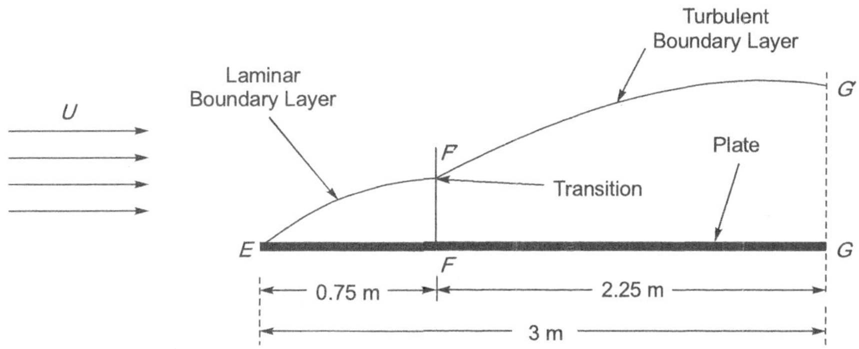

(a) **Drag due to laminar boundary (i.e., from E to F)**

$$F_{EF} = \frac{1}{2}\rho A U^2 \times C_D$$

where C_D is given by Blasius solution for laminar boundary layer as

$$C_D = \frac{1.328}{\sqrt{R_{ex}}}, \text{ where } R_{ex} = 5 \times 10^5$$

$$= \frac{1.328}{\sqrt{5 \times 10^5}} = 1.878 \times 10^{-3}$$

$$A = \text{Area of plate upto laminar boundary layer}$$

$$= 0.75 \times b = 0.75 \times 0.3 = 0.225 \text{ m}^2$$

$$\rho = 1.24 \text{ kg/m}^3$$

$$\therefore \quad F_{EF} = \frac{1}{2} \times 1.24 \times 0.225 \times 10^2 \times 0.001878 = 0.0262 \text{ N}$$

(b) **Drag force due to turbulent boundary layer from F to G**

= Drag force due to turbulent boundary layer from E

to G – Drag force due to turbulent Boundary Layer from E to F

$$= (F_{EG})_{turb.} - (F_{EF})_{turb.}$$

Now, $$(F_{EG})_{turb.} = \frac{1}{2}\rho AU^2 \times C_D$$

where C_D for turbulent boundary layer is given by equation as

$$C_D = \frac{0.074}{(R_{e_L})^{1/5}}$$

But $$R_{e_L} = \frac{U \times L}{\nu} = \frac{10 \times 3}{0.15 \times 10^{-4}} = 20 \times 10^5$$

$\therefore$ $$C_D = \frac{0.074}{(20 \times 10^5)^{1/5}} = 0.00406$$

$\therefore$ $$(F_{EG})_{turb.} = \frac{1}{2}\rho AU^2 \times C_D$$

$$= \frac{1}{2} \times 1.24 \times (3 \times 0.3) \times 10^2 \times 0.00406$$

$$= 0.2265 \text{ N}$$

Now, $$(F_{EF})_{turb.} = \frac{1}{2}\rho \times A_{EF} \times U^2 \times C_D$$

where, $$A_{EF} = \text{Area of plate upto } EF = EF \times b$$

$$= 0.75 \times 0.3 = 0.225 \text{ m}^2$$

and $$C_D = \frac{0.074}{\left[(R_e)_{EF}\right]^{1/5}} = \frac{0.074}{(5 \times 10^5)^{1/5}} = 0.00536$$

$\therefore$ $$(F_{EF})_{turb.} = \frac{1}{2} \times 1.24 \times 0.225 \times 10^2 \times 0.00536 = 0.0747 \text{ N}$$

$\therefore$ Drag force due to turbulent boundary layer from F to G

$$= (F_{EG})_{turb.} - (F_{EF})_{turb.} = 0.2265 - 0.0747 = 0.1518 \text{ N}$$

$\therefore$ Total drag force when flow is parallel to the length of the plate

= Drag due to laminar boundary layer upto F + Drag due to turbulent boundary layer from F to G

$$= 0.0262 + 0.1518 = 0.178 \text{ N} \quad \text{...(A)}$$

(ii) **Drag when flow is parallel to the width of the plate**

We have already calculated that upto the length of 0.75 m from the leading edge, the boundary layer is laminar. As the width of the plate is only 0.3 m, hence when flow is parallel to the width of the plate, only laminar boundary layer will be formed.

$\therefore$ Drag force on the plate

$$= \frac{1}{2}\rho AU^2 \times C_D$$

where C_D from Blasius solution for laminar boundary layer is given as

$$C_D = \frac{1.328}{\sqrt{R_{e_x}}}, \text{ here } x = \text{width of plate} = 0.3 \text{ m}$$

$$\therefore \quad R_{e_x} = \frac{U \times x}{\nu} = \frac{10 \times 0.3}{0.15 \times 10^{-4}} = 2 \times 10^5$$

$$\therefore \quad C_D = \frac{1.328}{\sqrt{2 \times 10^5}} = 0.00297$$

$$A = \text{Area of plate} = 3 \times 0.3 = 0.9 \text{ m}^2$$

$$\rho = 1.24 \text{ kg/m}^3$$

$\therefore$ Total drag force on the plate

$$= \frac{1}{2} \times 1.24 \times 0.9 \times 10^2 \times 0.00297$$

$$= 0.1657 \text{ N} \qquad \text{...(B)}$$

$\therefore$ Ratio of two total drags given by equations (A) and (B) becomes

$$\frac{\text{Total drag when flow is parallel to the length of the plate}}{\text{Total drag when flow is parallel to the width of the plate}} = \frac{\text{Equation (A)}}{\text{Equation (B)}}$$

$$= \frac{0.1738}{0.1657} = 1.05$$

8. Laminar Flow

Level-1

8.1 A flow of 420 *l*/min of oil of specific gravity 0.91 and viscosity 1.24 poise is pumped through a pipe line 75 mm diameter having a length of 62 m and whose outlet is 3 m higher than its inlet. Estimate the power required for the pump if its efficiency is 60%. Also find positive head loss through the pipe.

[10 Marks]

Sol:

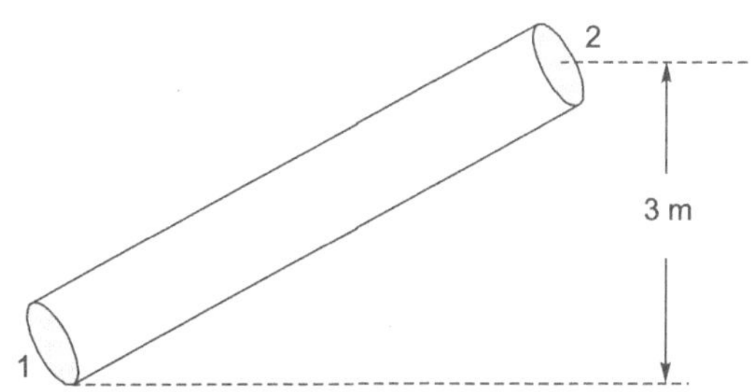

Given data : $Q = 420$ *l*/min $= 7 \times 10^{-3}$ m^3/sec; Velocity, $V = \frac{Q}{A} = \frac{7 \times 10^{-3}}{\frac{\pi}{4} \times (0.075)^2} = 1.584$ m/sec

$$\text{Reynold's number, } R_e = \frac{\rho VD}{\mu} = \frac{0.9 \times 1000 \times 1.584 \times 0.075}{1.24 \times 0.1} = 862.25$$

$$R_e < 2000$$

$\Rightarrow$ Laminar flow

$$\text{Head loss due to friction } h_f = \frac{32\mu \bar{V} L}{\rho g D^2}$$

$$h_f = \frac{32 \times 1.24 \times 0.1 \times 1.584 \times 62}{0.91 \times 9810 \times (0.075)^2} = 7.76 \text{ m}$$

$$\text{Total head required for pump, } H = 3 + \frac{V^2}{2g} + h_f = 3 + \frac{(1.584)^2}{2 \times 9.81} + 7.76 = 10.88 \text{ m}$$

$$\text{Power req.} = \frac{0.91 \times 9810 \times 7 \times 10^{-3} \times 10.88}{0.6} = 1134 \text{ Watt}$$

$$P = 1.134 \text{ kW}$$

Level-2

8.2 Laminar flow takes place between parallel plates 10 mm apart. The plates are inclined at 45° with the horizontal. For oil of viscosity 0.9 kg/ms and mass density of 1260 kg/m³. The pressure at two point 2 m vertically apart are 80 kN/m² and 250 kN/m² when the upper plate moves at 2 m/sec velocity relative to the lower plate but in opposite direction to flow. Determine

(i) Velocity distribution

(ii) Maximum velocity

(iii) Shear stress on top plate

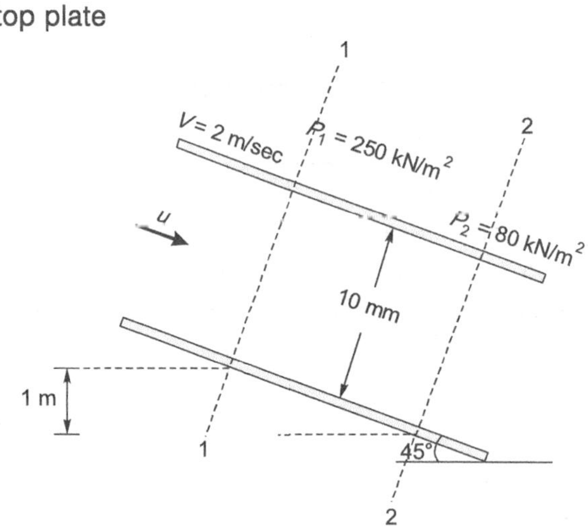

[20 Marks]

Sol:

Piezometric head, difference between section 1-1 and 2-2.

$$h_1 - h_2 = \left(\frac{P_1}{\gamma} + Z_1\right) - \left(\frac{P_2}{\gamma} + Z_2\right)$$

$$= \left(\frac{250 \times 10^3}{9.81 \times 1260} + 1\right) - \left(\frac{80 \times 10^3}{9.8 \times 1260}\right) = 21.234 - 6.475 = 14.759 \text{ m}$$

$$\text{Actual distance between both sections} = \frac{1}{\sin 45°} = 1.414 \text{ m}$$

$$\frac{\partial h}{\partial x} = \frac{-14.759}{1.414} = -10.438$$

$$\frac{\partial p}{\partial x} = \gamma \frac{\partial h}{\partial x} = -9.81 \times 1260 \times (-10.438) = -128.96 \text{ kN/m}^2$$

Now using equation for Coutte flow

$$u = \frac{1}{2\mu}\left(\frac{-\partial p}{\partial x}\right)\left(By - y^2\right) + \frac{y}{B} \times V$$

$$= \frac{1}{2 \times 0.9}\left(+128.96 \times 10^3\right)\left(0.01y - y^2\right) + \frac{y}{0.01} \times 2$$

$$= 71.644 \times 10^3 (0.01y - y^2) + 200y$$

(ii) For maximum velocity

$$\frac{\partial u}{\partial y} = 0$$

$$\frac{\partial u}{\partial y} = 71.644 \times 10^3 (0.01 - 2y) + 200 = 0$$

Solving, $y = 6.4 \times 10^{-3}$ m

$$u_{max} = 71.644 \times 10^3 \left(0.01 \times 6.4 \times 10^{-3} - (6.4)^2 \times 10^{-6} + 200 \times 6.4 \times 10^{-3}\right)$$

$$= 2.93 \text{ m/sec}$$

9. Turbulent Flow

Level-1

9.1 **Water is flowing steadily in a 5 cm diameter horizontal pipe made of stainless steel at a rate of 6 *l*/sec. Determine the pressure drop, head loss, and the power input for 60 m long section of pipe. Take ρ = 1000 kg/m^3, μ = 1.138 x 10^{-3} kg/m-sec, k_s = Relative surface roughness = 0.002 mm.**

[12 Marks]

Sol:

Given data, $Q = 6 \times 10^{-3}$ m^3/sec, V = Velocity of flow $= \frac{Q}{A} = \frac{6 \times 10^{-3}}{\frac{\pi}{4} \times (0.05)^2} = 3.06$ m/sec

R_e = Reynold number $= \frac{\rho VD}{\mu} = \frac{1000 \times 3.06 \times 0.05}{1.0138 \times 10^{-3}} = 1.34 \times 10^5 > 2000$

Hence, the flow is turbulent.

Now, using
$$\frac{1}{\sqrt{f}} = 2\log_{10}\left(\frac{D}{2k_s}\right) + 1.73 = 2\log_{10}\left(\frac{5 \times 10^{-2}}{2 \times 2 \times \times 10^{-6}}\right) + 1.73$$

f = friction factor = 0.01

Now from Darcy's equation, head loss is given by

$$h_f = \frac{hLV^2}{2gd}$$

l = Length of section = 60 m

$$h_f = \frac{0.01 \times 60 \times 3.06^2}{2 \times 9.81 \times 0.05} = 5.73 \text{ m}$$

$$\text{Pressure drop} = \rho g \times h_f = 1000 \times 9.81 \times 5.73 = 56.18 \text{ kN/m}^2$$

$$\therefore \quad \text{Power input required} = \dot{m} g h_f$$

$$= \rho \times Q \times g \times h_f = 10^3 \times 6 \times 10^{-3} \times 9.81 \times 5.73 = 337.27 \text{ Watt} \quad \textbf{(Ans)}$$

9.2 Water is flowing through a smooth pipe of 100 mm diameter at rate of 0.036 m^3/s. Determine

(i) Darcy's friction factor
(ii) Shear velocity
(iii) Shear stress at wall
(iv) Normal thickness of viscous sub layer

Take kinematic viscosity = 10^{-6} m^2/s
Density, ρ = 1000 kg/m^3 and

$$f \text{ (Darcy's friction factor)} = 0.0032 + \frac{0.221}{R_e^{0.237}}$$

[12 Marks]

Sol:

For given flow,

Diameter, $D = 100 \text{ mm} = 0.1 \text{ m}$

Discharge, $Q = 0.036 \text{ m}^3/\text{s}$

Velocity, $$V = \frac{0.036}{\frac{\pi}{4} \times (0.1)^2} = 4.58 \text{ m/s}$$

Reynold's number, $$R_e = \frac{VD}{\nu} = \frac{4.58 \times 0.1}{10^{-6}} = 458000$$

(Turbulent flow)

(i) Friction factor, $$f = 0.0032 + \frac{0.221}{Re^{0.237}} = 0.0032 + \frac{0.221}{(458000)^{0.237}} = 0.01326$$

(ii) For smooth pipe,

Shear velocity, $$V_* = V\sqrt{\frac{f}{8}} = 4.58\sqrt{\frac{0.01326}{8}} = 0.1865 \text{ m/s}$$

(iii) Wall shear stress, $$\tau_0 = \rho V_*^2 = 1000 \times (0.1865)^2 = 34.77 \text{ N/m}^2$$

(iv) Laminar sub-layer thickness, $$\delta' = \frac{11.6}{V_*} = \frac{11.6 \times 10^{-6}}{0.1865} = 6.22 \times 10^{-5} \text{ m}$$

Level-2

9.3 The wind velocities were measured above a smooth plane ground surface at 2 m^2 4 m heights as 2 and 2.3 m/s respectively. Compute the shear velocity. Assume Karman's constant as 0.40 and kinematic viscosity of air as 0.145 centi stoke. What is the velocity at 8 m above the ground surface? Calculate the probably laminar sublayer thickness.

[20 Marks]

Sol:

Given : Karman's constant, $K = 0.40$; $n = 0.145 \times 10^{-6}$ m²/s; At $y_1 = 2$ m, $u_1 = 2$ m/s; At $y_2 = 4$ m, $u_2 = 2.3$ m/s

By Reynold's theory:
$$\tau = \rho\left(l \cdot \frac{du}{dy}\right)^2$$

$$\sqrt{\frac{\tau}{\rho}} = l \cdot \frac{du}{dy}$$

Let shear velocity be V^*
$$V^* = \sqrt{\frac{\tau}{\rho}}$$

By Karman's experiment
$$l = k \cdot y$$
$$l = (0.4)\,y$$

$\therefore$
$$V^* = (0.4)y \cdot \frac{du}{dy}$$
$$du = 2.5\ V^*\ dy$$

Integrating both sides
$$\int du = 2.5\ V^*\ dy$$
$$u = 2.5\ V^*\ lny + C$$

Keeping values in above equation $u_1 = 2.5\ V^*\ lny_1 + C$
$$2 = 2.5\ V^*\ ln(2) + C \quad \text{... (i)}$$
$$u_2 = 2.5\ V^*\ ln(y_2) + C$$
$$2.3 = 2.5\ V^*\ ln(4) + C \quad \text{... (ii)}$$

Equation (ii) — (i)
$$0.3 = 2.5\ V^*\ \{ln(4) - ln\,(2)\}$$
$$0.3 = 2.5\ V^* ln\,2$$
$$V^* = \frac{0.3}{2.5 \times ln(2)} = 0.1731 \text{ m/s}$$

$\therefore$ Shear velocity = 0.1731 m/s

Keeping value of V^* in equation (i)

$\therefore$ $2 = 2.5 \times 0.1731 \times ln\,(2) + C$

$\Rightarrow$ $C = 1.7$

$\therefore$ $u = 2.5 \times 0.1731\ ln\,(y) + C$

$$u = 0.43275\ ln\,(y) + 1.7$$

At $y = 8$ m
$$u = 0.43275\ ln\,(8) + 1.7 = 2.5999 \text{ m/s}$$

$\therefore$ Velocity at 8 m above the ground,
$$v \simeq 2.6 \text{ m/s}$$

Using Nikuradse's result
$$\delta' = \frac{11.6 \times \nu}{V^*} = \frac{11.6 \times 0.145 \times 10^{-6}}{0.1731} = 9.7169 \times 10^{-6} \text{ m}$$

$\therefore$ Probable laminar sublayer thickness = 9.7169×10^{-3} mm

10. Dimensional Analysis

Level-1

10.1 A 1 : 49 scale model of a proposed dam is used to predict prototype flow conditions. The design flood discharge near the spillway is 15000 m^3/s. Establish the relation between velocities in the model and prototype. What flow rate should be established in the model to simulate this flow? If a velocity of 1.2 m/s is measured at a point in the model what is the velocity at a corresponding point in the prototype?

[8 Marks]

Sol:

Given, Scale ratio

$$L_r = \frac{L_p}{L_m} = 49.1$$

Discharge $Q_p = 15000$

Then scale ratio for discharge

$$Q_r = (49)^{2.5}$$

$$\Rightarrow \quad \frac{Q_p}{Q_m} = (L_r)^{2.5}$$

$$\Rightarrow \quad Q_m = \frac{15000}{(49)^{2.5}} = 0.89\ m^3/sec$$

For velocity

$$\frac{V_p}{V_m} = \sqrt{L_r} = 7$$

Given $V_m = 1.2$ m/sec

$\Rightarrow$ $V_p = 7 \times 1.2 = 8.4$ m/sec

10.2 A 1 : 25 scale model of a ship has a submerged surface area of 6 m^2, a length of 5 m and experiences a total drag of 25 N when towed through water with a velocity of 1.2 m/sec. Estimate the total drag on the prototype when cruising at the corresponding speed. The skin friction can be estimated by $F_s = \frac{C_f A \rho V^2}{2}$ where the coefficient $C_f = \frac{0.0735}{(R_e)^{1/5}}$. [Assume $\mu_{water} = 1 \times 10^{-3}$ Pa sec and $\rho_{water} = 1030\ kg/m^3$, for both model and the prototype].

[12 Marks]

Sol:

For model:

Step 1 : Calculate the surface friction (F_{SD})

$$R_{eL} = \frac{\rho VL}{\mu} = \frac{1030 \times 1.2 \times 8}{10^{-3}} = 6.18 \times 10^6$$

$$C_{fL} = \frac{0.0735}{(61.8 \times 10^5)^{1/5}} = 3.22 \times 10^{-3}$$

$$(F_{SD})_m = \frac{1}{2} \times 3.22 \times 10^{-3} \times 1030 \times 6 \times 1.2^2 = 14.34\ N$$

Hence, Wave drag , $F_{wd} = 25 - (14.34) = 10.66$ N

For prototype

$$(F_{WD})_r = \rho_r L_r^3 = \left(\frac{1}{25}\right)^3$$

$$(F_{WD})_P = (25)^3 \times 10.66 = 166.6 \text{ kN}$$

$$(F_{WD})_P = ?$$

$$(R_e)_L = \frac{1030 \times (5 \times 1.2) \times 25 \times 5}{10^{-3}} = 7.73 \times 10^8$$

$$C_f = \frac{0.0735}{(7.73 \times 10^8)^{1/5}} = 1.226 \times 10^{-3}$$

$$(F_{SD})_P = \frac{1}{2} \times 1.226 \times 10^{-3} \times 1030 \times (6 \times 25^2) \times (1.2 \times 5)^2 = 85.3 \text{ kN}$$

$$\text{Total, } F_T = 85.3 + 166.6 = 251.9 \text{ kN}$$

10.3 (i) The aerodynamic drag of a new sports car is to be predicted at a speed of 80 km/hr at an air temperature of 25°C engineering build a one-fifth scale model of the car to test in a wind tunnel. It is winter and the wind tunnel air having a temperature of 5°C. Determine how fast the engineers should run the wind tunnel in order to achieve similarty between model and prototype?

(ii) Suppose engineers run wind tunnel of 354 km/hr to achieve similarly, calculate drag force on prototype. If drag force on model is 94.3 N.

Given data :

At 25°C ⇒ $\mu_{air} = 1.849 \times 10^{-5}$ kg/m.sec, $\rho_{air} = 1.184$ kg/m^3

At 5°C ⇒ $\rho_{air} = 1.269$ kg/m^3, $\mu_{air} = 1.754 \times 10^{-5}$ kg/m.sec

[8 Marks]

Sol:

(i) As only viscous effects are predominant:

$$R_{em} = (R_e)_p$$

$$\frac{\rho_m V_m L_m}{\mu_m} = \frac{\rho_p V_p L_p}{\mu_p}$$

$$\frac{1.269 \times V_m \times \left(\frac{1}{5}\right)}{1.754 \times 10^{-5}} = \frac{1.184 \times 80 \times 1}{1.849 \times 10^{-5}}$$

Solving, $V_m = 354$ km/hr

(ii) Using

$$F_r = \rho_r V_r^2 L_r^2$$

$$\frac{F_m}{F_p} = \left(\frac{1.269}{1.184}\right)\left(\frac{354}{80}\right)^2 \times (5)^2$$

Solving using $F_m = 94.3$ N

$$F_p = 112 \text{ N}$$

Level-2

10.4 An air jet discharging vertically upwards supports a ball placed in the jet in a stable position as shown in figure. The height of the ball h above the nozzle is known to depend on the diameter of the ball D, the exit diameter of the nozzle is d, the velocity of jet V, the air density ρ. The viscosity μ and weight of ball W. Use Buckingam π theory to develop the appropriate relationship bet them.

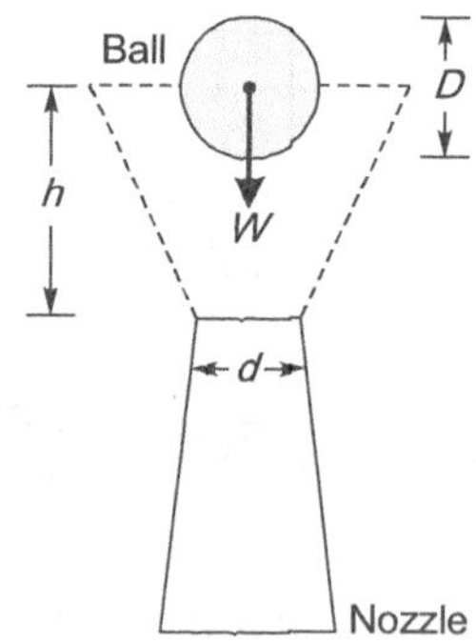

[20 Marks]

Sol:

Given data : Height of ball $(h) = f^n(D, d, v, \rho, \mu, w)$

From variables

Total number of variables $(m) = 7$

Number of fundamental dimensions $(n) = 3\ (M, L, T)$

Number of π-terms $= 7 - 3 = 4$

(i) First term

$$\pi_1 = d\left[D^a V^b \rho^c\right]$$

$$M^0L^0T^0 = L^1\left[\left(L^1\right)^a\left(L^1T^{-1}\right)^b\left(M^1L^{-3}\right)^c\right]$$

Solving, $c = 0$, $b = 0$, $a = -1$

$$\pi_1 = \frac{d}{D}$$

(ii) Second term

$$\pi_1 = \mu\left[D^a V^b \rho^c\right]$$

$$M^0L^0T^0 = M^1L^1T^{-1}\left((L)^a\left(L^1T^{-1}\right)^b\left(M^1L^{-3}\right)^c\right)$$

Solving $c = -1$, $a = -1$, $b = -1$

$$\pi_2 = \frac{\mu}{Dv\rho}$$

(iii) Third term

$$\pi_3 = W(D^a V^b \rho^c)$$

$$M^0L^0T^0 = M^1L^1T^{-2}\left(\left(L^1\right)^a\left(L^1T^{-1}\right)^b\left(M^1L^{-3}\right)^c\right)$$

Solving, $a = -2$. $b = -2$, $c = -1$

$$\pi_3 = \frac{W}{D^2V^2\rho}$$

(iv) Fourth term

$$\pi_4 = h(D^aV^b\rho^c)$$

$$M^0L^0T^0 = L^1\left(\left(L^1\right)^a\left(L^1T^{-1}\right)^b\left(M^1L^{-3}\right)^c\right)$$

Solving, $c = 0$, $b = 0$, $a = -1$

$$\pi_4 = \frac{h}{D}$$

Now,

$$\pi_1 = f^n\left(\frac{d}{D}, \frac{\mu}{D\rho v}, \frac{w}{D^2v^2\rho}\right)$$

$$h = Df^n\left(\frac{d}{D}, \frac{\mu}{D\rho v}, \frac{w}{D^2v^2\rho}\right)$$

10.5 **In order to predict lift and drag forces on a scale model of an aircraft during a section of operational envelope, involves see level flight at 100 ms^{-1}, where the speed of sound may be taken as 340 ms^{-1}. It is proposed to utilise cryogenic wind tunnel with Nitrogen at 5 atm of pressure and a temperature of –90°C (ρ = 7.7 kg/m^3) (Viscosity 1.2 × 10^{-5} Ns/m^2 for Nitrogen). The speed of sound in Nitrogen at this temperature is 295 m/s. Determine the wind tunnel flow velocity, the scale of model to ensure full dynamic similarity and the ratio of forces acting on the model and prototype. Mass density of air 1.2 kg/m^3 and viscosity 1.8 × 10^{-5} Ns/m^2.**

[15 Marks]

Sol:

Given : For prototype: Velocity of aircraft, V_p = 100 m/s; Speed of sound, C_p = 340 m/s; Density of air, ρ = 1.2 kg/m^3; Dynamic viscosity of air, μ_p = 1.8 × 10^{-5} Ns/m^2; For model → Fluid : Nitrogen Pressure P_m = 5 atm; Temperature, T_m = –90° C; Density, ρ_m = 7.7 kg/m^3; Dynamic viscosity, μ_m = 1.2 × 10^{-5} Ns/m^2; Speed of sound, C_m = 295 m/s

Applying Mach Model law,

$$(M)_{model} = (M)_{prototype}$$

$$\frac{V_m}{C_m} = \frac{V_p}{C_p}$$

$$\frac{V_m}{295} = \frac{100}{340}$$

$$\Rightarrow \quad V_m = 86.7647 \text{ m/s}$$

$\therefore$ Wind tunnel flow velocity, V_m = 86.7647 m/s

For full dynamic similarity

$$(R_e)_{model} = (R_e)_{prototype}$$

$$\frac{\rho_mV_mL_m}{\mu_m} = \frac{\rho_pV_pL_p}{\mu_p}$$

$$\frac{7.7\times 86.7647\times L_m}{1.2\times 10^{-5}} = \frac{1.2\times 100\times L_p}{1.8\times 10^{-5}}$$

$$\frac{L_p}{L_m} = 8.3511$$

$\Rightarrow$ $L_r = 8.3511$ i.e. scale model

$$\text{Drag force} = \text{Mass} \times \text{Acceleration} = \rho L^3 \times \frac{V}{t} = \rho L^2 V^2$$

Ratio of Drag forces,

$$\frac{F_p}{F_m} = \frac{(\rho L^2 V^2)}{(\rho L^2 V^2)_m} = \frac{\rho_p}{\rho_m} \times \left(\frac{L_p}{L_m}\right)^2 \times \left(\frac{V_p}{V_m}\right)^2$$

$$= \frac{1.2}{7.7} \times (8.3511)^2 \times \left(\frac{100}{86.7647}\right)^2 = 14.4375$$

10.6 **On the apron of spillway model built to a scale of 1 : 30, a hydraulic jump of 100 mm is formed and time required to open the spillway gate is 100 seconds. If the discharge in the prototype spillway is 1500 cumecs, make calculations for the corresponding**

(i) height of the jump in prototype

(ii) time in the prototype to produce the same surge effects in model.

(iii) discharge over the model **[15 Marks]**

Sol:

In case of hydraulic jump over a spillway, gravity forces are predominant, hence Froude model law is applied.

(i) For geometric similarity,

$$L_r = d_r$$

$$\Rightarrow \quad \frac{L_m}{L_p} = \frac{d_m}{d_p}$$

$$\Rightarrow \quad \frac{1}{30} = \frac{100}{d_p}$$

$$\Rightarrow \quad d_p = 3000 \text{ mm} = 3 \text{ m}$$

Thus a hydraulic jump of 3 m will be formed in the prototype.

(ii) For kinematic similarity, velocity scale ratio is given by

$$V_r = \frac{L_r}{T_r}$$

But according to Froude model law,

$$V_r = \sqrt{L_r}$$

$$\therefore \quad \sqrt{L_r} = \frac{L_r}{T_r}$$

$$\Rightarrow \quad T_r = \sqrt{L_r}$$

$$\Rightarrow \quad \frac{T_m}{T_p} = \sqrt{\frac{L_m}{L_p}}$$

$$\Rightarrow \quad \frac{100}{T_p} = \sqrt{\frac{1}{30}}$$

$$\Rightarrow \quad \frac{(100)^2}{T_p^2} = \frac{1}{30}$$

$$\Rightarrow \quad T_p^2 = 30 \times 10^4$$

$$\Rightarrow \quad T_p = 547.72 \text{ seconds.}$$

Time in the prototype to produce the same surge is 547.72 seconds.

(iii) We know that

$$Q = AV$$

$$\therefore \quad Q_r = A_r \times V_r$$

$$\Rightarrow \quad Q_r = (L_r)^2 \times V_r$$

But according to Froude model law,

$$V_r = \sqrt{g_r L_r}$$

But $g_r = 1$

$$\therefore \quad V_r = \sqrt{L_r}$$

$$\Rightarrow \quad Q_r = (L_r)^2 \times (L_r)^{1/2}$$

$$\Rightarrow \quad Q_r = (L_r)^{5/2}$$

$$\Rightarrow \quad \frac{Q_m}{Q_p} = \left(\frac{L_m}{L_p}\right)^{5/2}$$

$$\Rightarrow \quad \frac{Q_m}{1500} = \left(\frac{1}{30}\right)^{5/2}$$

$$\Rightarrow \quad Q_m = 1500 \times 2.029 \times 10^{-4}$$

$$\Rightarrow \quad Q_m = 0.3043 \text{ cumecs}$$

Thus, the discharge over the model is 0.3043 cumecs.

10.7 **Determine the speed and torque required to drive an agitator of 675 mm diameter rotating in air if similar agitator of diameter 225 mm, rotating at 23 rev/s in water requires a torque of 1.1 Nm. Given $\rho_{air} = 1.23$ kg/m^3, $\rho_{water} = 1000$ kg/m^3, $\mu_{air} = 1.86 \times 10^{-5}$ Pa-s, $\mu_{water} = 1.1 \times 10^{-3}$ Pa-s. Obtain your result after arriving at the relevant group of dimensionless terms and the corresponding similarity law, considering that power required as a function of the speed, diameter of agitator, the fluid density and viscosity of fluid in which agitator rotates.**

[20 marks]

Sol:

For an agitator

Power (P) = f(speed (N), Diameter (D), Viscosity (μ), Density (ρ))

Using Buckingham's π theorem

$$f(P, N, D, \mu, \rho) = c$$

$$\text{Number of } \pi \text{ terms} = 5 - 3 = 2$$

Using N, D, ρ as repeating variable

$$\phi(\pi_1, \pi_2) = c$$

$$\pi_1 = N^{a_1} D^{b_1} \rho^{c_1} P$$

$$\Rightarrow \quad [M^0 L^0 T^0] = [T^{-1}]^{a_1} [L^1]^{b_1} [M^1 L^{-3}]^{c_1} [M^1 L^2 T^{-3}]$$

$$\therefore \quad c_1 + 1 = 0$$

$$\Rightarrow \quad c_1 = -1$$

$$b_1 + 2 - 3c_1 = 0$$

$$\Rightarrow \quad b_1 = -5$$

$$-a_1 - 3 = 0$$

$$\Rightarrow \quad a_1 = -3$$

$$\pi_1 = \frac{P}{\rho D^5 N^3}$$

$$\pi_2 = N^{a_2} D^{b_2} \rho^{c_2} \mu$$

$$[M^0 L^0 T^0] = [T^{-1}]^{a_2} [L^1]^{b_2} [M^1 L^{-3}]^{c_2} [M^1 L^{-1} T^{-1}]$$

$$c_2 + 1 = 0$$

$$\Rightarrow \quad c_2 = -1$$

$$b_2 + (-3c_2) - 1 = 0$$

$$\Rightarrow \quad b_2 = -2$$

$$-a_2 - 1 = 0$$

$$\Rightarrow \quad a_2 = -1$$

$$\pi_2 = \frac{\mu}{ND^2\rho}$$

i.e.,
$$\phi\left(\frac{P}{\rho D^5 N^3}, \frac{\mu}{ND^2\rho}\right) = \text{Constant}$$

Here π_2 looks similar to inverse of Reynold's number, so Reynold's model law follows.
For similar pumps

$$\frac{\mu_{air}}{N_1 D_1^2 \rho_{air}} = \frac{\mu_{water}}{N_2 D_2^2 \rho_{water}}$$

$$\Rightarrow \quad \frac{1.86 \times 10^{-5}}{N_1 \times 675^2 \times 1.23} = \frac{1.1 \times 10^{-3}}{23 \times (225)^2 \times 1000}$$

$$\Rightarrow \quad N_1 = 35.13 \text{ rev/s} \qquad \text{...(i)}$$

Also
$$\frac{P_1}{\rho_{air} D_1^5 N_1^3} = \frac{P_2}{\rho_{water} D_2^5 N_2^3}$$

$$\Rightarrow \quad \frac{P_1}{1.23 \times 675^5 \times (35.13)^3} = \frac{P_2}{1000 \times 225^5 \times 23^3}$$

$$\Rightarrow \quad \frac{P_1}{P_2} = 1.065$$

But
$$P = \frac{2\pi NT}{60}$$

$$\therefore \quad N_1 T_1 = 1.065\, N_2 T_2$$

$$\Rightarrow \quad 35.13\, T_1 = 1.065 \times 23 \times 1.1$$

$$\Rightarrow \quad T_1 = 0.767 \text{ Nm}$$

10.8 A river stretch 10 km long and 920 m wide is to be modeled for flood studies in a hydraulic laboratory where the available flume size is 2 m in width, 20 m in length and 70 cm in height. It is decided to use a depth scale of 10. If the peak flood in the river is 3000 cumec at a depth of 6 m, find the suitable horizontal scale to be adopted for construction of model. Maximum available flow is 150 litres/s in the laboratory. The horizontal scale should be a multiple of 10 such that the model is as big as possible. Determine the model dimensions, depth and discharge. Assume that hydraulic depth is same as the depth of flow in both the model and prototype.

[20 marks]

Sol:

Given : for prototype → $L_p = 10$ km; $B_p = 920$ m; $Q_p = 3000$ m/s; $H_p = 6$ m; for model → $Q_m \leq 150$ *l*ps; $B_m \leq 2$ m; $L_m \leq 20$ m; $H_m < 70$ cm

Depth scale, $(L_r)_v = 10$

$$\therefore \quad \frac{H_p}{H_m} = 10$$

$$H_m = \frac{6}{10} = 0.6 \text{ m} = 60 \text{ cm} < 70 \text{ cm} \quad \text{(Hence OK)}$$

Using Froude Model law

$$(F_e)_m = (F_e)_p$$

$$\frac{V_m}{\sqrt{g(L_m)_v}} = \frac{V_p}{\sqrt{g(L_p)_v}}$$

$$\frac{V_p}{V_m} = \sqrt{\left(\frac{L_p}{L_m}\right)} = \sqrt{(L_r)_v}$$

$$V_r = \sqrt{(L_r)_v}$$

$$\frac{Q_p}{Q_m} = \frac{A_p}{A_m} \times \frac{V_p}{V_m}$$

$$Q_r = (L_r)_H \times (L_r)_v \times \sqrt{(L_r)_v}$$

$$Q_r = (L_r)_H \times (L_r)_v^{3/2}$$

$$\therefore \quad \frac{3000}{150 \times 10^{-3}} = (L_r)_H \times 10^{3/2}$$

$$(L_r)_H = 632.455$$

As given, horizontal scale should be multiple 10

∴ Let us assume, $(L_r)_H = 640$

Checking for discharge,

$$Q_r = (L_r)_H \times (L_r)_v^{3/2}$$

$$\frac{Q_p}{Q_m} = 640 \times 10^{3/2}$$

$$Q_m = \frac{3000}{640 \times 10^{3/2}} = 0.14823 \text{ m}^3/\text{s}$$

$$= 148.23 \; l\text{ps} < 150 \; l\text{ps}$$

∴ okay

Checking for length

$$(L_r)_H = \frac{L_p}{L_m}$$

$$L_m = \frac{10 \times 10^3}{640} = 15.625 \text{ m} < 20 \text{ m}$$

Checking for width

$$(L_r)_H = \frac{B_p}{B_m}$$

$$B_m = \frac{920}{640} = 1.4375 \text{ m} < 2 \text{ m}$$

So we can safely take $(L_r)_H = 640$

Hence,
$$(L_r)_V = 10$$
$$(L_r)_H = 640$$
$$Q_m = 148.23\ lps$$
$$H_m = 60\text{ cm}$$
$$L_m = 15.625\text{ m}$$
$$B_m = 1.4375\text{ m}$$

11. Drag & Lift

Level-1

11.1 **Find the power required to tow lengthwise a plate 1.2 m wide and 3 m long at a velocity of 2.4 m/sec in water at 23°C. Make allowance for the felt that the boundary layer will change from laminar to turbulent over the plate, *V* for water at 23°C is 0.9 × 10⁻⁶ m²/sec and ρ = 1000 kg/m³.**

[15 Marks]

Sol:

$$R_{eL} = \frac{VL}{\nu} = \frac{2.4\times 3}{0.9\times 10^{-6}} = 8\times 10^{6}$$

$$C_D = \frac{0.074}{(R_{eL})^{1/5}} - \frac{1700}{R_{eL}}$$

$$C_D = \left[\frac{0.074}{(8\times 10^{6})^{1/5}} - \frac{1700}{8\times 10^{6}}\right] = 2.87\times 10^{-3}$$

The drag force is given by

$$F_D = 2\times\left[\frac{1}{2}\times C_D\times\rho\times A\times V_\infty^2\right]$$

$$= 2\times\left[\frac{1}{2}\times 2.87\times 10^{-3}\times 1.2\times 3\times 1000\times 2.4^2\right] = 59.51\text{ N}$$

Power required $= 59.51\times 2.4 = 143$ W

11.2 **A small air bubble of diameter 1.0 mm rise with a steady velocity of 1.5 cm/sec through an oil. Estimate the viscosity of the oil if its specific gravity is 0.9.**

[10 Marks]

Sol:

Neglecting the weight of the bubble, the drag and buoyancy force should keep the bubble in equilibrium

$$F_B = F_D$$

$$\frac{\pi d^3}{6}\rho g = 3\pi\mu v d$$

Dynamic viscosity, $\mu = \dfrac{\rho g d^2}{18\nu}$

$$= \frac{(1000\times 0.9)\times 9.81\times(0.001)^2}{18\times 0.0018}$$

$$= 32.4\times 10^{-3}\text{ Ns/m}^2$$

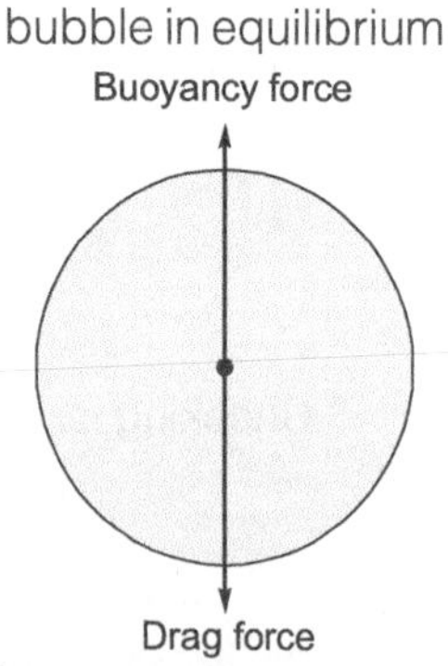

Check :

$$R_e = \frac{\rho VD}{\mu} = \frac{0.015 \times 0.001 \times 1000 \times 0.9}{32.4 \times 10^{-3}}$$

$$= 0.41 < 1 \qquad \text{(OK)}$$

The expression used for drag force is ok.

Level-2

11.3 To improve the fuel efficiency, the design of side rearview mirrors has been changed from circular plate to a stream lined shape.

Determine the amount of fuel and money saved per year as a result of replacement.

Assume the car is driven 15000 km is a year at an average speed of 60 km/hr, Density of gasoline = 0.75 kg/L, Price of gasoline = Rs. 75 /litre, Heating value of gasoline = 44000 kJ/kg, Overall efficiency of engine = 30%, Density of air = 1.2 kg/m^3

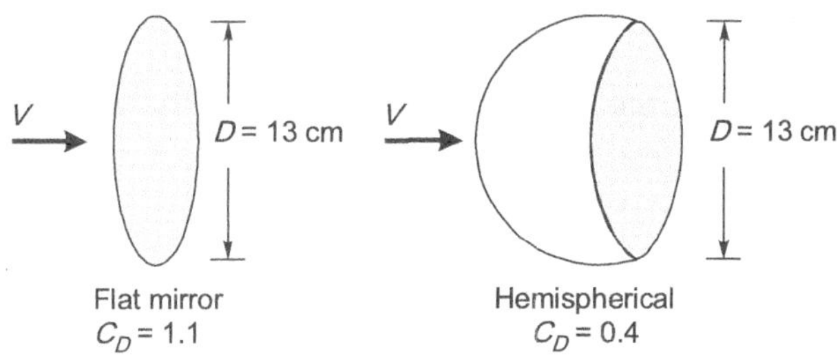

[20 Marks]

Sol:

$$F_1 = \text{Drag force on flat mirror} = \frac{1}{2} \times C_{D_1} \times \rho \times A \times U_\infty^2$$

$$A = \frac{\pi}{4} \times (0.13)^2,$$

$$U_\infty = 60 \text{ km/hr}$$

$$F_2 = \text{Drag force on hemisphere}$$

$$= \frac{1}{2} C_{D_2} \times \rho \times A \times U_\infty^2$$

$$\text{Reduction of drag force} = \Delta F = F_1 - F_2$$

$$= \frac{1}{2} \times \left(C_{D_1} - C_{D_2}\right) \times \rho \times A \times U_\infty^2$$

$$\Delta F = \frac{1}{2} \times (1.1 - 0.4) \times 1.2 \times \frac{\pi}{4} \times 0.13^2 \times 16.67^2 = 1.55 \text{ N}$$

$$\text{Total distance travelled in a year} = 15000 \text{ km}$$

$$\text{Work saved} = \Delta F \times \text{Distance travelled}$$

$$= 1.55 \times 15000 \times 10^3$$

$$= 23.25 \times 10^6 \text{ Joule}$$

$$= 23250 \text{ kJ}$$

$$\text{Gasoline saved} = \frac{23250}{44000} = 0.5284 \text{ kg}$$

$$\text{Given density of gasoline} = 0.75 \text{ kg/L}$$

$$\text{Volume saved} = \frac{0.5284}{0.75} = 0.704 \text{ litre}$$

$$\text{Since overall efficiency} = 0.3$$

$$\text{Hence actual reduction} = 0.704 \times 0.3 = 0.211 \text{ litre}$$

$$\text{Saving in price} = 0.211 \times 75 = \text{Rs. } 15.85$$

$$\text{For both mirrors} = \text{Rs. } 31.71$$

11.4 A 2 m high, 4 m wide rectangular advertisement panel is attached to a 4 m wide, 0.15 m high rectangular concrete block (density = 2300 kg/m^3) by two 10 cm diameter 4 m high (exposed) poles, as shown in figure. If the sign is to withstand 120 km/hr winds from any direction. Determine

(a) Maximum drag force on the panel.

(b) Drag force on poles.

(c) Minimum length L of concrete block required to resist the winds.

Given : Density of air = 1.3 kg/m^3, C_D for rectangular part = 1.1, C_D for poles = 1.0

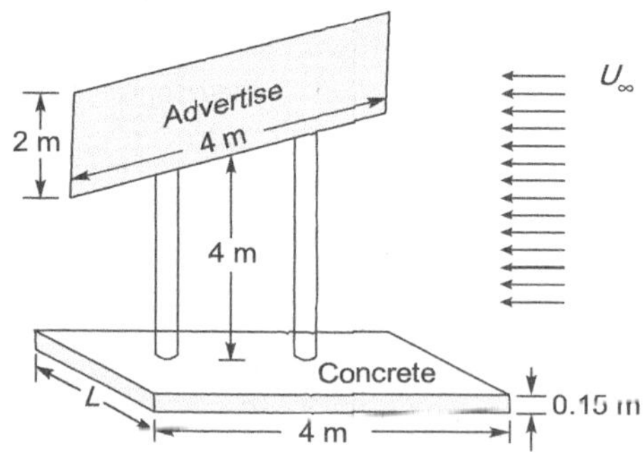

[20 Marks]

Sol:

(a) Using $F_D = \frac{1}{2} C_{D_1} \rho A \times U_\infty^2$

$$\text{Exposed area, } A = 4 \text{ m} \times 2 \text{ m} = 8 \text{ m}^2$$

$$\rho = 1.3 \text{ kg/m}^3$$

$$U_\infty = 120 \text{ kmph} = 33.33 \text{ m/sec}$$

$$F_{D_1} = \frac{1}{2} \times 1.1 \times 1.3 \times 8 \times 33.33^2 = 6354.28 \text{ N}$$

(b) $$\text{Drag on poles} = 2 \times \left[\frac{1}{2} C_{D_2} \times \rho \times A \times U_\infty^2\right]$$

$$\text{Projected area, } A = 0.1 \times 4 = 0.4 \text{ m}^2$$

$$F_{D_2} = 1.0 \times 1.3 \times 0.4 \times (33.33)^2 = 577.66 \text{ N}$$

(c) For reg. length of concrete block

Calculate overturning moment about rear edge

$$M_0 = F_{D_1} \times L_{A_1} + F_{D_2} \times L_{A_2}$$

$$L_{A_1} = \text{Lever Arm } 1 = 4\text{ m} + 0.15\text{ m} + 1\text{ m} = 5.15\text{ m}$$

$$L_{A_2} = \text{Lever Arm } 2 = 0.15\text{ m} + 2\text{ m} = 2.15\text{ m}$$

$$M_0 = 6354.28 \times 5.15 + 577.66 \times 2.15 = 33966.52\text{ Nm}$$

$$\text{Resting moment} = \text{Self weight} \times L_{A_3}$$

$$= 2300 \times [L \times 0.15 \times 4] \times \frac{L}{2} = 690L^2$$

Equating, $33966.52 = 690\,L^2$

Solving, $L = 7.02\text{ m}$

12. Open Channel Flow

Level-1

12.1 **A hydraulic jump occurs in a wide, rectangular channel with initial and sequent depths of 0.5 m and 2.0 m respectively. Calculate**

(a) the discharge in m³/s per metre width

(b) the possible critical depth for this discharge

(c) the energy loss in the jump, (in metres head) and

(d) sketch the critical depth line on the jump profile

[10 Marks]

Sol:

Alternate depth, $y_1 = 0.5$ m, $y_2 = 2$ m

(a) Discharge in m³/sec per metre width 'q' is given by

$$\frac{2q^2}{g} = y_1 y_2 (y_1 + y_2)$$

or
$$q^2 = \frac{y_1 y_2 (y_1 + y_2)}{2} g$$

$$q^2 = \frac{0.5 \times 2 \times (0.5 + 2)}{2} \times 9.81$$

$\Rightarrow$ $q = 3.5$ m³/sec per metre width

(b) Critical depth 'y_c' for this discharge is given by

$$y_c = \left(\frac{q^2}{g}\right)^{1/3} = \left(\frac{3.5^2}{9.81}\right)^{1/3} = 1.077\text{ m}$$

(c) Energy loss in the jump (in metre head) is given by

$$\Delta E = \frac{(y_2 - y_1)^3}{4y_1 y_2} = \frac{(2 - 0.5)^3}{4 \times 2 \times 0.5} = 0.84\text{ m}$$

12.2 A channel has vertical walls 1.5 m apart and a semicircular invert. If the centre-line depth is 1 m and the bed slope is 1 in 2500, what would be the value of "C" in the Chezy's formula if the discharge is 0.60 m^3/s?

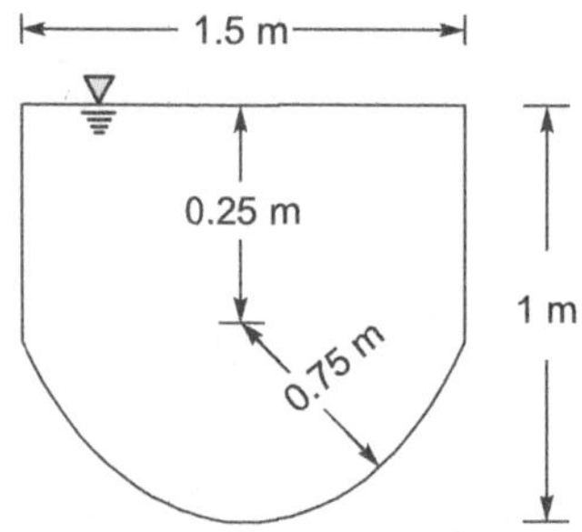

[10 marks]

Sol:

$$\text{Bed slope, } S = \frac{1}{2500}; \text{ Discharge, } Q = 0.60 \text{ m}^3/\text{s}$$

$$\text{Area of flow, } A = (1.5\times 0.25) + \left[\frac{\pi}{2}\times(0.75)^2\right] = 1.2586 \text{ m}^2$$

$$\text{Wetted perimeter, } P = 2\times 0.25 + \frac{1}{2}[2\pi\times(0.75)] = 2.8562 \text{ m}$$

$$R = \frac{A}{P} = \frac{1.2586}{2.8562} = 0.4407$$

Using Chezy's formula,

$$Q = AC\sqrt{RS}$$

$$\Rightarrow \quad 0.60 = 1.2586\times C\sqrt{0.4407\times\frac{1}{2500}}$$

$$\Rightarrow \quad C = 35.9056$$

12.3 A rectangular channel 3 m wide and 1.8 m deep with discharge 20 m^3/sec, suddenly has its discharge reduced to 12 m^3/s at downstream end. Calculate the speed and height of surge moving downstream.

[12 marks]

Sol:

For sudden reduction in discharge (due to closing of gate) a negative surge will move downstream.

Let, suffix 1 refers to flow conditions before the gate closure and suffix 2 after the passage of negative wave.

So,

$$\text{Velocity, } V_1 = \frac{20}{3\times 1.8} = 3.704 \text{ m/s}$$

$$\text{Depth, } y_1 = 1.8 \text{ m}$$

$$\text{Discharge, } Q_1 = 20 \text{ m}^3/\text{s}$$

After passage of wave,

$$\text{Discharge, } Q_2 = 12 \text{ m}^3/\text{s}$$

$$\text{Discharge per unit width, } q_2 = \frac{12}{3} = 4 \text{ m}^2/\text{s}$$

and, $$q_2 = V_2 y_2$$

$\Rightarrow$ $$4 = V_2 y_2 \quad \text{...(i)}$$

Also, for negative surge,

$$V_2 = V_1 + 2\sqrt{gy_2} - 2\sqrt{gy_1}$$

$$= 3.704 + 2\sqrt{9.81 \times y_2} - 2\sqrt{9.81 \times 1.8}$$

$$= 6.264\sqrt{y_2} - 4.70 \quad \text{...(ii)}$$

From equation (i) and (ii),

Solving by trial and error,

$$y_2 = 1.43 \text{ m}$$

$$V_2 = 2.797 \text{ m/s}$$

$$\text{Height of surge, } H = 1.8 - 1.43 = 0.37 \text{ m}$$

$$\text{Velocity of wave at crest, } C_1 = \sqrt{gy_1} + V_1$$

$$= \sqrt{9.81 \times 1.8} + 3.704 = 7.906 \text{ m/s}$$

$$\text{Velocity of wave at trough, } C_2 = \sqrt{gy_2} + V_2$$

$$= \sqrt{9.81 \times 1.43} + 2.797 = 6.542 \text{ m/s}$$

12.4 **A sluice gate discharges water into a horizontal rectangular channel with a velocity of 10 m/s and depth of flow of 1 m. Determine the depth of flow after the jump and consequent loss in total head.**
[8 marks]

Sol:

Given: Velocity of flow before hydraulic jump,

$$V_1 = 10 \text{ m/s}$$

Depth of flow before hydraulic jump,

$$y_1 = 1 \text{ m}$$

Discharge per unit width,

$$q_1 = V_1 y_1 = 10 \times 1 = 10 \text{ m}^2/\text{s}$$

$\therefore$ $$F_1^2 = \frac{V_1^2}{gy_1} = \frac{10^2}{9.81 \times 1} = 10.194$$

The depth of flow after the jump is given by

$$\frac{y_2}{y_1} = \frac{1}{2}\left[-1 + \sqrt{1 + 8F_1^2}\right]$$

$\Rightarrow$ $$y_2 = \frac{1}{2} \times \left[-1 + \sqrt{1 + 8 \times 10.194}\right] \times 1$$

$\Rightarrow$ $$y_2 = 4.043 \text{ m}$$

Loss in total head, $$h_L = \frac{(y_2 - y_1)^3}{4y_1 y_2} = \frac{(4.043 - 1)^3}{4 \times 4.043 \times 1} = 1.74 \text{ m}$$

12.5 A spillway has its crest at elevation of +144.0 m and a horizontal apron at an elevation of +105.0 m on the downstream side. Find the tail water elevation required to form a hydraulic jump when the elevation of energy line above the crest is +146.5 m. The C_d for the flow can be assumed as 0.73. The energy loss over the spillway face may be neglected.

[12 Marks]

Sol:

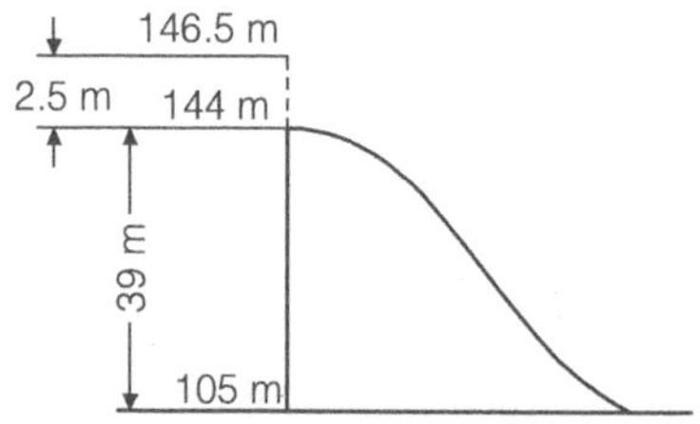

(For unit length)

$$Q = \frac{2}{3}C_d L\sqrt{2g}\,h^{3/2}$$

$$= \frac{2}{3}\times 0.73\times 1\times\sqrt{2\times 9.81}\times 2.5^{3/2}$$

$$= 8.52\,\text{m}^3/\text{sec}$$

Average fall of water

$$h_{av} = 39+\frac{2.5}{2} = 40.25$$

Theoretical velocity

$$= \sqrt{2\times g\times h_{av}} = \sqrt{2\times 9.81\times 40.25} = 28.10\text{ m/sec}$$

Actual velocity

$$= 0.9\times 28.10$$
$$= 25.30\text{ m/sec}$$

Depth of flow at foot of spillway

$$Y_1 = \frac{Q}{V} = \frac{8.52}{25.30} = 0.3370\text{ m}$$

$$F_1 = \frac{V_1}{\sqrt{gY_1}} = \frac{25.30}{\sqrt{9.81\times 0.3370}} = 13.91$$

$$Y_2 = \frac{Y_1}{2}\left[\sqrt{1+8F_1^2}-1\right] = \frac{0.3370}{2}\left[\sqrt{1+8\times 13.9^2}-1\right] = 6.46\text{ m}$$

12.6 Water flows in a rectangular channel at a depth of 1.20 m and a velocity of 2.4 m/sec. What would be the effect of a local rise in the channel bed of 0.60 m on the water surface?

[12 Marks]

Sol:

Froude number, of section (1)-(1)

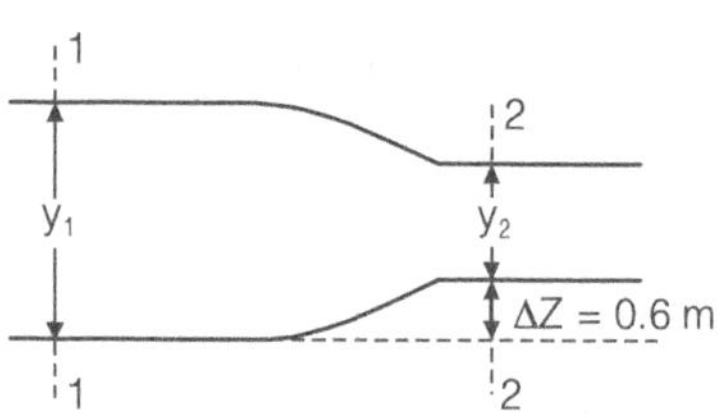

$$F_{r_1} = \frac{V}{\sqrt{gy_1}} = \frac{2.4}{\sqrt{9.81\times 1.2}}$$

$$= 0.69 < 1.0$$

Apply energy equation at points (i) and (ii)

$$\Rightarrow \quad y_1+\frac{V_1^2}{2g} = (y_2+\Delta Z)+\frac{V_2^2}{2g}$$

$$\Rightarrow \quad 1.2+\frac{(2.4)^2}{2\times 9.81} = (y_2+0.6)+\frac{V_2^2}{2g}$$

$$\left[\because q = y_1V_1 = y_2V_2 \Rightarrow y_2 = \frac{2.4\times1.2}{V_2}\right]$$

$$\Rightarrow \quad 1.2+\frac{(2.4)^2}{2\times9.81} = (y_2+0.6)+\frac{(2.88/y_2)^2}{2\times9.81}$$

$$\Rightarrow \quad 1.49 = y_2+0.6+\frac{0.423}{y_2^2}$$

$$\Rightarrow \quad 0.89y_2^2 = y_2^3+0.423$$

$$\Rightarrow \quad y_2 = 0.72\text{ m}$$

$$\Rightarrow \quad F_{r_2} = \frac{V_2}{\sqrt{gy_2}} = \frac{\frac{2.4\times1.2}{0.72}}{\sqrt{9.81\times0.72}} = 1.51 > 1$$

The basic effect in the flow is that, the flow will change from supercritical stage to subcritical stage.

Level-2

12.7 A 20 m wide rectangular channel has a normal depth of 2.5 m when bottom slope is 0.0006 and manning's, n is 0.015. Two piers of 2 m width each and 4 m long with rounded nose are constructed in the channel. What will be the water depth upstream of piers?

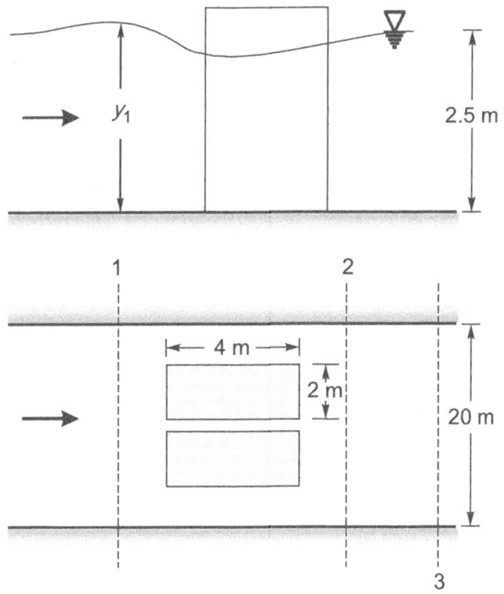

[20 Marks]

Sol:

Under normal condition,

$$\text{Hydraulic radius, } R = \frac{A}{P} = \frac{20\times2.5}{25} = 2\text{ m}$$

$$V_1 = \frac{1}{n}R^{2/3}S^{1/2} \qquad \text{(Mannings)}$$

$$= \frac{1}{0.015}\times(2)^{2/3}\times(0.0006)^{1/2} = 2.529\text{ m/sec}$$

$$Q = 20\times2.5\times2.592 = 129.614\text{ m}^3\text{/sec}$$

At section 2, downstream of piers,

$$16v_2y_2 = 129.614$$

$$v_2y_2 = 8.1$$

Applying momentum equation between 2 and 3

$$\dot{m}(V_3 - V_2) = \frac{1}{2}\gamma_B\left(y_2^2 - y_3^2\right)$$

$$\rho Q(V_3 - V_2) = \frac{1}{2}\gamma_B\left(y_2^2 - y_1^2\right)$$

$$\frac{2\times 129.614}{9.81\times 20}\left(2.592 - \frac{8.1}{y_2}\right) = y_2^2 - 2.5^2$$

Solving,

$$y_2^2 - 9.677y_2 + 10.708 = 0$$

$$y_2 = 2.185$$

$$V_2 = \frac{8.1}{2.185} = 3.707 \text{ m/sec}$$

Now apply energy equation between section 1 and 2
(No energy loss since no wake region)

$$y_1 + \frac{V_1^2}{2g} = y_2 + \frac{V_2^2}{2g}$$

$$y_2 + \left(\frac{129.614}{20\times y_1}\right)^2 \times\frac{1}{2g} = 2.185 + 0.701 = 2.886$$

Solving,

$$y_1 = 2.532 \text{ m}$$

$$\text{Rise} = 2.532 - 2.5 = 0.032 \text{ m}$$

12.8 A 3.5 m wide rectangular channel carries a discharge of 10 m³/sec at a depth of 1.75 m. If the width of the channel is reduced to 2.25 m and the bed level is lowered by 0.97 m, determine the difference in water level elevations between upstream and contracted sections. Assume no energy loss.

[15 Marks]

Sol:

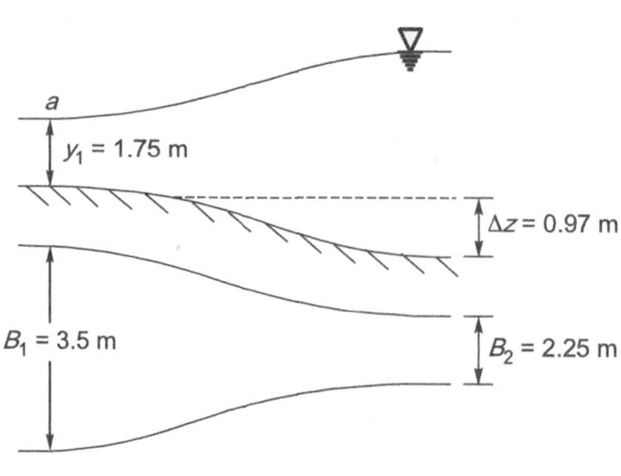

$$V_1 = \frac{Q}{B_1\times y_1} = \frac{10}{3.5\times 1.75} = 1.633 \text{ m/sec}$$

Using

$$E_1 = \text{Specific energy at section 1} = 1.75 + \frac{1.633^2}{2\times 9.81} = 1.866 \text{ m}$$

$$q_2 = \frac{10}{2.25} = 4.44 \text{ m}^2/\text{sec}$$

$$2 + E_1 = E_2$$

$$0.97 + 1.866 = y_2 + \frac{4.44^2}{2 \times 9.81 \times y_2^2}$$

Solving, $y_2 = 2.72$ m

12.9 Sketch the water surface profile in a long rectangular channel (n = 0.014). If the channel is 3 m wide, flow rate is 10 m³/sec and there is an abrupt change in slope from 0.0016 to 0.0150.

[15 Marks]

Sol:

$$\text{Critical depth, } y_c = \left(\frac{q^2}{g}\right)^{1/3} = \left(\frac{(10/3)^2}{9.81}\right)^{1/3} = 1.0424 \text{ m}$$

$$Q = \frac{1}{n} A R^{2/3} S^{1/2} \qquad \text{(Manning's equation)}$$

In upstream segment

$$10 = \frac{1}{0.014}(3y_{n_1})\left[\frac{3y_{n_1}}{3+2y_{n_1}}\right]^{2/3}(0.0016)^{1/2}$$

Solving

Normal depth, $y_{n_1} = 1.435$ m

In downstream segment

$$10 = \frac{1}{0.014}(3y_{n_2})\left(\frac{3y_{n_2}}{3+2y_{n_2}}\right)^{2/3}(0.015)^{1/2}$$

$y_{n_2} = 0.647$ m

Hence the flow is subcritical before the break in grade and supercritical after the break.

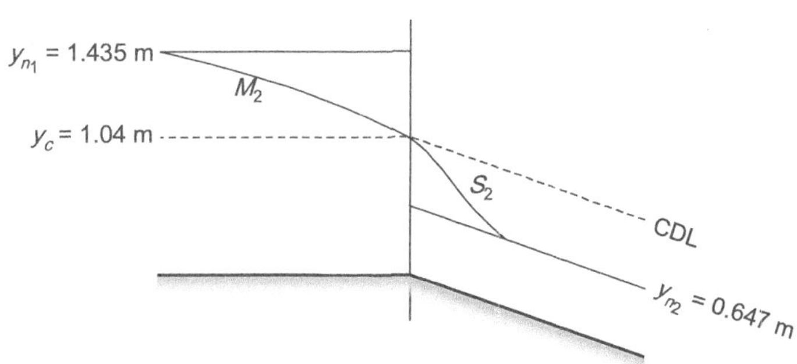

12.10 Show that for critical flow in a circular conduit of diameter D

$$\frac{Q^2}{gD^5} = \frac{\theta - \sin\theta\cos\theta}{64\sin\theta}$$

Where 2θ is the angle subtended by Water surface at the centre.

[15 Marks]

Sol:

From basic equation for critical condition.

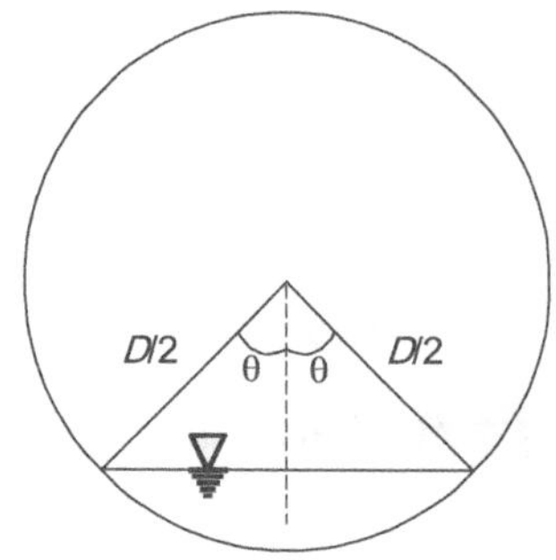

$$\frac{Q^2}{g} = \frac{A^3}{T}$$

where,
$$\frac{a}{A} = \frac{2\theta}{360} - \frac{\sin 2\theta}{2\pi}$$

a = Actual wetted area, A = Total area

$$a = \frac{\pi}{4} \times D^2 \left[\frac{\theta}{180^\circ} - \frac{\sin 2\theta}{2\pi}\right]$$

$$T = \text{Top width} = 2\left[\frac{D}{2}\sin\theta\right] = D\sin\theta$$

$\Rightarrow$
$$\frac{Q^2}{g} = \frac{\left(\frac{\pi}{4}\right)^3 \times D^6 \left[\frac{\theta}{180^\circ} - \frac{\sin 2\theta}{2\pi}\right]^3}{D\sin\theta}$$

$$\frac{Q^2}{gD^5} = \frac{\pi^3\left[\frac{\theta}{180} - \frac{\sin\theta\cos\theta}{\pi}\right]^3}{64\sin\theta} \quad [\theta_1 = \text{in degrees}, \theta = \text{in radians}]$$

$$= \frac{\left[\theta \times \frac{\pi}{180} - \sin\theta\cos\theta\right]^3}{64\sin\theta}$$

$$= \frac{[\theta - \sin\theta\cos\theta]^3}{64\sin\theta}$$ (Hence proved) $[\theta_1 = \theta = \text{in radian}]$

12.11 Water flows in a channel whose bottom slope is 0.002 and whose cross section is shown in figure. The dimensions and manning's coefficients for the surfaces of different section are given in the figure. Determine the flow rate through the channel and effective manning's coefficient for the channel.

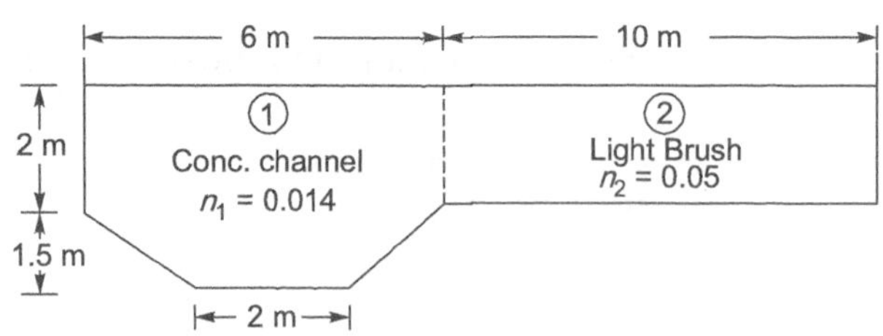

[15 Marks]

Sol:

For section (i)

$$\text{Wetted perimeter, } P_1 = 2 + 2\sqrt{2^2 + 1.5^2} + 2 \text{ m} = 9 \text{ m}$$

$$\text{Area, } A_1 = 6 \times 2 + \frac{1}{2} \times [2+6] \times 1.5$$

$$= 12 + 6 = 18 \text{ m}^2$$

$$R_1 = \frac{A_1}{P_1} = \frac{18}{9} = 2 \text{ m}$$

$$Q_1 = \frac{1}{n_1} R_1^{2/3} A S_0^{1/2} = \frac{1}{0.014} \times (2)^{2/3} \times (0.002)^{1/2} \times 18 = 91.27 \text{ m}^3\text{/sec}$$

For section (2)

$$\text{Wetted perimeter, } P_2 = 10 + 2 = 12 \text{ m}$$

$$\text{Area, } A_2 = 10 \times 2 = 20 \text{ m}^2$$

$$P_2 = \frac{A_2}{P_2} = \frac{20}{12} = 1.63 \text{ m}$$

Now using manning's equation

$$Q_2 = \frac{1}{n} R_2^{2/3} S^{1/2} \times A_2$$

$$= \frac{1}{0.05} \times (1.067)^{2/3} (0.002)^{1/2} \times 20 = 25.18 \text{ m}^3\text{/sec}$$

$$\text{Total discharge} = Q_1 + Q_2 = 91.27 + 25.18$$

$$= 116.45 \text{ m}^3\text{/sec}$$

For effective coefficient,

$$\frac{1}{n_{eff}} \times \left(R_{eq}\right)^{2/3} \times (S)^{1/2} \times A = 116.48$$

where, $A = A_1 + A_2 = 18 + 20 = 38 \text{ m}^2$

$P = P_1 + P_2 = 9 + 12 = 21$ m

$$R_{eq} = \frac{38}{21} = 1.81 \text{ m}$$

Putting,

$$\frac{1}{n_{eff}} \times (1.81)^{2/3} (0.002)^{1/2} \times 38 = 116.45$$

Solving, $n_{eff} = 0.0216$

12.12 A horizontal rectangular channel 3 m wide and 2 m water depth conveys water at 18 m^3/sec. If the flow rate is reduced to 2/3 or the original value. Compute the magnitude and speed of the upstream surge. Assume the front of the surge rectangular and friction in the channel is neglected.

[15 Marks]

Sol:

Before surge,

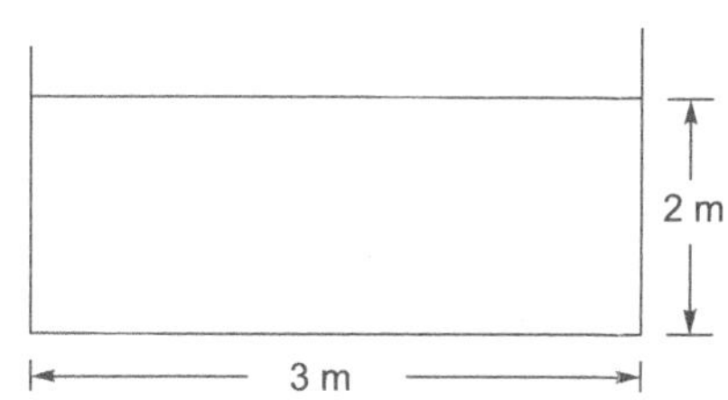

$$Q_1 = 18 \text{ m}^3\text{/sec}$$

$$V_1 = \frac{18}{6} = 3 \text{ m/sec}$$

$$A_1 = 3 \text{ m} \times 2 \text{ m} = 6 \text{ m}^2$$

After surge,

$$Q_2 = \frac{2}{3} \times 18 = 12 \text{ m}^3\text{/sec}$$

$$B = 3 \text{ m}$$

$$V_2 = \frac{12}{3 \times y_2} = \frac{y}{y_2}$$

Since the surge will travel upstream from continuity equation

$$y_1(V_1 + V_w) = y_2(V_2 + V_w)$$

$$V_w = \frac{y_2V_2 - y_1V_1}{y_1 - y_2} = \frac{4 - 3\times 2}{2 - y_2} = \frac{-2}{2 - y_2}$$

Also using, $$C = \sqrt{\frac{9y_2}{2}\left(\frac{y_2}{y_1}+1\right)}$$ [from momentum equation]

c = Velocity of wave w.r.t. water = $V_w + V_1$

$$V_w + V_1 = \sqrt{\frac{9.81}{2}\times y_2\left(\frac{y_2}{2}+1\right)}$$

$$\frac{-2}{2-y_2} + 3 = \sqrt{\frac{9.81}{2}\times y_2\left(\frac{y_2}{2}+1\right)}$$

Solving for y_2

$$y_2 = 2.75 \text{ m}$$

Hence, $$V_w = \frac{-2}{2-2.75} = 2.67 \text{ m/sec}$$

12.13 A trapezoidal channel having bottom width 6 m, side slopes 2H : 1V and bottom slope 0.0016, carries a discharge of 10 m³/sec. A weir placed across the channel backs up the water to a depth of 2 m. Calculate how for upstream the depth i.e. 1.5. [Take manning's n = 0.0025]

[20 Marks]

Sol:

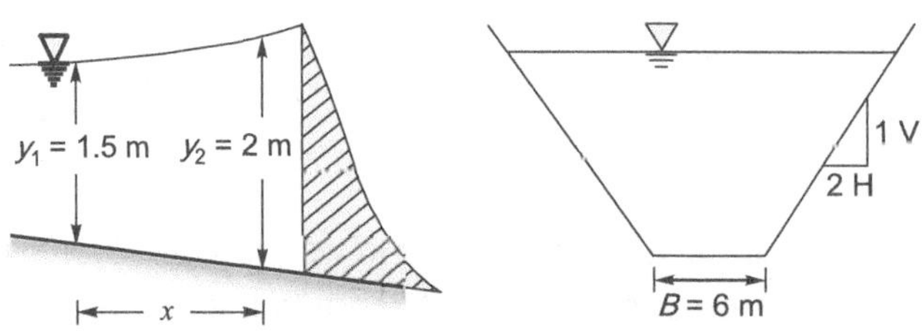

At section (i)-(i)

$$y_1 = 1.5 \text{ m}, B = 6 \text{ m}$$

$$A_1 = y_1(B + my) = 1.5(6 + 2\times 1.5) = 13.5 \text{ m}^2$$

$$P_1 = B + 2y\sqrt{1+m^2} = 6 + 2\times 1.5\sqrt{1+2^2} = 12.71 \text{ m}$$

$$Q = \frac{1}{n}R^{2/3}S_{f_1}^{1/2}A$$ [From manning's]

Putting

$$10 = \frac{1}{0.025}\times\left(\frac{13.5}{12.71}\right)^{2/3}(s_{f_1})^{1/2}\times 13.5$$

Solving for $$s_{f_1} = 3.164\times 10^{-4}$$

At section (2)-(2)

$$y_2 = 2 \text{ m}$$

$$A_2 = 2[6 + 2\times 2] = 20 \text{ m}^2$$

$$P_2 = 6 + 2\times 2\sqrt{5} = 14.94 \text{ m}$$

From Manning's

$$10 = \frac{1}{0.025} \times \left(\frac{20}{14.94}\right)^{2/3} \times S_{f_2}^{1/2} \times 20$$

Solving $\quad s_{f_2} = 1.06 \times 10^{-4}$

Hence, $\quad \bar{s}_f = \frac{s_{f_1} + s_{f_2}}{2} = \frac{3.164 \times 10^{-4} + 1.06 \times 10^{-4}}{2} = 2.112 \times 10^{-4}$

Also, $\quad E_1 = \text{Specific energy at 1-1} = y_1 + \frac{Q^2}{2gA_1^2} = 1.5 + \frac{10^2}{2g \times 13.5^2} = 1.53 \text{ m}$

E_2 = specific energy at 2-2

$$E_2 = y_2 + \frac{Q^2}{2gA_2^2} = 2 + \frac{10^2}{2 \times 9.81 \times 20^2} = 2.01 \text{ m}$$

Now using $\quad \frac{E_2 - E_1}{s_0 - \bar{s}_f} = \Delta x$

$$\Delta x = \frac{2.01 - 1.53}{1.6 \times 10^{-4} - 2.112 \times 10^{-4}} = 346.2 \text{ m} \qquad \text{(Ans)}$$

12.14 An overflow spillway is 40 m high. Water flows down the spillway with a head of 2.5 m over the spillway crest. The spillway discharge coefficient $C_d = 0.738$. Show that the water depth at the toe of the spillway would be 0.3 m. Determine the sequent depth required for the formation of the hydraulic jump and the loss of head in the jump.

[15 Marks]

Sol:

Given data

Height of spillway, $\quad H = 40$ m

Head, $\quad H = 2.5$ m

Spillway discharge coefficient, $C_d = 0.738$

Water depth of the toe of spillway = 0.3 m

$\because \quad Q = \frac{2}{3} C_d B \sqrt{2g} H^{3/2}$

$\therefore \quad q = \frac{Q}{B} = \frac{2}{3} C_d \sqrt{2g} (H)^{3/2}$

$$= \frac{2}{3} \times 0.738 \times \sqrt{2 \times 9.81} \times (2.5)^{3/2}$$

$= 8.614 \text{ m}^3\text{/sec/m}$

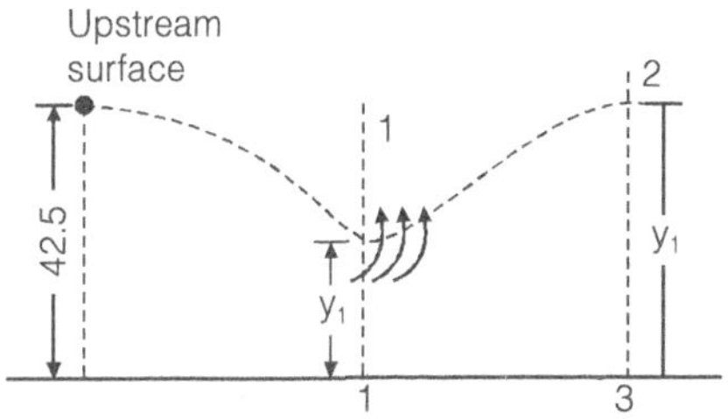

Apply Bernoulli's equation between upstream level and section (1)-(1)

$$\Rightarrow \quad 42.5 = y_1 + \frac{V^2}{2g}$$

$$\Rightarrow \quad 42.5 = y_1 + \frac{q^2}{2gy_1^2}$$

$$\Rightarrow \quad 42.5 = y_1 + \frac{(8.614)^2}{2\times 9.81\times y_1^2}$$

$$\Rightarrow \quad 42.5 = y_1 + \frac{3.78}{y_1^2}$$

$$\Rightarrow \quad y_1^3 - 42.5\,y_1^2 + 3.78 = 0$$

$$\Rightarrow \quad y_1 = 0.29 \approx 0.3 \text{ m}$$

$\therefore \quad y_1 = 0.3$ m **Hence proved.**

Now,
$$V_1 = \frac{q}{y_1} = \frac{8.614}{0.3} = 28.71 \text{ m/s}$$

Froude numbers,

$$F_{r_1} = \frac{V_1}{\sqrt{gy_1}} = \frac{28.71}{\sqrt{9.81\times 0.3}} = 16.74$$

$$\because \quad \frac{y_2}{y_1} = \frac{1}{2}\left[-1+\sqrt{1+8F_{r_1}^{\,2}}\right]$$

$$\Rightarrow \quad y_2 = \frac{0.3}{2}\left[-1+\sqrt{1+8\times(16.74)^2}\right] = 6.95 \text{ m}$$

$\therefore$ Loss of head in the jump

$$= \frac{(y_2-y_1)^3}{4y_1y_2} = \frac{(6.95-0.3)^3}{4\times 6.95\times 0.3} = 35.26 \text{ m}$$

12.15 A sluice gate in a 2.0 m wide horizontal rectangular channel is discharging freely as shown in figure. If the depths at a small distance upstream (y_1) and downstream (y_2) are 2.5 m and 0.20 m respectively, estimate the discharge in the channel (i) by neglecting energy losses at the gate and (ii) by assuming the energy loss at the gate to be 10% of the upstream depth y_1.

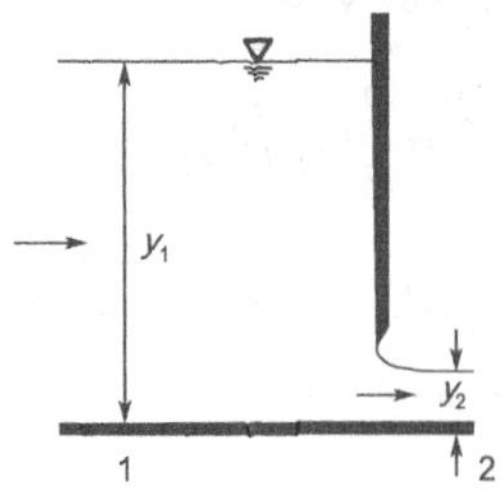

[15 marks]

Sol:

Given, $\quad y_1 = 2.5$ m and $y_2 = 0.20$ m

$$\therefore \quad Q = By_1V_1 = By_2V_2$$

$$\Rightarrow \quad V_2 = \frac{y_1}{y_2}\times V_1 = \frac{2.5}{0.20}\times V_1 = 12.5V_1$$

(i) When there is no energy loss

$$Z_1 + y_1 + \frac{V_1^2}{2g} = Z_2 + y_2 + \frac{V_2^2}{2g}$$

Since the channel is horizontal, so $Z_1 = Z_2$ and thus

$$\frac{V_2^2}{2g} - \frac{V_1^2}{2g} = (y_1 - y_2)$$

$$\Rightarrow \quad \frac{V_1^2}{2g}\left[(12.5)^2 - 1\right] = 2.50 - 0.20 = 2.30$$

$$\Rightarrow \quad \frac{V_1^2}{2g} = \frac{2.30}{155.25} = 0.01481 \Rightarrow V_1 = 0.539 \text{ m/s}$$

$\therefore$ Discharge, $Q = By_1V_1 = 2.0 \times 2.5 \times 0.539 = 2.695 \text{ m}^3\text{/s}$

(ii) When there is energy loss

Given H_L = Energy loss = $0.10\, y_1 = 0.25$ m

$$\therefore \quad y_1 + \frac{V_1^2}{2g} = y_2 + \frac{V_2^2}{2g} + H_L$$

$$\frac{V_2^2}{2g} - \frac{V_1^2}{2g} = (y_1 - y_2 - H_L)$$

$$\Rightarrow \quad \frac{V_1^2}{2g}\left[(12.5)^2 - 1\right] = 2.50 - 0.20 - 0.25 = 2.05$$

$$\Rightarrow \quad \frac{V_1^2}{2g} = \frac{2.05}{155.25} = 0.0132 \Rightarrow V_1 = 0.509 \text{ m/s}$$

$\therefore$ Discharge, $Q = By_1V_1 = 2.0 \times 2.5 \times 0.509 = 2.545 \text{ m}^3\text{/s}$

13. Turbines

Level-1

13.1 A Pelton wheel installation which receiver water from a reservoir through a penstock has a gross head at the wheel of 495 m. A nozzle supplies 2.0 m³/sec or water to the turbine wheel the jet could be taken as the third or the gross head. Considering speed ratio = 0.45, c_v = 0.98 and overall efficiency as 0.85 determine the (i) power generated by the Pelton wheel and (ii) the hydraulic efficiency. Take blade friction coefficient k = 0.95.

[10 Marks]

Sol:

Given data : Efficiency, η_o = 85% = 0.85, Q = 2 m³/s, Cross head = 4.95 m, Loss of head at penstock $= \frac{1}{3} \times 495 = 165$ m, Net head at turbine =495 – 165 = 330 m, Power, $P = \eta_0\,(\rho gQH) = 0.85 \times 9.81 \times 2 \times 330 = 5530.4$ kW, $\rho g = 9.81$ kN/m³

$$V_1 = c_V\sqrt{2gh} = 0.98\sqrt{2 \times 9.81 \times 330} = 78.88 \text{ m/sec}$$

$$u = k_u\sqrt{2gh} = 0.45\sqrt{2\times9.81\times330} = 36.21 \text{ m/sec}$$

$$H_e = \frac{1}{g}(V-u)u(1+k\cos\beta)$$

$$= \frac{1}{9.81}\times 36.21\times(78.86-36.21)(1+0.95\times\cos 55°) = 301.89 \text{ m}$$

$$\eta_h = \frac{H_e}{H} = \frac{301.89}{330} = 0.915$$

13.2 The following data were obtained from a test on Pelton wheel

(i) Head at the base of nozzle = 32 m

(ii) Discharge of the nozzle = 0.18 m^3/sec

(iii) Area of jet = 7500 mm^2

(iv) Power available at the shaft = 44 kW

(v) Mechanical efficiency = 94%

Calculate the power lost:

(a) in the nozzle

(b) in the runner

(c) in the mechanical friction

[12 Marks]

Sol:

$$\text{Power of jet} = \frac{1}{2}\dot{m}v^2 = \frac{1}{2}\times10^3\times0.18\times\left(\frac{0.18}{7500\times10^{-6}}\right)^{1/2}$$

$$= 51840 \text{ Watt} = 51.84 \text{ kW}$$

$$\text{Power at base of nozzle} = \dot{m}gh$$

$$= 10^3\times0.18\times9.81\times32 = 56.50 \text{ kW}$$

$$\text{Power loss in nozzle} = 56.50 - 51.84 = 4.66 \text{ kW}$$

$$\text{Runner power} = \frac{44}{0.94} = 46.80 \text{ kW}$$

$$\text{Power loss in the runner} = 51.84 - 46.80 = 5.03 \text{ kW}$$

$$\text{Power loss in mechanical friction} = 46.80 - 44 = 2.80 \text{ kW}$$

13.3 A vertical water jet is issuing upwards from a nozzle with a velocity of 10 m/s. The nozzle exit diameter is 60 mm. A flat horizontal plate with a total of 25 N is supported by the impact of the jet. Determine the equilibrium height of the plate above the nozzle exit. Neglect all losses and take unit weight of water as 1000 kg/m^3

[10 Marks]

Sol:

Let equilibrium height of plate above nozzle exit = h m.

Now, force with which the jet strikes the horizontal plate = $2\rho aV^2$.

For equilibirum of the plate, this force must be equal to force on plate i.e.,

$$2\rho aV^2 = 25$$

$$\Rightarrow \quad 2\times1000\times\frac{\pi}{4}\times(0.06)^2\times V^2 = 25$$

$$\Rightarrow \quad V = 2.1 \text{ m/s}$$

∴The final velocity of jet when it strikes the plate,

$$V = 2.1 \text{ m/s}$$

Using equation of motion,

$$V^2 - u^2 = 2gh$$

$\Rightarrow (2.1)^2 - (10)^2 = -2 \times 9.81 \times h$

Where, u = initial velocity of jet = 10 m/s

$\Rightarrow$ $h = 4.87$ m

∴The equilibrium height of plate above nozzle exit = 4.87 m

13.4 A Francis turbine has an inlet diameter of 2.5 m and an outlet diameter of 1.5 m. The breadth of the blade is constant at 0.20 m. The runner rotates at a speed of 300 rpm with a discharge of 10.0 m^3/s. The vanes are radial at the inlet and discharge is radially outwards at the outlet. Calculate the angle of guide vane to be set at the inlet and the blade angle at the outlet.

[10 Marks]

Sol:

Given: $D_1 = 2.5$ m; $B_1 = B_2 = 0.2$ m; $D_2 = 1.5$ m; N =300 rpm

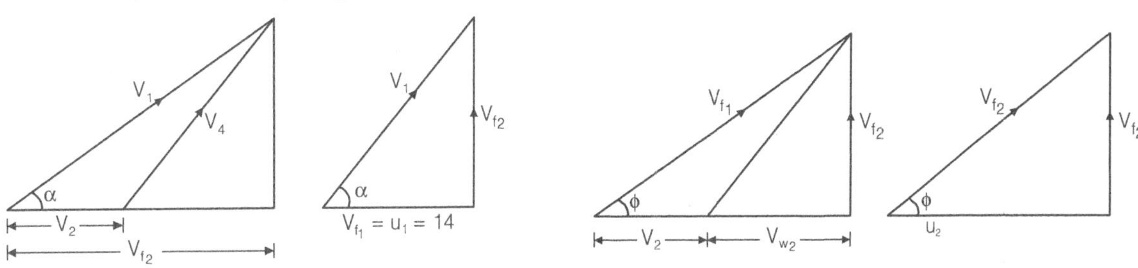

$$u_1 = \frac{\pi D_1 N}{60} = \frac{\pi \times 2.5 \times 300}{60} = 39.27$$

$$u_2 = \frac{\pi D_2 N}{60} = \frac{\pi \times 1.5 \times 300}{60} = 23.56$$

$$V_{f1} = 6.366 \text{ m/s}$$

$$V_{f2} = 10.62 \text{ m/s}$$

$$\tan \alpha = \frac{V_{f1}}{U_1} = \frac{6.366}{39.27}$$

or

$$\alpha = \frac{9.21°}{V_{f2}}$$

$$\tan \phi = \frac{V_{f2}}{u_2} = \frac{10.62}{23.56} = 0.45$$

$\Rightarrow$ $\phi = 24.26°$

13.5 A jet or water having velocity of 30 m/sec strikes a series of radial curved valves mauled on a wheel which is rotating at 200 rpm. The jet makes an angle of 20° with the tangent to the wheel at inlet and leaves the wheel with the velocity of 5 m/sec at an angle of 130° to the tangent to the wheel at exit. Water is flowing from outward in radial direction. Outer and inner radii of wheel are 0.5 m and 0.25 m respectively. Find

(a) The angle of vanes tips.

(b) Work done per second per unit weight of water striking per second and efficiency.

[12 Marks]

Sol:

$$u_1 = \frac{\pi D_1 N}{60} = \frac{\pi \times 1 \times 200}{60} = 10.46 \text{ m/sec}$$

$$u_2 = 5.23 \text{ m/sec}$$

$$V_{w_1} = 30 \cos 20° = 28.19 \text{ m/sec}$$

$$V_{f_1} = 30 \sin 20° = 10.26 \text{ m/sec}$$

$$\theta = \tan^{-1}\left(\frac{V_{f_1}}{V_{w_1} - 10.4719}\right) = 30.07°$$

$$V_{w_2} = 5 \cos 50° = 3.21 \text{ m/sec}$$

$$V_{f_2} = 5 \sin 50° = 3.83 \text{ m/sec}$$

$$\phi = \tan^{-1}\left(\frac{V_{f_2}}{V_{w_2} + 5.2359}\right) = 24.41°$$

$$\frac{RP}{\dot{m}g} = \frac{\dot{W}}{\dot{m}g} = \frac{1}{g}\left(V_{w_1}u_1 + V_{w_2}u_2\right)$$

$$= \frac{1}{9.81} \times (28.19 \times 10.46 + 3.21 \times 5.23) = 31.77 \text{ m}$$

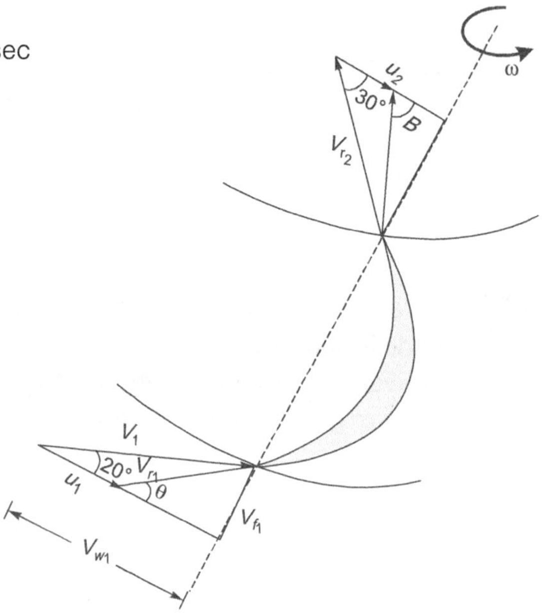

13.6 A inward flow reaction turbine works under an average head of 160 m with a discharge of 80 m^3/s. The inlet and outlet diameters of the turbine are 4 m and 2 m respectively. The runner blade angle at the inlet is 120°. Radial discharging velocity at the outlet is 15 m/sec. Assuming constant breadth of the wheel and 90% hydraulic efficiency. Determine the HP produced in MW and rpm of the machine.

[12 Marks]

Sol:

Given data : $H = 160$ m, $Q = 80$ m^3/sec, $D_1 = 4$ m, $D_2 = 2$ m, $\alpha = 120°$, $V_2 = 18$ m/sec (radial)
$\beta = 90°$ $(V_{w_2} = 0)$ $(V_2 = V_{f_2})$, $\beta_1 = \beta_2$

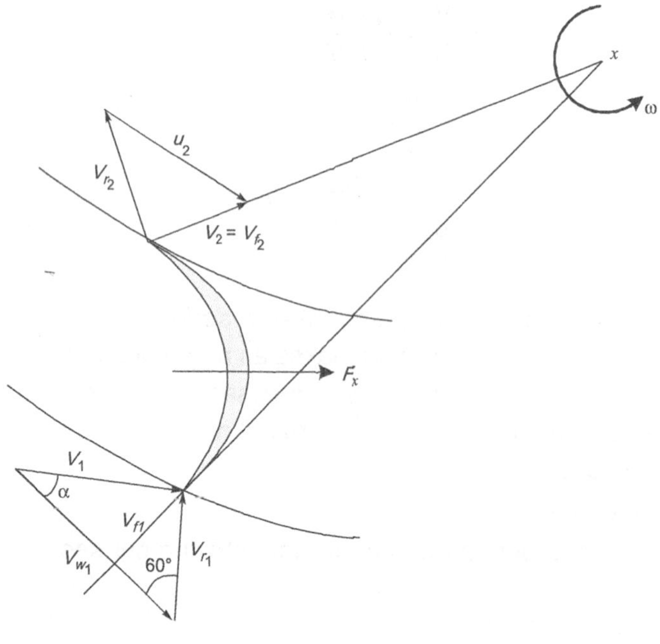

$$\eta_h = 0.9 = \frac{RP}{WP} = \frac{\dot{m}V_{w_1}u_1}{\dot{m}gH}$$

$$V_{w_1}u_1 = 1412.64 \quad \text{...(i)}$$

Continuity equation,

$$(\pi D_1 B_1)V_{f_1} = (\pi D_2 B_2)V_{f_2}$$

$$V_{f_1} = 7.5 \text{ m/sec}$$

$$V_{w_1} = u_1 - V_{f_1}\cot 60°$$

Put in (i) get $u_1 = 39.812$ m/sec

$$u_1 = \frac{\pi D_1 N}{60}$$

$$\Rightarrow \quad N = 190.083 \text{ rpm}$$

$$RP = \dot{m}V_{w_1}u_1 = (1000 \times 80) \times 1412.64$$

$$= 113011.2 \text{ kW}$$

$$RP = 113.011 \text{ MW}$$

13.7 A conical draft tube having inlet and exit diameters of 1 m and 1.5 m discharges water at the outlet with a velocity of 2.5 m/sec. The total length of the tube is 6 m and the 1.2 of the length of tube is inside of water. If the head loss due to friction in the true is the 20% of the exit kinetic head of the true then determine:

(i) Pressure head at the inlet of the tube.

(ii) Efficiency of the tube.

[10 Marks]

Sol:

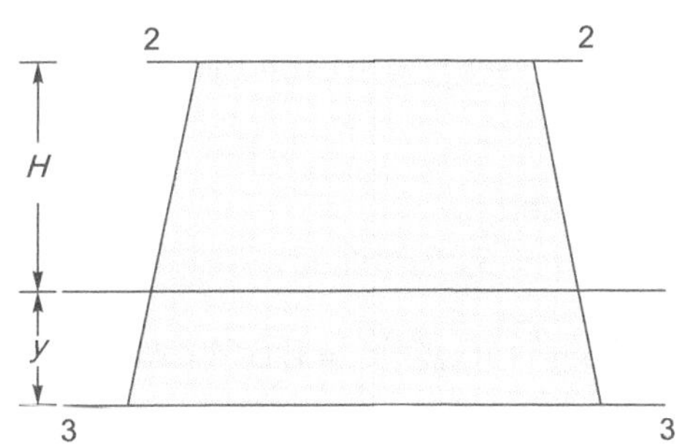

Given data : $D_2 = 1$ m, $D_3 = 1.5$ m, $V_3 = 2.5$ m/sec

$$H + y = 6 \text{ m}$$

$$y = 1.2 \text{ m}$$

$$H = 4.8 \text{ m}$$

$$h_L = 0.2\frac{V_3^2}{2y} = 0.0637 \text{ m}$$

$$\left(\frac{\pi}{4} \times 1^2 \times V_2 = \frac{\pi}{4} \times 1.5^2 \times 2.5 \Rightarrow V_2 = 5.625 \text{ m/sec}\right)$$

$$\frac{P_2}{\rho g} = 10.3\left[\frac{5.625^2 - 2.5^2}{2g} + 4.2 - 0.0637\right]$$

$$= 4.2695 \text{ m} > 2.5 \text{ m} \quad \text{(No vapour)}$$

$$h_{DT} = \frac{\frac{V_2^2 - V_3^2}{2g} - h_L}{\frac{V_2^2}{2g}} = 0.7629 = 76.29\%$$

Level-2

13.8 A jet of water having a velocity of 15 m/sec strikes a curved surface vane which is moving with a velocity of 5 m/sec. The vane is symmetrical and so shaped that the jet is deflected through 120°. Find the angle of jet of inlet so that there is no shock at inlet. What is the velocity or jet at outlet in magnitude and direction and the work done per second per unit weight of the water sticking per second. Assume the vane to be smooth.

[20 Marks]

Sol:

Given data : $V_1 = 15$ m/sec, $u_1 = u_2 = u = 5$ m/sec, $\delta = 120°$, $\phi + \phi = 180° - 120° = 60°$

Symmetrical vane, $\phi = \theta = 30°$

Applying sine rule,

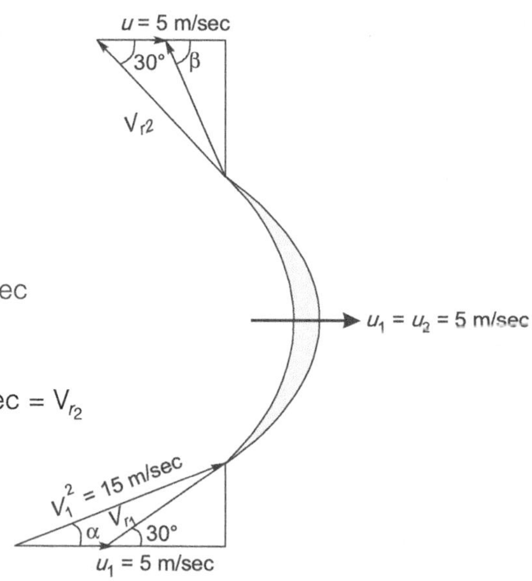

$$\frac{15}{\sin 150°} = \frac{5}{\sin\gamma}$$

$$\gamma = 9.594°$$

$$\alpha = 180° - (150° + 9.594) = 20.405°$$

$$V_{w_1} = V_1 \cos\alpha = 15\cos\alpha = 14.05 \text{ m/sec}$$

$$V_{f_1} = 15\sin\alpha = 5.22 \text{ m/sec}$$

$$V_{r_1} = \sqrt{V_{f_1}^2 + (V_{w_1} - 5)^2} = 10.454 \text{ m/sec} = V_{r_2}$$

$$V_{w_2} = V_{r_2}\cos 30° - 5 = 4.053 \text{ m/sec}$$

$$V_{f_2} = V_{r_2}\sin 30° = 5.227 \text{ m/sec}$$

$$V_2 = \sqrt{(4.053)^2 + (5.227)^2} = 6.614 \text{ m/sec}$$

$$\beta = \tan^{-1}\left(\frac{5.227}{4.053}\right) = 52.21°$$

For TFR

$$\Rightarrow \quad \frac{W}{\dot{m}g} = \frac{1}{g}(V_{w_1} + V_{w_2})u = 9.23 \text{ metre}$$

13.9 (i) A Francis turbine is designed for a hydroelectric dam, engineers decide to geometrically scale up a previously designed hydroelectric turbine. The existing turbine (A) has diameter $D_A = 2.05$ m, $N_A = 120$ rpm, $\theta = 350$ m³/sec, $H_A = 75$ m, $SP = 242$ MW.

For new turbine, assume, $N_B = 120$ rpm, $H_B = 104$ m. Calculate $D_B = ?$ $Q_B = ?$ $S_{PB} = ?$

(ii) Also calculate the dimensionless specific speed for both turbines.

[15 Marks]

Sol:

(i) For similar turbines using

$$DN \propto \sqrt{H}$$

$$\frac{D_A}{D_B} = \sqrt{\frac{H_A}{H_B}} = \sqrt{\frac{75}{104}}$$

$$D_B = 2.41 \text{ m}$$

Also,

$$Q \propto D^2\sqrt{H}$$

$$\frac{Q_A}{Q_B} = \left(\frac{D_A}{D_B}\right)^2 \sqrt{\frac{H_A}{H_B}}$$

$$\frac{350}{Q_B} = \left(\frac{2.05}{2.41}\right)^2 \sqrt{\frac{75}{164}}$$

Solving $Q_B = 570 \text{ m}^3/\text{sec}$

Also using $P \propto D^2 H^{3/2}$

$$\frac{242}{P_B} = \left(\frac{2.05}{2.41}\right)^2 \left(\sqrt{\frac{75}{104}}\right)^{3/2}$$

Solving $P = 547 \text{ MW}$

(ii) For dimensionless specific speed,

$$N_s = \frac{N\sqrt{P/\rho}}{(gH)^{5/4}}$$

For turbine A

$$N_s = \frac{120\sqrt{\dfrac{242\times 10^3}{10^3}}}{(9.81\times 75)^{5/4}} = 0.487$$

For turbine B

$$N_s = \frac{120\sqrt{\dfrac{547\times 10^3}{10^3}}}{(9.81\times 104)^{5/4}} = 0.487$$

Hence, both are equal and are true homogeneous products.

13.10 A Kaplan turbine runner is to be designed to develop 7357.5 kW shaft power. The net available head is 5.50 m. Assume that the speed ratio is 2.09 and flow ratio is 0.68, and the overall efficiency is 60%. The diameter of the boss is $\frac{1}{3}$rd of the diameter of the runner. Find the diameter of the runner, its speed and its specific speed.

[20 marks]

Sol:

Given :

Shaft Power, $P = 7357.5 \text{ kW}$

Head, $H = 5.50 \text{ m}$

Speed ratio, $\frac{u_1}{\sqrt{2gH}} = 2.09$

$\therefore \quad u_1 = 2.09 \times \sqrt{2 \times 9.81 \times 5.50} = 21.71 \text{ m/s}$

$$\text{Flow ratio} = \frac{V_{f1}}{\sqrt{2gH}} = 0.68$$

$\therefore \quad V_{f1} = 0.68 \times \sqrt{2 \times 9.81 \times 5.50} = 7.064 \text{ m/s}$

Overall efficiency, $\eta_o = 60\% = 0.60$

Diameter of boss, $D_b = \frac{1}{3} \times D_o$

Now, overall efficiency, $\eta_o = \frac{\text{Shaft power}}{\text{Water power}} = \frac{7357.5}{\frac{\rho \times g \times Q \times H}{1000}}$

$$\Rightarrow \quad 0.60 = \frac{7357.5 \times 1000}{\rho \times g \times Q \times H} = \frac{7357.5 \times 1000}{1000 \times 9.81 \times Q \times 5.5}$$

$$\Rightarrow \quad Q = \frac{7357.5 \times 1000}{1000 \times 9.81 \times 5.5 \times 0.60} = 227.27 \text{ m}^3\text{/s}$$

The discharge through the runner

$$Q = \frac{\pi}{4}\left(D_o^2 - D_b^2\right) \times V_{f1}$$

$$\Rightarrow \quad 227.27 = \frac{\pi}{4}\left[D_o^2 - \left(\frac{D_o}{3}\right)^2\right] \times 7.064 \qquad \left(\because D_b = \frac{D_o}{3}\right)$$

$$\Rightarrow \quad 227.27 = \frac{\pi}{4} \times \frac{8}{9} D_o^2 \times 7.064 = 4.9316 D_o^2$$

$$\Rightarrow \quad D_o = \sqrt{\frac{227.27}{4.9316}} = 6.789 \text{ m}$$

$$\therefore \quad D_b = \frac{1}{3} \times 6.789 = 2.263 \text{m}$$

Now, peripheral velocity at inlet

$$u_1 = \frac{\pi D_o N}{60}$$

$$\therefore \quad N = \frac{60 \times u_1}{\pi D_o} = \frac{60 \times 21.71}{\pi \times 6.789} = 61.07 \text{ rpm}$$

The specific speed (N_s) of turbine is given by,

$$N_s = \frac{N\sqrt{P}}{H^{5/4}} = \frac{61.07 \times \sqrt{7357.5}}{5.50^{5/4}} \approx 622 \text{ (SI units)}$$

14. Pump

Level-1

14.1 A centrifugal pump can deliver a discharge of 0.1 m^3/sec of water to a head of 30 m. The critical cavitation number σ_c for the pump is found to be 0.12. The pump is to be installed at a location where the barometric pressure is 96 kPa (abs) and the vapour pressure is 3 kPa (abs). Assuming the intake pipe friction of 0.3 m. Determine the value of NPSHR. What would be the maximum allowable elevation above the pump-water surface of which the pump can be installed? Assume a value of 2.0 m for the merging of NPSHR. Use g = 9.79.

[10 Marks]

Sol:

$$\sigma_c = \frac{NPSHR}{H_m}$$

$$0.12 = \frac{NPSHR}{30}$$

$$\text{NPSHR} = 3.6 \text{ m}$$

Allowing margin of 2 m

$$\text{NPSHA} = 3.6 + 2 = 5.6 \text{ m}$$

$$= H_a - H_s - H_{Ls} - H_v$$

$$5.6 = \frac{96}{9.79} - H_s - 0.3 - 0.306$$

$$H_s = 3.6 \text{ m}$$

14.2 A single acting reciprocating pump has the following characteristics:

Cylinder diameter = 15 cm, Stroke length = 30 cm, Suction head = 2.5 m, Diameter of suction pipe = 5 cm, Suction pipe length = 5 m

Assume atmospheric pressure = 10 m of water (abs.) and cavitation occurs of 2.0 m water (abs.), determine the maximum speed at which the pump can be run without cavitations.

[12 Marks]

Sol:

Given, l_s = 5 m; d = 5 cm; D = 15 cm; Stroke length = 30 cm

Maximum acceleration, head, $H_{asm} = \frac{l_s}{g}\left(\frac{A}{a}\right)\omega^2 r$

$$\omega = \frac{2\pi N}{60} = 0.1047 \text{ N}, \quad r = 0.15 \text{ m}$$

$$H_{asm} = \frac{5}{9.81}\left(\frac{15}{5}\right)^2 \times (0.1047N)^2 \times 0.15 = 0.007543\, N^2$$

For critical condition

$$H_{asm} + H_v + H_s = H_{atm}$$

$$H_{asm} + 2 + 2.5 = 10 \text{ m}$$

$$H_{asm} = 5.5 \text{ m}$$

Equating, $0.007543\, N^2 = 5.5$

Solving, $N = 270$ rpm

Level-2

14.3 A centrifugal pump having an impeller of 35 cm outside diameter rotates at 1000 rpm. The vanes are radial at exit and are 7 cm wide. The velocity of radial flow through the impeller is 3 m/sec. The velocity in the suction and delivery pipes are 2.5 m/s and 1.5 m/s respectively. Neglecting frictional losses, determine the height through which pump lifts the water and the horse-power of the pump. [20 marks]

Sol:

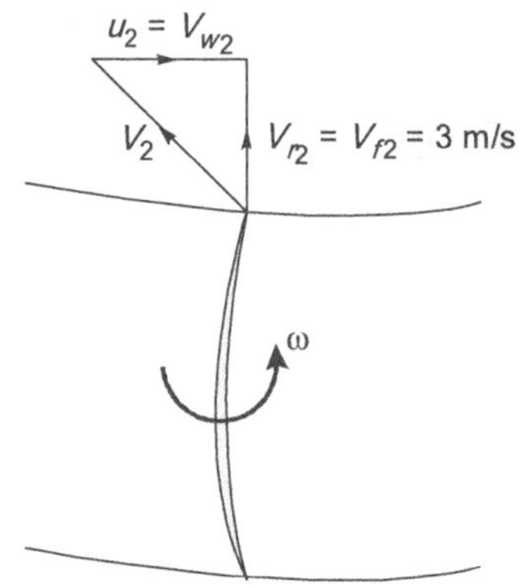

Given data:

Outside diameter of impeller,

$$D_2 = 35 \text{ cm} = 0.35 \text{ m}$$

$$\text{Speed, } N = 1000 \text{ rpm}$$

$$\text{Width of outlet, } B_2 = 7 \text{ cm} = 0.07 \text{ m}$$

$$\text{Velocity of flow, } V_{f_2} = 3 \text{ m/sec}$$

$$\text{Velocity in suction pipe, } V_s = 2.5 \text{ m/s}$$

$$\text{Velocity in delivery pipe, } V_d = 1.5 \text{ m/s}$$

Now, tangential velocity of impeller at outlet,

$$u_2 = \frac{\pi D_2 N}{60} = \frac{\pi \times 0.35 \times 1000}{60} = 18.326 \text{ m/sec}$$

As the vanes are radial at exit,

$$V_{w_2} = u_2 = 18.326 \text{ m/s}$$

As we know, the head developed will be given by,

$$H_m = \frac{V_{w_2} u_2}{g} \eta_{mano}$$

Assuming losses of the pmp to be zero, head developed will be

$$H_m = \frac{V_{w_2} u_2}{g} = \frac{u_2^2}{g} = \frac{(18.326)^2}{9.81} = 34.235 \text{ m}$$

Now, $$H_m = h_s + h_d + h_{fs} + h_{fd} + \frac{V_d^2}{2g}$$

where, $$h_s + h_d = H_s$$

which is known as the static head which is the net vertical height through which the liquid is lifted by the pump.

$\because$ Frictional losses are to be neglected

$$\therefore \quad H_s = H_m - \frac{V_d^2}{2g} = 34.235 - \frac{1.5^2}{2 \times 9.81} = 34.12 \text{ m}$$

The power of the pump is given by,

$$H_P = \rho Q g H_s (\text{Watt}) = Q g H_s (\text{kW}) = \frac{Q g H_s}{0.746} (HP)$$

where, $$Q = \pi B_1 D_i V_{f_1} = \pi \times 0.35 \times 0.07 \times 3 = 0.231 \text{ m}^3/\text{s}$$

$$1 \text{ Horse power} = 745.7 \text{ Watt} \simeq 0.746 \text{ kW}$$

$\therefore$ Horse power of pump, $H_P = \dfrac{0.231 \times 34.12 \times 9.81}{0.746} = 103.604$ hp

14.4 A centrifugal pump with 1.2 m diameter runs at 200 r.p.m. and pumps 1880 litres/s, the average lift being 6 m. The angle which the vanes make at exit with the tangent to the impeller is 26° and the radial velocity of flow is 2.5 m/s. Determine the manometric efficiency and the least speed to start pumping against a head of 6 m, the inner diameter of the impeller being 0.6 m.

[15 Marks]

Sol:

Classsification of jumps: The hydraulic jumps in horizontal rectangular channels are classified into five categories based on the Froude number F_1 of the supercritical flow.

1. **Undular jump $1.0 < F_1 \leq 1.7$:** The water surface is undulating with a very small ripple on the surface. The sequent-depth ratio is very small and E_L/E_1 is practically zero. A typical undular jump is shown in fig. (a).
2. **Weak jump $1.7 < F_1 \leq 2.5$:** The surface roller makes its appearance at $F_1 \approx 1.7$ and gradually increases in intensity towards the end of this range, i.e., $F_1 \approx 2.5$. The energy dissipation is very small, is E_L/E_1 about 5 per cent at $F_1 = 1.7$ and 18 per cent at $F_1 = 2.5$. The water surface is smooth after the jump [Fig. (b)].
3. **Oscillating jump $2.5 < F_1 \leq 4.5$:** This category of jump is characterised by an instability of the high-velocity flow in the jump which oscillates in a random manner between the bed and the surface. These oscillations produce large surface waves that travel considerable distances downstream [Fig. (c)].

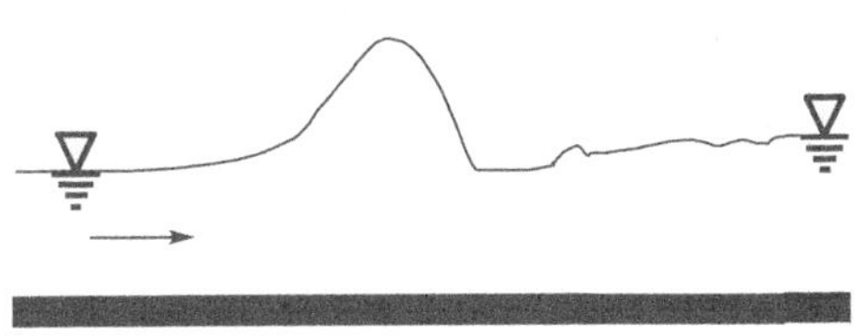

(a) Undular jump, $1.0 < F_1 \leq 1.7$

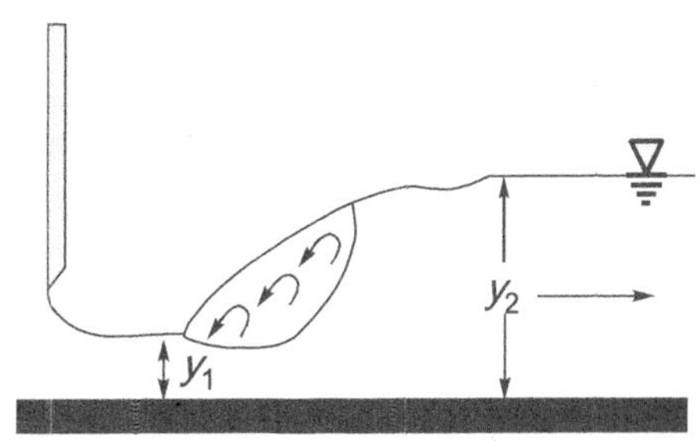

(b) Weak jump, $1.7 < F_1 \leq 2.5$

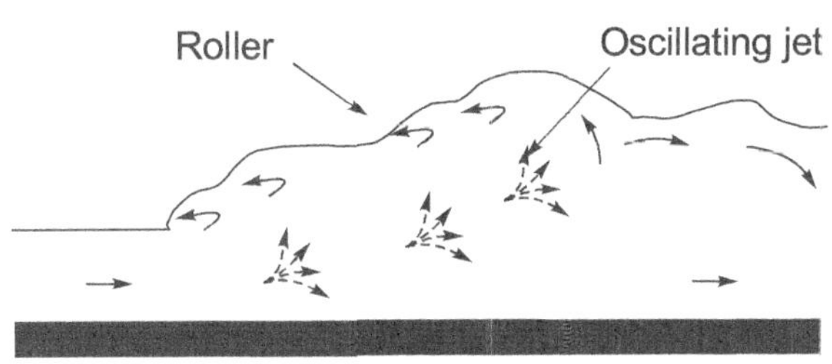

(c) Oscillating jump, $2.5 < F_1 \leq 4.5$

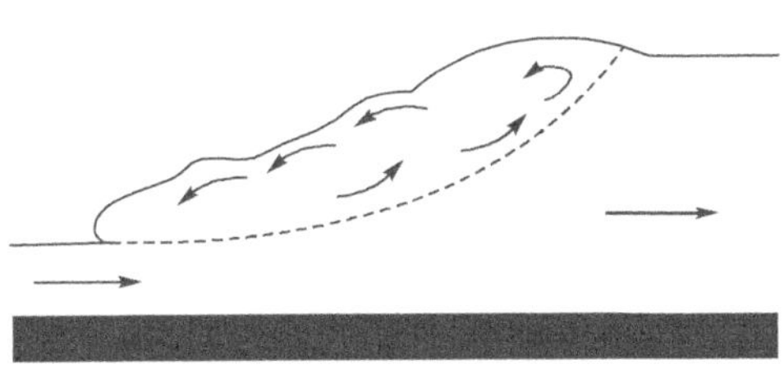

(d) 'Steady' jump, $4.5 < F_1 \leq 9$

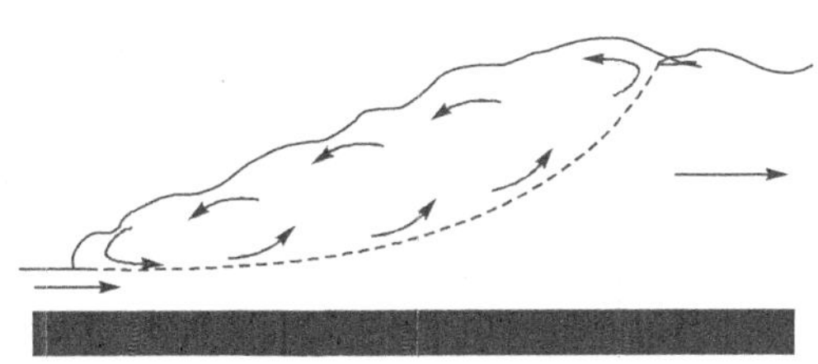

(e) Strong or Choppy jump, $F_1 > 9$

Special care is needed to suppress the waves in stilling basins having this kind of jump. Energy dissipation is moderate i.e. about E_L/E_1 = 45 per cent at F_1 = 4.5.

4. **'Steady' jump 4.5 < F_1 ≤ 9.0:** In this range of Froude numbers, the jump is well-established, the roller and jump action is fully developed to cause appreciable energy loss (fig. (d)). The relative energy loss E_L/E_1 ranges from 45 percent to 70 percent in this, class of jump. The 'steady jump' is least sensitive in terms of the toe-position to small fluctuations in the tailwater elevation.
5. **Strong or Choppy Jump F_1 > 9.0:** In this class of jump the water surface is very rough and choppy. The water surface downstream of the jump is also rough and wavy (Fig. (e)). The sequent-depth ratio is large and the energy dissipation is very efficient with E_L/E_1 values greater than 70 per cent.

14.5 **A centrifugal pump is required to discharge 600 *l*/sec and develop a head of 15 m when the impeller rotates at 750 rpm. The manometric efficiency is 80%. The less of head in the pump due to fluid resistance being assumed to be 0.027 V² m of water is have *V* is the velocity with which water leaves the impeller without shock or whirl and the velocity or flow is 3.2 m/sec. Determine**

(i) The impeller diameter and outlet area.

(ii) The blade angle at the outlet edge.

[20 Marks]

Sol:

Given data : Q = 0.6 m³/sec, H = 15 m, N = 750 rpm, $n_{mano.}$ = 80%, loss in pump = 0.027 V², $V_{f_1} = V_{f_2}$ = 3.2 m/s

$$n_{mano} = 0.8 = \frac{H}{\frac{1}{g}V_{w_2}u_2}$$

$$\frac{1}{g}V_{w_2u_2} = \frac{H}{0.8} = \frac{15}{0.8} = 18.78 \text{ m}$$

$$H = 15 \text{ m} = \frac{1}{g}V_{w_2}u_2 - \text{loss in casing}$$

$$15 = \frac{1}{g}V_{w_2}u_2 - 0.027V_2^2$$

$$V_2 = 11.785 \text{ m/sec} \qquad \left[\because \frac{1}{g}V_{w_2}u_2 = 18.75\right]$$

$$V_{w_2} = \sqrt{11.785^2 - 3.2^2} = 11.342 \text{ m/sec}$$

$$u_2 = 16.216 \text{ m/sec} = \frac{\pi D_2 \times 750}{60}$$

$$D_2 = 0.4129 \text{ m}$$

$$Q = (\pi D_2 B_2) V_{f_2}$$

$$A = .1875 \text{ m}^2$$

$$\phi = \tan^{-1}\left(\frac{3.2}{16.216 - 11.342}\right) = 33.286°$$

■■■■

02 Engineering Hydrology

1. Precipitation and General Aspects of Hydrology

Level-1

1.1 What are the objectives of integrated development of surface and ground water resources? Explain. [12 marks]

Sol:

The objectives of integrated development of surface and ground water resources are listed below:

1. Improving security and reliability of supply.
2. Managing impacts of factors other than ground water and surface water.
3. Maintaining water quality to acceptable levels.
4. Improving efficiency of integrated system and resource conservation.
5. Storing and delivering water where and when it is needed.
6. Ground water and surface water and related to each other as seepage from ground water gives rise to contribution in surface water. Their integrated development will lead to balanced dependability of human on both of them.
7. It will also protect either of source from overuse and exploitation.
8. It will also enable developers to counter and resort to suitable remedy for problem associated with both resources.
9. Conjunctive/Integrated development will the economy as water as a resource in a pillar of every country's economy.
10. To cover and fill the gap between needs of urban sector and rural sector. Because urban sector of population primarily depends upon surface water resources while rural sector satisfies its needs though ground water resource (wells).
11. Their conjunctive development will also encourage sustainable development which is the main avenue in road map for country's growth.

1.2 A pond has a water surface elevation of 110 m above datum at the beginning of certain month of 30 days. In that month, the pond received an average inflow of 10 m^3/sec from surface runoff. In the same period out flow from the pond has an average value of 8 m^3/sec. Further in that month pond received a rainfall of 152 mm. The evaporation from the pond surface was estimated as 8.3 cm. Write water budget equation for the pond, and calculate water 4surface elevation of river, at the end of month. The average pond surface area is 5500 ha. Assume no contribution to or from Groundwater.

[10 Marks]

Sol:

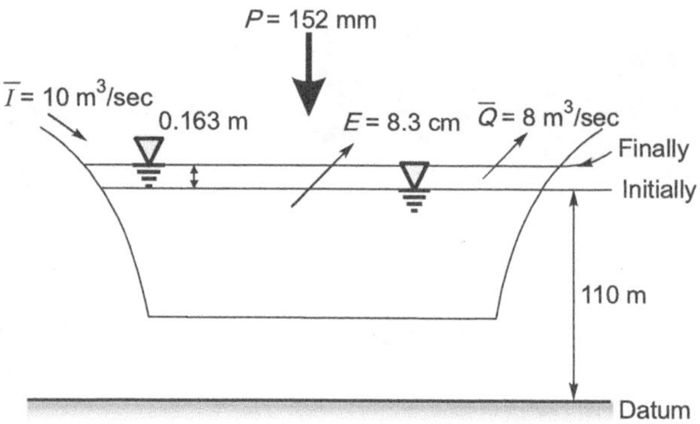

Given:

$\bar{I}$ (inflow rate) = 10 m³/sec

$\bar{Q}$ (outflow rate) = 8 m³/sec

Total rainfall (P) = 152 mm

Evaporation from pond (E) = 8.3 cm

Pond area (A) = 5500 ha

By hydrological water budget equation,

$$V_i - V_0 = \Delta S$$

V_i = Inflow volume

V_0 = Outflow volume

ΔS = Change in storage

$$\left(\bar{I}\Delta t + PA\right) - \left(\bar{Q}\Delta t + EA\right) = \Delta S$$

Assume, Δt = 30 days

$= 30 \times 24 \times 60 \times 60 = 2.592 \times 10^6$ sec

$$\text{Now,}\left(\frac{10\times 2.592\times 10^6}{10^6}+\frac{152\times 10^{-3}\times 5500\times 10^4}{10^6}\right)\text{Mm}^3-\left(\frac{8\times 2.592\times 10^6}{10^6}+\frac{8.3\times 10^{-2}\times 5500\times 10^4}{10^6}\right)\text{Mm}^3 = \Delta S$$

$$\Delta S = (34.28 - 25.30) = 8.98 \text{ Mm}^3$$

$$\therefore \text{ Rise in water surface elevation} = \frac{\Delta S}{A} = \frac{8.98\times 10^6}{5500\times 10^4} = 0.163$$

Now, water surface elevation at the end of this month

$= 110 + 0.163 = 110.163$ m

1.3 The normal annual precipitation of five raingauge stations P, Q, R, S, T are respectively 125, 102, 76, 113 and 137 cm. During a particular storm, the precipitation recorded by station P, Q, R and S are 13.2, 9.2, 6.8 and 10.22 cm respectively. The raingauge at station T was inoperative during the storm. Estimate the rainfall at station T during that storm.

[10 Marks]

Sol:

Raingauge	Normal precipitation (cm)	Precipitation (cm)
P	125	13.2
Q	102	9.2
R	76	6.8
S	113	10.22
T	137	?

At raingauge station *T*, precipitation is missing,

N_x = 137 cm

$0.9N_x = 0.9 \times 137 = 123.3$ cm

$1.1N_x = 1.1 \times 137 = 150.7$ cm

$$N_S, N_Q, N_R \not\in [0.9N_x, 1.1N_x]$$

Hence, normal ratio method is used,

$$P_x = \frac{N_x}{(n-1)}\left[\frac{P_P}{N_P}+\frac{P_Q}{N_Q}+\frac{P_R}{N_R}+\frac{P_S}{N_S}\right]$$

$$= \frac{137}{(5-1)}\left[\frac{13.2}{125}+\frac{9.2}{102}+\frac{6.8}{76}+\frac{10.22}{113}\right] = 12.868 \text{ cm}$$

$$P_x \simeq 12.9 \text{ cm}$$

∴ Precipitation at the raingauge station, $T = 12.9$ cm

1.4 (a) Rainfall due to storms in a basin is shown using isohyets. Find the average depth of rainfall in basin.

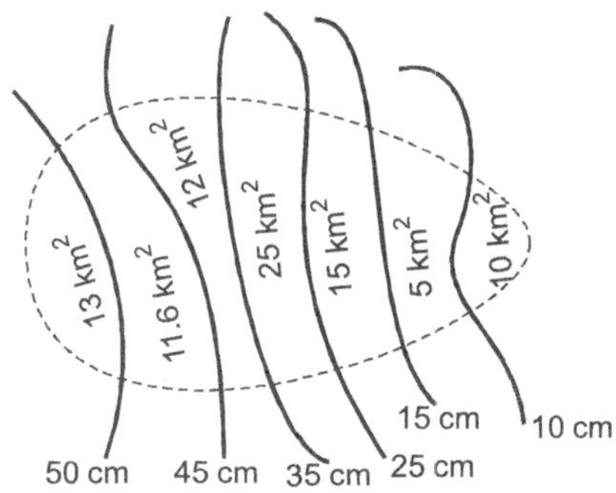

(b) What are the limitations of the isohyetal method?

[10+5 Marks]

Sol:

Isohyet interval (cm)	Avg. value of isohyet (cm)	Area (km^2)	Area × Avg. osohyet (km^2-cm)
< 10	10	10	100
10 - 15	12.5	5	62.5
15 - 25	20	15	300
25 - 35	30	25	750
35 - 45	40	12	480
45 - 50	47.5	11.6	551
> 50	50	13	650
		$\Sigma = 91.6$ km^2	$\Sigma = 2893.5$ km^2-cm

$$P_{avg} = \frac{2893.5}{91.6} = 31.588 \text{ cm}$$

Limitations of isohyetal methods are as follows:

(a) It is a time consuming process.

(b) Drawing isohyets is a difficult and cumbersome process.

(c) New isohyets have to drawn to account for changes in rainfall pattern over time.

1.5 (i) Explain with the help of neat sketch the hydrology cycle indicating its various phases.

[10 Marks]

Sol:

(i) Hydrological cycle:

- It is a global sun driven process in which water is transported from oceans to the atmosphere then to the land and then back to the sea.
- It is a continuous process with no definite starting point. A convenient starting point to describe the cycle is taken as oceans.

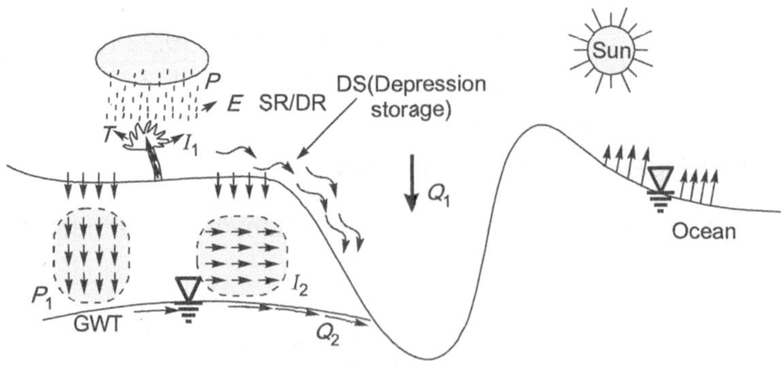

- Due to solar radiation falling on ocean surface water evaporates and mixes with the dry air above making it moist. Moist air being lighter than dry air rises and in the process cools down thereby increasing its relative humidity.
- Relative humidity subsequently reaches 100% full saturation, any further rise in elevation of moist air causes condensation followed by precipitation.

Component of hydrological cycle:

Evaporation (E) - Change of water from liquid to gaseous state.

Precipitation (P) - The deposition of water on Earth surface as rain, snow, hail etc.

Interception (I_1) - Short term retention of rain water by vegetation, roof tops, pavement etc.

Infiltration (I) - Movement of water into the soil at the surface.

Percolation (P_1) - Movement of water from one soil zone to a lower soil zone.

Transpiration (T) - It is the water absorbed from the ground and evaporated into atmosphere through leaves.

Interflow (I_2) - It is the groundwater flowing horizontally above the GWT. It is also known as subsurface flow.

Depression storage (DS) - Rainwater accumulated in small depressions and ditches above the Earth surface.

Surface Runoff(SR) - It is the part of rain which reaches the stream immediately after the rainfall flowing over the surface.

(Q_2) - It is the discharge obtained in the stream due to GWT (Ground water table) through seepage.

- Also known as dry weather flow or effluent seepage.

(Q_1) - It is the discharge obtained in the stream due to SR/DR.

1.6 Discuss different forms of precipitation.

[10 Marks]

Sol:

Different forms of precipitation: Some of the common forms of precipitation are rain, snow, drizzle, glaze, sleet and hail.

1. **Rain:**
   - It is the principal mode of precipitation in India.
   - The term rainfall is used to describe precipitations in the form of water drops of size larger than 0.5 mm.
   - The maximum size of raindrop is about 6 mm.
   - On the basis of the intensity, rainfall is classified as follows:

Type	Intensity
Light rain	Trace to 2.5 mm/h
Moderate rain	2.5 mm/h to 7.5 mm/h
Heavy rain	> 7.5 mm/h.

2. Snow:
   - Snow consists of ice crystals which usually combine to form flakes.
   - When fresh, snow has an initial density varying from 0.06 to 0.15 g/cm^3 and it is usual to assume an average density of 0.1 g/cm^3.
3. Drizzle:
   - A fine sprinkle of numerous water droplets of size less than 0.5 mm and intensity less than 1 mm/h is known as drizzle. In this, the drops are so small that they appear to float in the air.
4. Glaze:
   - When water droplets comes in contact with cold surface < 0°C. It freezes an ice coating known as Glaze.
5. Sleet:
   - It is frozen raindrops of transparent grains which forms when rain falls through air at subfreezing temperature.
6. Hail:
   - It is a showery precipitation in the form of irregular pallets or lumps of ice of more than 8 mm.
   - Hails occur in violent thunder storms in which vertical currents are very strong.

Level-2

1.7 **(a) Explain the optimum number of raingauge stations?**

(b) A catchment area has seven raingauges stations. In a year, the annual rainfall recovered by raingauges are as follows:

Station	*P*	*Q*	*R*	*S*	*T*	*U*	*V*
Rainfall (cm)	130	142.1	118.2	108.5	165.2	102.1	146.9

(i) Determine the standard error in the estimation of mean rainfall in the existing set of raingauges.

(ii) for a 5% error in the estimation of rainfall, calculate the minimum number of additional raingauges stations to be established in the catchment.

[20 Marks]

Sol:

(a) **Optimum number of raingauge stations:** Statistics has been used in determining the optimum number of raingauges required to be installed in a given catchment.

- The reason behind such statistical calculation is that a certain number of raingauges stations are necessary to give average rainfall with a certain percentage of error. If the allowable percentage error is more, lesser number of gauges would be required.

Hence, it is the number of raingauge required to calculate the average rainfall value with a certain acceptable percentage of error.

- Based upon this statistical principle, the optimum number of raingauges (N) can be obtained by the following equation:

$$N = \left(\frac{C_V}{\varepsilon}\right)^2$$

Water, C_V = Coefficient of variation of rainfall based on existing raingauge stations.

ε = Allowable percentage error in the estimation of basic mean rainfall.

C_V can be computed as follows:

$$C_V = \frac{\sigma_{n-1}}{\overline{P}} \times 100$$

σ_{n-1} = Standard deviation,

$\overline{P}$ = Mean precipitation of existing number of raingauges

$$\sigma_{n-1} = \sqrt{\frac{\sum_{1}^{n}(p_i - p_m)^2}{n-1}}$$

n = Number of existing raingauge

$$P_m = \frac{\Sigma P}{n}$$

(b)

$$P_m = \frac{\Sigma P}{n}$$

$$P_m = \frac{130 + 142.1 + 118.2 + 108.5 + 165.2 + 102.1 + 146.9}{7}$$

$$P_m = 130.43 \text{ cm}$$

Station	P	Q	R	S	T	U	V
Rainfall (cm)	130	142.1	118.2	108.5	165.2	102.1	146.9

$$\sigma_{n-1} = \sqrt{\frac{\varepsilon(P_i - P_m)^2}{n-1}}$$

$$\sigma_{n-1} = \sqrt{\frac{(130-130.43)^2 + (130-142.1)^2 + (130-118.2)^2 + (130-108.5)^2 + (165.2-130)^2 + (130-102.1)^2 + (146.9-130)^2}{7-1}}$$

$$\sigma_{n-1} = 22.55$$

$$C_V = \frac{\sigma_{n-1}}{p_m} \times 100$$

$$C_V = \frac{22.55}{130.43} \times 100$$

Now, $C_V = 17.3\%$

(i) Error in the estimation of mean rainfall in the existing set of raingauges.

$$N = \left(\frac{C_V}{\varepsilon}\right)^2$$

$$7 = \left(\frac{17.3}{\varepsilon}\right)^2$$

$$\varepsilon = \frac{17.3}{(7)^{1/2}} = 6.53\%$$

(ii) Given, $\varepsilon = 5\%$

$$N = \left(\frac{C_V}{\varepsilon}\right)^2 = \left(\frac{17.3}{5}\right)^2 = 11.97 \simeq 12$$

$\therefore$ Additional raingauge station required = 12 – 7 = 5

1.8 A semicircle of diameter 20 km with an equilateral triangle of side 20 km below its diameter is a close approximation to a river basin. The position coordinates of 5 raingauge stations A, B, C, D and E located within the basin with respect to a coordinate axes system whose X-axis and origin are coincident with diameter and centre of the circle are (5, 5), (–5, 5), (–5,–5), (5, –5) and (0, 0) km respectively. If the rainfall recorded at these raingauge are 85, 92, 77, 80 and 105 mm respectively determine the average depth of rainfall using Thiessen polygon method.

[20 Marks]

Sol:

Given, diameter of semicircle = 20 km

The area of the basin is as shown in figure with the locations of the raingauge stations A, B, C, D and E marked on it

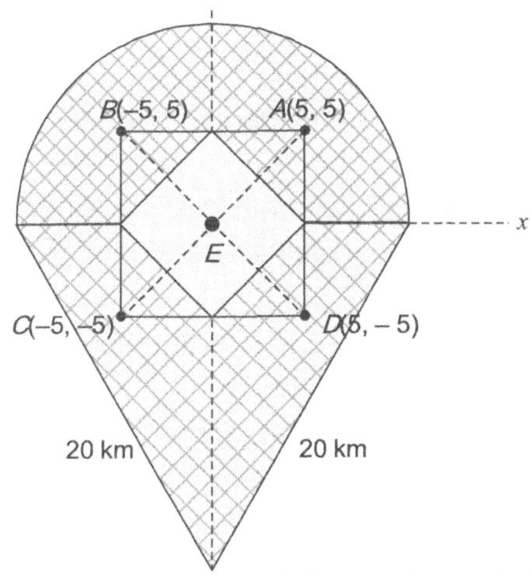

$$\text{Area of the basin} = \text{Area of semicircle} + \text{Area of triangle}$$

$$\text{Area of semicircle} = \frac{\pi R^2}{2} = \frac{\pi}{2}\times\left(\frac{20}{2}\right)^2 = \frac{\pi}{2}\times(10)^2 = 157.08 \text{ km}^2$$

$$\text{Area of triangle} = \frac{1}{2}\times\text{Base}\times\text{height}$$

$$\text{Height(H)} = \sqrt{(20)^2-(10)^2} = \sqrt{300}\text{ km}$$

$$\therefore \quad \text{Area of triangle} = \frac{1}{2}\times 20\times\sqrt{300} = 173.205 \text{ km}^2$$

Now, $$\text{Area of basin} = 157.08 + 173.205 = 330.285 \text{ km}^2$$

It can be seen from the Theissen polygons constructed for the raingauge stations that the polygon of station 'E' is a square of side $\left(5\sqrt{2}\text{ km}\right)$.

$$\therefore \quad \text{Area of polygon for station, } E = \left(5\sqrt{2}\right)^2 = 50 \text{ km}^2$$

Area of polygon for raingauge A and B are equal and each is given by

$$= \frac{1}{2}(\text{Area of semicircle}) - \frac{1}{4}(\text{Area of square of side } 5\sqrt{2})$$

$$= \frac{1}{2}\times(157.08) - \frac{1}{4}(50) = 66.04 \text{ km}^2$$

$$\therefore \quad A_a = A_b = 66.04 \text{ km}^2$$

Again, area of the polygons for C and D are equal and each is given by

$$= \left[\frac{1}{2}(\text{Area of triangle}) - \frac{1}{4}(\text{Area of squqare})\right]$$

$$= \left[\frac{1}{2}(173.205) - \frac{1}{4}(50)\right] = 74.1\,\text{km}^2$$

$\therefore \qquad A_c = A_d = 74.1\ \text{km}^2$

Now, as per Theissen polygon method,

$$\text{Average depth of rainfall, } P_{avg} = \frac{\Sigma P_A}{A} = \frac{P_aA_a + P_bA_b + P_cA_c + P_dA_d + P_eA_e}{A}$$

$$= \frac{85 \times 66.04 + 92 \times 66.04 + 77 \times 74.1 + 80 \times 74.1 + 105 \times 50}{330.285}$$

$$P_{avg} = 86.51\ \text{mm}$$

1.9 Describe the methods of measuring rainfall and snowfall of a basin. How is inconsistency of rainfall data determined? Why does inconsistency, arise? How is consistency adjusted?

[15 Marks]

Sol:

- Rainfall is measured and collected in a raingauge. A raingauge is a cylindrical vessel assembly kept in open to collect rain.
- Precipitation is expressed in terms of the depth to which rainfall water would stand on an area of all the rainwater collected on it. Thus, 1 cm of rainfall over a catchment area of 1 km^2 represents a volume of water equal to 10^4 m^3.
- Terms such as pelvimeter, ombrometer and hytometer are also sometimes used to designate a raingauge.

Raingauge can be broadly classified into two categories:

(i) Non-recording raingauges.

(ii) Recording raingauges.

- **Non Recording raingauges:** The non-recording raingauge used in India is the symon's raingauge. It essentially consist of a circular collecting area of 12.7 cm (5.0 inch) diameter connected to a tunnel. The rim of the collector is set in a horizontal plane at a height of 30.5 cm above the ground level. The tunnel discharges the rainfall catch into a receiving vessel. The tunnel and receiving vessel are housed in a metallic container.
- Figure shows the details of the installation water contained in the receiving vessel is measured by a suitably graduated measuring glass, with an accuracy upto 0.1 mm.

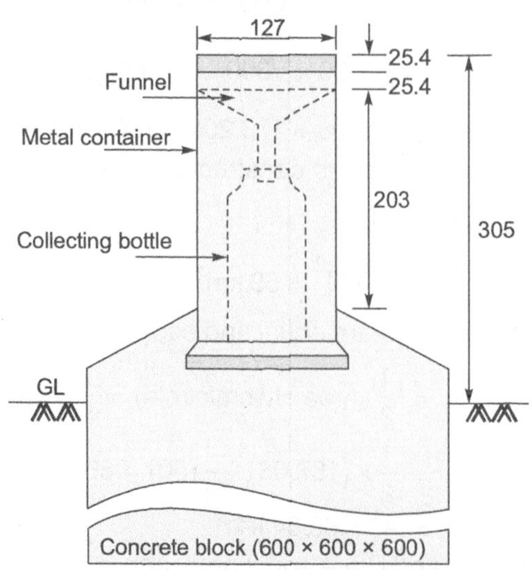

Non-recording raingauges (Symon's gauge)

- **Recording raingauge:** Recording gauges produce a continuous plot of rainfall against time and provide valuable data of intensity and duration of rainfall for hydrological analysis of storms. The following are some of the commonly used recording raingauges.

(i) Tipping-Bucket type

(i) Weighing-Bucket type

(i) Natural-Siphon type

- **Telemetering raingauge:** Recording type raingauge contain electronic units to transmit data on rainfall to a base station used in mountainous and inaccessible locations.
- **Radar measurement of rainfall:** Good degree of accuracy, used to measure rainfall over large areas. Even Doppler type radar measure velocity and distribution of raindrops.
- To measure rainfall from raingauge, it has to be setup via adopting standard settings which are as follows:
  (a) The ground must be level and in open and instrument must present a horizontal catch surface.
  (b) The gauge must be set as near the ground as possible to reduce wind effects but should be adequately high enough to avoid flooding, splashing etc.
  (c) The instrument must be surrounded by an open faced area of atleast 5.5 m × 5.5 m. No object should be nearer to instrument then 30 m or twice the height of obstruction.
- **Snow fall:** Snowfall as a form of precipitation differs from rainfall in that it may accumulate over a surface for sometimes before it melts and causes runoff. Further, evaporation from the surface of accumulated snow surface is a factor to be considered in analysis dealing with snow. Water equivalent of snowfall is included in the total precipitation amounts of a station to prepare seasonal and annual precipitation records.

  [Water equivalent of snowfall ≃ Depth of water that would result in melting of a unit of snow.]

  It should obtained/measured in 2 ways:
- **Snow gauges:** Larger cylindrical receiver (without tunnel and collecting bottle), 203 mm in diameter used to collect snow as it fells. After collecting it is brought in a warm room and snow is melted by adding premeasured quantity of hot water.
- **Snow tubes:** A set of telescopic metal tubes provided with a cutter edge for easy penetration as well as to enable extraction of core sample, 40 mm diameter in normal size.
- To extract a sample, the tube is driven into the snow deposit till it reaches the bottom of the deposit and then and then twisted and turned to cut a core. The core is extracted carefully and studies for its physical properties and then is melted to obtain water equivalent of the snow.
- The checking for inconsistency of a record is done by the double mass curve technical. The technique is based on the principle that when each recorded data comes from some parent population, they are consistent.

Incosistency in records arises due to following reasons

(i) Shifting of raingauge station to a new location.

(ii) The neighbourhood of the station undergoing a marked change.

(iii) Change in ecosystem due to calamities, such as forest fires, landslides etc.

(iv) Occurrence of observational error from a certain date.

- Let there be η raingauges surroundings the raingauge station X for which consistency is to be checked the accumulated rainfall at station X and the accumulated rainfall of average of surrounding raingauge station (ΣP_{avg}) is calculated in river chronology. i.e. starting from the latest record.
- The value of ΣP_x as ordinate is plotted against (ΣP_{avg}) as abscissa to obtain a curve known as double mass curve.

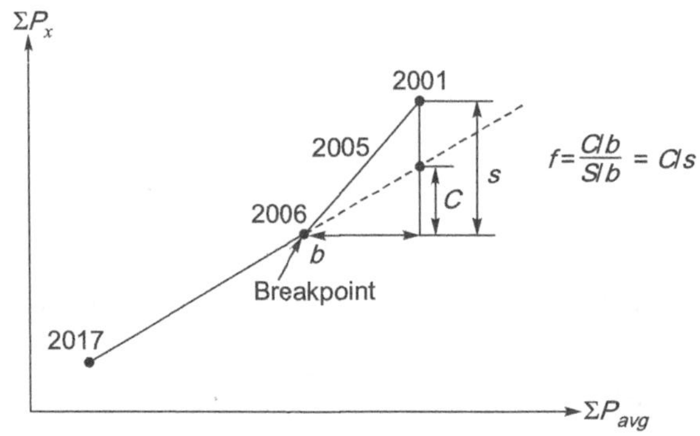

- If the curve obtained is a straight line then station 'X' is said to be consistent and if a change in slope is observed then station 'X' is said to be inconsistent beyond the point slope changes that point is called breakpoint.
- The values of stations for station X for years beyond the breakpoint are corrected by multiplying them by a correction factor 'f'.

$$f = \frac{\text{Slope of line before breakpoint}}{\text{Slope of line after breakpoint}}$$

- According to figure, the corrected values of precipitation for years 2005-2001,

$$(P_x)_{\text{corrected}} = (P_x)_{\text{recorded}} \times f$$

1.10 **The accumulated rain falls at a raingauge station for various durations are listed below. Construct maximum intensity duration curve. Comment on nature of curve.**

Time (in mm)	0	5	10	15	20	25	30	35	40	45	50	60	80	100	120
Accumulated rainfall (cm)	0	0.15	0.38	1.78	3.05	3.86	4.32	4.83	5.23	5.64	6.05	6.86	8.71	10.54	11.05

[15 Marks]

Sol:

Time since start (min)	Cumulative rainfall (mm)	Rainfall (mm) in any possible ΔT equal to					
		5 min	10 min	15 min	20 min	25 min	30 min
0	0						
5	1.5	1.5	3.8	17.8			
10	3.8	2.3	16.3	29	30.5	38.6	
15	17.8	12.7	26.7	34.8	37.1	41.7	43.2
20	30.5	8.1	20.8	25.4	39.4	44.5	46.8
25	38.6	4.6	12.7	17.8	30.5	34.5	48.5
30	43.2	5.1	9.7	13.7	21.8	25.9	38.6
35	48.3	4	9.1	13.2	17.8		
40	52.3	4.1	8.1				
45	56.4						

Maximum intensity in 5 mm interval $= \frac{14}{5} = 2.8$ mm/ min

Maximum intensity in 10 mm interval $= \frac{26.7}{10} = 2.67$ mm/ min

Maximum intensity in 15 mm interval $= \frac{34.8}{15} = 2.32$ mm/ min

Maximum intensity in 20 mm interval $= \frac{39.4}{20} = 1.97$ mm/ min

$$\text{Maximum intensity in 25 mm interval} = \frac{44.5}{25} = 1.78 \text{ mm/ min}$$

$$\text{Maximum intensity in 30 mm interval} = \frac{48.5}{30} = 1.62 \text{ mm/ min}$$

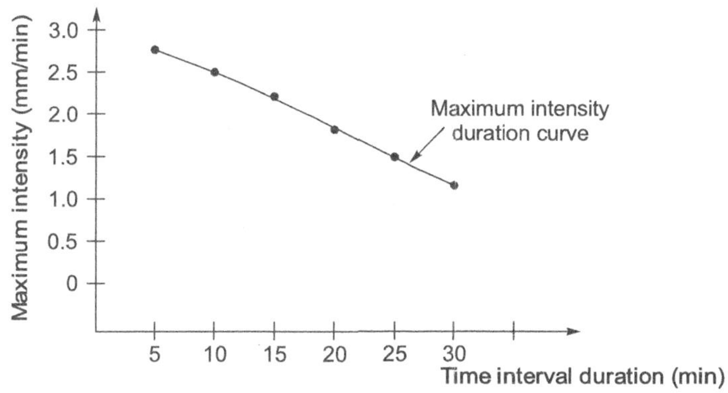

1.11 The following are the raingauge observations during storms. Construct the hyetograph.

Time since commencement of storm (min)	Accumulated rainfall (cm)
5	0.1
10	0.2
15	0.8
20	1.5
25	1.8
30	2
35	2.5
40	2.7
45	2.9
50	3.1

[15 Marks]

Sol:

Time since commencement of storm (min)	Accumulated rainfall (cm)	Incremental rainfall (cm)	Rainfall intensity (cm/min)
5	0.1	0.1	0.1/5 = 0.02
10	0.2	0.1	0.02
15	0.8	0.6	0.12
20	1.5	0.7	0.14
25	1.8	0.3	0.06
30	2	0.2	0.04
35	2.5	0.5	0.1
40	2.7	0.2	0.04
45	2.9	0.2	0.04
50	3.1	0.2	0.04

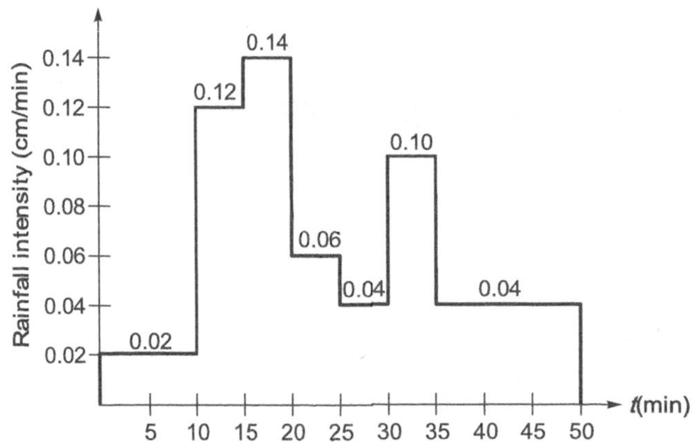

2. Evaporation, Transpiration and Stream Flow Measurement

Level-1

2.1 What are the factors affecting evaporation process? Also discuss the method to reduce evaporation losses.

[10 Marks]

Sol:

The evaporation losses from water surface depend upon the following factors:

(i) **Vapour pressure or Humidity:** The rate of evaporation is proportional to the difference between the saturation vapour pressure at the water temperature (ρ_w) and the actual vapour pressure in the air (ρ_a) thus,

$$E_L = c(\rho_w - \rho_a)$$

The above equation, also known as Dalton's law of evaporation,

Where, E_L = Rate of evaporation (mm/day)

ρ_w and ρ_a are in mm of mercury.

Evaporation continues till $\rho_w = \rho_a$; if $\rho_w < \rho_a$ then condensation takes place.

(ii) **Temperature:** If the temperature is more, the saturation vapour pressure increases and the evaporation increases. Thus, in summer season or in hot countries the evaporation will be more as compared to that in the winter season or in cold countries.

(iii) **Wind:** Wind aids in removing the evaporated water from the zone of evaporation and consequently create greater scope for evaporation.

- If the wind velocity is large enough to remove all the evaporated water vapour, a further increase in wind velocity does not influence the evaporation.

(iv) **Atmospheric pressure:** Other factors remaining same rate of evaporation decreases with increase in atmospheric pressure.

(v) **Quality of water:** The quality of water in the water body also affects the rate of evaporation. Since the presence of any dissolved salts in water reduces the saturated vapour pressure of water, which consequently reduces the rate of evaporation.

- Turbidity may also affect the rate of the evaporation by affecting the head transfer within the depth of the water body.
- Evaporation of sea water is around 2 to 3% less than that of fresh water.

(vi) **Area of water surface:** The amount of the evaporation is directly proportional to the area of evaporation. If the exposed area is large, the evaporation will be more and vice versa.

(vii) **Depth of water:** The depth of water influences the evaporation considerably more depth reduces the summer, evaporation and increases the winter evaporation. In summer, the temperature of shallow water bodies rises rapidly and hence more evaporation in shallow water bodies.

Various methods available for reduction of evaporation losses can be considered as:

- By providing mechanical surface covers.
- By opting for deep reservoirs as against shallow reservoir for storing same volume of water.
- By adding chemical like cetyl alcohol (Hexa decanol) or stearyl alcohol (Octa decanol).

Properties of chemicals to be added

- Density should be less than that of water.
- Rate of evaporation should be less than that of water.
- Non toxic.

2.2 **Enlist the formulae used to assess evaporation. Explain any one method to assess potential evapotranspiration.**

[15 Marks]

Sol:

Following formulae are used to calculate evapotranspiration.

(i) Pan evaporation method:

$$ET_0 = k \times E_{pan}$$

where, ET_0 is reference crop evapotranspiration (mm/day), k is non-dimensional coefficient. E_{pan} is pan evaporation (mm/day).

(ii) Blaney-Criddle method:

$$ET_0 = \frac{kp}{40}(1.8t_m + 32)$$

k = Crop factor

p = Monthly percentage of hours of bright sunshine in the year.

t_m = Mean monthly temperature (°C)

(iii) Penman's equation: Penman developed a theoretical formula based on principles of both energy budget and mass transfer approaches to compute potential evapotranspiration using following equations:

$$E_t = \frac{AH + \alpha E_a}{A + \alpha}$$

where, E_t = Evapotranspiration (mm/day)

α = Psychrometric constant (0.49 mm Hg/°C)

A = Slope of the curve between saturated vapour pressure (ρ_a) and temperature at mean air temperature.

E_a = Drying power of air which includes wind velocity and saturation deficit and is given by the relation.

$$E_a = 0.002187\,(160 + U_2)\,(\rho_s - \rho_a)$$

where, U_2 = Mean wind velocity (km/day) measured 2 m above the ground.

ρ_s = Saturation vapour pressure at mean air temperature.

H = Daily net radiation in mm of H_2O

Given by,
$$H = R_a(1-r)\left(0.29\cos\phi + \frac{0.55n}{N}\right) - \sigma T_a^4\left(0.56 - 0.092\sqrt{\rho_a}\right)\left(0.1 + \frac{0.9n}{N}\right)$$

where,
R_A = Mean monthly extra – Terrestial radiation in mm of H_2O/day
ϕ = Latitude of place
r = Reflection coefficient of surface
n = Actual duration of bright sunshine
N = Maximum possible hours
σ = Stefan Boltzman constant
T_a = Mean air temperature (k)

2.3 **(a) What volume of water is required to be added to an evaporation pan of 1.22 m diameter placed in a catchment area experiencing 10 mm rainfall, total evaporation on that day being 5 mm?**

(b) Determine the amount of evapotranspiration from an area if the total rainfall precipitated during a storm is 10 mm. Given the antecedent moisture at the root in the soil was 5 mm, the loss of water due to seepage 2.5 mm, losses due to percolation 2.00 mm. Surface runoff 3 mm and the moisture retained in the soil is 1.00 mm.

[5 + 5 Marks]

Sol:

(a) Given, Rainfall, P = 10 mm
Pan evaporation, E = 5 mm

$$\text{Area of pan} = \frac{\pi}{4}\times(1.22)^2 = 1.169\text{ m}^2$$

As rainfall is greater than evaporation water has to be taken out of pan,

$$\text{Volume of water taken out, } V = A(P-E)$$
$$= 1.169\times(10-5)\times 10^4 \times \frac{1}{10} = 5845\text{ cm}^3$$

(b) As we know, from water budget equation,

$$P–R–G–E–T = \Delta S$$

Total rainfall, P = 10 mm
Antecedent moisture at root in the soil = 5 mm
Loss of water due to seepage = 2.5 mm
Loss due to percolation = 2 mm
Surface runoff = 3 mm
Moisture retained in soil = 1 mm

∴ From water budge equation,

$$10 + 5 - 2.5 - 2 - 3 - T = 1$$
$$7.5 - T = 1$$
$$T = 6.5\text{ mm}$$

∴ Amount of evapotranspiration = 6.5 mm

2.4 **Describe briefly a method of estimating the evaporation.**

[15 Marks]

Sol:

The amount of water evaporated from a water surface is estimated by the following methods:

- Using evaporimeter data.
- Using empirical evaporation equations.
- Analytical methods.
- **Using evaporimeter data:** Evaporimeter are water containing pans which are exposed to the atmosphere, and the loss of water by evaporation in them is measured at regular intervals.
  Type of evaporimeter:
  (i) Class A evaporation pan (US Weather Bureau)
  (ii) ISI standard Pan (used in India)
  (iii) Colorado sunken pan (Pan is sunk below ground such that water level in pan is at ground level)
  (iv) Us Geological survey floating pan.
- **Pan coefficient, C_P:**
  (i) Evaporation pans are not exact models of large reservoirs.
  (ii) In view of the above, the evaporation observed from a pan has to be corrected to obtain the value of evaporation from a lake under similar climatic and exposure conditions.
  (Lake evaporation = C_P × Pan evaporation)
  C_P = Pan coefficient

The value of C_P in the for different pond are given in table.

S.No.	Types of pan	Average value	Range
1	Class A pan	0.70	0.60 - 0.80
2	ISI pan (modified class A)	0.80	0.64 - 1.10
3	Colorado sunken pan	0.78	0.75 - 0.86
4	US GS floating pan	0.80	0.70 - 0.52

- **Using empirically evaporation equations:** A large number of empirical equations are available to estimate lake evaporation using commonly available meteorological data. Most formulae are based on the Delton type equation and can be expressed as

$$E_L = k.f(u).(\rho_w - \rho_a)$$

E_L = Lake evaporation (mm/day)
ρ_w = Saturated vapour pressure of water surface temperature (in mm of mercury)
ρ_a = Actual vapour pressure of overlying air at a specific height (in mm of mercury)
$f(u)$ = Wind speed correction function
k = a constant

Two commonly used formulae are:

(i) Meyer's formula:

$$E_L = k_m(\rho_w - \rho_a)\left(1+\frac{V_9}{16}\right)$$

V_9 = Monthly mean wind velocity (km/h) at 9 m above the ground.
k_M = Coefficient accounting for various other facing with a value of 0.36 for large deep water and 0.50 for small, shallow waters.

(ii) Rhower's Formula: Rhower's formula considers a correction for the effect of pressure in addition to the wind speed effect and is given by

$$E_L = 0.771\,(1.465 - 0.000732\,P_a)\,(0.44 + 0.073340 \times V_0)\,(\rho_w - \rho_a)$$

P_a = Mean barometric reading (in mm of Hg)
V_0 = Mean wind velocity (in km/h) at ground the velocity at 0.6 m height above ground.

- **Analytical methods:** The analytical method for the determination of lake evaporation can be broadly classified as:
  1. **Water budget method:** In involves writing the hydrological continuity equation and determining the evaporation from a knowledge or estimation of other variables. Continuity equation is written as:

$$P + V_{is} + V_{ig} = V_{os} + V_{og} + E_L + \Delta S + T_L$$

P = Daily precipitation

V_{is} = Daily surface inflow into the lake

V_{ig} = Daily groundwater inflow

V_{os} = Daily surface outflow from lake

V_{og} = Daily seepage outflow

E_L = Daily lake evaporation

ΔS = Increase in lake storage in a day

T_L = Daily transpiration loss

$$E_L = P + (V_{is} - V_{os}) + (V_{ig} - V_{os}) - T_L - \Delta S$$

  2. **Energy budget method:** The energy budget method uses the law of conservation of energy. The energy available for evaporation is determined by considering the incoming energy, outgoing energy and energy stored in the water body over a known time interval.

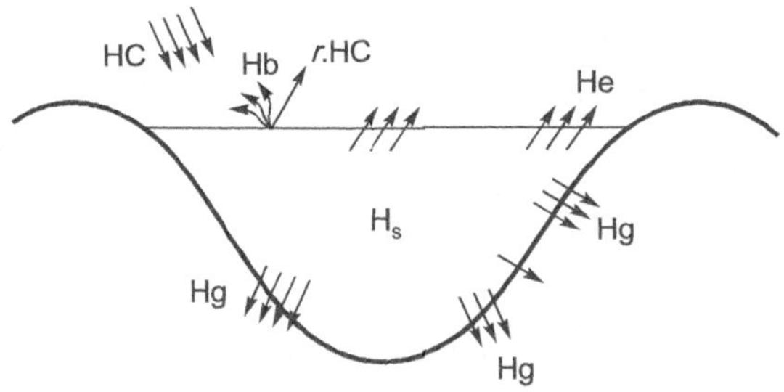

H_C – Total heat incident on water body

rH_C = Heat reflected

r = Reflection coefficient

H_b = Back radiation

H_s = Net heat energy absorbed by water body

$= H_C - rH_C - H_b = H_C(1 - r) - H_b$.

from conservation of energy we have,

$$H_\eta = H_a + H_e + H_g + H_i + H_s$$

H_a = Sensible heat transfer to the atmosphere

It is calculated using Bowen's ratio (β)

$$\beta = \frac{H_a}{H_e}$$

$H_a = \beta H_e$

H_e = Latent heat transfer to the atmosphere or energy used in evaporation

$H_e = P.LE_L$

ρ = Density of water

L = Latent heat of vaporisation

E_L = Rate of evaporation

H_g = Heat flux into the ground

H_i = Advected heat i.e. heat lost in seepage

$$H_s = \text{Heat energy stored}$$
$$H_a + H_e = H_\eta - H_g - H_i - H_s$$
$$\beta H_e + H_e = H_\eta - H_g - H_i - H_s$$
$$(1+\beta)L\rho E_L = H_\eta - H_g - H_i - H_s$$
$$E_L = \frac{H_\eta - H_g - H_i - H_s}{(1+\beta)\rho L}$$

3. **Mass transfer method:** This method is based on theories of turbulent mass transfer in boundary layer to calculate mass vapour transfer from surface to the surrounding atmosphere.

2.5 **The data pertaining to a stream-gauging operation at a gauging site are given below. The rating equation of the current meter is $v = 0.51\, N_s + 0.03$ m/s where N_s = revolutions per second. Calculate the discharge in the stream.**

Distance from left water edge (m)	0	1.0	3.0	5.0	7.0	9.0	11.0	12.0
Depth (m)	0	1.1	2	2.5	2.0	1.7	1.0	0
Revolutions of a current meter kept at 0.6 depth	0	39	58	112	90	45	30	0
Duration of observation (s)	0	100	100	150	150	100	100	0

[12 Marks]

Sol:

The calculations are performed in a tabular form.

For the first and last sections,

$$\text{Average width, } \bar{W} = \frac{\left(1+\frac{2}{2}\right)^2}{2\times 1} = 2.0 \text{ m}$$

For the rest of segments,

$$\bar{W} = \left(\frac{2}{2}+\frac{2}{2}\right) = 2.0 \text{ m}$$

Since the velocity is measured at 0.6 depth, the measured velocity is the average velocity at that vertical $(\bar{v})$. The calculate of discharge by the mid-section method is shown in tabular form below.

Distance from left water edge (m)	Average width $\bar{W}$ (m)	Depth y (m)	N_s = Reve./s	Velocity $\bar{v}$ (m/s)	Segmental discharge ΔQ_i (m^3/s)
0	0	0			0.0000
1	2	1.10	0.390	0.2289	0.5036
3	2	2.00	0.580	0.3258	1.3032
5	2	2.50	0.747	0.4110	2.0549
7	2	2.00	0.600	0.3360	1.3440
9	2	1.70	0.450	0.2595	0.8823
11	2	1.00	0.300	0.1830	0.3660
12	0	0.00			0.0000
				Sum =	6.45393

Discharge in the stream = 6.454 m^3/s

2.6 (a) A reservoir with a surface area of 250 hectares has saturation vapour pressure at water surface = 17.54 mm of Hg and actual vapour pressure of air = 7.02 mm of Hg wind velocity of 1 m above ground surface = 16 km/h. Estimate the average daily evaporation from the lake using Meyer's formula.

(b) An ISI standard evaporation pan at the site indicated a pan coefficient of 0.80 on the basis of calibration against controlled water budgeting method. If this pan indicated an evaporation of 72 mm in the week.

(i) Estimate the accuracy, if Meyer's method relative to the pan evaporation measurements.

(ii) Also estimated the volume of water evaporated from the lake in that week.

[6 + 6 Marks]

Sol:

(a)

$\rho_w = 17.54$ mm of Hg, $\rho_a = 7.02$ mm of Hg

Wind speed at a height of 9 m above ground

$$V_1 = a(h_1)^{1/7}$$

$$a = 16$$

$$V_9 = a(h_9)^{1/7}$$

$$16 = a(1)^{1/7}$$

$$= 16 \times (9)^{1/7} = 21.9 \text{ km/h}$$

Now, using Meyer's formula

$$E_L = 0.36(\rho_w - \rho_a)\left(1 + \frac{V_9}{16}\right)$$

$$= 0.36(17.54 - 7.02)\left(1 + \frac{21.9}{16}\right)$$

$$E_L = 8.97 \text{ mm/day}$$

(b)(i)

Daily evaporation as per pan evaporimeter

$$= 0.8 \times \frac{72}{7} = 8.23 \text{ mm}$$

$\therefore$ Error by Meyer's formula = (8.23 – 8.97) = –0.74 mm

Hence, Meyer's method overestimates the evaporation relative to the pan.

Percentage overestimation by Meyer's formula

$$= \left(\frac{0.74}{8.23}\right) \times 100 = 9\%$$

(ii) Considering the pan measurements as the basis, volume of water evaporated from the lake in 7 days.

$$= \left(7 \times \frac{8.23}{1000} \times 250 \times 10^4\right) = 144025 \text{ m}^3$$

Level-2

2.7 Compute the stream flow for the measurement data given below:

Distance from left (m) water edge	0	0.6	1.2	1.8	2.4	3.0	3.6	4.2	4.8	5.4	6.0	6.6
Depth (m)	0	0.3	1.29	2.16	2.55	2.22	1.68	1.41	1.05	0.63	0.42	0
Velocity (m/sec) at 0.2*d*	0	0.42	0.57	0.78	0.87	0.81	0.75	0.69	0.63	0.54	0.45	0
Velocity (m/sec) at 0.8*d*	0	0.21	0.36	0.54	0.60	0.30	0.51	0.45	0.39	0.33	0.30	0

[20 Marks]

Sol:

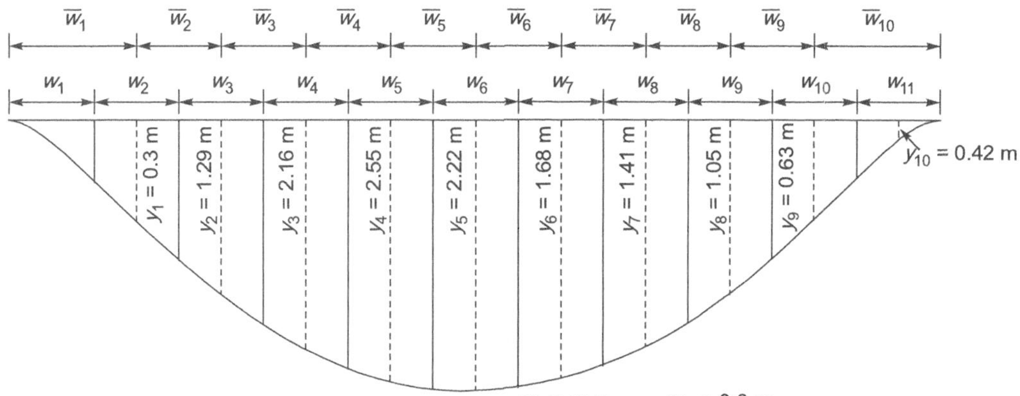

Average width for first and last segment $= \dfrac{\left(W_1 + \dfrac{W_2}{2}\right)^2}{2W_1} = \dfrac{\left(0.6 + \dfrac{0.6}{2}\right)^2}{2\times 0.6} = 0.675\text{m}$

Average width of each intermediate segment = (2nd to 10th) = 0.6 m

S.No.	Distance from one edge (b) (in m)	Avg. effective width $(\overline{w})_m$	Depth (y) m	Velocity (in m/sec)			Segmental discharge (m^3/sec) $q = \overline{w} \times \overline{y} \times \overline{v}$
				at 0.2d	at 0.8d	Mean velocity $(\overline{v})$	
1	2	3	4	5	6	7 = 5+6/2	8 = (3) × (4) × (7)
0	0	0	0	0	0	0	0
1	0.6	0.675	0.3	0.42	0.21	0.315	0.064
2	1.2	0.6	1.29	0.57	0.36	0.465	0.360
3	1.8	0.6	2.16	0.78	0.54	0.66	0.855
4	2.4	0.6	2.55	0.87	0.60	0.735	1.125
5	3.0	0.6	2.22	0.81	0.30	0.555	0.739
6	3.6	0.6	1.68	0.75	0.51	0.645	0.650
7	4.2	0.6	1.41	0.69	0.45	0.57	0.482
8	4.8	0.6	1.05	0.63	0.39	0.51	0.321
9	5.4	0.6	0.63	0.54	0.33	0.435	0.164
10	6.0	0.675	0.42	0.45	0.30	0.375	0.106
11	6.6	0	0	0	0	0	0

∴ The sum of total all the segmental discharge, computed in column (8) of table works out to 4.867 m^3/sec as the total discharge of the stream.

2.8 In order to compute the flood discharge, in a stream by slope area method, the following data have been obtained.

	Upstream section	Middle section	Downstream section
Area (m^2)⟶	108.6	103.1	99.80
Wetted perimeter (m)⟶	65.3	60.7	59.40
Gauge reading (m)⟶	316.8	-	316.55

Determine the flood discharge assuming Manning, n = 0.029 and length between downstream section and upstream section as 250 m.

[20 Marks]

Sol:

Let us the subscripts 1, 2 and 3 for U/S, D/S and middle section.

Average area of reach, $A_{eq} = \frac{A_1 + 2A_3 + A_2}{4} = \frac{108.6 + 2 \times 103.1 + 99.80}{4} = 103.65 \text{ m}^2$

Average wetted perimeter of reach,

$$P_{eq} = \frac{P_1 + 2P_3 + P_2}{4} = \frac{65.3 + 2 \times 60.7 + 59.4}{4} = 61.525 \text{ m}^2$$

Hydraulic radius, $R_{eq} = \frac{A_{eq}}{P_{eq}} = \frac{103.65}{61.525} = 1.68 \text{ m}$

$$k_{eq} = \frac{1}{n} A_{eq} \times R_{eq}^{2/3} = \frac{1}{0.029} \times (103.65)(1.68)^{2/3}$$

$$= 5051 \text{ m}$$

Assume, $V_1 = V_2$

$$h_L = h_1 - h_2 = 316.8 - 316.55 = 0.25 \text{ m}$$

Energy line slope, $s_f = \frac{h_L}{L} = \frac{0.25}{250} = \frac{1}{1000}$

$\therefore$ $Q = k_{eq}\sqrt{S} = 5051 \times \sqrt{\frac{1}{1000}}$

$$= 159.72 \text{ m}^3/\text{sec}$$

(2nd trial), $V_1 = \frac{Q}{A_1} = \frac{159.72}{108.6} = 1.47 \text{ m/ sec}$

$$V_2 = \frac{Q}{A_2} = \frac{159.72}{99.80} = 1.6 \text{ m/ sec}$$

Now, $h_1 + \frac{V_1^2}{2g} = h_2 + \frac{V_2^2}{2g} + h_L$

$$h_L = (h_1 - h_2) + \frac{V_1^2 - V_2^2}{2g} = 0.25 + \frac{(1.47)^2 - (1.6)^2}{2 \times 9.81} = 0.23 \text{ m}$$

$$Q = k_{eq}\sqrt{S} = 5051 \times \sqrt{\frac{h_L}{L}} = 5051 \times \sqrt{\frac{0.23}{250}}$$

$$= 153.2 \text{ m}^3/\text{sec}$$

(Next Trial) $V_1 = \frac{Q}{A_1} = \frac{153.2}{108.6} = 1.41 \text{ m/ sec}$

$$V_2 = \frac{Q}{A_2} = \frac{153.2}{99.8} = 1.53 \text{ m/ sec}$$

$$h_L = (h_1 - h_2) + \frac{V_1^2 - V_2^2}{2g} = 0.25 + \frac{(1.41)^2 - (1.53)^2}{2 \times 9.81} = 0.232 \text{ m}$$

$$Q = k_{eq}\sqrt{S} = 5051 \times \sqrt{\frac{0.232}{250}}$$

$= 153.8 \text{ m}^3/\text{sec}$

Thus, value is closer to the value of Q in previous trial and hence we should stop here.

The final value of Q can be taken as 153.8 m^3/sec.

2.9 Explain briefly the tracer method of flow measurement. A 25 g/lit of a fluorescent tracer, was discharged into a stream at a constant rate of 10 cm^3/sec. The background concentration of the tracer in the stream water was found to be zero. At a downstream section sufficiently for away, the day was found to reach on equilibrium concentration of 5 parts per billion. Estimate the stream discharge.

[20 Marks]

Sol:

- In this method a solution of a stable chemical such as common salt or sodium dichromate or a radioactive chemical, known as tracer is injected into the stream at either a constant rate or all at once.
- When the tracer is introduced into the stream at constant rate, it is called plateau method and when it is introduced all at once, it is called the 'gulp method'.
- The solution will be diluted by the discharge of the stream. The concentration of the tracer at a cross-section downstream (known as sampling section) of the injection point is determined.

Gulp method: Let C_o be the small initial concentration of the tracer in the stream flow. At section 1 a small quantity (volume, V_1) of high concentration C_1 of this tracer is added all at once. The concentration profile at section-2, which is far away from 1 is schematically shown in the figure.

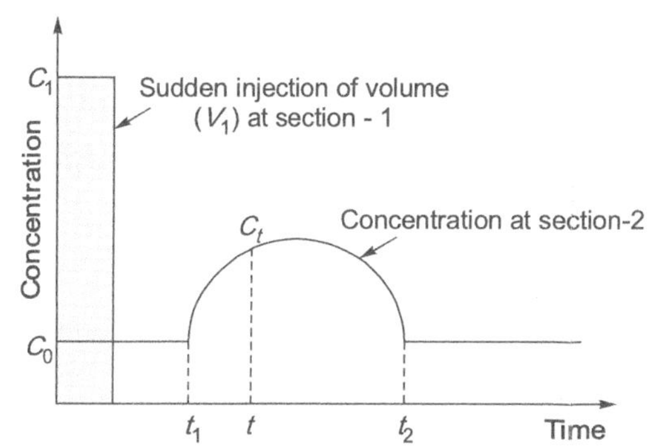

By continuity equation,

Mass of tracer at section 1 = Mass of tracer in the original discharge of stream (Q) at section '2' + Mass of tracer in the additional volume, V_1

$$\Rightarrow \quad V_1C_1 = \int_{t_1}^{t_2} Q(C_t - C_0)dt + \int_{t_1}^{t_2} \frac{V_1(C_t - C_0)}{(t_2 - t_1)}dt$$

2nd terms is negligible,

$$Q = \frac{V_1C_1}{\int_{t_1}^{t_2}(C_t - C_0)dt}$$

Discharge of the stream.

Plateau method: When tracer is injected continuously. Let concentration C_1 is continuously added at constant rate Q_1 at section 1. Let the initial concentration in tracer in stream be C_0 and at section 2 after sometime the concentration becomes constant at C_{eq} then,

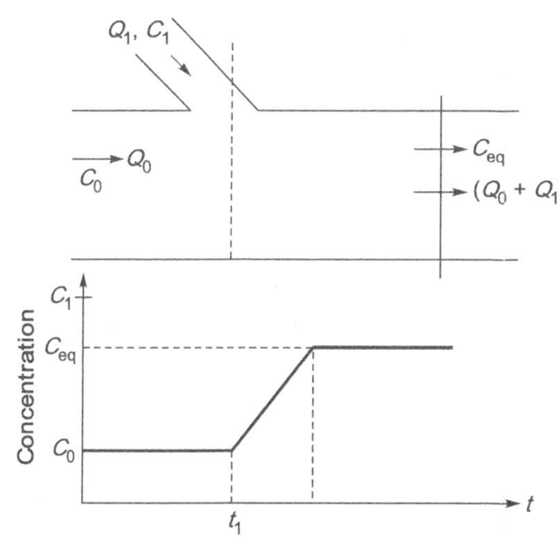

Applying mass conservation for tracer between section (i) and (ii) after equilibrium is reached, we have

$$(Q_0 \times C_0) + (Q_1C_1) = (Q_0 + Q_1).\ C_{eq}$$

$$C_{eq} = \frac{Q_0C_0 + Q_1C_1}{Q_0 + Q_1}$$

$$Q_0 = \frac{Q_1(C_1 - C_{eq})}{(C_{eq} - C_0)}$$

Discharge of the stream.

(2nd pert)

Since tracer is added at constant rate

Given,
$$C_1 = 25 \text{ gm}/l$$
$$Q_1 = 10 \text{ cm}^3/\text{sec}$$
$$C_0 = 0$$
$$C_{eq} = 5 \text{ ppb} = 5 \times 10^{-6} \text{ gm}/l$$

Now, Stream discharge,
$$Q_0 = \frac{Q_1(C_1 - C_{eq})}{(C_{eq} - C_0)} = \frac{10(25 - 5\times10^{-6})}{(5\times10^{-6} - 0)}$$
$$= 49.99999 \times 10^6 \text{ cm}^3/\text{sec}$$
$$Q_0 = 50 \text{ m}^3/\text{sec}$$

2.10 **During a flood the water surface at a section in a river was found to increase at a rate of 11.2 cm/h. The slope of the river is $\frac{1}{3600}$ and the normal discharge for the river stage read from a steady flow rating curve was 160 m^3/sec. If the velocity of the flood wave can be assumed as 2 m/sec. Determine the actual discharge.**

[10 Marks]

Sol:

It may be noted that, at the same stage, more discharge passes through the river during rising stages than in falling ones.

If Q_n is the normal discharge at a given stage under steady uniform flow and Q_m in the measured (actual) unsteady flow, the two are related as

$$\frac{Q_m}{Q_n} = \sqrt{1 + \frac{1}{V_\omega S_o} \times \frac{dh}{dt}}$$

Where,
S_o = Channel slope = Water surface slope at uniform flow.

$\frac{dh}{dt}$ = Rate of change of stage

V_ω = Velocity of the flood wave

Given,
$$\frac{dh}{dt} = 11.2 \text{ cm/h} = \frac{11.2}{100 \times 3600} \text{ m/sec}$$
$$S_0 = \frac{1}{3600}$$
$$Q_n = 160 \text{ m}^3/\text{sec};\ V_\omega = 2 \text{ m/sec}$$
$$\frac{Q_m}{160} = \sqrt{1 + \frac{1\times3600}{2} \times \frac{11.2}{100\times3600}} = 1.02762$$
$$Q_m = 164.42 \text{ m}^3/\text{sec}$$

2.11 Enumerate the various methods used for the measurements of discharge in a river. Explain the velocity area method, and also the factors which influence this method.

[15 Marks]

Sol:

Various method of estimating discharge in a river are as following:

(a) **Direct methods:**

(i) Velocity area method
(ii) Dilution method
(iii) Ultrasonic method
(iv) Moving boat method
(v) Electromagnetic method

(b) **Indirect methods:**

(i) Slope area method
(ii) Use of discharge measuring devices.

Velocity area method: In this method whole cross-section of river is divided into number of sections and sectional discharge is calculated and total discharge is equal to summation of sectional discharges.

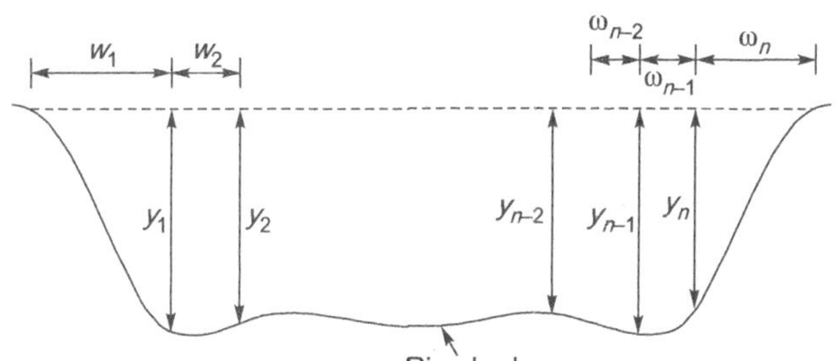

Cross section of river

In the above figure, discharge is calculated in three parts and added together to get total discharge.

$$Q_{\text{total}} = Q_{\text{first section}} + Q_{\text{last section}} + Q_{\text{remaining section}}$$

$$Q_{\text{first section}} = \frac{\left(W_1 + \frac{W_2}{2}\right)^2}{2W_1} \times y_1 \times \bar{V}_1$$

Here average velocity can be calculate as

$$\bar{V} = \frac{V_{0.2y} + V_{0.8y}}{2}$$

where, $V_{0.2y}$ and $V_{0.8y}$ is the velocity of river at 0.2y and 0.8y depth.

$$\text{Average width of first section,} = \frac{\left(W_1 + \frac{W_2}{2}\right)^2}{2W_1}$$

$$Q_{\text{last section}} = \frac{\left(W_n + \frac{W_{n-1}}{2}\right)^2}{2W_n} \times y_{n-1} \times \bar{V}_{n-1}$$

$$Q_{\text{remaining section}} = \sum_{i=2}^{n-2}\left(\frac{W_i}{2} + \frac{W_{i+1}}{2}\right) \times y_i \times \bar{V}_i$$

Average width for sections other than first and last section will be = $\frac{W_i}{2} + \frac{W_{i+1}}{2}$

Number of segment '*n*' as above has to be chosen based on the following criteria:

1. The segment width should not be greater than $\frac{1}{15}$ and $\frac{1}{20}$ of the width of river.
2. The discharge in each segment would be less than 10% of the total discharge.
3. The difference of velocities in adjacent segments should not be more than 20%.

Factors affecting velocity area method are:

(i) Number of segments and width of segment taken.
(ii) Geometry of the river.
(iii) Actual velocity distribution of the flow.
(iv) Type of measurement of velocity.

3. Infiltration, Runoff and Hydrograph

Level-1

3.1 **Direct runoff from a flood over a basin of area 50 hectare was observed as 25000 m^3 during a storm. The equivalent depth of rain fall over the basin was 97 mm. The time distribution of the rain fall during the storm is given below. Determine the f-index for the storm.**

Time(hr.)	0	1	2	3	4	5	6	7
Rainfall(mm)	0.0	6.0	11.0	34.0	28.0	12.0	6.0	0.0

[10 Marks]

Sol:

Time	Rainfall (mm)
0	0
1	6.0
2	11.0
3	34.0
4	28.0
5	12.0
6	6.0
7	0.0

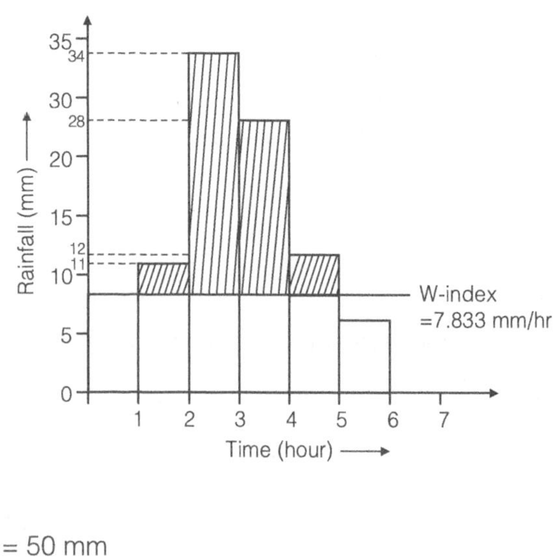

Total precipitation,

$$P = 97 \text{ mm}$$

Total runoff,

$$Q = \frac{25000}{50 \times 10^4} = 0.05 \text{ m} = 50 \text{ mm}$$

$\therefore$ $$\text{W-index} = \frac{P-Q}{t} = \frac{97-50}{6} = 7.833 \text{ mm/hr}$$

$\because$ $$\phi\text{-index} \geq \text{W-index}$$

$$\phi\text{-Index} = \frac{\text{(Total infiltration + infiltration during period when no excess rain occurs)}}{\text{Period of excess rainfall}}$$

$$= \frac{47-6-6}{4} = 8.75 \text{ mm/hr}$$

3.2 Measured infiltration rates, f, in cm/hour, for every hour from t = 0, when the rainfall just commenced to t = 8 hours are given in the table below. The rainfall lasts over 8 hours. Calculate the total infiltration quantity during 8 hours using HORTON constant of k = 4 (day^{-1})

Time(hr)	0	1	2	3	4	5	6	7
f(cmhr)	2.00	1.10	0.75	0.65	0.55	0.50	0.50	0.50

[5 Marks]

Sol:

According to HORTON'S equation

$$f_{ct} = f_{cf} + (f_{co} - f_{cf})e^{-k_h t}$$

f_{ct} = infiltration capacity at any time 't'
f_{co} = initial infiltration capacity at t = 0
f_{cf} = final steady state value
t_d = duration of rainfall
k_h = constant

Given,

$$f_{co} = 2 \text{ cm/hour}$$
$$f_{cf} = 0.5 \text{ cm/hour}$$
$$k_h = 4(day^{-1}) = 0.1666 \ (hour^{-1})$$
$$t = 8 \text{ hours}$$

$\therefore$
$$f_{ct} = 0.5 + (2 - 0.5)e^{-0.1666 \times 8}$$
$$= 0.8954 \text{ cm/hour}$$

3.3 What do you understand by infilteration index? How do you determine it?

[12 Marks]

Sol:

In hydrological calculations involving floods. It is found convenient to use a constant value of infilteration rate for the duration of the storm. The average infiltration rate is called infiltration index and two types of indices, in this regards are in common use.

(i) **ϕ - index:** It is on approximate infiltration rate so defined that the area of hytograph above ϕ-index gives depth of SR/DR/ER/Rainfall Excess. Hence, area of hyetograph below ϕ-index approximately gives depth of infiltration.

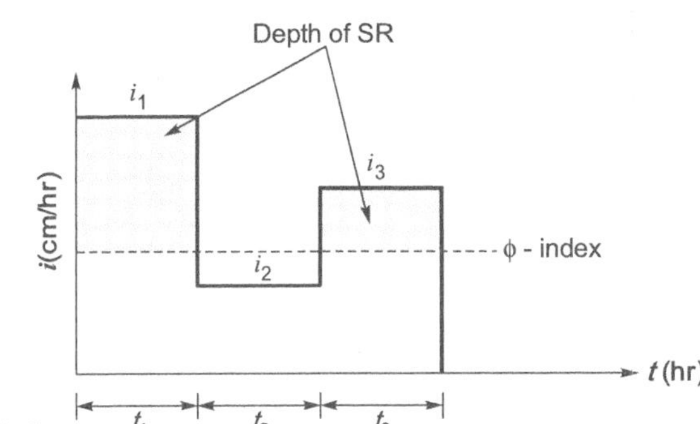

$(t_1 + t_3) \rightarrow$ Effective duration of rainfall

(ERH)

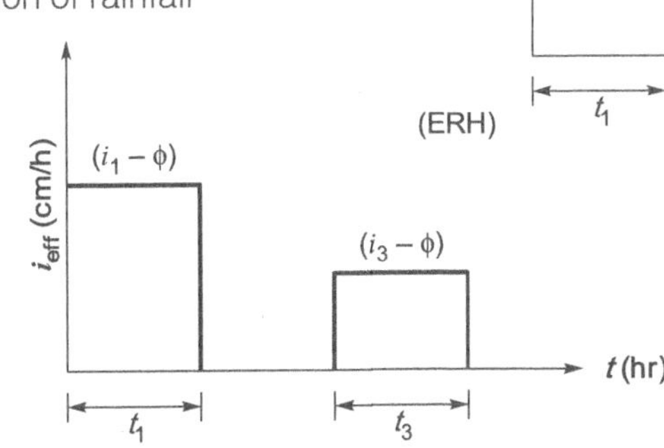

$(\phi \times t)$ - will give depth of infiltration only when ponding condition exist i.e. $(i > \phi)$; in that duration.

$$\phi\text{-index} = \left(\frac{P-R}{t}\right)_{\text{eff. duration}}$$

P = Precipitation in effective duration
R = Runoff
t = Effective duration of rainfall.

Steps to calculate ϕ-index:

Step 1: Drew hyetograph

Step 2: Assume ϕ - to be lower than the lowest intensity of rainfall.

Step 3: Calculate ϕ-index from the definition

$$\Sigma(i > \phi) \times t = SR$$

If the calculated ϕ - support our assumption, then it is correct otherwise move to the next step.

Step 4: Assume ϕ - index greater than the one calculated above and lower than the next higher intensity of rainfall.

If the calculated ϕ - support our assumption, then it is correct otherwise keep this process repeating.

(ii) ***W* - index:** The W-index is a refined version of ϕ - index. It excludes the depression storage and interception from the total losses. It is the average infiltration rate during the time rainfall intensity exceeds the capacity rate.

$$W\text{- index} = \left(\frac{P-R-\text{Minor losses}}{t}\right)_{\text{eff. duration}}$$

$$= \left(\frac{P-R}{t}\right)_{\text{Total duration}}$$

3.4 **The following is the set of observed data for successive 15 min periods to 105 minutes storm in a catchment,**

Duration (min)	15	30	45	60	75	90	105
Rainfall (cm/hr)	2.0	2.0	8.0	7.0	1.25	1.25	4.5

If the value of ϕ - index is 3 cm/hr, estimate the net runoff, the total rainfall and the value of *W*-index.

[12 Marks]

Sol:

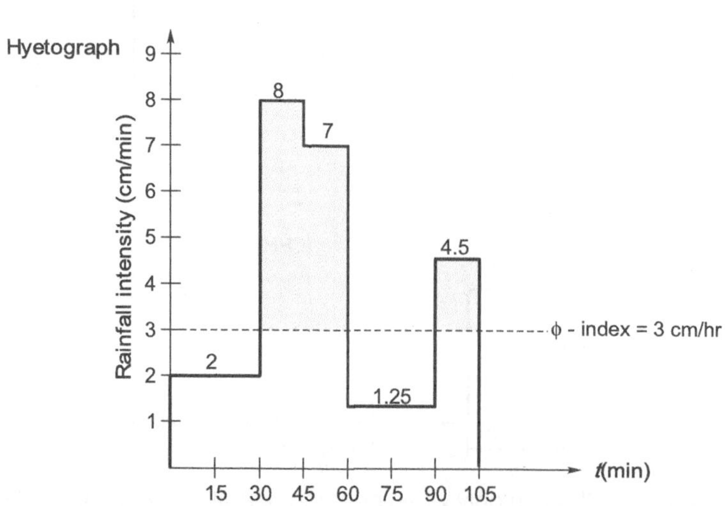

Given ϕ - index = 3 cm/hr

$$\text{Net runoff, } R = (8-3)\times\frac{15}{60} + (7-3)\times\frac{15}{60} + (4.5-3)\times\frac{15}{60} = 2.625 \text{ cm}$$

$$\text{Total precipitation, } P = 2\times\frac{15}{60} + 2\times\frac{15}{60} + 8\times\frac{15}{60} + 7\times\frac{15}{60} + 1.25\times\frac{30}{60} + 4.5\times\frac{15}{60} = 6.5 \text{ cm}$$

$$W\text{-index} = \left(\frac{P-R}{t}\right)_{\text{Total duration}} = \left(\frac{6.5-2.625}{105/60}\right) = 2.21 \text{ cm/hr}$$

3.5 An isolated 3 hr storm over a basin occurs as given below, estimate the runoff from the catchment due to the storm:

		Rainfall(cm)		
Percentage of catchment	φ-index (cm/hr)	1st hr	2nd hr	3rd hr
20	1	0.8	2.3	1.5
30	0.75	0.7	2.1	1
50	0.5	1	2.5	0.8

[10 Marks]

Sol:

Calculation of rainfall excess from different catchment areas:

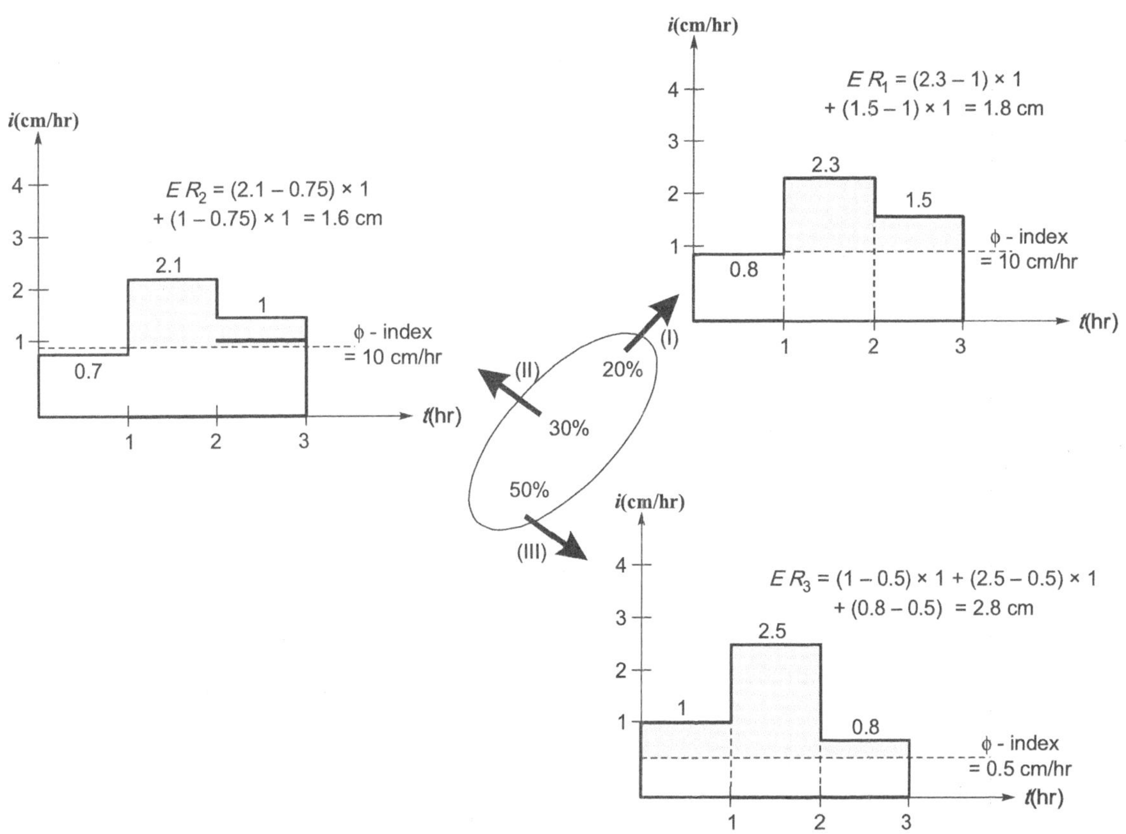

Now, net runoff from the catchment,

$$ER = \frac{(ER_1)\times(0.2A)+(ER_2)\times(0.3A)+(ER_3)\times(0.5A)}{A}$$

$$= \frac{\{(1.8\times 0.2)+(1.6\times 0.3)+(2.8\times 0.5)\}A}{A} = 2.24 \text{ cm}$$

3.6 The intensity frequency duration relationship for the catchment area is given by

$$I = \frac{800T_r^{0.2}}{(t+15)^{0.7}}$$ in which,

I = Rainfall intensity (mm/hr)

T_r = Return period (in years)

t = Rainfall duration (in min)

The catchment is divided into five sub areas as given below:

Sub area (km^2)	1.5	3	2	1	2.5
Runoff coefficient	0.7	0.6	0.4	0.55	0.9

Calculate the 25 years flood using the rational method. Assume the concentration time of 35 min.

[10 Marks]

Sol:

Given,: T_r = 25 years, t = 35 min

$$I = \frac{800\times(25)^{0.2}}{(35+15)^{0.7}} = 98.49 \text{ mm/ hr}$$

Equivalent runoff coefficient of the catchment,

$$C_{eq} = \frac{C_1A_1 + C_2A_2 + C_3A_3 + ...}{A_1 + A_2 + A_3 + ...}$$

$$= \frac{(0.7\times 1.5)+(3\times 0.6)+(2\times 0.4)+(1\times 0.55)+(2.5\times 0.9)}{1.5+3+2+1+2.5}$$

$$= 0.645$$

Using rational formula, $$Q = \frac{1}{36}kP_CA$$

$$P_C(\text{in cm/hr}) = \frac{98.49}{10} = 9.849 \text{ cm/ hr}$$

A (in ha) = 10 km^2 = 10 × 100 ha = 1000 ha

k = Equivalent runoff coefficient = 0.645

$$Q = \frac{1}{36}\times 0.645\times 9.849\times 1000 = 176.46 \text{ m}^3\text{/sec}$$

3.7 State and discuss the various factors affecting runoff from a drainage basin.

[10 Marks]

Sol:

Runoff from a drainage basin depends upon several factors, which are further classified in 2 categories:

(i) Physical characteristics of basin.

(ii) Climatic characteristics.

(a) Physical factors:

(i) **Shape:** Fan shaped/semicircular catchment gives rise to faster movement of runoff, low time base:
While elongated catchment slows down movement of runoff, quickness/fastness of runoff is determined by time taken by it to reach the basin outlet.

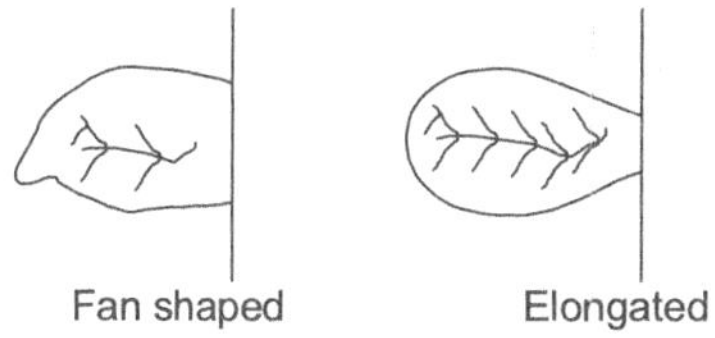

(ii) **Slope:** Higher the slope, runoff will move quickly, discharge of runoff will be more.

(iii) **Area:** More the area of drainage basin, more will be the runoff discharge.

(iv) **Stream density/drainage density:** Higher will be their value, more rapid will be the surface runoff.

(v) **Presence of any storage (Lake swamps):** It will decreases both magnitude and discharges of runoff.

(vi) **Vegetation/soil cover:** More the vegetation and soil cover, more will be infiltration thus, runoff volume will be decreased.

(b) Climatic factors:

(i) **Precipitation:** For a fixed duration of rainfall, runoff volume will increase with increase in rainfall; magnitude. Also for a fixed intensity of rainfall, more the duration, more will be the runoff yield.

(ii) **Evaporation:** More will be evaporation loss, less will be volume of runoff. It depends on seasons. In summer loss in runoff volume due to evaporation is more than as compared to that occur in winter season.

(iii) **Interception:** More the interception, more will be decrease in runoff volume due to it.

3.8 **Given below are the ordinates of 6 hr unit hydrograph for a catchment. Calculate the ordinates of the flood hydrograph due to a rainfall excess of 4.5 cm occurring in 6 hr. Assuming base flow as 10 m³/sec.**

Time (hr)	0	3	6	9	12	15	18	24	30	36
UH ordinate (m^3/s)	0	20	40	85	115	110	70	40	22	0

[12 Marks]

Sol:

Time (hr)	6 hr UH	6 hr 4.5 cm DRH	Base flow	fH
0	0	0	10	10
3	20	20 × 4.5 = 90	10	100
6	40	40 × 4.5 = 180	10	190
9	82	85 × 4.5 = 382.5	10	392.5
12	115	115 × 4.5 = 517.5	10	527.5
15	110	110 × 4.5 = 495	10	505
18	70	70 × 4.5 = 315	10	325
24	40	40 × 4.5 = 180	10	190
30	22	22 × 4.5 = 99	10	109
36	0	0	10	10

3.9 **The following are the ordinates of the hydrograph of flow from a catchment area of 740 km² due to 6 hr rainfall. Derive the ordinates of 6 hr UH. Make suitable assumption regarding base flow.**

Time from beginning of storm (hr)	0	6	12	18	24	30	36	42	48	54
Discharge (m^3/s)	30	55	200	340	390	330	250	190	130	30

[12 Marks]

Sol:

Assuming base flow of 30 m^3/s from straight line method.

Time (hr)	fH	Base flow	DRH = FH – Base flow	6 hr UH = DRH/ER(cm)
0	30	30	0	0
6	55	30	25	25/4.8 = 5.21
12	200	30	170	170/4.8 = 35.41
18	340	30	310	64.58
24	390	30	360	75
30	330	30	300	62.5
36	250	30	220	45.83
42	190	30	160	33.83
48	130	30	100	20.83
54	30	30	0	0

$$\text{Area of DRH} = (ER) \times A_{catchment}$$

$$(0 + 25 + 170 + 310 + 360 + 300 + 220 + 160 + 100 + 0) \times 6 \times 3600 = \{ER(cm)\} \times 10^{-2} \times 740 \times 10^6$$

$$ER = \frac{1645 \times 6 \times 3600}{740 \times 10^4} = 4.8 \text{ cm}$$

Level-2

3.10 Rainfall over a basin in 3-consecutive hrs are 4 cm, 5 cm and 3 cm respectively. Estimate the surface runoff from the basin assuming negligible surface retention and evaporation losses. The infiltration losses can be estimated using Horton's equation, $f = 1.2 + 4.2\,e^{-2.5t}$, f (in cm/hr), t (in hr);

[20 Marks]

Sol:

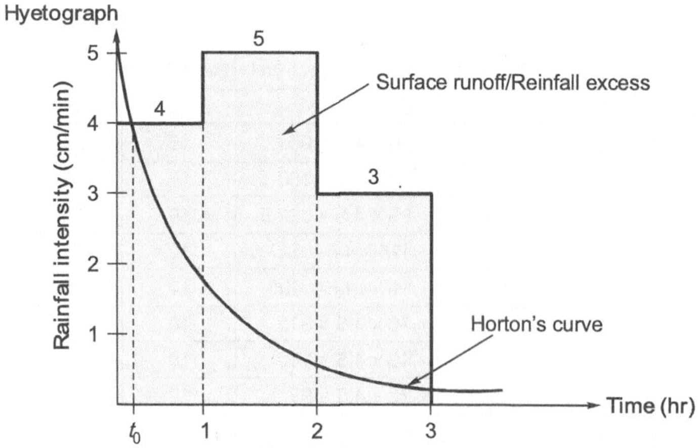

Given, Horton's infiltration equation:

$$f = 1.2 + 4.2\,e^{-2.5t}$$

At, $t = 0$ $\quad f = 1.2 + 4.2 = 5.4$ cm/hr

At, $t = 1$ hr $\quad f = 1.2 + 4.2\,e^{-2.5 \times 1} = 1.544$ cm/hr

Area bellow Horton's curve will given depth of infiltration only when ponding condition exist.

From 0 to t_0 (No ponding condition exist)

$$f = i = 4 \text{ cm/hr}$$

$$4 = 1.2 + 4.2e^{-2.5t_0}$$

$$2.8 = 4.2e^{-2.5t_0}$$

$$\frac{2}{3} = e^{-2.5t_0}$$

Taking log both sides, $\ln\left(\frac{2}{3}\right) = -2.5\ t_0$

$$t_0 = 0.162 \text{ hr}$$

Now, total depth of infiltration, $f = \int_{0.162}^{3} \left(1.2 + 4.2e^{-2.5t}\right) dt + 0.162 \times 4$

$$= 1.2[t]_{0.162}^{3} + 4.2\left[\frac{e^{-2.5t}}{-2.5}\right]_{0162}^{3}$$

$$= 1.2(3 - 0.162) + 4.2\left[\frac{e^{-2.5\times3}}{-2.5} + \frac{e^{-2.5\times0.162}}{2.5}\right]_{0162}^{3}$$

$$f = 4.525 + 0.648 = 5.173 \text{ cm}$$

$\therefore$ $P = 5 + 3 + 4 = 12$ cm

$\therefore$ Rainfall excess, $ER = P - F = 12 - 5.173 = 6.827$ cm

3.11 (a) Measured infiltration rates, f (in cm/hr), for every hour from $t = 0$, when the rainfall just commenced to $t = 7$ hrs are given in the table below. The rainfall last over 8 hrs. Calculate the total infiltration quantity during 8 hrs using Horton constant $k = 4$ day^{-1}.

(b) The infiltration capacity in a basin is represented by Horton's equation as $f = 4.7 + e^{-4.2t}$ where f is in (cm/hr) and t is (in hrs). Assuming the infiltration to take place at capacity rates in storms of 60 minutes duration, estimate the depth of infiltration in

(i) The first 30 min.

(ii) The second 30 min. of storm

Time (hrs.)	f (cm/hr)
0	2.00
1	1.10
2	0.75
3	0.65
4	0.55
5	0.50
6	0.50
7	0.50

[10+10 =20 Marks]

Sol:

(a) Given: Horton's constant, $k_h = 4 \text{ day}^{-1} = \frac{4}{24} = \frac{1}{6}\text{hr}^{-1}$

From table, $f_0 = 2$ cm/hr

$f_c = 0.50$ cm/hr

$\therefore$ Horton's infiltration equation, $f = f_c + (f_0 - f_c)e^{-k_h t} = 0.5 + (2 - 0.5)e^{-1/6t}$

$$= 0.5 + 1.5\ e^{-t/6}$$

$\therefore$ Total infilteration (in 8 hrs), $F = \int_0^8 f \cdot dt = \int_0^8 \left(0.5 + 1.5e^{-t/6}\right) dt = 10.628$ cm

(b) Given, Hortons' infiltration equation

$$f = 4.7 + e^{-4.2t}$$

(i) Infiltration depth (in first 30 min) = $\int_0^{1/2} \left(4.7 + e^{-4.2t}\right) dt = 2.559$ cm

(ii) Infiltration depth (in second 30 min of storm) = $\int_{1/2}^{1} f(t)dt = \int_{1/2}^{1} (4.7 + e^{-4.2t})dt = 2.375$ cm

3.12 A storm with a 16.33 cm precipitation produced a direct runoff and 6.8 cm. The time distribution of the storm is as follows.

Time from start (min)	30	60	90	120	150	180	210	240
rainfall (cm)	4.6	7	2.46	5.4	2.2	5.6	3.4	2

Draw the hyetograph and estimate ϕ - index.

[20 Marks]

Sol:

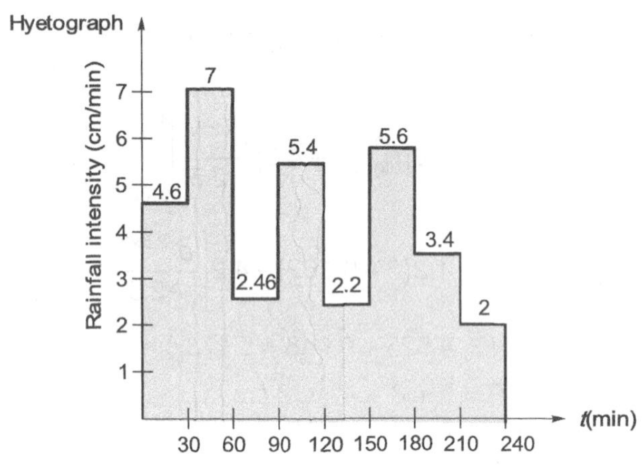

Calculation of ϕ - index:

Assuming ϕ - index < 2 cm/hr.

$$S(i-\phi)\times t = SR$$

$$(4.6-\phi)\times\frac{1}{2}+(7-\phi)\times\frac{1}{2}+(2.46-\phi)\times\frac{1}{2}+(5.4-\phi)\times\frac{1}{2}+(2.2-\phi)\times\frac{1}{2}+(5.6-\phi)\times\frac{1}{2}+(3.4-\phi)\times\frac{1}{2}$$
$$+(2-\phi)\times\frac{1}{2}=6.8$$

$$16.33-4\phi = 6.8$$

$$\phi = 2.3825 \text{ cm/h}$$

Hence, our assumption is wrong.

Now, assuming ϕ - index

$$2.3825 \le \phi$$

$$(4.6-\phi)\times\frac{1}{2}+(7-\phi)\times\frac{1}{2}+(2.46-\phi)\times\frac{1}{2}+(5.4-\phi)\times\frac{1}{2}+(5.6-\phi)\times\frac{1}{2}+(3.4-\phi)\times\frac{1}{2}=6.8$$

$$14.23-3\phi = 6.8$$

$$\phi = 2.47 \text{ cm/h}$$

Hence, our assumption is wrong.

Now, assuming ϕ - index

$$2.47 < \phi < 3.4 \text{ cm/h}$$

$$(4.6-\phi)\times\frac{1}{2}+(7-\phi)\times\frac{1}{2}+(5.4-\phi)\times\frac{1}{2}+(5.6-\phi)\times\frac{1}{2}+(3.4-\phi)\times\frac{1}{2}=6.8$$

$$13-2.5\phi = 6.8$$

$$\phi = 2.48 \text{ cm/h}$$

Supports our assumption

Hence, ϕ - index of the storm = 2.48 cm/hr

3.13 The ordinates of a 4h U.H. of a basin of area 300 km^2 measured at 1 h intervals are 6, 36, 66, 91, 106, 93, 79, 68, 58, 49, 41, 34, 27, 23, 17, 17, 13, 9, 6, 3 and 1.5 m^3/s respectively. Obtained the ordinates of a 3h U.H for the basin using the S-curve technique.

[20 Marks]

Sol:

Time t hours	Ordinates of 4 h U.H. U(t) m^3/s	S-curve additions S(t−4) m^3/s	S-curve ordinates S(t) m^3/s	S-curve lagged by 3h, S(t−3) m^3/s	Difference graph m^3/s	Ordinates of 3h U.H.
(1)	(2)	(3)	(4) = (2) + (3)	(5)	(6) = (4) − (5)	(7) = (6) × 4/3
0	0		0		0	0
1	6		6		6	8.0
2	36		36		36	48.0
3	66		66	0	66	88.0
4	91	0	91	6	85	113.3
5	106	6	112	36	76	101.3
6	93	36	129	66	63	84.0
7	79	66	145	91	54	72.0
8	68	91	159	112	47	62.7
9	58	112	170	129	41	54.7
10	49	129	178	145	33	44.0
11	41	145	186	159	27	36.0
12	34	159	193	170	23	30.7
13	27	170	197	178	19	25.3
14	23	178	201	186	15	20.0
15	17	186	203	193	10	13.3
16	13	193	206	197	9	12.0
17	9	197	206.5*	201	5.5	7.3
18	6	201	207.0	203	4.0	5.3
19	3	203	207.5*	206	1.5	2.0
20	1.5	206	208.0*	206.5	0	0
21	0	206.5	208.5*	207.0	0	

3.14 From the topographical map of a drainage basin of following quantities are measured:

Area of catchment = 3500 km^2

Length of main stream = 150 km

Distance of centroid from outlet = 80 km

The coefficient C_t and C_p of the catchment are assumed to be 1.8 and 0.65. Derive the 4 hr UH using Synder's method.

[20 Marks]

Sol:

Given, A = 3500 km^2, L = 150 km, L_{ca} = 80 km, C_t = 1.8, C_p = 0.65

Now, as per Synder;

$$t_L(\text{leg time}) = C_t(L.L_{ca})^{0.3}$$

$$= 1.8(150 \times 80)^{0.3} = 30.131 \text{ hr}$$

$$\text{Standard duration of SUH, } t_r = \frac{t_L}{5.5} = \frac{30.131}{5.5} = 5.478 \text{ hr}$$

$$D' = 4 \text{ hr}$$

$$\therefore \quad t_L' = t_L + \frac{D'-D}{4} = 30.131 + \frac{4-5.478}{4} = 29.76 \text{ hr}$$

$$\text{Base period } (T) = 72 + 3t_L' = 72 + 3 \times 29.76 = 161.28 \text{ hr}$$

$$\text{Peak discharge, } (Q_p) = 2.78\frac{C_pA}{t_L'} = 2.78 \times \frac{0.65 \times 3500}{29.76} = 212.52 \text{ m}^3/\text{sec}$$

Width of SUH of 50% of peak discharge,

$$W_{50} = \frac{5.87}{\left(\frac{Q_P}{A}\right)^{1.08}}$$

$$= \frac{5.87}{\left(\frac{212.52}{3500}\right)^{1.08}} = 120.96 \text{ hr}$$

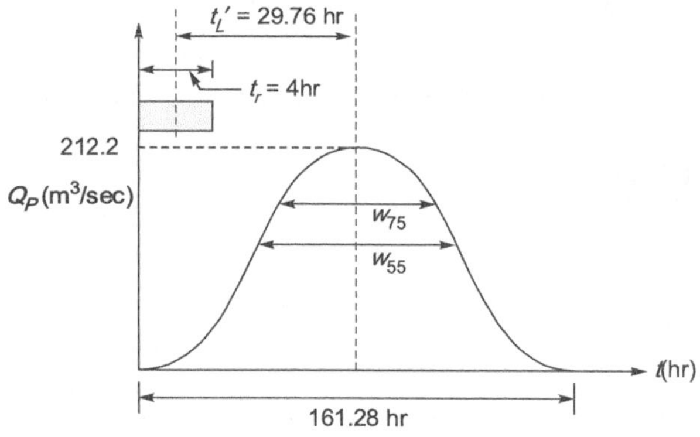

Width of SUH at 75% of peak discharge:

$$W_{75} = \frac{W_{50}}{1.75} = \frac{120.56}{1.75} = 69.12 \text{ hr}$$

3.15 A 4 hr storm in a basin producing a 10 cm of runoff results in the following flow in a stream

Time (hr)	0	2	4	6	8	12	16	20
Flow (m³/s)	0	2.044	8.1	13.5	11.34	6.75	2.70	0

Calculate the peak flow and the time of its occurrence of the flood created by 8 hr storm in the basin which produces 5 cm of runoff during first 4 hrs and 7.5 cm of runoff during the second 4 hrs. Assume base flow is negligible.

[20 Marks]

Sol:

Time (hr) 1	4hr DRH flow (m³/sec) 2	4hr UH (m³/sec) 3	Col. 3 lagged by 4hr 4	8hr UH (Col.3 + col.4)/2 5	Col.5 × 5 6	Col. 5 lagged by 4hr × 7.5 7	8 hr (DRH) 8
0	0	0	-	0	0	-	0
2	2.044	0.2044	-	0.1022	0.511	-	0.511
4	8.1	0.81	0	0.405	2.025	0	2.025
6	13.5	1.35	0.2044	0.7772	3.886	0.7665	4.6525
8	11.34	1.134	0.81	0.972	4.86	3.0375	7.8975
10	9.045	0.9045	1.35	1.127	5.635	5.829	11.464
12	6.75	0.675	1.134	0.9045	4.5225	7.29	11.8125
14	4.725	0.4725	0.9045	0.6885	3.4425	8.4525	11.895
16	2.70	0.270	0.675	0.4725	2.3625	6.7837	9.1462
18	1.35	0.135	0.4725	0.30375	1.5187	5.1637	6.682
20	0	0	0.270	0.135	0.675	3.5437	4.218
			0.135	0.0675	0.3375	2.275	2.6155
			0	0	0	1.0125	1.0125
						0.506	0.506
						0	0

Peak flow = 11.895 m³/sec

and time of its occurrence = 14 hr

3.16 The 6 hr UH of a basin is triangular in shape with peak of 100 m³/sec occurring at 24 hr from start. The base period is 72 hr.

(i) What is area of the catchment represented by this unit hydrograph.

(ii) Calculate the flood hydrograph due to storm of rainfall excess of 2 cm during first 6 hrs and 4.0 cm during the second 6 hrs. interval. The base flow can be assumed as 25 m³/sec. Constant throughout.

[10+10=20 Marks]

Sol:

(i)

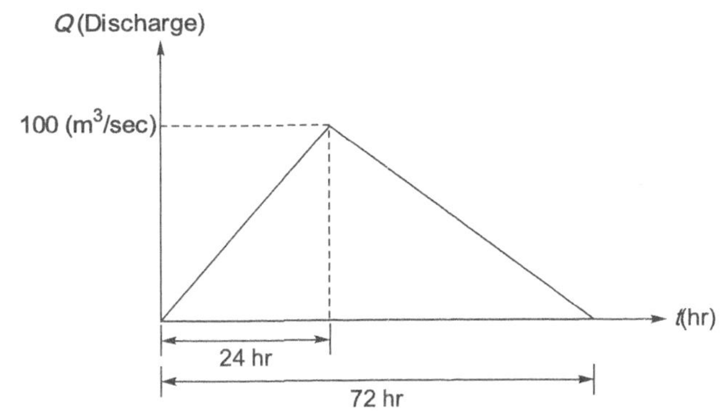

$$\text{Area of UH} = 1\text{ cm} \times A_{\text{cathment}}$$

$$\frac{1}{2} \times 72 \times 3600 \times 100 = 1 \times 10^{-2} \times A(\text{km}^2) \times 10^6$$

$$A = 1296\text{ km}^2$$

(ii)

Time (hr)	6hr UH ordinate (m^3/sec)	Direct runoff due to ER = 2 cm Col.2 × 2	Direct runoff due to ER = 4 cm Col.2 lagged by 6hr × 4	DRH (Col.3 + col.4)	Base flow	Flood hydrograph
1	2	3	4	5	6	7
0	0	0	-	0	25	25
6	25	50	0	50	25	75
12	50	100	100	200	25	225
18	75	150	200	350	25	375
24	100	200	300	500	25	525
30	87.5	175	400	575	25	600
36	75	150	350	500	25	525
42	62.5	125	300	425	25	450
48	50	100	250	350	25	375
54	37.5	75	200	275	25	300
60	25	50	150	200	25	225
66	12.5	25	100	125	25	150
72	0	0	50	50	25	75
78			0	0	25	25

3.17 A small rural catchment had rapid urbanization over a period. The impact of urbanization on the urban hydrology of the area is reflected in the change occurring in the 1-hour unit hydrograph of the area as follows:

Before Urbanization : Properties of 1-hour unit hydrograph:

Shape : Triangle

Peak : 3.6 m^3/s occurring 3 hours from the start

Base : 10 hours

ϕ-index : 7 mm/hour

After Urbanization : Properties of 1-hour unit hydrograph:

Shape : Triangle

Peak : 6.0 m^3/s occurring 1.5 hours from the start

Base : 6 hours

ϕ-index : 4 mm/hour

If a design storm has intensity of 40 mm/h, estimate the percentage increase in the peak storm runoff and in flood volume due to urbanisation.

[20 marks]

Sol:

Design rainfall = 40 mm/h

Excess design rain of 1 hr duration in 1st case (i.e. before urbanization)

$$= (40\text{ mm} - 7\text{ mm}) = 33\text{ mm} = 3.3\text{ cm}$$

Excess design rain of 1 hr duration in 2nd case (i.e. after urbanization)

$$= (40\text{ mm} - 4\text{ mm}) = 36\text{ mm} = 3.6\text{ cm}$$

Peak discharge in 1st case = Peak ordinate of UH × Excess rain

Ordinates of UH always imply rainfall excess depth of 1 cm

$$= \left[3.6\text{m}^3/\text{s} \times 3.3\right] = 11.88\text{ m}^3/\text{s} \quad \text{...(i)}$$

$$\text{Peak discharge in 2nd case} = \left[6.0\text{m}^3/\text{s} \times 3.6\right] = 21.6\text{ m}^3/\text{s} \quad \text{...(ii)}$$

$$\text{Percentage increase in peak storm runoff} = \frac{21.6 - 11.88}{11.88} \times 100\% = 81.82\%$$

Volume of water contained in 1st flood hydrograph

$$= \frac{1}{2} \times 11.88\text{ m}^3/\text{s} \times 10\text{ h}; \quad (\text{i.e. area of } \Delta)$$

$$= \frac{1}{2} \times 11.88 \times (10 \times 3600)\text{ m}^3 = 213840\text{ m}^3 \quad \text{...(iii)}$$

Volume of water contained in 2nd flood hydrograph

$$= \frac{1}{2} \times 21.6\text{ m}^3/\text{s} \times 6\text{h} \qquad [\textit{i.e.}\text{ Area of } \Delta]$$

$$= \frac{1}{2} \times 21.6 \times (6 \times 3600)\text{m}^3 = 233280\text{ m}^3 \quad \text{...(iv)}$$

$$\text{Percentage increase in flood volume} = \frac{233280 - 213840}{213840} \times 100\% = 9.09\%$$

3.18 **A 2 hour unit hydrograph in a catchment is given below :**

Time (hr)	0	2	4	6	8	10	12
Discharge (m^3/sec)	0	54	175	127	58	25	0

Compute and tabulate the ordinates of 1 hour UH.

[20 marks]

Sol:

The ordinates at 1 hr. interval of the 2 hr. UH is to be interpolated.

Time (hr) (1)	Ordinates of 2-h UH (m^3/s) (2)	S_2-curve addition (m^3/s) (3)	S_2-curve ordinates Col(2) + Col(3) (m^3/s) (4)	S_2-curve lagged by 1 hr (5)	Col(4) – Col(5) DRH of (1/2) = 0.5 cm (6)	1-h UH ordinates $= \frac{Col(6)}{(0.5)}$ (7)
0	0	–	0	–	0	0
1	27	–	27	0	27	54
2	54	0	54	27	27	54
3	114.5	27	141.5	54	87.5	175
4	175	54	229	141.5	87.5	175
5	151	141.5	292.5	229	63.5	127
6	127	229	356	292.5	63.5	127
7	92.5	292.5	385	356	29	58
8	58	356	414	385	29	58
9	41.5	385	426.50	414	12.5	25
10	25	414	439	426.50	12.5	25
11	12.5	426.50	439	439	0	0
12	0	439	439	439	0	0

3.19 Determine using the 'rational formula' with runoff co-efficient equal to 0.80, the peak storm runoff to be carried out by the underground drainage pipeline *AB* for the drainage basin area of 0.2 ha as shown in the given figure.

Take the length of pipeline as 100 m and the level difference between *A* and *B* as 0.5 m. Assume initially a pipe diameter of 250 mm. Consider Darcy-Weisbach friction factor $f = 0.015$. The rainfall duration and rainfall intensity is furnished in tabular form below. Consider time of entry at end *A* as 2 min.

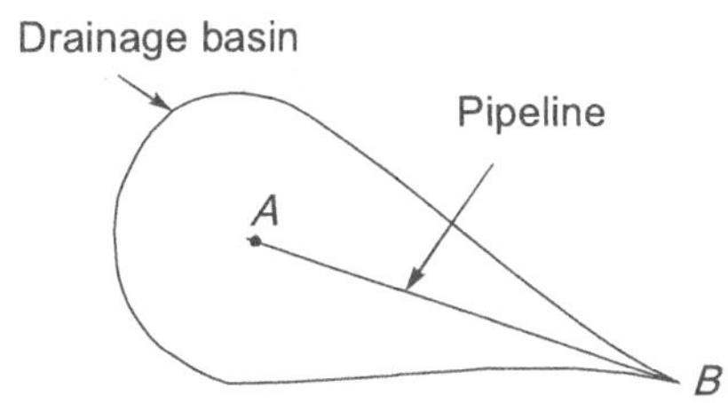

Duration (mins)	Rainfall intensity (mm/hr)
2.0	122
2.5	108
3.0	97
3.5	89
4.0	82
4.5	77

Check whether the pipe size is adequate or not.

[20 Marks]

Sol:

$K = 0.8$; $A = 0.2$ ha; $L = 100$ m; $H_L = 0.5$ m; $D = 250$ mm = 25 cm = 0.25 m; $f = 0.015$

$$H_L = \frac{f.LV^2}{2gD}$$

$$\Rightarrow \quad 0.5 \text{ m} = \frac{0.015 \times 100 \times V^2}{2 \times 9.81 \times 0.25}$$

$$\Rightarrow \quad V = 1.279 \text{ m/s}$$

$$\text{Time of flow} = T_f = \frac{\text{Length of pipe}}{\text{Vel. of flow}} = \frac{100}{1.279 \text{ m/s}} = 78.2 \text{ sec} = 1.3 \text{ min}$$

$\therefore$ Time for concentration, $T_c = T_i + T_f = 2$ min. + 1.3 min = 3.3 min

Rainfall intensity (as per given table):

For T_i = 3.0 min., i = 97 mm/hr

and for T_2 = 3.5 min., i = 89 mm/hr

Interpolating, we get, p_c for T_c of 3.3 min.

$$i = \left[97 - \frac{97-89}{0.5} \times 0.3\right] = 92.2 \text{ mm/h} = 9.22 \text{ cm/h}$$

Now, $$Q_P = \frac{1}{36} Kp_c.A = \left[\frac{1}{36} \times 0.80 \times 9.22 \times 0.2\right] = 0.041 \text{ m}^3\text{/s}$$

The capacity of the 0.25 m dia pipe is

$$Q = V.A = 1.279 \times \left(\frac{\pi}{4} \times 0.25^2\right) = 0.062 \text{ m}^3\text{/s}$$

Since, capacity of pipe (0.062 m^3/s) is more than the discharge (0.041 m^3/s), and thus the pipe dia. is adequate.

3.20 The following rainfall hyetograph and the corresponding direct runoff are recorded in a watershed. Compute the one-hour unit hydrograph ordinates for the first four hours. Assume f-index = 0.50 cm/hr.

Time(hr)	1	2	3	4	5	6	7	8	9	10
Rainfall(cm)	2.8	5.2	4.7	0.0	0.0	0.0	0.0	0.0	0.0	0.0
Direct Runoff (m^3/sec)	64.2	288.4	794.5	1369.6	1593.7	1175.1	588.1	286.9	170.5	110.0

[20 Marks]

Sol:

Time	Rainfall	DRH	R_1u	R_2u	R_3u	$R_1u+R_2u+R_3u=M$
1.	2.8	$64.2 = M_1$	$2.3u_1$	–	–	$27.91 = u_1$
2.	5.2	$288.4 = M_2$	$2.3u_2$	$4.7u_1$	–	$68.36 = u_2$
3.	4.7	$794.5 = M_3$	$2.3u_3$	$4.7u_2$	$4.2u_1$	$154.78 = u_3$
4.	0	$1369.6 = M_4$	$2.3u_4$	$4.7u_3$	$4.2u_2$	$154.36 = u_4$
5.	0	$1593.7 = M_5$	$2.3u_5$	$4.7u_4$	$4.2u_3$	$94.84 = u_5$
6.	0	$1175.1 = M_6$	$2.3u_6$	$4.7u_5$	$4.2u_4$	$34.46 = u_6$
7.	0	$588.1 = M_7$	$2.3u_7$	$4.7u_6$	$4.2u_5$	$12.09 = u_7$
8.	0	$286.9 = M_8$	$2.3u_8$	$4.7u_7$	$4.2u_6$	$37.10 = u_8$
9.	0	$170.5 = M_9$	$2.3u_9$	$4.7u_8$	$4.2u_7$	$0 = u_9$
10	0	$110.0 = M_{10}$	$2.3u_{10}$	$4.7u_9$	$4.2u_8$	$0 = u_{10}$
				$4.7u_{10}$	$4.2u_9$	$0 = u_{11}$
					$4.2u_{10}$	

Given, ϕ-index = 0.50 c/hr

$$\text{Rainfall excess in first hour } (R_1) = 2.8 - 0.50 \times 1 = 2.3 \text{ cm}$$
$$\text{Rainfall excess in second hour } (R_2) = 5.2 - 0.50 \times 1 = 4.7 \text{ cm}$$
$$\text{Rainfall excess in third hour } (R_3) = 4.7 - 0.50 \times 1 = 4.2 \text{ cm}$$
$$2.3u_1 = M_1$$
$$\Rightarrow \quad 2.3u_1 = 64.2$$
$$\Rightarrow \quad u_1 = 27.91 \text{ m}^3/\text{sec}$$
$$2.3u_2 + 4.7\,u_1 = M_2$$
$$\Rightarrow \quad 2.3u_2 + 4.7 \times 27.91 = 288.4$$
$$\Rightarrow \quad u_2 = 68.36 \text{ m}^3/\text{sec}$$
$$2.3\,u_3 + 4.7u_2 + 4.2u_1 = M_3$$
$$\Rightarrow 2.3 \times u_3 + 4.7 \times 68.36 + 4.2 \times 27.91 = 794.5$$
$$\therefore \quad u_3 = 154.78 \text{ m}^3/\text{sec}$$
$$2.3\,u_4 + 4.7\,u_3 + 4.2\,u_2 = M_4$$
$$\Rightarrow \quad 2.3u_4 + 4.7 \times 154.78 + 4.2 \times 68.36 = 1369.6$$
$$\Rightarrow \quad u_4 = 154.36 \text{ m}^3/\text{sec}$$
$$2.345 + 4.7u_4 + 4.2u_3 = 1513.7$$
$$\Rightarrow \quad u_5 = 94.84$$
$$2.3u_6 + 4.7u_5 + 4.2u_4 = 1175.1$$
$$\Rightarrow \quad u_6 = 34.46 \text{ m}^3/\text{sec}$$
$$2.3u_7 + 4.7u_6 + 4.2u_5 = 588.1$$
$$\Rightarrow \quad u_7 = 12.09 \text{ m}^3/\text{sec}$$
$$2.3u_8 + 4.7u_7 + 4.2u_6 = 586.9$$
$$\Rightarrow \quad u_8 = 37.10 \text{ m}^3/\text{sec}$$
$$2.3u_9 + 4.7u_8 + 4.2u_7 = 170.5$$
$$\Rightarrow \quad u_9 = 0$$
$$\Rightarrow \quad 2.3u_{10} + 4.7u_9 + 4.2u_8 = 110.0$$
$$\Rightarrow \quad u_{10} = 0$$
$$u_{11} = 0$$

$u_1, u_2 \ldots u_{10}$ are the ordinate of unit hydrograph.

3.21 A 6 hour unit hydrograph of a watershed is given below. Calculate 18 hour unit hydrograph using S-curve method and tabulate the results.

Time(hr)	0	6	12	18	24	30	36	42
6 hr unit hydrograph $\left[m^3/(scm)\right]$	0	1.8	30.9	85.6	41.8	14.6	5.5	1.8

[12 Marks]

Sol:

Time	Ordinate of 6 hours hydrograph	Offset ordinate	S-curve ordinate	offset S-curve	Δy	Ordinate of 18 hr UH=Δy/3
(1)	(2)	(3)	(4)	(5)	(6)	(7)
0	0	–	0	–	0	0
6	1.8	0	1.8	–	1.8	0.6
12	30.9	1.8	32.7	–	32.7	10.9
18	85.6	32.7	118.3	0	118.3	39.43
24	41.8	118.3	160.1	1.8	158.3	52.77
30	14.6	160.1	174.7	32.7	142	47.3
36	5.5	174.7	180.2	118.3	61.9	20.63
42	1.8	180.2	182	160.1	21.9	7.3
		182	182	174.7	7.3	2.43
		182	182	180.2	2.2	0.73
		182	182	182	0	0
				182	0	0

3.22 Given below are the monthly rainfall, 'P' and the corresponding runoff values, R, for a period of 10 months for a catchment develop a correlation between R and P.

Month no.	1	2	3	4	5	6	7	8	9	10
P(cms)	4	22	28	15	12	8	4	15	10	5
R(cms)	0.2	7.1	10.9	4.0	3.0	1.3	0.4	4.1	2.0	0.3

[15 Marks]

Sol:

The relationship between rainfall and the resulting runoff is quite complex and is influenced by a host of factors relating the catchment and climate. One method is to fit a linear regression line between R and P. The calculations are tabulated below:

Month no.	1	2	3	4	5	6	7	8	9	10
P(cms)	4	22	28	15	12	8	4	15	10	5
R(cms)	0.2	7.1	10.9	4.0	3.0	1.3	0.4	4.1	2.0	0.3
P^2	16	484	784	225	144	64	16	225	100	25
PR	0.8	156.2	305.2	60	36	10.4	1.6	61.5	20.0	1.5

The equation for straight line regression between runoff R and rainfall P is:

$$R = aP + b$$

where,

$$a = \frac{N\Sigma PR - \Sigma P \cdot \Sigma R}{N \cdot \Sigma P^2 - (\Sigma P)^2}$$

$$b = \frac{\Sigma R - a\Sigma P}{N}$$

N = Number of data set of runoff = 10

$$a = \frac{10 \times 653.2 - (123 \times 33.3)}{10 \times 2083 - (123)^2} = 0.427$$

$$b = \frac{33.3 - 0.427 \times 123}{10} = -1.922$$

$$\therefore \quad R = 0.427P - 1.922$$

3.23 The 4-hour unit hydrograph (UH) for a catchment having an area of 536 km^2 is shown in the Table below. Find the peak discharge when a 3-hour period of rainfall excess with intensity of 6 mm/hr was realized in the catchment. Assume that there is no base flow.

Time (hours)	UH ordinate (m^3/sec)	Time (hours)	UH ordinate (m^3/sec)
0	0	11	73
1	10	12	59
2	60	13	48
3	120	14	36
4	170	15	28
5	200	16	20
6	180	17	13
7	150	18	8
8	124	19	3
9	104	20	0
10	88	21	0

[15 Marks]

Sol:

1 Time	2 UH 1 m^3/sec	3 S-curve addition	4 S-curve ordinate	5 Lagged by 3hr	3hr UH Col.4-col.5 (4/3) m^3/sec
0	0	–	0	0	0
1	10	–	10	0	13.3
2	60	–	60	0	80
3	120	–	120	0	160
4	170	0	170	10	213.33
5	200	10	210	60	200
6	180	60	240	120	160
7	150	120	270	170	133.33
8	124	170	294	210	112
9	104	210	314	240	98.67
10	88	240	328	270	77.33
11	73	270	343	294	65.33
12	59	294	353	314	52
13	48	314	362	328	45.33
14	36	328	364	343	28
15	28	343	371	353	24
16	20	353	373	362	19.67
17	10	362	375	364	17.33
18	8	364	375	371	5.33
19	3	371	375	373	2.67
20	0	373	375	375	0
21	0	375	–	–	0
22	–	375	–	–	0
23	–	375	–	–	–
24	–	375	–	–	–

In 3 hr

Rainfall = $3 \times 6 = 18$ mm = 1.8 cm

Peak discharge = $213.33 \times 1.8 = 384$ m^3/sec

4. Floods, Flood Routing and Flood Channel

Level-1

4.1 Write short note on the following:

(i) Prism storage

(ii) Wedge storage

Also state difference between these two.

[12 marks]

Sol:

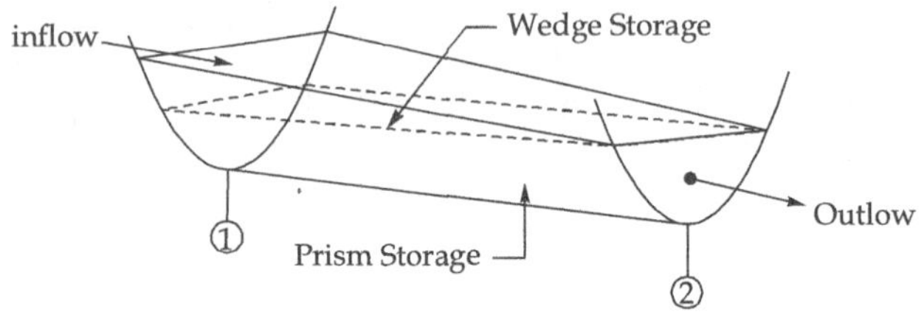

Storage in a channel reach

(i) **Prism Storage:** It is the volume which exist if uniform flow occurs at the downstream depth. It means it is the volume found by an imaginary plane parallel to the channel bottom drawn at the outflow section of the water surface.

(ii) **Wedge Storage:** It is the triangular volume enclosed between the actual water surface profile and the top of the prism storage.

Difference between prism storage and wedge storage

- At fixed depth at a downstream section of a river reach, the prism storage is constant while the wedge storage changes from a positive value at an advancing flood (in this case depth increases) to a negative value during receding flood (in this case depth decreases).
- The prism storage is steady so it is similar to reservoir storage and is expressed as a function of outflow discharge

$$S_p = f(Q).$$

Wedge storage is a function of inflow discharge.

It can be expressed as $s_w = f(I)$.

- The total storage is expressed as;

$$S = K[xI^m + (1-x)Q^m]$$

where,

x = Weighing factor

$0 < x \leq 0.5$

$0.1 \leq x \leq 0.3$ for natural channel

K = Storage time constant which has dimension of time

m = Constant = 1 for natural channel

4.2 **What is meant by flood routing? Explain briefly, Muskingham method of flood routing in a river channel.**

[10 Marks]

Sol:

Flood routing: It is a procedure in which shape of flood hydrograph at a particular location on the stream is calculated from a known or assumed flood hydrograph at some location upstream.

Muskingham method of flood routing: A common method of channel routing is Muskingham's method, in which the channel storage is expressed as a function of both inflow and outflow storage.

$$S = k[xI^m + (1-x)Q^m]$$

S = Storage

I = Inflow discharge

Q = Outflow discharge

m = Constant, having value 1 for natural channel.

Hence, for natural channel, $S = k[xI + (1-x)Q]$

k is a coefficient known as storage time coefficient.

- It has a dimension of time.
- It is approximately the time taken by flood wave of cross the channel reach.
- It depends on the length of channel reach and channel roughness characteristics. x is weighing factor, its values lies between (0 – 0.5) for ΔT, we have

$$(S_2 - S_1) = k[x(I_2 - I_1) + (1-x)(Q_2 - Q_1)]$$

from Muskingham equation,

$$(S_2 - S_1) = \left(\frac{I_1 + I_2}{2}\right)\Delta t - \left(\frac{Q_1 + Q_2}{2}\right)\Delta t$$

from continuity equation, eliminating S_1 and S_2 from above two equations and rearranging we have

$$Q_2 = C_0 I_2 + C_1 I_1 + C_2 Q_1$$

$$C_0 = \frac{-kx + 0.5\Delta t}{k - kx + 0.5\Delta t}$$

$$C_1 = \frac{kx + 0.5\Delta t}{k - kx + 0.5\Delta t}$$

$$C_2 = \frac{k - kx - 0.5\Delta t}{k - kx + 0.5\Delta t}$$

where, $C_0 + C_1 + C_2 = 1$

$$2kx < \Delta t < k$$

$$Q_n = C_0 I_n + C_1 I_{n-1} + C_2 Q_{n-1}$$

4.3 **Analysis of 30 hr flood data at a point on a river yields mean flood of 1200 m^3/sec with standard deviation of 650 m^3/sec, for what discharge would you design the structure at this point to provide 95% assurance that the structure would not fail in next 50 years. Use Gumbel's method.**

[10 Marks]

Sol:

Given: n = 30 years, $\bar{x}$ (mean flood) = 1200 m^3/sec, Standard deviation (σ_{n-1}) = 650 m^3/sec

95% assurance that structure would not fail in next 50 years.

$\Rightarrow$ $Re = 95\%$

$$\left(1 - \frac{1}{T}\right)^{50} = 0.95$$

$$\left(1-\frac{1}{T}\right) = 0.95^{1/50}$$

$$\frac{1}{T} = 1-0.998974$$

$$T = 975.348 \text{ years}$$

Assuming, infinite sample size

$$\bar{y}_n = 0.577$$

$$S_n = 1.2825$$

$$y_T = -\ln\ln\left(\frac{T}{T-1}\right) = -\ln\ln\left(\frac{975.348}{975.348-1}\right) = 6.882$$

$$k = \frac{y_T - \bar{y}_n}{S_n} = \frac{6.882-0.577}{1.2825} = 4.916$$

Design flood,

$\therefore$

$$x_T = \bar{x} + k\sigma_{n-1} = 1200 + (4.916 \times 650)$$

$$= 4395.4 \text{ m}^3\text{/sec}$$

4.4 **Data covering a period of 92 years for the river Ganga at Raiwala yielded the mean and standard derivation of the annual flood series as 6437 and 2951 m^3/s respectively. Using Gumbel's method estimate the flood discharge with a return period of 500 years. What are the (a) 95% and (b) 80% confidence limits for this estimate.**

Sol:

For $N = 92$ years, $\bar{y}_n = 0.5589$ and $S_n = 1.2020$

$$Y_{500} = -[\ln \ln (500/499)] = 6.21361$$

$$K_{500} = \frac{6.21361-0.5589}{1.2020} = 4.7044$$

$$x_{500} = 6437 + (4.7044 \times 2951) = 20320 \text{ m}^3\text{/s}$$

From eq.

$$b = \sqrt{1+1.3k_T+1.1k_T^2}$$

$$b = \sqrt{1+1.3(4.7044)+1.1(4.7044)^2} = 5.61$$

$$S_e \text{ probable error} = 5.61 \times \frac{2951}{\sqrt{92}} = 1726$$

(a) For 95% confidence probability $f(c) = 1.96$

$$x_{1/2} = 20320 \pm (1.96 \times 1726)\, x_1$$

$$= 23703 \text{ m}^3\text{/s and } x_2$$

$$= 16937 \text{ m}^3\text{/s}$$

Thus estimated discharge of 20320 m^3/s has a 95% probability of lying between 23700 and 16940 m^3/s.

(b) For 80% confidence probability, $f(c) = 1.282$

$$x_{1/2} = 20320 \pm (1.282 \times 1726) \text{ giving } x_1 = 22533 \text{ m}^3\text{/s and } x_2 = 18107 \text{ m}^3\text{/s}$$

Thus estimated discharge of 20320 m^3/s has a 80% probability of lying between 22530 and 18110 m^3/s.

4.5 On the basis of isopluvial maps of 50 years. 24 hr maximum rainfall at Bangalore is found to be 16 cm. Determine the probability of a 24hr rainfall of magnitude 16 cm occurring at Bangalore.

(i) Once in ten successive years.

(ii) Twice in ten successive years.

(iii) At least once in ten successive years.

[15 Marks]

Sol:

Given: Return period (T) = Probability of occurrence,

$$\therefore \quad P = \frac{1}{T} = \frac{1}{50} = 0.02$$

Probability of non occurrence, $q = 1 - p = 1 - 0.02 = 0.98$

(i) Once in ten successive years,

$$n = 10 \text{ years}$$

$$p = n_{C_r} p^r . q^{n-r}$$

$$= 10_{C_1} (0.02)^1 \times (0.98)^{10-1} = 10 \times 0.02 \times (0.98)^9 = 0.166 \simeq 16.6\%$$

(ii) Twice in ten successive years, n= 10 years.

$$p = 10_{C_2} (0.02)^2 \times (0.98)^{10-2} = 0.0153 \simeq 1.53\%$$

(iii) At least once in ten successive years.

$$\text{Risk (R)} = 1 - q^n = 1 - (1-p)^n = 1 - (1 - 0.02)^{10} = 18.3\%$$

4.6 The records of annual rainfall at a raingauge station, expressed in mm, in chronological sequence from 1981 to 1992 are given below:

360, 285, 557, 815, 273, 229, 707, 478, 309, 176, 285, 651. Using Hazen's method estimate the maximum annual rainfall which has a recurrence interval of 10 years.

(10 Marks)

Sol:

The probability of occurrence or exceedance (p) of a rainfall in a year is calculated empirically as below:

By Hazen's method,

$$p = \frac{m - 0.5}{N}$$

where,

p = Probability of occurrence or exceedance of an event with rank 'm' in a year.

N = Total number of data available

m = Rank of an event obtained by arranging all the available data's in decreasing order and ranking them from 1 (highest event) to N (lowest event)

Rainfall	815	707	651	557	478	360	309	285	285	273	229	176
Rank	1	2	3	4	5	6	6	–	9	10	11	12
Probability $P = \frac{m-0.5}{N}$	0.041	0.125	0.208	0.291	0.375	0.458	0.541	–	0.70	0.791	0.875	0.958
Return period T = 1/p (years)	24	8	4.8	3.42	2.67	2.18	1.84	–	1.41	1.26	1.14	1.04

By interpolation maximum annual rainfall which has a recurrence interval of 10 years

$$= 707 + \frac{815 - 707}{24 - 8} \times (10 - 8) = 720.5 \text{ mm}$$

4.7 **Calculate the time duration for a cofferdam to remain in a river with an even chance of not being overtopped if it is designed to be secured against a flood having a recurrence interval of 10 years.**

(10 Marks)

Sol:

Given, Even chance = 50% survival

$\therefore$ Risk = 50% = 0.5

Recurrence interval (T) = 10 years.

$$\therefore \quad p = \frac{1}{T} = \frac{1}{10} = 0.1$$

$$\text{Risk(R)} = 1-(1-p)^n$$

$$0.5 = 1-(1-0.1)^n$$

$$0.5 = 0.9^n$$

Taking log both sides, $\log 0.5 = n \log 0.9$

$$n = \frac{\log 0.5}{\log 0.9} = 6.58$$

Coffer dam will here 50% survival chance (even chance) during next 6.58 years.

4.8 **What information is needed to rout a flood through a reservoir? Discuss the salient features of outflow hydrographs as it differs from inflow hydrograph.**

[10 Marks]

Sol:

For flood routing through reservoir, the following data, graphs, tables have to known.

1. Storage volume versus elevation for reservoir.
2. Water surface elevation versus outflow avoid hence storage Vs. outflow discharge.
3. Inflow hydrograph
4. Initial values of S, I and Q at $t = 0$;

Inflow and outflow hydrograph

- Inflow hydrograph has the time base which is enlarged in outflow hydrograph.
- Peak of flood attenuated due to storage effect, peak of inflow hydrograph is greater than that of outflow hydrograph.
- The peak of outflow hydrograph occurs after to that of inflow hydrograph.

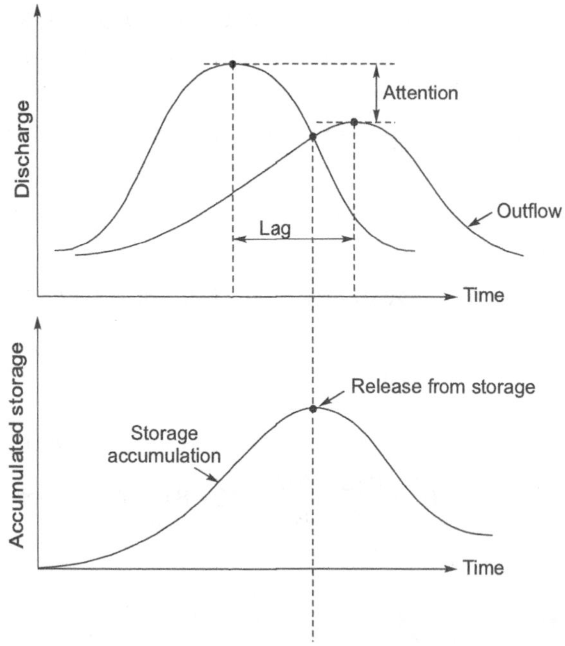

- When the inflow and outflow hydrographs intersect each other, that point is the point of maximum storage.
- Before the point of maximum storage, ordinate of inflow hydrograph > Ordinate of outflow hydrograph
- Storage increases before point of inflection.
- Storage decreases after point of inflection.

4.9 **Write a brief note on kinematic flood routing.**

[10 Marks]

Sol:

Kinematic flood routing: Kinematic flood routing is a generic term which identifies broad class of numerical methods used to route flood waves through a channel or water way. Flood waves which behaved in kinematic fashion are observed in various rivers. Ex:

(i) Lag and route method.

(ii) Muskingham flood routing or reservoir routing.

In kinematic wave modelling, following equations are used:

- Continuity equation
- Momentum equation
- Saint venant equation

Kinematic wave method assumes that weight or gravity force of flowing water is simply balanced by resistive forces of bed friction. This method can be used to derive flow hydrographs which can be added to produce collector or channel hydrograph and eventually, as stream or channel hydrograph. This method is the combination of continuity equation and simplified form of St. Venant equations.

Level-2

4.10 **A storage type hydropower plant has an output of 7.5 MW when working under a net head of 25 m and with a plant efficiency of 85%. Inflow in the reservoir during a year is as given below:**

Month	Jan	Feb	March	April	May	June
Inflow, 10^6 m^3	105	111	93	84	75	90
Month	July	Aug	Sep	Oct	Nov	Dec
Inflow, 10^6 m^3	120	90	96	102	103	99

Find the minimum capacity of reservoir to satisfy the uniform quantity of water. Assume reservoir is full at beginning of January.

[20 marks]

Sol:

$$\text{Hydro power developed, } P = \rho QgH\eta$$

But, $P = 7.5$ MW $= 7.5 \times 10^3$ kW $= 7.5 \times 10^6$ W

$$\therefore \quad 10^3 \times 9.81 \times 0.85 \times Q \times 25 = 7.5 \times 10^6$$

$$\Rightarrow \quad Q = \frac{7.5 \times 10^6}{10^3 \times 9.81 \times 0.85 \times 25} = 35.98 \text{ m}^3/\text{s}$$

$$= (35.98 \times 3600 \times 24) \text{ m}^3/\text{day}$$

$$= 3.1087 \text{ Mm}^3/\text{day}$$

$$\therefore \quad \text{Uniform water demand} = 3.1087 \text{ Mm}^3/\text{day}$$

The computations of cumulative deficit are carried out in table below.

Month	Inflow Mm^3	Outflow Mm^3	Surplus Mm^3	Deficit Mm^3	Cumulative surplus	Cumulative deficit
Jan	105	$3.1087 \times 31 = 96.370$	8.6370	•		
Feb	111	$3.1087 \times 28 = 87.044$	23.956		32.586	
March	93	96.370	•	3.370		
April	84	93.261	•	9.261		
May	75	96.370	•	21.370		
June	90	93.261	•	3.261		37.262
July	120	96.370	23.630	•	23.630	
August	90	96.370	•	6.370		6.370
September	96	93.261	2.739	•		
October	102	96.370	5.630	•		
November	103	93.261	9.739			
December	99	96.370	2.630		20.738	

The minimum storage required to meet the demand is equal to the maximum of the cumulative deficit = 37.262 Mm^3.

4.11 **The inflow hydrograph for a river reach is given below. Route this flood to a downstream point of the river using Muskingham method of flood routing. Assume that the initial flood discharge at the downstream point is equal to 100 m^3/s. For the river reach, Muskingham coefficient (K) = 18 hours and weighing factor (x) = 0.3. Use a routing period of 12 hours.**

Time (hr)	0	12	24	36	48
Inflow (m^3/s)	100	750	780	470	270

[20 Marks]

Sol:

$$Q_{initial} = 100 \text{ m}^3/\text{s}$$
$$k = 18 \text{ hours}$$
$$x = 0.3$$
$$\Delta t = 12 \text{ hrs}$$

Time(hr)	0	12	24	36	48
Inflow(m^3/s)	100	750	780	470	270

Using Muskingham equation

$$C_0 = \frac{-kx + 0.5\,\Delta t}{k(1-x) + 0.5\,\Delta t} = \frac{-18 \times 0.3 + 0.5 \times 12}{18(1-0.3) + 0.5 \times 12} = 0.0322$$

$$C_1 = \frac{kx + 0.5\,\Delta t}{k(1-x) + 0.5\,\Delta t} = \frac{18 \times 0.3 + 0.5 \times 12}{18(1-0.3) + 0.5 \times 12} = 0.613$$

$$C_2 = \frac{k - kx - 0.5\,\Delta t}{k(1-x) + 0.5\,\Delta T} = \frac{18(1-0.3) - 0.5 \times 12}{18(1-0.3) + 0.5 \times 12} = 0.355$$

Now, $C_0 + C_1 + C_2$

$$= 0.0322 + 0.613 + 0.355 = 1$$

Initial flood discharge, Q = 100 m^3/sec.

The outflow ordinates are worked out in the table using the general equation.

$$Q_n = C_0 I_n + C_1 I_{n-1} + C_2 Q_{n-1}$$
$$Q_2 = C_0 I_2 + C_1 I_1 + C_2 Q_1$$

Time from start (h) (1)	Inflow (m^3/s) (2)	$C_0I_2 = 0.032I_2$ (3)	$C_1I_1 = 0.613I_1$ (4)	$C_2Q_1 = 0.355Q_1$ (5)	Q in cumecs (6)
0	100	–	–	–	100
12	750	$0.032 \times 750 = 24$	$0.613 \times 100 = 61.3$	$0.355 \times 100 = 35.5$	120.8
24	780	24.96	459.75	42.88	527.59
36	470	15.04	478.14	187.29	680.47
48	270	8.64	288.11	241.566	538.31

The above routing shows that the peak which occurred at t = 24 h at upstream point of river reach, now occurs at t = 36 hr at downstream point i.e., at a lag of 12 hr.

The peak discharge also reduced from 780 m^3/s to 680.47 m^3/sec.

4.12 Route the following flood hydrograph through a river reach for which K = 12.0 h and x = 0.20. At the start of the inflow food, the outflow discharge is 10 m^3/s.

Time (h)	0	6	12	18	24	30	36	42	48	54
Inflow (m^3/s)	10	20	50	60	55	45	35	27	20	15

[20 marks]

Sol:

Since K = 12 h and $2Kx = 2 \times 12 \times 0.2 = 4.8$ h, Δt should be such that 12 h > Δt > 4.8 h. In the present case Δt = 6 h is selected to suit the given inflow hydrograph ordinate interval.

Coefficient C_0, C_1 and C_2 are calculated as

$$C_0 = \frac{-Kx + 0.5\Delta t}{K - Kx + 0.5\Delta t} = \frac{-12 \times 0.20 + 0.5 \times 6}{12 - 12 \times 0.2 + 0.5 \times 6} = \frac{0.6}{12.6} = 0.048$$

$$C_1 = \frac{Kx + 0.5\Delta t}{K - Kx + 0.5\Delta t} = \frac{12 \times 0.2 + 0.5 \times 6}{12.6} = 0.429$$

$$C_2 = \frac{K - Kx - 0.5\Delta t}{K - Kx + 0.5\Delta t} = \frac{12 - 12 \times 0.2 - 0.5 \times 6}{12.6} = 0.523$$

For the first time interval, 0 to 6 h,

$$I_1 = 10.0 \qquad C_1 I_1 = 4.29$$
$$I_2 = 20.0 \qquad C_0 I_2 = 0.96$$
$$Q_1 = 10.0 \qquad C_2 Q_1 = 5.23$$

Using equation $Q_2 = C_0 I_2 + C_1 I_1 + C_2 Q_1 = 10.48$ m^3/s

For the next time step, 6 to 12 h, Q_1 = 10.48 m^3/s.

$\Delta t = 6$h

Time (h)	I(m³/s)	0.048 I_2	0.429 I_1	0.523 Q_1	Q(m³/s)
(1)	(2)	(3)	(4)	(5)	(6)
0	10				10.00
		0.96	4.29	5.23	
6	20				10.48
		2.40	8.58	5.48	
12	50				16.46
		2.88	21.45	8.61	
18	60				32.94
		2.64	25.74	17.23	
24	55				45.61
		2.16	23.60	23.85	
30	45				49.61
		1.68	19.30	25.95	
36	35				46.93
		1.30	15.02	24.55	
42	27				40.87
		0.96	11.58	21.38	
48	20				33.92
		0.72	8.58	17.74	
54	15				27.04

$$\text{Attenuation} = 60 - 49.61 = 10.39 \text{ m}^3/\text{sec}$$

$$\text{Peak lag} = 30 - 18 = 12 \text{ hrs}$$

4.13 A and B are two concentration points along a river in a catchment of area 3.5 km². The concentration point A is 1 km downstream of concentration point B. The base flow in the river reach AB is 10 m³/s. A 2-hour rainfall event occurs on this catchment having rainfall intensities of 4 cm/hr during the first hour and 2 cm/hr during the second hour. A 2 hr UH at concentration point B is given below.

Time (hr)	0	3	6	9	12	15	18	21
UH ordinate(m³/s)	0	25	50	160	110	45	8	0

Compute the ordinates of the flood hydrograph at concentration point A resulting from the 2-hr rainfall event. (Assume ϕ-index = 2 cm/hr. Muskingum routing constant K = 6 hours and x = 0.2)

[20 Marks]

Solution:

$$A = 3.5 \text{ km}^2$$

$$\text{Base flow} = 10 \text{ m}^3/\text{sec}$$

$$\text{Rainfall intensity} = 4 \text{ cm/hr in first hour}$$

$$= 2 \text{ cm/h second hour}$$

$$C_0 = \frac{-Kx + 0.5\Delta t}{K - Kx + 0.5\Delta t} = \frac{-6 \times 0.2 + 0.5 \times 3}{6 - 6 \times 0.2 + 0.5 \times 3} = 0.0476$$

$$C_1 = \frac{Kx + 0.5\Delta t}{K - Kx + 0.5\Delta t} = \frac{6 \times 0.2 + 0.5 \times 3}{63} = 0.428$$

$$C_2 = \frac{K - Kx - 0.5\Delta t}{K - Kx + 0.5\,\Delta t} = \frac{6 - 6 \times 0.2 - 0.5 \times 3}{63} = 0.524$$

$$\text{Excess rainfall} = 4 + 4 - 4 = 4 \text{ cm}$$

Time (i)	UH ordinate (ii)	Ordinate ×4 cm (iii)		C_0I_2 0.048I	C_1I_1 0.431I_1	C_2Q_1 0.524	Q	Remarks
0	0	0	0	4.8	0	5.24	10	
3	25	100	100	9.6	43.0	5.26	10.04	
6	50	200	200	30.72	86.0	27.56	52.6	
9	160	640	640	21.12	275.2	61.16	116.72	
12	110	440	440	8.64	189.2	155.27	296.32	
15	45	180	180	1.54	77.4	103.67	197.84	
18	8	32	32	0	13.76	41.36	78.94	
21	0	0	0	—	—	7.21	13.76	

4.14 Route the following flood through a reach for which k = 22 hr and x = 0.25. Plot the inflow and outflow hydrograph and determine the peak leg and attenuation. At t = 0, the outflow discharge is 40 m^3/sec.

Time (hr)	0	12	24	36	48	60	72	84	96	108	120	132	144
In flow (m^3/s)	40	65	165	250	240	205	170	130	115	85	70	60	54

[20 Marks]

Sol:

Given: $k = 22$ hr, $x = 0.25$

Muskingham constant

$$\Delta t = 12 \text{ hr}$$

$$2kx = 2 \times 22 \times 0.25 = 11 \text{ hr}$$

$$k = 22 \text{ hr}$$

$$2kx < \Delta t < k \quad \text{(OK)}$$

$$C_0 = \frac{-kx + 0.5\Delta t}{k - kx + 0.5\Delta t} = \frac{-22 \times 0.25 + 0.5 \times 12}{22 - 22 \times 0.25 + 0.5 \times 12} = 0.022$$

$$C_1 = \frac{kx + 0.5\Delta t}{k - kx + 0.5\Delta t} = \frac{22 \times 0.25 + 0.5 \times 12}{22 - 22 \times 0.25 + 0.5 \times 12} = 0.511$$

where, $C_0 + C_1 + C_2 = 1$

$$C_2 = 1 - (C_0 + C_1) = 1 - (0.022 + 0.511) = 0.467$$

Time (hr)	Time (hr)	C_0I_2	C_1I_1	C_2Q_1	Outflow $Q_2 = C_0I_1 + C_1I_1 + C_2Q_1$ (m^3/sec)
0	40	-	-	-	40 ($Q_1 = I_1$)
12	65	0.022 × 65	0.511 × 40	0.467 × 40	40.55
24	165	0.022 × 165	0.511 × 65	0.467× 40.55	55.78
36	250	0.022 × 250	0.511 × 165	0.467 × 55.78	115.86
48	240	0.022 × 240	0.511 × 250	0.467 × 115.86	187.14
60	205	0.022 × 205	0.511 × 240	0.467 × 187.14	214.54 ⟶ Peak of
72	170	0.022 × 170	0.511 × 205	0.467 × 214.54	208.68 outflow
84	130	0.022 × 130	0.511 × 170	0.467 × 208.68	187.18 hydrograph
96	115	0.022 × 115	0.511 × 130	0.467 × 187.18	156.37
108	85	0.022 × 85	0.511 × 115	0.467 × 156.37	133.66
120	70	0.022 × 70	0.511 × 85	0.467 × ??	107.39
132	60	0.022 × 60	0.511 × 70	0.467 × 107.39	87.24
144	54	0.022 × 54	0.511 × 60	0.467 × 87.24	72.58

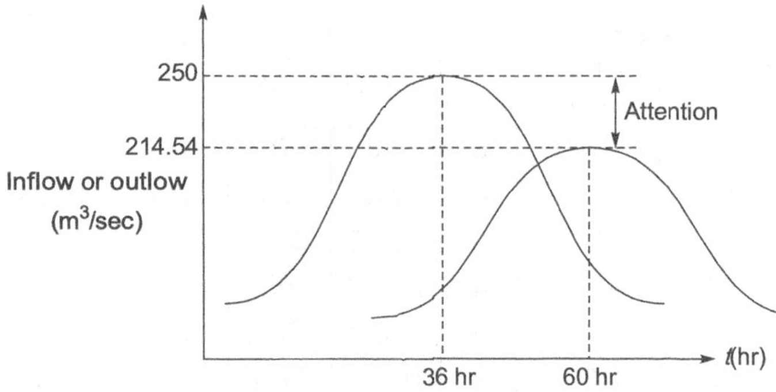

Attenuation = (Peak of inflow hydrograph) – (Peak of outflow hydrograph)
= 250 – 214.54 = 35.46 m^3/sec

(Peak lag) = Time lag between occurrence of peak of outflow – inflow hydrograph
= 60 – 36 = 24 hr.

4.15 **(a)** **Annual maximum flood data in a river at a station have been processed to estimate the maximum flood for different return periods using Gumbel's method. If the estimated maximum flood for return periods 100 and 50 years are 450 m^3/sec and 400 m^3/sec, respectively estimate the flood discharge for return period of 500 years.**

(b) **Annual flood data of a river for 30 years yielded mean annual discharge of 30000 m^3/sec and a standard deviation of 15000 m^3/sec. For a proposed bridge on this river, it is decided to have an acceptable risk of 10% and service life of 50 years.**

1. **Estimate the flood discharge for design of bridge.**
2. **If the actual flood value adopted in design is 118000 m^3/sec, what is the safety margin for maximum flood discharge.**

Given that, reduced mean and standard deviation for Gumbel distribution for 30 years are 0.5380 and 1.119 respectively.

[10+10 =20 Marks]

Sol:

(a) Given:

As per Gumbel's method,

Flood discharge	Return period
450 m^3/sec	100 years
400 m^3/sec	50 years

$$x_T = \bar{x} + k\sigma_{n-1}$$

$$x_{100} = \bar{x} + k_{100}\sigma_{n-1}$$

$$x_{50} = \bar{x} + k_{50}\sigma_{n-1}$$

$$k = \frac{y_T - \bar{y}_\eta}{S_\eta} = \frac{y_T}{S_\eta} - \frac{\bar{y}_\eta}{S_\eta}\text{(constant)}$$

$$(x_{100} - x_{50}) = (k_{100} - k_{50})\,\sigma_{n-1} = (y_{100} - y_{50})\frac{\sigma_{n-1}}{S_\eta}$$

$$y_{100} = -\ln\ln\left(\frac{100}{100-1}\right) = 4.6$$

$$y_{50} = -\ln\ln\left(\frac{50}{50-1}\right) = 3.9$$

$$(450-400) = (4.6-3.9)\frac{\sigma_{n-1}}{S_\eta}$$

$$\frac{\sigma_{n-1}}{S_\eta} = 71.42$$

$$x_{100} = \bar{x} + k_{500}\sigma_{n-1}$$

$$(x_{500} - x_{100}) = (y_{500} - y_{100})\frac{\sigma_{n-1}}{S_\eta}$$

$$y_{500} = -\ln\ln\left(\frac{500}{500-1}\right) = 6.213$$

$$(x_{500} - 450) = (6.213 - 4.6) \times 71.42 = 565.2 \text{ m}^3/\text{sec}$$

(b) Given, x = 30000 m3/sec, $\bar{y}_\eta$ = 0.5380, σ_{n-1} = 15000 m^3/sec, S_η = 1.119

Risk = 10%

$$R = 0.1 = 1 - \left(1 - \frac{1}{T}\right)^n$$

$$0.1 = 1 - \left(1 - \frac{1}{T}\right)^{50} = \left(1 - \frac{1}{T}\right)^{50} = 0.9$$

T = 475 yres.

$\therefore$
$$y_T = -\ln\ln\left(\frac{T}{T-1}\right)$$

$$y_T = -\ln\ln\left(\frac{475}{475-1}\right) = 6.16$$

$$k = \frac{y_T - \bar{y}_n}{S_n} = \frac{6.16 - 0.538}{1.119} = 5.024$$

(i) Now, flood design discharge

$$x_T = \bar{x} + k\sigma_{n-1} = 30000 + 5.024 \times 15000 = 105360 \text{ m}^3/\text{sec}$$

(ii) Actual value taken for design = 1180000 m^3/sec

Estimated design value for bridge = 105360 m^3/sec

Now, Safety margin = Design value – estimated value

= 118000 – 105360 = 12640 m^3/sec

4.16 The record of annual rainfall at station A covering a period of 22 years is given below:

(a) Estimate the annual rainfall with return periods of 10 years and 5 years.

(b) What would be the probability of an annual rainfall of magnitude equal to or exceeding 100 cm occurring at station A?

(c) What is the 75% dependable annual rainfall at station A?

Years	Annual rainfall (cm)	Years	Annual rainfall (cm)
1960	130	1971	90
1961	84	1972	102
1962	76	1973	108
1963	89	1974	60
1964	112	1975	75
1965	96	1976	120
1966	80	1977	160
1967	125	1978	85
1968	143	1979	106
1969	89	1980	83
1970	78	1981	95

Use Weibull's method.

[20 marks]

Sol:

The data are arranged in descending order and the rank numbers are assigned to the recorded events.

m	Annual rainfall (cm)	Probability $P = \frac{m}{N+1}$	Return period ($T = 1/P$ Year)
1	160	0.043	23.256
2	143	0.087	11.5
3	130	0.130	7.69
4	125	0.174	5.75
5	120	0.217	4.6
6	112	0.261	3.831
7	108	0.304	3.289
8	106	0.348	2.874
9	102	0.391	2.558
10	96	0.435	2.3
11	95	0.478	2.092
12	90	0.522	1.916
13	89	0.565	1.77
14	89	0.609	1.642
15	85	0.652	1.534
16	84	0.696	1.437
17	83	0.739	1.353
18	80	0.783	1.277
19	78	0.826	1.211
20	76	0.87	1.150
21	75	0.913	1.095
22	60	0.957	1.045

(a) For $T = 10$ years

Interpolating between values between $T = 11.5$ years and $T = 7.69$ years

143	11.5
R_{10}	10
130	7.69

$$R_{10} = 130 + \left(\frac{143-130}{11.5-7.69}\right)(10-7.69)$$

$$\Rightarrow \quad R_{10} = 137.91 \text{ cm}$$

Similarly for $T = 5$ years, interpolating values between $T = 5.75$ years and $T = 4.6$ years

125	5.75
R_5	5
120	4.6

$$R_5 = 120 + \left(\frac{125-120}{5.75-4.6}\right)(5-4.6)$$

$$\Rightarrow \quad R_5 = 121.74 \text{ cm}$$

(b) For 100 cm rainfall

Rainfall (cm)	P
102	0.391
100	P_{100}
96	0.435

$$P_{100} = 0.435 + \left(\frac{0.435 - 0.391}{96 - 102}\right)(100 - 96)$$

$\Rightarrow$ $P_{100} = 0.406$

(c) 75% dependable rainfall refers to probability of 0.75.

So, interpolating between proper limits,

Rainfall	P
83	0.739
$R_{75\%}$	0.75
80	0.783

$$R_{75\%} = 83 - \left(\frac{83 - 80}{0.783 - 0.739}\right)(0.75 - 0.739)$$

$\Rightarrow$ $R_{75\%} = 82.25 \text{ cm}$

■■■■

03 Irrigation Engineering

1. Water Requirement of Crops

Level-1

1.1 **Enumerate advantages and disadvantages of drip irrigation.**

[8 marks]

Sol:

Advantages of drip irrigation:

- Water requirement is very less (since water is delivered directly to the plant).
- Evaporation loss and wind loss are almost negligible.
- There is least wetting of soil surface due to direct water supply to the plants and hence loss of water is minimum.
- Suitable for all types of soils, especially for coarse textured soils.
- Highest rate of vegetative growth.
- Land levelling is not required.
- Roots stay within the moist zone.
- Reduction of deleterious effects of salts.
- Low labour requirement, as most drip systems are permanent setups.
- Field operations are easier to manage.

Disadvantages of drip irrigation:

- Requires regular flushing (to clear off the dirt collected near the ends of the drip lines) and supervision.
- High skill is required in the design, installation, operation and maintenance.
- Plastic drip-lines and sub-mains may be attacked by rodents and small animals.

1.2 **Find the time required to cover an area of 0.1 hectares when a tube well is discharging at the rate of 0.03 cumecs for irrigating Rabi crops. Average depth of flow is expected to be 7.5 cm, average infiltration rate for the soil may be taken as 5 cm/hr. Also, find the maximum area that can be irrigated by the available discharge of 0.03 cumecs.**

[5 Marks]

Sol:

q = 0.03 cumecs = 0.0108 hectare-m/hr, y = 7.5 cm = 0.075 m, I = 5 cm/hr = 0.05 m/hr, A = 0.1 hectare

$\because$ Time required to irrigate the area = $t = 2.303\frac{y}{I}\log_{10}\left(\frac{Q}{Q-IA}\right)$

$$= 2.303\times\frac{0.075}{0.05}\log_{10}\left(\frac{0.0108}{0.0108-0.05\times 0.1}\right)\text{hr} = 0.9327 \text{ hr} \simeq 56 \text{ min}$$

Maximum area that can be irrigated = $A_{max} = \frac{Q}{I} = \frac{0.0108}{0.05} = 0.216$ hectare.

1.3 An irrigation canal has gross commanded area of 80000 hectares out of which 85% is culturable irrigable. The intensity of irrigation for Kharif season is 30% and for Rabi season 60%. Find the discharge required at the head of the canal if the duty at its head is 800 hectares/cumecs for Kharif season and 1700 hectares cumec for Rabi season.

[5 Marks]

Sol:

$$\text{Culturable irrigation area} = 80000 \times 0.85 = 68000 \text{ ha}$$
$$\text{Area under Kharif season} = 68000 \times 0.30 = 20400 \text{ ha}$$
$$\text{Area under Rabi season} = 68000 \times 0.60 = 40800 \text{ ha}$$

Water required at head of the canal to irrigate the land under:

$$\text{Kharif season} = \frac{20400}{800} = 25.5 \text{ cumecs}$$
$$\text{Rabi season} = \frac{40800}{1700} = 24 \text{ cumecs}$$

$\because$ The water requirement during Kharif is more than that in Rabi. Hence the canal must be designed to carry a discharge of 25.5 cumecs. Actually the canal must be designed to carry the Peak discharge and not the average discharge calculated based on the average duty. Peak requirements may be taken to be equal to 20-25% higher than than avg. requirements.

1.4 A water course has a CCA of 1200 ha. The intensity of irrigation for crop A is 40% and for B is 35% both the crops being Rabi crops, crop A has kor period of 20 days and crop B has kor period 15 days. Calculate the discharge of the water course if the depth for crop A is 10 cm and for B it is 16 cm.

[10 Marks]

Sol:

(a) For crop A:

$$\text{Area of irrigation} = 1200 \times 0.40 = 480 \text{ ha}$$
$$\text{Kor period, } b = 20 \text{ days}$$
$$\text{Kor depth, } d = 10 \text{ cm} = 0.1 \text{ m}$$
$$\text{Duty} = \frac{8.64 \times b}{d} = \frac{8.64 \times 20}{0.1} = 1728 \text{ ha/cumec}$$

So,
$$\text{Discharge required} = \frac{\text{Area under irrigation}}{\text{Duty(D)}} = \frac{480}{1728}$$
$$= 0.278 \text{ cumec}$$

(b) For crop B:

$$\text{Area of irrigation} = 1200 \times 0.35 = 420 \text{ ha}$$
$$\text{Kor period, } b = 15 \text{ days}$$
$$\text{Kor depth, } d = 16 \text{ cm} = 0.16 \text{ m}$$
$$\text{Duty} = \frac{8.64 \times b}{d} = \frac{8.64 \times 15}{0.16} = 810 \text{ ha/cumec}$$

So,
$$\text{Discharge required} = \frac{420}{810} = 0.518 \text{cumec}$$

So, design discharge of water course $= 0.278 + 0.518$

$$= 0.796 \approx 0.8 \text{ cumec}$$

1.5 Table below gives data about crop, their duty and area under each crop, commanded by a canal taking off from a storage tank. Take a time factor for the canal to be $\frac{13}{20}$, calculate the discharge required to the head of the canal. If the capacity factor is 0.8. Determine the design discharge.

Crops	Base period (days)	Area (ha)	Duty at the head of the canal (ha/cumec)
Sugar-cane	320	850	580
Overlap for sugar-cane in hot weather	90	120	580
Wheat (Rabi)	120	600	1600
Bajri (Monsoon)	120	500	2000
Vegetable (Hot weather)	120	360	600

[12 Marks]

Sol:

Discharge required for crops:

$$\text{For sugar-cane} = \frac{850}{580} = 1.465 \text{ cumecs}$$

$$\text{For overlap sugar-cane} = \frac{120}{580} = 0.207 \text{ cumecs}$$

$$\text{For wheat} = \frac{600}{1600} = 0.375 \text{ cumecs}$$

$$\text{For Bajri} = \frac{500}{2000} = 0.25 \text{ cumecs}$$

$$\text{For Vegetables} = \frac{360}{600} = 0.60 \text{ cumecs}$$

Since sugarcane has a base period of 320 days, it will require water in Rabi, Mansoon and Hot weather.

$\therefore$ Discharge required in Rabi = 1.465 + 0.375 = 1.84 cumecs

Discharge required in Monsoon = 1.465 + 0.25 = 1.715 cumecs.

Discharge required in Hot weather = 1.465 + 0.207 + 0.6 = 2.272 cumecs.

Thus, the maximum demand of 2.272 cumecs is in the hot weather.

$\therefore$ Full supply discharge at the head of canal will be

$$\frac{\text{Max. demand}}{TF} = \frac{2.272}{(13/20)} = 3.50 \text{ cumecs}$$

$$\therefore \quad \text{Design discharge} = \frac{\text{Full supply discharge}}{\text{Capacity factor}} = \frac{3.50}{0.8} = 4.375 \text{ cumecs}$$

1.6 The base period, IOI, duty of various crops under a canal system are given in table below. Find the reservoir capacity if the canal losses are 20% and reservoir losses are 12%.

Crops	Base period (days)	Duty at the field (ha/cumec)	Area under the crop (ha)
Wheat	120	1800	4800
Sugar-cane	360	800	5600
Cotton	200	1400	2400
Rice	120	900	3200
Vegetable	120	700	1400

[15 Marks]

Sol:

(i) Wheat:

$$\text{Discharge required} = \frac{4800}{1800} \text{ cumecs} = 2.67 \text{ cumecs}$$

$\therefore$ Volume of water required = $2.67 \times 120 = 320.4$ cumec-days

(ii) sugar-cane:

$$\text{Discharge required} = \frac{5600}{800} \text{ cumecs} = 7 \text{ cumecs}$$

$\therefore$ Volume of water required = $7 \times 360 = 2520$ cumec-days

(iii) Cotton:

$$\text{Discharge required} = \frac{2400}{1400} = 1.71 \text{ cumecs}$$

$\therefore$ Volume of water required = $1.71 \times 200 = 342$ cumec-days

(iv) Rice:

$$\text{Discharge required} = \frac{3200}{900} = 3.55 \text{ cumecs}$$

$\therefore$ Volume of water required = $3.55 \times 120 = 426$ cumec-days

(v) Vegetables:

$$\text{Discharge required} = \frac{1400}{700} = 2 \text{ cumecs}$$

$\therefore$ Volume of water required = $2 \times 120 = 240$ cumec-days

$\therefore$ Total volume of water required on field for all crops

$= 320.4 + 2520 + 342 + 426 + 240 = 3848.4$ cumec-days.

$\therefore$ Total volume of water required on field per day

$= 3848.4 \times 86400 \times 10^{-4}$ ha-m.

$= 33250.176$ ha-m.

Since losses in the canal system are 20% the volume of water required at the head of the canal

$$= 33250.176 \times \frac{100}{80} = 41562.72 \text{ ha-m}$$

Allowing 12% reservoir loss, storage capacity of reservoir

$$= \frac{41562.72}{0.88} = 47230.36 \text{ ha-m}$$

1.7 Using the table given below with details for a certain crop, determine:

(i) Consumptive use

(ii) Consumptive irrigation requirement

(iii) Field irrigation requirement, if water application efficiency is 0.7.

The latitude of the place is 30°N. Use Blaney-cradle equation and a crop factor $k = 0.75$.

Month	Monthly temp. (Avg.)°C	Monthly % of day time hrs. of the year	Useful Rainfall (cm)
November	19.0	7.19	-
December	16.0	7.15	1.2
January	12.5	7.30	0.8
February	13.0	7.03	-

[8 Marks]

Sol:

Using Blarey -Criddle equation

$$C_u = k.\Sigma f$$

where, $f = \frac{p}{40}[1.8t + 32]$

In tabular form:

Month	Temp. (°C)	Percentage	f
November	19.0	7.19	11.9
December	16.0	7.15	10.9
January	12.5	7.30	9.9
February	13.0	7.03	9.7
			$\Sigma = 42.4$

$\therefore$ $C_u = k.\Sigma f = 0.75 \times 42.4 = 31.8$ cm

Efficiency rainfall = 1.2 + 0.8 = 2.0 cm

$$\text{CIR} = C_u - R_e = 31.8 - 2.0 = 29.8 \text{ cm}$$

NIR = CIR ($\because$ on water is used for deep percolation)

$$\text{FIR} = \frac{NIR}{\eta_a} = \frac{29.8}{0.7} = 42.57 \text{ cm}$$

1.8 **A certain crop is grown in an area of 3000 ha, which is fed by a canal system. The data pertaining to irrigation are: Field capacity of soil = 26%, optimum moisture = 12%, PWP = 10%, efficiency depth of RZ = 80 cm, Apparent relative density of soil = 1.4. If the frequency of irrigation is 10 days and the overall irrigation efficiency is 22%. Find:**

(i) Daily consumptive uses Cu

(ii) Water discharge in m³/s required in the canal feeding the area.

[10 Marks]

Sol:

Flow diagram:

Moter discharge in canal ⟶ (Total irrigation required) × Area ⟶ Given

↓

$\text{NIR}/\eta_{\text{irrigation}}$ → Given

↓

Daily water consumed by plant

↓

$\frac{\gamma_d \times d}{\gamma_\omega}[\text{FC} - \text{OMC}]$ in 10 days

Depth of water used by plants for growth, which is supplemented by irrigation after every 10 days.

$$= \frac{\gamma_d}{\gamma_\omega} \times d \times [FC - OMC]$$

$$= 1.4 \times 0.8 \times [0.26 - 0.12]$$

$$= 0.1568 \text{ m} = 15.68 \text{ cm}$$

(i) Daily water consumed by plants, i.e. daily consumptive use

$$\frac{15.68}{10} \text{ cm} = 1.568 \text{ cm}$$

(ii) total irrigation water required i.e. including losses in field and conveyance

$$\frac{NIR}{\eta_{\text{irrigation}}} = \frac{1.568}{0.22} \text{ cm/ day} = 7.127 \text{ cm/d}$$

Field area = 3000 ha = 3000×10^4 m²

$$\therefore \text{Discharge required in irrigation canal} = \frac{7.127}{100} \times \frac{3000 \times 10^4}{24 \times 3600} = 24.75 \text{ m}^3/\text{s}$$

1.9 After how many days will you supply water to soil (clay 10 cm) in order to ensure efficient irrigation of the given crop, if.

(i) FC of soil = 27% (ii) PWP = 14%

(iii) Density of soil = 1.5 g/cc (iv) Efficient depth of Root Zone = 75 cm

(v) Daily consumptive use for the given crop = 11 mm

[10 Marks]

Sol:

$$\text{Available moisture} = FC - PWP = 27 - 14 = 13\%$$

Assuming, readily available moisture (RAM) be 80% of the available moisture.

$$\therefore \quad RAM = 13 \times 0.8 = 10.4\%$$

$\therefore$ Optimum moisture is 27 to 16.6%

Hence, when irrigation water is applied, moisture is raised from 16.6 to 27%.

$\therefore$ Depth of water stored in Root Zone, during each watering:

$$= \frac{\gamma_d \times d}{\gamma_\omega}[FC - \text{Optimum water content}]$$

$$= \frac{1.5 \times 0.75}{1}[0.27 - 0.166] = 0.117 \text{ m} = 11.7 \text{ cm}$$

Thus, depth of water available for evapo-transpiration = 11.7 cm

Daily consumptive use of water = 1.1 cm

$$\therefore \quad \text{Watering frequency} = \frac{11.7}{1.1} = 10.67 \approx 10 \text{ days (Lower side value)}$$

Hence, water should be applied after every 10 days.

1.10 (a) Elaborate the concept of frequency of Irrigation?

(b) List the factors affecting duty of water?

[6 + 6 = 12 Marks]

Sol:

(a)

- Frequency of irrigation is the time cycle during which moisture present in the soil decreases from field capacity to minimum allowable moisture content.
- As the moisture content of soil reaches to lower limit of readily available moisture, irrigation water needs to be applied and moistured content again raised to field capacity of the soil.

$$\text{Frequency of irrigation } (f_c) = \frac{\text{Depth of water to be applied}}{\text{Consumptive use}}$$

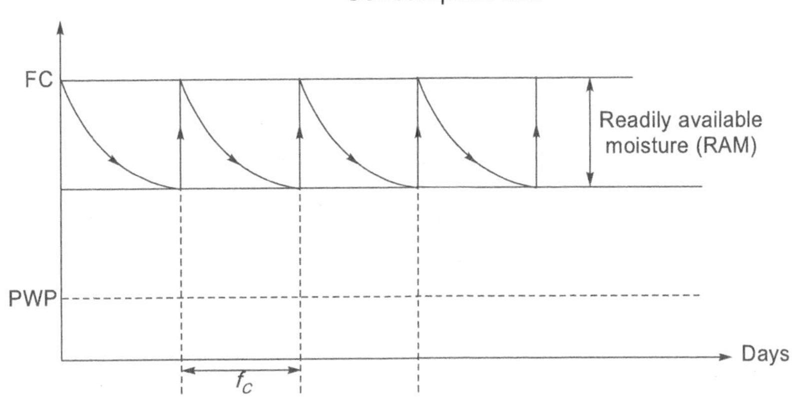

(b)

Factors affecting duty of water are:

(i) Type of crops - Crops, having large delta will have lesser duty of water.

(ii) Temperature - High temperature leads lesser duty.

(iii) Wind velocity - Higher, duty lesser

(iv) Humidity - Higher, duty higher

(v) Precipitation - More, duty more

(vi) Base period - More, duty lesser

(vii) Permeability of soil - More, duty lesser

(viii) Topography of land - A levelled land results in economical use of water leading to higher duty.

(ix) Method of irrigation - Duty of water is high for sprinkler and drop irrigation method. Over surface irrigation method.

(x) Frequency of cultivation - Frequent cultivation of land reduces the loss of moisture hence more duty.

1.11 **A Farmer has 250 ha of agriculture land. He does the cultivation, in Kharif season with the help of rain water only while during Rabi season he irrigates the field from his well. Water requirement for his cultivation is tabulated as follows:**

Months	Wheat (cm)	Potatoes (cm)	Mustard (cm)
October	12.4	-	18.4
November	8.6	4.2	6.2
December	4.4	8.1	8.1
January	10.2	9.1	10.9
February	6.4	4.3	6.2
March	-	12.8	14.6
Area of cultivation (ha)	100	80	40

Determined:

(i) Maximum yield from well

(ii) Horse power of pump required to lift water from 12 m deep well. If pump efficiency is 56%.

[12 Marks]

Sol:

Months	Water required (ha-m)			Monthly discharge (ha.m/day)
	Wheat	Potatoes	Mustard	
October	12.4	-	7.36	0.637
November	8.6	3.36	2.48	0.481
December	4.4	6.48	3.24	0.455
January	10.2	7.28	4.36	0.705
February	6.4	3.44	2.48	0.44
March	-	10.24	5.84	0.518

(i) Maximum yield = 0.705 ha-m/day

$$= \frac{0.705 \times 10^4}{3600 \times 24} = 0.0816\, m^3/s = 81.6\ l/s$$

(ii) Volume of water lifted/sec = $0.0816\ m^3/s$

Weight of water lifted/sec = $0.0816 \times 9.81 = 0.8$ kN/sec

Lift of pump = Static lift = 12 m

Work done by pump = $0.8 \times 12 = 9.6$ kN-m/sec

1HP = 7.5 watt [Metric horse power] (N-m/sec = watt)

$$\text{Output power required} = \frac{9.60}{0.735} = 13.06 \text{ HP}$$

$$\text{Input power of pump} = \frac{\text{Output}}{\eta} = \frac{13.06}{0.56} = 23.32 \text{ HP}$$

1.12 **The depth of penetrations along the length of a border strip at point 30 m apart were probed. Their observed values are 2.0, 1.9, 1.8, 1.6 and 1.5 m. Compute the water distribution efficiency.**

[10 Marks]

Sol:

Given data: The observed depths at five station are

$x_1 = 2.0$, $x_2 = 1.9$, $x_3 = 1.8$, $x_4 = 1.6$, $x_5 = 1.5$ m.

$$\eta_d = \left(1 - \frac{y}{d}\right) \times 100$$

$$d = \text{Mean depth} = \frac{\Sigma x_i}{n};$$

$$y\text{(Average deviation)} = \frac{\Sigma \text{ Absolute value of } (x_i - d)}{n}$$

$$\text{Mean depth, } d = \frac{2.0 + 1.9 + 1.8 + 1.6 + 1.5}{5} = \frac{8.8}{5} = 1.76 \text{ m}$$

Deviations from mean are:

(2 – 1.76), (1.9 – 1.76), (1.8 – 1.76), (1.6 – 1.76), (1.5 – 1.76) i.e. 0.24, 0.14, 0.04, –0.16 and –0.26.

The absolute values of these deviations from the mean are 0.24, 0.14, 0.04, 0.16 and 0.26.

Average value of these absolute values of deviations from the mean,

$$y = \frac{0.24 + 0.14 + 0.04 + 0.16 + 0.26}{5} = \frac{0.84}{5} = 0.168 \text{ m}$$

$$\text{The water distribution efficiency} = \left(1 - \frac{d}{D}\right) \times 100 = \left(1 - \frac{0.168}{1.76}\right) \times 100 = (1 - 0.095) \times 100$$

$\therefore$ Water distribution efficiency = 90.5%

1.13 **For 5000 ha of Rabi crop, an irrigation canal runs as follows:**

Months	Oct.	Nov.	Dec.	Jan.	Feb.
No. of running days	14	28	21	21	14
Discharge (m^3/s)	2	2.5	2.25	2.25	2

Calculate overall duty and depth of water used.

[6 Marks]

Sol:

Months	Running days	Discharge (m^3/s)	Volume = 86400 × Running day × discharge/10^4 (ha-m)
October	14	2	241.92
November	28	2.5	604.8
December	21	2.25	408.24
January	21	2.25	408.24
February	14	2	241.92

Total volume = 1905.12 ha-m

Depth, $\Delta = \dfrac{\text{Volume}}{\text{Area}} = \dfrac{1905.12}{5000} = 0.381\,\text{m}$

Base period, $B = 31 + 30 + 31 + 31 + 28 = 151$ days.

$$\text{Overall duty, } D = 8.64\frac{B}{\Delta} = \frac{8.64 \times 151}{0.381}$$

$$= 3424.25 \text{ ha/cumec.}$$

1.14 In a farm land irrigated by a system of pumps from wells, the area irrigated is 50 hectares. Water pumped from wells is conveyed through a canal to the farms. Efficiency of water conveyance is 85% and pumps work at 12 hours/day. Available moisture holding capacity of the soil is 20 cm per metre depth and the average root zone depth is 1 m. Water application efficiency is 80%. Irrigation is started when moisture extraction level of 50% of available moisture is reached. Peak rate of moisture use by plants is 5 mm. Calculate the irrigation period in days and total pumping capacity required in litres/minute.

[10 Marks]

Sol:

Given data,

Area to be irrigated = 50×10^4 m^2

Water conveyance efficiency, $\eta_c = 0.85$

Pumps working hours = 12 hours/day

Moisture holding capacity of soil = 20 cm/m depth

Average root zone depth = 1 m

Water application efficiency, $\eta_a = 0.8$

Moisture content extraction = 50% of available moisture content

Peak rate of moisture use = 5 mm = 0.5 cm

∴ Moisture holding capacity of soil

= 20 cm per m depth × 1 m depth of root zone = 20 cm

Moisture already available at the start of irrigation

$$= \frac{50 \times 20}{100} = 10 \text{ cm}$$

∴ Additional water required

$$= 20 - 10 = 10 \text{ cm}$$

∴ Irrigation period

$$= \frac{10}{0.5} \text{days} = 20 \text{ days}$$

Now, water application efficiency = 80%

Water required through irrigation = 10 cm

∴ Pumping capacity required

$$= \frac{10}{0.8} = 12.5 \text{ cm}$$

∴ Required pumping capacity

$$= \frac{12.5 \times 10^{-2} \times 50 \times 10^4 \times 10^3}{20 \times 24 \times 60} = 2170.14 \text{ litres/min}$$

1.15 An area of 1 hectare is irrigated through a stream of 0.03 m^3/s. Depth of the root zone is 1 m and available moisture holding capacity is 16 cm/m. Irrigation water is supplied when 50% of the available moisture is depleted. Water application efficiency is 60%. Determine the storage efficiency.

[8 Marks]

Sol:

Insufficient data, as time for irrigation is missing and run off losses.

Assuming it to be 8 hrs and runoff losses = 15%

Then, water applied

$$= 0.6 \times 0.03 \times 8 \times 60 \times 60 = 518.4 \text{ m}^3$$

Let runoff losses

$$= 0.15 \times 518.4 = 75.6 \text{ m}^3$$

⇒ Water stored

$$= 518.4 - 75.6 = 442.8 \text{ m}^3$$

Now, moisture holding capacity

$$= 16 \times 1 = 16 \text{ cm}$$

Additional water required

$$= 8 \text{ cm } (50\%) = \frac{8}{100} \times 10^4 = 800 \text{m}^3$$

$$\Rightarrow \qquad \eta_{ws} = \frac{442.8}{800} \times 100 = 55.35\%$$

Level-2

1.16 A reservoir is proposed to be constructed to command an area of 1,20,000 hectares. The area has a monsoon rainfall of about 100 cm per year. It is anticipated that sugar and rice each would be equal to 20% of the command area and wheat equal to 50% of the command area, making a total of annual irrigation equal to 90% of the command area.

(a) Work out the storage required for the reservoir, assuming the water requirements given below, canal losses as 25% of the head discharge and reservoir evaporation and dead storage losses as 20% of the gross capacity of the reservoir.

(b) Determine also the full supply discharge of the canal at the head of the canal.

Crop	Showing time	Harvesting time	Total water depth in cm	"Kor" period in weeks	"Kor" watering in cm
Transplanted rice	July	Nov.	150	2.5	19.0
Sugar cane	Feb-Mar	Dec.-Mar Next year	90	4	16.5
Wheat	October	Mar.-Apr	37.5	4	13.5

Note that $\Delta = \frac{864B}{D}$ in which Δ = depth of water in cm, B = base period in days, and D = duty of water in hectares/cumec.

[20 Marks]

Sol:

Given data:

Total command area,

$$A = 1,20,000 \text{ Ha}$$

Area of rice and sugarcane

$$= 20\% \text{ of command area}$$

$$= 120000 \times \frac{20}{100} = 24000 \text{ Ha}$$

Area for wheat

$$= 1,20,000 \times \frac{50}{100} = 60000 \text{ Ha}$$

Delta for rice = 150 cm = 0.15 m

Delta for sugarcane = 90 cm = 0.9 m

Delta for wheat = 37.5 cm = 0.375 m

Quantity of water required for rice

$$= 24000 \times 1.5 = 36000 \text{ Ha-m}$$

Quantity of water required for sugarcane

$$= 24000 \times 0.9 = 21,600 \text{ Ha-m}$$

Quantity of water required for wheat

$$= 60.000 \times 0.375 = 22500 \text{ Ha-m}$$

∴ Total quantity of water required for crops

$$= 36000 + 21600 + 22500$$
$$= 80100 \text{ Ha-m}$$

Given, total loss

$$= 25\% + 20\% = 45\%$$

∴ Total loss of water

$$= 80100 \times \frac{45}{100} = 36045 \text{ Ha-m}$$

Reservoir capacity

$$= 80100 + 36045$$
$$= 116145 \text{ Ha-m}$$

We know that, outlet factor is the duty at the head of water course.

∴ Outlet factor for rice

$$= \frac{8.64 \times 17.5}{0.19}$$
$$= 795.8 \text{ Ha/cumec}$$

Outlet factor for sugarcane

$$= \frac{8.64 \times 28}{0.16}$$

Crop	Total water depth (cm)	kor period in days(B)	kor watering (cm)	Outlet factor $D = \frac{8.64B}{\Delta}$ (Ha/cumec)
Rice	150	$2.5 \times 7 = 17.5$	0.19	795.8
Sugarcane	90	28	0.165	1466.18
Wheat	37.5	28	0.135	1792

∴ Discharge of outlet $= \dfrac{\text{Free irrigational water}}{\text{Outlet factor}}$

$\therefore$ Discharge for Rice $= \dfrac{24000}{795.8} = 30.2\ m^3/sec$

Discharge for Sugarcane $= \dfrac{24000}{1466.18} = 16.37\ m^3/sec$

Discharge for wheat $= \dfrac{60000}{1792} = 33.5\ m^3/sec$

$\therefore$ Full supply discharge of canal

$= \mathbf{33.5\ m^3/sec}$

1.17 **A river valley consists of 14 hectare of irrigatable land to be used for growing a crop whose peak consumptive use is 5 mm/day. If the average daily precipitation on the valley is 20 mm/day of which 2.0 mm is available for the crop and the irrigation system is operated for 12 hrs a day, determine the discharge in the irrigation stream if field irrigation efficiency is 50%.**

[12 Marks]

Sol:

Given data:

Area of irrigable land $= 14 \times 10^4\ m^2$

Consumptive use rate, $C_u = 5$ mm/day

Average daily precipitation = 20 mm/day

Duration of effective rainfall = 2 mm per day

Duration of operation of irrigation system = 12 hr

Field irrigation efficiency, $\eta_a = 50\%$

We know that

Consumptive irrigation requirement,

$$\text{C.I.R.} = C_u - P_{eff}$$
$$= 5 - 2 = 3.0 \text{ mm/day}$$

Net irrigation requirement,

N.IR = C.I.R + leaching requirement

$$= 3.0 + 0$$
$$= 3.0 \text{ mm/day}$$

Field irrigation requirement,

$$\text{FIR} = \frac{\text{NIR}}{\eta_a} = \frac{3.0}{0.5} = 6 \text{ mm/day}$$

Discharge,
$$Q = \frac{14 \times 10^4 \times 6 \times 10^{-3}}{86400} = 9.722 \times 10^{-3}\ m^3/sec$$

Now if irrigation system is operated for 12 hour in a day then discharge,

$$Q = 2 \times 9.722 \times 10^{-3} = 0.019\ m^3/sec$$

Volume of water required per day

$$= 3 \times 10^{-3} \times 14 \times 10^4 = 420\ m^3/day$$

Total discharge required

$$= \frac{420}{0.5} = 840\ m^3/day$$

$\because$ $\eta_a = 50\%$

$\therefore$ Discharge of irrigation system

$$= 2 \times \text{discharge required}$$
$$= 2 \times 840 = 1680\ m^3/day$$

1.18 Calculate the reservoir capacity for irrigating an area of GCA = 50000 ha. The cropping pattern is as follows:

Crops	Period	Base period (days)	Outlet factor (ha/cumec)	Intensity of cropping (%)	Crop ratio
Sugar-cane	Oct-Sep	360	1000	60	4
Cotton	May-Nov	180	1200	60	3
Wheat	Dec-April	120	1500	80	2
Gram	Dec-April	120	2000	80	3

The area has 40% non-culturable are:

Assume:

(i) Time factor = $\frac{7}{10}$

(ii) Extra allowance for period of peak water use = 20%

(iii) Conveyance losses = 20%

(iv) Reservoir losses = 10%

Also calculate design discharge and capacity factor of main canal.

[20 Marks]

Sol:

$$\text{GCA} = 50000 \text{ ha}$$

$$\text{CCA} = (1-0.4)\times 50000 = 30000 \text{ ha}$$

Sugar-cane: Area under irrigation (A_s) = Intensity × Crop ratio × CCA

$$= 0.6\times\frac{4}{12}\times 30000 = 6000 \text{ ha}$$

$\because$
$$\Delta = \left(\frac{8.64B}{D}\right)$$

D = duty (outlet factor) (ha/cumec)

B = Base period (days)

Δ = Delta (m)

$$\Delta_s = \frac{8.64\times 360}{1000} = 3.11\text{m}$$

$\therefore$ Volume required (V_s) = $A_s\times\Delta_s$ = 6000 × 3.11 = 18660 ha-m

Cotton: Area under irrigation (A_C) = $0.6\times\frac{3}{12}\times 30000$ = 4500 ha

$$\Delta_C = 8.64\frac{B}{D} = 8.64\times\frac{180}{1200} = 1.296 \text{ m}$$

Volume required, $V_C = A_C\times\Delta_C$

= 4500 × 1.296 = 5832 ha-m

Wheat: Area under irrigation (A_w) = $0.8\times\frac{2}{12}\times 30000$ = 4000 m

$$A_w = 8.64\times\frac{120}{1500} = 0.691\text{m}$$

$$V_w = A_w\times\Delta_w = 4000\times 0.691 = 2764 \text{ ha-m}$$

Gram: Area under irrigation (A_g) = $0.8\times\frac{3}{12}\times 30000$ = 6000 m

$$\Delta_G = 8.64 \times \frac{120}{2000} = 0.518 \text{ m}$$

$$\therefore \quad V_g = A_g \times \Delta_g = 6000 \times 0.518 = 3108 \text{ ha-m}$$

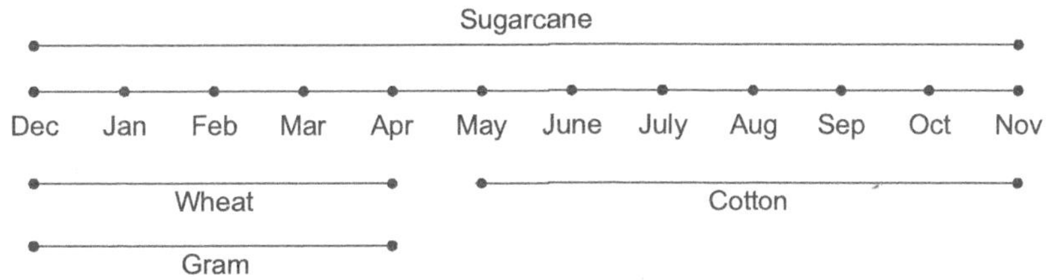

$$\text{Peak water use} = \text{Maximum } (V_s + V_C, V_C + V_s + V_g)$$

$$= \text{Maximum } (18660 + 5832 = 24492, 18660 + 2764 + 3108 = 24532)$$

$$= 24532 \text{ has-m}$$

$$\text{Due to 20\% extra allowance}(V_{reqd}) = 1.2 \times 24532 = 29438.4 \text{ ha-m}$$

Given:

$$\eta_{conv.} = 0.8$$

$$\eta_{reservoir} = 0.9$$

$$\therefore \text{ Gross reservoir capacity needed} = \frac{29438.4}{0.8 \times 0.9} = 40886.67 \text{ ha-m}$$

Design discharge calculations:

$$q_{sugarcane} = \frac{A_s}{D_s} = \frac{6000}{1000} = 6 \text{ cumec}$$

$$q_{cotton} = \frac{A_c}{D_c} = \frac{4500}{1200} = 3.75 \text{ cumec}$$

$$q_{wheat} = \frac{A_w}{D_w} = \frac{4000}{1500} = 2.67 \text{ cumec}$$

$$q_{gram} = \frac{A_g}{D_g} = \frac{6000}{2000} = 3 \text{ cumec}$$

$$Q_{design} = \frac{\text{Max}[\text{Dec-April, May-Nov}]}{\text{Time factor}}$$

$$= \frac{\text{Maximum } (q_s + q_w + q_g, q_s + q_c)}{\text{Time factor}}$$

$$= \frac{\text{Maximum } (11.67, 9.75)}{0.7} = \frac{11.67 \text{ cumec}}{0.7}$$

$$Q_{design} = 16.67 \text{ cumec}$$

$$\text{Capacity factor} = \frac{Q_{mean}}{Q_{max}}$$

$$Q_{mean} = \frac{11.67 + 9.75}{2} = 10.71 \text{ cumec}$$

$$\therefore \quad CF = \frac{10.71}{11.67} = 0.918$$

1.19 A stream of 150 lps was delivered from a canal and 110 lps were delivered to the field. An area of 2.2 ha was irrigated in 8 hrs. The effective depth of root zone was 1.5 m. Runoff loss in the field was 445 m^3. Depth of water penetration varied linearly from 1.5 m at the head end of the field to 1.1 m at the tail end. Available moisture holding capacity of the soil is 200 mm per meter depth of soil. Determine the water conveyance efficiency, water application efficiency, water storage efficiency and water distribution efficiency. Given that irrigation was started at a moisture extraction level of 50%.

[20 Marks]

Sol:

$$Q_{canal} = 150\ l/s;$$
$$Q_{field} = 110\ l/s$$
$$\text{Area, } A_{field} = 2.2\ \text{ha},$$
$$T = 8\ \text{hr}$$
$$d_h = 1.5\ \text{m}$$
$$d_t = 1.1\ \text{m}$$
$$\text{Moisture holding capacity} = 200\ \text{mm/m}$$
$$\text{Runoff loss} = 455\ \text{m}^3$$

$$\text{Water conveyance efficiency, } \eta_c = \frac{W_f}{W_r} \times 100 = \frac{110}{150} \times 100 = 73.33\%$$

$$\text{Water application efficiency, } \eta_a = \frac{W_s}{W_f} \times 100$$

$$\text{Water delivered to the plot} = 110 \times 10^{-3} \times (8 \times 3600) = 3168\ \text{m}^3$$
$$\text{Water stored in rout zone, } w_s = 3168 - 445 = 2723\ \text{m}^3$$

$$\therefore \quad \eta_a = \frac{2723}{3168} \times 100 = 85.95\%$$

$$\text{Water storage efficiency, } \eta_s = \frac{W_s}{W_h} \times 100$$

$$\text{Water holding capacity of soil} = 200 \times 1.5 = 300\ \text{mm}$$

$$\text{Moisture required in the root zone} = \left(300 - \frac{300 \times 50}{100}\right) = 150\ \text{mm}$$

$$\eta_s = \frac{2723}{3300} \times 100 = 82.52\%$$

$$\text{Water distribution efficiency, } \eta_d = \left(1 - \frac{y}{d}\right) \times 100$$

$$\text{Average depth of water stored in Root zone } (d) = \frac{1.5 + 1.1}{2} = 1.3\ \text{m}$$

Deviations from depth of Penetration

$$\text{At the head end} = 1.5 - 1.3 = 0.2\ \text{m}$$
$$\text{At the tail end} = 1.3 - 1.1 = 0.2\ \text{m}$$

Average numerical deviation in depth of water stored,

$$y = \frac{0.2 + 0.2}{2} = 0.2\ \text{m}$$

$$\eta_d = \left(1 - \frac{0.3}{1.3}\right) \times 100 = 76.92\%$$

1.20 A reservoir is proposed to be constructed for a commanded area of 1.20 lakh hectares. It is presumed that ultimately sugarcane and paddy will be irrigated equal to 20% of the commanded area for each and wheat equal to 50% of the commanded area, making a total annual irrigation equal to 90% of the commanded area. The water requirements, average duties, capacity duties etc. for three crops prevailing in the area are given below:

Crop	Total water depth required (cm)	Avg. duty in (ha/cumecs)	Duty on capacity i.e. max. demand duty in (ha/cumecs)
Sugarcane	90	2600	600
Paddy	120	864	637
Wheat	40	3464	1728

(a) Determine the storage capacity for the proposed reservoir assuming canal losses = 25% of the head discharge and reservoir evaporation. Dead storage losses = 20% of gross capacity. [20 Marks]

Sol:

(a) Total commanded area = 120000 ha

Area under Sugarcane = $0.20 \times 120000 = 24000$ ha

Area under Paddy = $0.20 \times 120000 = 24000$ ha

Area under Wheat = $0.50 \times 120000 = 60000$ ha

The quantity of water required for all crops is worked out as.

The volume of water required at the outlet:

$$= 24000 \times 0.9 + 24000 \times 1.2 + 60000 \times 0.4 = 74400 \text{ ha-m}$$

$$= 74400 \times 10^4 \text{ m}^3 = 744 \text{ Mm}^3$$

Canal losses = 25% of head discharge

$\therefore$ Total volume of water required at the head of canal system.

$$= \frac{744}{0.75} \text{ Mm}^3 = 992 \text{ Mm}^3$$

or, Net reservoir storage = 992 Mm^3

Given, reservoir evaporation and dead storage loss = 20% of gross storage

$\therefore$ Gross reservoir storage $= \frac{992}{0.80} = 1240 \text{ Mm}^3$.

(b) Channel capacity: Capacity discharge required for each crop:

Crop (1)	Duty (2)	Area under cultivation (3)	Discharge reqd. in cumec = col(3)/col(2) (4)
Sugarcane	600	24000	40
Paddy	637	24000	37.7
Wheat	1728	60000	34.72

$\therefore$ Discharge required for Kharif season = (40 + 37.7) = 77.7 m^3/s.

Discharge required for Rabi season = (40 + 34.72) = 74.72 m^3/s.

Channel capacity at the output = 77.7 m^3/s.

Channel capacity in the head reaches $= \frac{77.70}{0.75} = 103.60 \text{ m}^3/\text{s}$.

2. Design of Stable Channels and Canals

Level-1

2.1 Design a trapezoidal shaped concrete lined channel to carry a discharge of 100 m^3/s at a slope of 25 cm/km. Side slopes of the channel are 1.5 : 1. The value of $N = 0.016$. Assume limiting velocity as 1.5 m/s.

[15 Marks]

Sol:

Using Manning's equation,

$$V = \frac{1}{N} \times R^{2/3} \times S^{1/2}$$

$$S = \frac{25}{1000 \times 100} = \frac{1}{4000};$$

$$V = 1.5 \text{ m/s}; \quad N = 0.016$$

$\therefore$ $$1.5 = \frac{1}{0.016} \times R^{2/3} \times \left(\frac{1}{4000}\right)^{1/2}$$

$\Rightarrow$ $$R = 1.87 \text{ m}$$

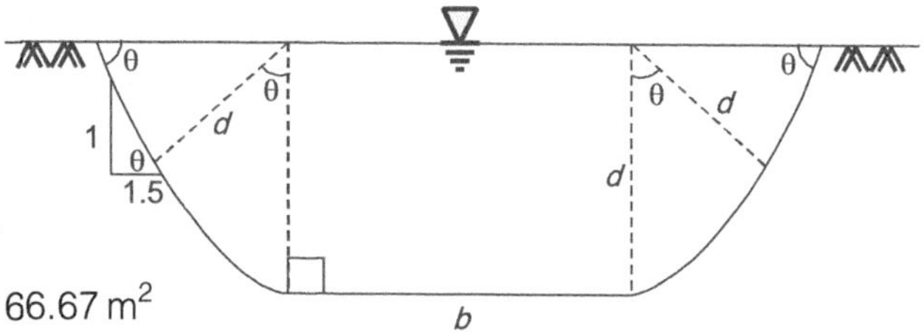

For side slope 1.5 : 1

$\Rightarrow$ $$\cot\theta = 1.5$$

$$\theta = 0.59 \text{ radian.}$$

$$\text{Area, } A = \frac{Q}{V} = \frac{100}{1.5} = 66.67 \text{ m}^2$$

$$\text{Perimeter, } P = \frac{A}{R} = \frac{66.67}{1.87} = 35.65 \text{ m}$$

For trapezoidal shaped channel,

$$P = b + 2d(\theta + \cot\theta)$$

$$35.65 = b + 2d\,(0.59 + 1.5)$$

$$35.65 = b + 4.18d$$

$$A = bd + d^2(\theta + \cot\theta)$$

and $$66.67 = bd + d^2(0.59 + 1.5)$$

$$66.67 = (35.65 - 4.18d)d + 2.09d^2$$

$$2.09d^2 - 35.65d + 66.67 = 0$$

$$d = 14.92 \text{ m}, 2.14 \text{ m}$$

$$\text{Adopt, } d = 2.14 \text{ m} \qquad \text{(Neglecting higher non-feasible values)}$$

$\therefore$ $$b = 35.65 - 4318 \times 2.14 = 26.70 \text{ m}$$

So, $$\text{Required bed width} = 26.70 \text{ m}$$

$$\text{Depth} = 2.14 \text{ m}$$

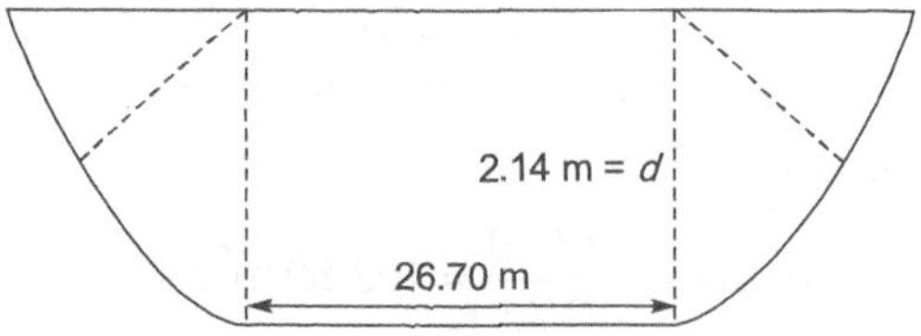

2.2 Design a concrete lined channel to carry a discharge of 45 m^3/s at a slope of 1 in 10000. The side of the channel are $1\frac{1}{4} : 1$ and Manning's N may be taken as 0.018.

[10 Marks]

Sol:

$\because$ $Q < 55$ m^3/s $\Rightarrow$ Adopt triangular rounded section of channel.

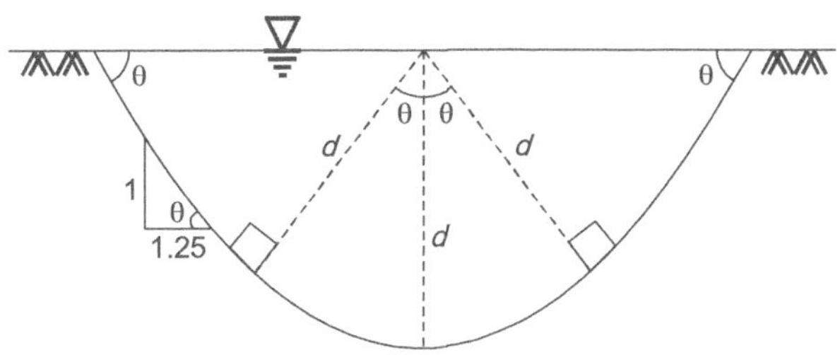

$$\text{Area, } A = d^2(\theta + \cot\theta)$$

Here, d = Depth of central portion,

$$\text{Perimeter, } P = 2d(\theta + \cot\theta)$$

$$\text{Hydraulic mean depth, } R = \frac{A}{P} = \frac{d}{2}$$

From Manning's equation,

$$V = \frac{1}{n} \times R^{2/3} \times S^{1/2}$$

$$= \frac{1}{0.018} \times \left(\frac{d}{2}\right)^{2/3} \times \frac{1}{\sqrt{10000}} = \frac{d^{2/3}}{2.86} \quad \text{... (i)}$$

$$\because \quad V = \frac{Q}{A} = \frac{45}{d^2(\theta + \cot\theta)}$$

$$\cot\theta = 1.25$$

$$\Rightarrow \quad \theta = 0.675 \text{ radian}$$

$$\therefore \quad V = \frac{45}{d^2(0.675 + 1.25)} = \frac{23.38}{d^2} \quad \text{... (ii)}$$

From (i) and (ii), we get

$$\frac{d^{2/3}}{2.86} = \frac{23.38}{d^2}$$

$$d = 4.84 \text{ m}$$

2.3 A rectangular irrigation channel has to be constructed in coarse alluvium gravel with D_{75} size of 50 mm. The channel has to be designed to carry 3.0 m^3/s of discharge at a longitudinal slope of 0.01 (1 in 100). The channel banks will be protected by gross against scouring. Calculate the minimum width of the channel using following assumptions:

Take hydraulic radius = Depth of flow in the channel.

No Scour should be there in the channel. Use stricker's formula to calculate Manning's rugosity coefficient.

[10 Marks]

Sol:

Given: $d = 50$ mm $= 0.05$ m (> 6 mm)

We know strickler's formula, $n = \frac{1}{24} d^{1/6}$ (d in 'm')

Now, for no scour condition,

$$d \geq 11 RS$$

$$R \leq \frac{d}{11S} = \left(\frac{0.05}{11 \times 0.01}\right)$$

$$R_{max} = 0.455 \text{ m}$$

$$n = \frac{1}{24} d^{1/6} = \frac{1}{24}(0.05)^{1/6} = 0.0253$$

Now, Using $V = \frac{1}{n} \times R^{2/3} \times S^{1/2}$

$$V_{max} = \frac{1}{0.0253} \times (0.455)^{2/3} \times (0.01)^{1/2} = 2.34 \text{ m/ s}$$

$\because$ $Q = AV = B \times y \times V$

$$B_{min} = \frac{Q}{Vy} = \frac{Q}{VR} = \frac{3.0}{0.455 \times 2.34} = 2.82 \text{ m} \approx 3 \text{ m}$$

$\therefore$ Canal bed width, $B = 3$ m

2.4 A semicircular channel carries uniform flow of 10 m³/s and runs full. Consider Chezy's coefficient, $C = 60$, bed slope = 1/3600. Determine the diameter of the channel. Also determine the cost of lining for 1.0 km long channel. Consider cost of lining at rate of ₹ 800 per m².

[8 Marks]

Sol:

For semi-circular channel:

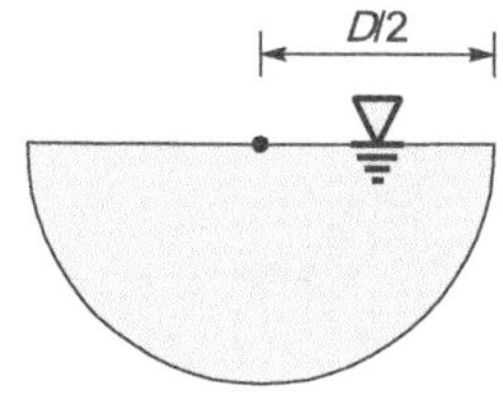

Discharge, $Q = 10$ m³/s

Bed slope, $S = \frac{1}{3600}$

$C = 60$

Assume diameter of channel as 'D'

$$\text{Area of channel, } A = \frac{\pi D^2}{4} \times \frac{1}{2} = \frac{\pi D^2}{8}$$

$$\text{Perimeter, } P = \frac{\pi D}{2}$$

$$\text{Hydraullic radius, } R = \frac{A}{P} = \frac{\pi D^2/8}{\pi D/2} = \frac{D}{4}$$

Use Chezy's equation, $Q = CA\sqrt{RS}$

$$10 = 60 \times \frac{\pi}{8} \times D^2 \times \sqrt{\frac{D}{4} \times \frac{1}{3600}}$$

$$D = 4.82 \text{ m}$$

$$\text{Cost of lining} = \frac{\pi D}{2} \times L \times 800 = \frac{\pi \times 4.82}{2} \times 1000 \times 800$$

$$= ₹\, 6056990$$

2.5 Design a channel transition at the location of an aqueduct using Mitra's approach. Normal width and flume width of channel are 15 m and 9 m respectively. The length of transition is 16 m.

[10 Marks]

Sol:

Given data : Normal width, $B_0 = 15$ m, Flume width, $B_f = 9$ m, Length of transition, $L_f = 16$ m

V_0, V_f, V_x - Velocities at original width, flumed width and at distance 'x' from the flumed section. And 'B_x' be the width at a distance 'x' from the flumed section.

Using Mitra's approach, rate of change of velocity per unit length of transition is kept constant.

$$\therefore \quad \frac{V_f - V_x}{x} = \frac{V_f - V_0}{L_f} \quad \text{... (i)}$$

Also water depth, D and discharge, Q are constant

So, $$Q = V_0 B_0 D = V_f B_f D = V_x . B_x . D \quad \text{... (ii)}$$

From (i) and (ii)

$$\frac{\left(\frac{Q}{B_f D} - \frac{Q}{B_x D}\right)}{x} = \frac{\left(\frac{Q}{B_f D} - \frac{Q}{B_0 D}\right)}{L_f}$$

$$\frac{(B_x - B_f) \times L_f}{B_x \times B_f} = \frac{(B_0 - B_f) \times x}{B_f \times B_0}$$

$$B_x = \frac{B_0 B_f L_f}{L_f B_0 - (B_0 - B_f)x}$$

$$B_x = \frac{B_0 B_f L_f}{L_f B_0 - (B_0 - B_f)x} = \frac{15 \times 9 \times 16}{16 \times 15 - (15 - 9) \times x}$$

$$= \frac{2160}{240 - 6x} = \left(\frac{360}{40 - x}\right)$$

x (m)	2	4	6	8	10	12	14
B_x(m)	9.47	10	10.59	11.25	12	12.86	12.85

So, transition width should be hyperbolic, as

$$B_x = \frac{360}{40 - x}$$

2.6 An area of 40000 ha. has to be irrigated by a canal for growing banana in 15000 ha. and 25000 ha. for growing paddy. The peak water requirements of banana and paddy are 12 cm/month and 16 cm/month respectively and these peak demands occur at the same month. Design a suitable canal section using Lacey's method. Adopt a side slope of 0.5 H : 1 V.

[10 Marks]

Sol:

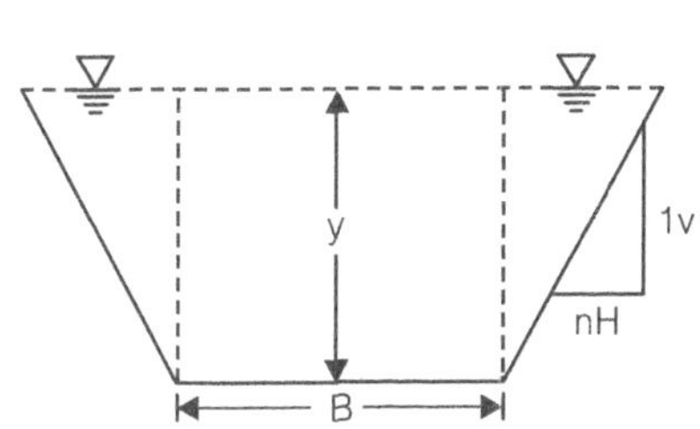

Maximum discharge required through canal

$Q = 15000 \times 10^4 \times 10^4 \times 12(cm^3/month) + 25000 \times 10^4 \times 10^4 \times 16\ (cm^3/month)$

$= (15 + 12 + 25 \times 16) \times 10^{11}\ cm^3/month$

$$= \frac{580 \times 10^{11} \times 10^{-6}}{30 \times 24 \times 60 \times 60} m^3/sec = 22.3\ m^3/sec \simeq 23\ m^3/sec$$

$$f = 1.05 \text{ (Assumed)}$$

Side slope 0.5 H : 1 V

$\Rightarrow$ $n = 0.5$

Now,

$$V = \left(\frac{Qf^2}{140}\right)^{1/6} = \left[\frac{23 \times (1.05)^2}{140}\right]^{116} = 0.752 \text{ m/s}$$

$$A = \frac{Q}{V} = \frac{23}{0.752} = 30.57 m^2$$

$$R = \frac{5}{2}\frac{V^2}{f} = \frac{5}{2} \times \frac{(0.752)^2}{1.05} = 1.35 \text{ m}$$

$$P = 4.75\sqrt{Q} = 4.75\sqrt{23} = 22.78 m$$

For a trapezoidal channel with $\frac{1}{2}H:1V$ slopes

$$P = B + 2y\sqrt{n^2+1}$$

$\Rightarrow$ $$P = B + 2y\sqrt{0.5^2+1}$$

$\Rightarrow$ $$P = B + y\sqrt{5}$$

Area $$A = (B + 0.5y)y$$

$\Rightarrow$ $$22.78 = B + \sqrt{5}y \quad ...(i)$$

and $$30.57 = (B + 0.5y)y \quad ...(ii)$$

Equation (i) and (ii)

$$B = 19.38 \text{ m}$$

$$y = 1.52 \text{ m}$$

2.7 **A suspended load sample taken at a height of 0.30 m from the bed indicated a concentration of 1000 ppm of sediment by weight. The stream is 5.0 m deep and has a bed slope of $\frac{1}{3800}$. The bed material can be assumed to be of uniform size with a fall velocity of 4 cm/s. Estimate the concentration of suspended load at its mid-depth.**

[10 marks]

Sol:

(ii) Rouse equation:

$$\frac{C}{C_a} = \left[\frac{a(D-y)}{y(D-a)}\right]^{\frac{W_o}{kV^*}} \quad ...(i)$$

Where total water depth (D) = 5 m

Bed slope, $S = \frac{1}{3800}$

y = distance from bed of channel, where sediment concentration 'C' is desired i.e., at mid-depth = $\frac{5}{2} = 2.5$ m

C_a = known sediment concentration = 1000 ppm, at a distance, a = 0.30 m from bed

w_0 = free fall velocity of sediment grain in still water

$\Rightarrow$ w_0 = 4 cm/s = 0.04 m/s

k = Von-Karman constant = 0.4 (assumed)

V^* = Shear friction velocity

$$= \sqrt{\frac{\tau_o}{\rho_w}}$$

$$= \sqrt{\frac{\gamma_w R.S}{\rho_w}} = \sqrt{gRS} \qquad [R \simeq D = 5 \text{ m (for wide channels)}]$$

$$= \sqrt{9.81 \times 5 \times \frac{1}{3800}}$$

$$= 0.1136 \text{ m/s}$$

From eq. (i)

$$\frac{C}{1000} = \left[\frac{0.3(5-2.5)}{2.5(5-0.3)}\right]^{\frac{0.04}{0.4 \times 0.1136}}$$

$$\Rightarrow \quad \frac{C}{1000} = \left[\frac{0.75}{11.75}\right]^{0.88} = 0.0888$$

$$\Rightarrow \quad C = 88.80 \text{ ppm}$$

Level-2

2.8 Design a trapezoidal shaped concrete lined channel to carry a discharge of 100 m^3/s at a slope of 25 cm/km. Side slopes of the channel are 1.5:1. Value of N is taken as 0.016. Given $\frac{b}{d} = 8$.

[15 Marks]

Sol:

For trapezoidal shaped channel,

$$A = bd + d^2(\theta + \cot\theta) \qquad \text{... (i)}$$

$$P = b + 2d(\theta + \cot\theta)$$

$$\cot\theta = 1.5$$

$$\Rightarrow \quad \theta = 0.59 \text{ radian}$$

$$\therefore \quad A = bd + 2.09d^2$$

$$P = b + 4.18d$$

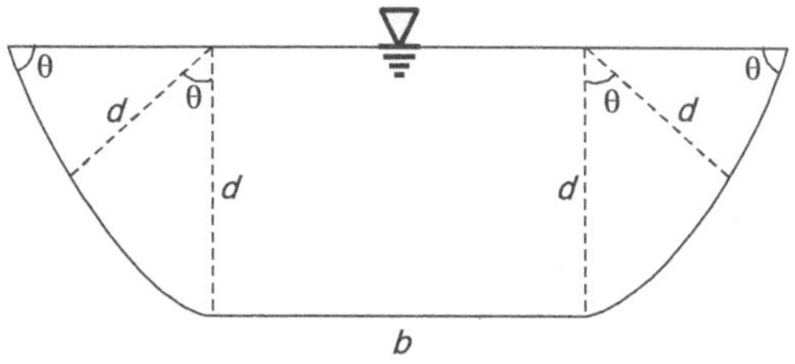

Velocity in the channel, $V = \dfrac{Q}{A} = \dfrac{100}{bd + 2.09d^2}$

Hydraulic mean depth, $R = \dfrac{A}{P}$

$$R = \frac{bd + 2.09d^2}{b + 4.18d}$$

Using Manning's equation, $V = \dfrac{1}{n} \times R^{2/3} \times S^{1/2}$

$$= \frac{1}{0.016} \times (R)^{2/3} \times \frac{1}{\sqrt{4000}} = \frac{R^{2/3}}{1.01}$$

$$\Rightarrow \quad R^{2/3} = 1.01\,V$$

$$\therefore \quad \left(\frac{bd + 2.09d^2}{b + 4.18d}\right)^{2/3} = 1.01 \times \left[\frac{100}{bd^2 + 2.09d^2}\right]$$

Put $b = 8d$

$$\left(\frac{8d^2 + 2.09d^2}{8d + 4.18d}\right)^{2/3} = 1.01 \times \left[\frac{100}{8d^2 + 2.09d^2}\right]$$

$$\left(\frac{10.09d}{12.18}\right)^{2/3} = 1.01\left(\frac{100}{10.09d^2}\right)$$

$$d^{8/3} = \frac{101}{10.09} \times \left(\frac{12.18}{10.09}\right)^{2/3} = 11.348$$

$$\therefore \quad d = (11.348)^{3/8} = 2.49 \text{ m} \approx 2.5 \text{ m}$$

$$\therefore \quad b = 8 \times 2.50 = 20 \text{ m}$$

Therefore, Required bed width = 20 m

Depth = 2.5 m

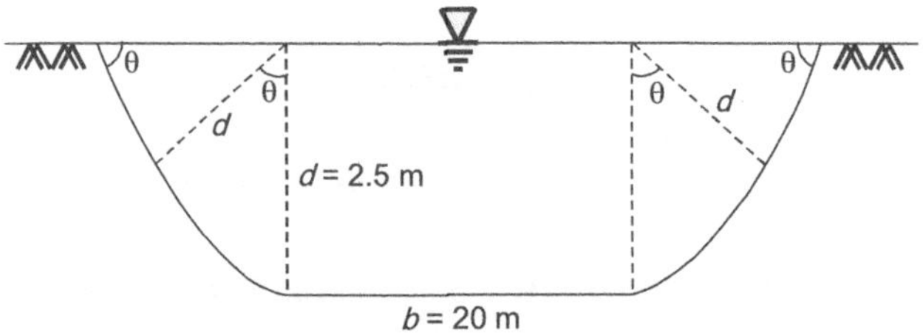

Channel section

2.9 A culturable command area for a distributary is 12000 hectares. The intensity of irrigation is 90 percent for Rabi and 80 percent for Kharif crop. The outlet factors for Rabi and Kharif crops are 720 ha/cumec and 775 ha/cumec, respectively. Design a lined canal having hydraulically efficient trapezoidal shape with 2 H : 1 V side slopes and a bed slope of 1/5000. Assume Manning's n to be 0.014.

[20 Marks]

Sol:

$$\text{Total area} = 12000 \text{ ha}$$
$$\text{Rabi} = 12000 \times 0.9 = 10800 \text{ ha}$$
$$\text{Kharif} = 12000 \times 0.8 = 9600 \text{ ha}$$
$$\text{Intensity of Rabi} = 720 \text{ ha/cumec}$$
$$\text{Intensity of Kharif} = 775 \text{ ha/cumec}$$
$$\text{Discharge for Rabi} = \frac{10800}{720} = 15 \text{ m}^3\text{/sec}$$

Discharge of Kharif

$$= \frac{9600}{775} = 12.39 \text{ m}^3\text{/sec}$$

Discharge of Rabi > Discharge of Kharif

Hence design should be taken as Rabi discharge

$$Q = 15 \text{ m}^3 \text{ sec}$$
$$\text{Slope} = \frac{1}{5000}$$
$$n = 0.014$$

From Manning's eq.

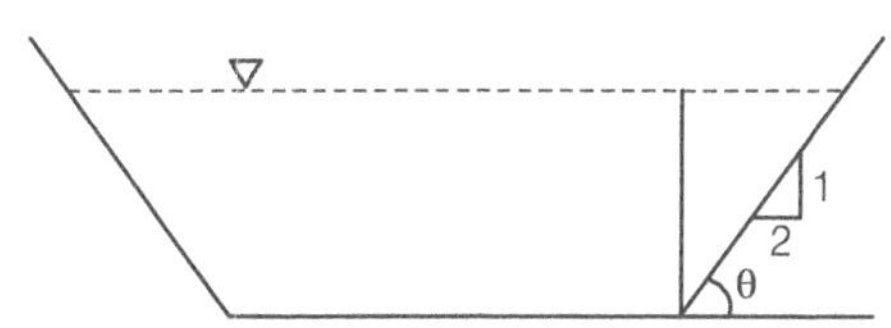

$$Q = \frac{A}{n}(R)^{2/3}(s)^{1/2}$$

Hydraulically efficient section is Trapezoidal

$$A_e = \left(2\sqrt{1+m^2} - m\right)y_c^2$$
$$= \left(2\sqrt{1+4} - 2\right)y_c^2 = 2.47\, y_c^2$$

and $$R_e = \frac{y_c}{2}$$

$\therefore$ $$Q = \frac{2.47y_c^2}{0.014}\left(\frac{y_c}{2}\right)^{2/3}\left(\frac{1}{5000}\right)^{1/2}$$

or $$15 = \frac{176.43y_c^{8/3}}{2^{2/3}} \times 0.014$$

or $$\frac{13 \times 1.587}{0.014 \times 176.43} = y_c^{8/3} = 9.64$$

or $$y_c = (9.64)^{3/8} = 2.34 \text{ m}$$
$$A = 2.47\, y_c^2$$
$$= 2.47 \times (2.34)^2 = 13.5 \text{ m}^2$$
$$B = 2y_c\left(\sqrt{1+m^2} - m\right) = 4.64\left(\sqrt{5} - 2\right) = 1.1 \text{ m}$$

$$\text{Width of channel} = 1.1 \text{ m}$$
$$\text{Area of channel} = 13.5 \text{ m}^2$$
$$\text{Depth of channel} = 2.34 \text{ m}$$
$$\text{Side slope} = 2 : 1$$

2.10 Design an irrigation channel to carry a discharge of 45 cumecs. Assume $N = 0.0225$ and $m = 1$. The channel has s bed slope of 0.16 m per km.

[15 Marks]

Sol:

Assuming trail depth, $y = 1.8$ m

Critical velocity, $V = 0.55\, my^{0.64}$

$$= 0.55 \times 1 \times 1.8^{0.64} = 0.80 \text{ m/s}$$

$$A = \frac{Q}{V} = \frac{45}{0.8} = 56.25 \text{ m}^2.$$

Assuming side slope of 2V : 1H

$$A = BD + 2\times\left(\frac{1}{2}\times D\times\frac{D}{2}\right) = BD + \frac{D^2}{2}$$

$$56.25 = B\times 1.8 + \frac{1.8^2}{2}$$

$$B = 30.35 \text{ m}$$

$$\text{Perimeter, } P = B + 2\sqrt{D^2 + \frac{D^2}{4}} = B + D\sqrt{5}$$

$$= 30.35 + 1.8\times\sqrt{5} = 34.37 \text{ m}$$

$$R = \frac{A}{P} = \frac{56.25}{34.37} = 1.636 \text{ m}$$

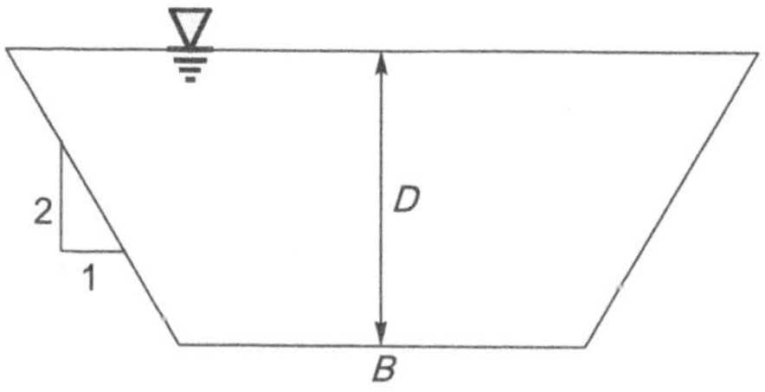

$\because$ $$V = C\sqrt{RS}$$

$$S = \frac{0.16}{1000}$$

and $$C = \frac{\frac{1}{N} + \left(23 + \frac{0.00155}{S}\right)}{1 + \left(23 + \frac{0.00155}{S}\right)\frac{N}{\sqrt{R}}}$$

$$= \frac{\frac{1}{0.0225} + \left(23 + \frac{0.00155}{0.16}\times 1000\right)}{1 + \left(23 + \frac{0.00155}{0.16}\times 1000\right)\frac{0.0225}{\sqrt{1.636}}} = 48.97 \simeq 49$$

$\therefore$ $$V = 49\sqrt{1.636\times\frac{0.16}{1000}} = 0.793$$

Ratio of velocities calculated above, $\frac{0.793}{0.80} = 0.991 \approx 1$

Assumed 'y' is satisfactory.

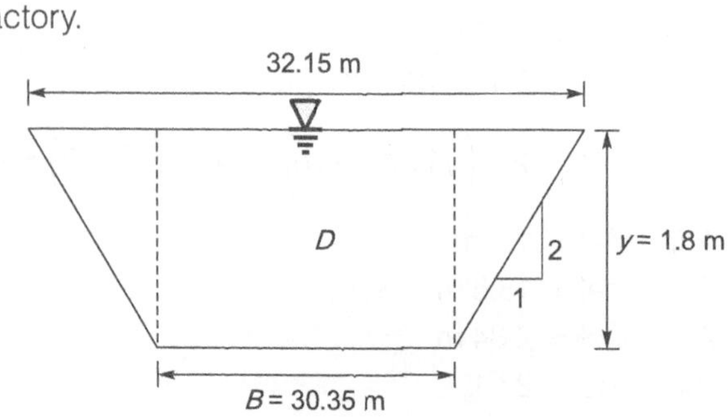

2.11 Design an irrigation channel by Kennedy's method, to carry 50 m^3/s of discharge of base width to depth ratio as 2.5. The critical velocity ratio is 1.1. Assume Kutter's velocity coefficient as 0.025 and side slopes of the channel as 0.5 H to 1 V.

[15 Marks]

Sol:

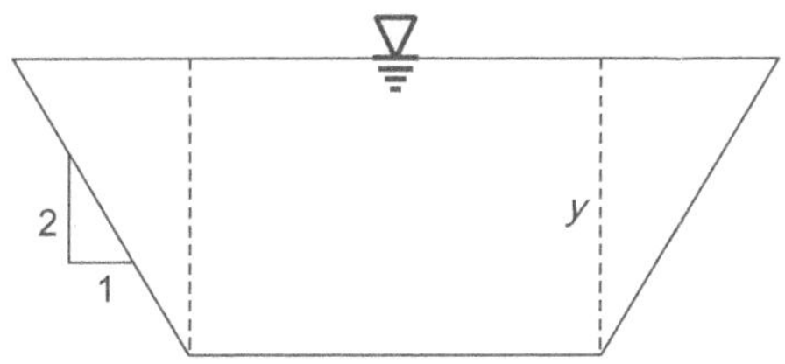

Given: $Q = 50\text{ m}^3/\text{s},\ \frac{B}{D} = 2.5,\ CVR\,(m) = 1.1,\ n = 0.025$

$$\text{Critical velocity, } V_0 = 0.55\, m\, y^{0.64}$$
$$= 0.55 \times 1.1 \times y^{0.64}$$
$$= 0.605\, y^{0.64}$$

$$A = By + \frac{y^2}{2} \quad \text{... (i)}$$

$$P = B + y\sqrt{5} = B + 2.236y \quad \text{... (ii)}$$

$$\because \quad Q = AV_0$$

$$50 = \left(B + \frac{y}{2}\right) \times y \times \left(0.605 \times y^{0.64}\right)$$

$$= \left(\frac{B}{y} + \frac{1}{2}\right) \times y^2 \times 0.605 \times y^{0.64} = (2.5 + 0.5) \times 0.605 \times y^{2.64}$$

$$y = 3.51\text{ m}$$

Now, $\quad$ Width, $B = 2.5 \times 3.51 = 8.78$ m

Putting values in equation (i)

$$A = 8.78 \times 3.51 + \frac{3.51^2}{2} = 36.98\text{m}^2$$

$$P = 8.78 + 2.236 \times 3.51 = 16.63\text{m}$$

$$\text{Hydrology mean depth, } R = \frac{A}{P} = \frac{36.98}{16.63} = 2.22\text{ m}$$

$$\therefore \quad V_0 = 0.605 \times 3.51^{0.64} = 1.351\text{ m/s}$$

Now, mean velocity using Kutter's formula,

$$V = \left[\frac{\frac{1}{N} + \left(23 + \frac{0.00155}{S}\right)}{1 + \left(23 + \frac{0.00155}{S}\right) \times \frac{n}{\sqrt{R}}} \times \sqrt{RS}\right]$$

Assume, $S = \frac{1}{3000}$

$$\therefore \quad V = \left[\frac{\frac{1}{0.025} + \left(23 + \frac{0.00155}{1/3000}\right)}{1 + \left(23 + \frac{0.00155}{1/3000}\right) \times \frac{0.025}{\sqrt{2.22}}}\right] \times \sqrt{2.22 \times \frac{1}{3000}}$$

$$= \frac{67.65}{1.464} \times 0.0272 = 1.257\text{m/s}$$

$$\Rightarrow \quad V < V_0$$

So, again assuming $S = \frac{1}{2600}$

$$V = \left[\frac{\dfrac{1}{0.025} + \left(23 + \dfrac{0.00155}{1/2600}\right)}{1 + \left(23 + \dfrac{0.00155}{1/2600}\right) \times \dfrac{0.025}{\sqrt{2.22}}}\right] \times \sqrt{2.22 \times \frac{1}{2600}}$$

$$= \frac{67.03}{1.45} \times 0.0292 = 1.35 \text{ m/s}$$

$$\therefore \quad V_0 = V \text{ at slope, } S = \frac{1}{2600}$$

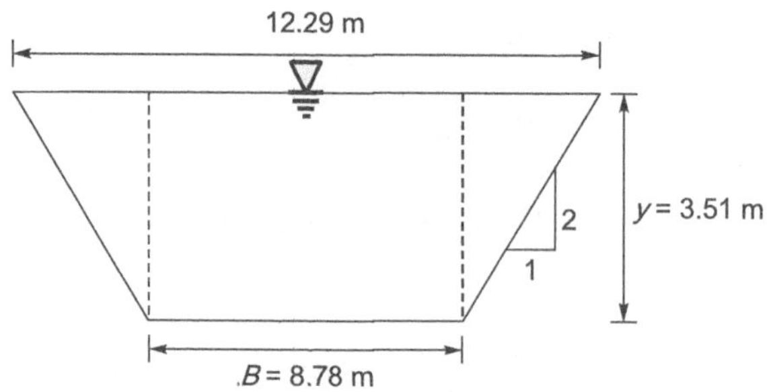

Design parameters:

$B = 8.78$ m,, $y = 3.51$ m, Bed slope, $S = \dfrac{1}{2600}$

2.12 **Design a regime channel for a discharge of 45 m³/s and silt factor of 0.8 by Lacey's theory, use side slopes as 2 V : 1 H.**

[15 Marks]

Sol:

$Q = 45$ m³/s, $f = 0.8$, S/S = 2V : 1 H

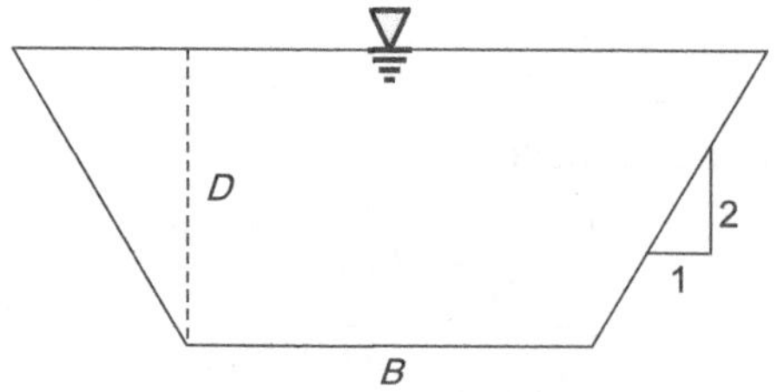

So for a regime channel, using Lacey's theory, we have

(i) Velocity, $V = \left(\dfrac{Qf^2}{140}\right)^{1/6} = \left(\dfrac{45 \times 0.8^2}{140}\right)^{1/6} = 0.768$ m/s

(ii) Hydraulic mean depth, $R = \dfrac{5}{2}\left(\dfrac{V^2}{f}\right) = \dfrac{5}{2} \times \left(\dfrac{0.768^2}{0.8}\right) = 1.843$ m

(iii) Area of channel, $A = \dfrac{Q}{V} = \dfrac{45}{0.768} = 58.594 \text{ m}^2$

(iv) Wetted perimeter, $P = 4.75\sqrt{Q} = 4.75\sqrt{45} = 31.86$ m

(v) Bed slope, $S = \dfrac{f^{5/3}}{3340 Q^{1/6}} = \dfrac{0.8^{5/3}}{3340 \times 45^{1/6}} = \dfrac{1}{9136.89} \simeq \dfrac{1}{9140}$

Now, for a trapezoidal channel,

$$A = By + \frac{y^2}{2}$$

$$58.594 = (B + 0.5y)y \quad \text{... (i)}$$

and $$P = B + y\sqrt{5}$$

$$31.86 = B + 2.236y$$

$\Rightarrow$ $$B = 31.86 - 2.236y \quad \text{... (ii)}$$

Putting value of of B in equation (i)

$\therefore$ $$58.594 = (31.86 - 2.236y + 0.5y)y$$

$$58.594 = (31.86 - 1.736y)y$$

Solving it, we get,

$$y = 16.28 \text{ m}, 2.07 \text{ m}$$

$\therefore$ $y = 2.07$ m (Neglecting higher non-feasible value)

$\therefore$ $$B = 31.86 - 2.236 \times 2.07 = 27.23 \text{ m}$$

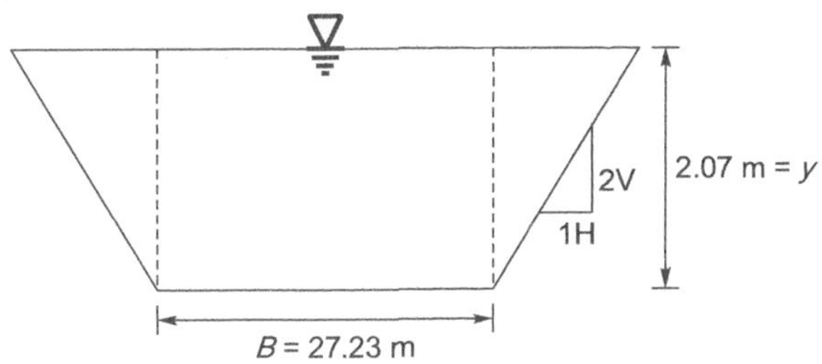

$$\text{Bed slope, } S = \frac{1}{9140}$$

2.13 Consider a trapezoidal rigid boundary channel of 3 m base width of side slopes 1 H : 0.5 V with depth of flow being 1.2 m, $n = 0.012$, with adequate free board. Sketch the shear stress distinction on wetted perimeter. Explain facts of zero and maximum, if both are so possible, shear stresses on each of side slopes and on bed.

[20 Marks]

Sol:

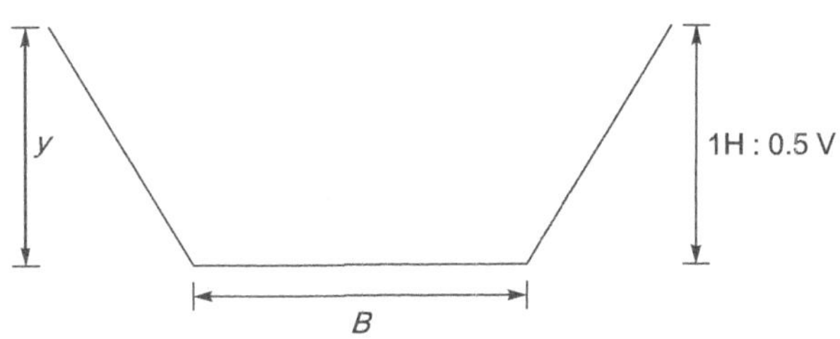

$$B = 3 \text{ m},$$

$$y = 1.2 \text{ m},$$

$$S = 1\text{H to } \frac{1}{2}\text{V} \ (m = 2)$$

$$P = B + 2y\sqrt{1+m^2} = B + 2y\sqrt{5}$$

$$= 3 + 2 \times 1.2 \times \sqrt{5} = 8.366 \text{ m}$$

$$A = (B + my)y = (3 + 2 \times 1.2) \times 1.2 = 6.48 \text{ m}^2$$

$\because$ $$S = \tan\theta \text{ (for small } Q, \sin\theta \simeq \tan\theta = S)$$

$$\tau\,(\text{Shear stress}) = \frac{(\gamma_\omega AL)\sin\theta}{\text{Wetted area}} = \frac{(\gamma_\omega AL)\times S}{PL}$$

$\Rightarrow$ $\tau = \gamma_\omega . \text{R.S.}$

$n = 0.012$

For rigid boundary channel, $n = \frac{1}{24}\times d^{1/6}\,(d \text{ in 'm'})$

$$0.012 = \frac{1}{24}\times d^{1/6}$$

$\Rightarrow$ $d = 5.71\times 10^{-4}$ m

$\because$ $d \geq 11\,RS$

$$5.71\times 10^{-4} = 11\times\left(\frac{6.48}{8.366}\right)\times S$$

$\Rightarrow$ $S = 0.67\times 10^{-4}$

$\tau_{\text{reqd.}}$ at bed $= \gamma_\omega . RS$

$$= 9.81\times 1000\times\left(\frac{6.48}{8.366}\right)\times 0.67\times 10^{-4} = 0.51\ \text{N/m}^2$$

τ_{avg} on sides $= 0.75\,\gamma_\omega . RS = 0.75\times 0.51 = 0.382\ \text{N/m}^2$

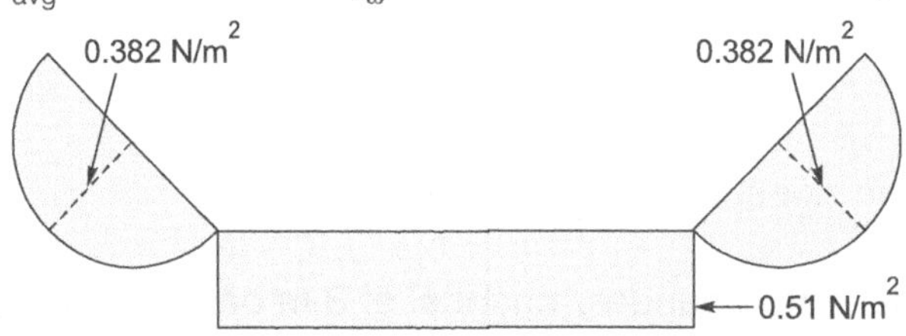

- Shear stress required to move a grain on side slopes is less than that shear stress required to move grain on canal bed.
- Shear stress will be zero at junction of bed and sides of section because it will be a sharp point.

2.14 **A canal is to be designed to carry a discharge of 56 m^3/s. The slope of the canal is 1 in 1000. The soil is coarse alluvium having a grain size of 5 cm. Assuming the canal to be unlined and of a trapezoidal section, determine a suitable section for the canal. Take ϕ = 37°, adopt side slopes at θ = 30°.**

[20 Marks]

Sol:

$$\frac{\tau_c'}{\tau_c} = \sqrt{1-\frac{\sin^2\theta}{\sin^2\phi}} = \sqrt{1-\frac{\sin^2 30^\circ}{\sin^2 37^\circ}} = \sqrt{1-\left(\frac{0.50}{0.60}\right)^2}$$

$\therefore$ $\frac{\tau_c'}{\tau_c} = 0.5527$

Therefore, minimum shear stress required to dislodge the grain on side slope,

$$\tau_c' = 0.5527\tau_c$$

For stability, shear stress generated, $\tau_c' \leq 0.5527\tau_c$ and actual shear stress to be generated on the side slopes of the channel, $\tau_c' = 0.75\,\gamma_\omega RS$

$\therefore$ $0.75\,\gamma_\omega RS \leq 0.5527\tau_c$

But, $\tau_c = \frac{\gamma_\omega d}{10.8}$

$$\therefore \qquad 0.75\,\gamma_\omega RS \le 0.5527 \times \frac{\gamma_\omega d}{10.8}$$

$$RS \le \frac{0.5527d}{0.75 \times 10.8}$$

$$\Rightarrow \qquad RS \le 0.068d \le (0.068 \times 0.05)\text{ m}$$

$$R \times \frac{1}{1000} \le (0.0688 \times 0.05)\text{ m}$$

$$R \le 3.4\text{ m}$$

or $$y \le 3.4\text{ m } (\because y \simeq R)$$

Adopt 20% FOS, $y = 0.8 \times 3.44 = 2.72 \approx 2.8$ m

Hence, chose depth as 2.8 m. Let us now choose the base width 'b' in such a way as to have a discharge of 56 cumec. Use hit and trial method to determine b,

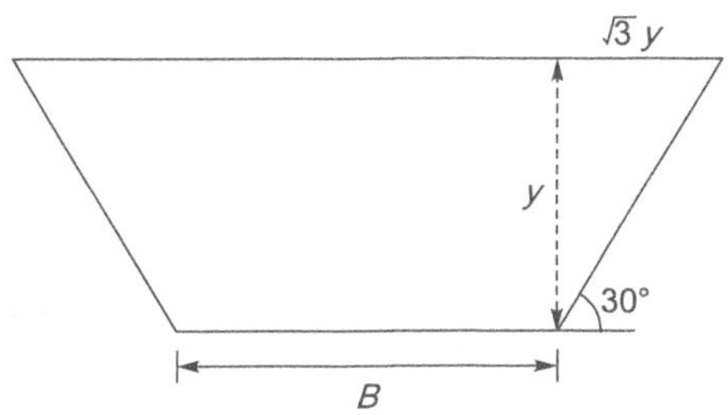

$$A = \left(B + \sqrt{3}y\right) \times y = 2.8\left(B + \sqrt{3} \times 2.8\right)$$

$$= 2.8\,(B + 4.85)$$

and $$P = B + 2 \times \sqrt{y^2 + 3y^2} = B + 2 \times 2y$$

$$= B + 4y = B + 4 \times 2.8 = B + 11.2$$

Now, $$n = \frac{1}{24} d^{1/6} = \frac{1}{24} \times (0.05)^{1/6} = 0.0253$$

$$V = \frac{1}{n} R^{2/3} S^{1/2} = \frac{1}{0.0253} \times R^{2/3} \sqrt{\frac{1}{1000}} = 1.25R^{2/3}$$

Hit and trial table:

B (in m)	$A = 2.8\,(B + 4.85)$ (m^2)	$P = (B + 11.2)$ (m)	$R = A/P$	$R^{2/3}$	$V = 1.25R^{2/3}$ (m/s)	$Q = AV$ (m^3/s)
3	22	14.2	1.549	1.338	1.67	36
4	24.8	15.2	1.632	1.386	1.673	42
5	27.6	16.2	1.704	1.427	1.777	48.3
6	30.4	17.2	1.767	1.462	1.825	54.8
6.15	30.8	17.35	1.77	1.466	1.832	56.4 or 56

So, use base width = 6.15 m

Depth = 2.8 m

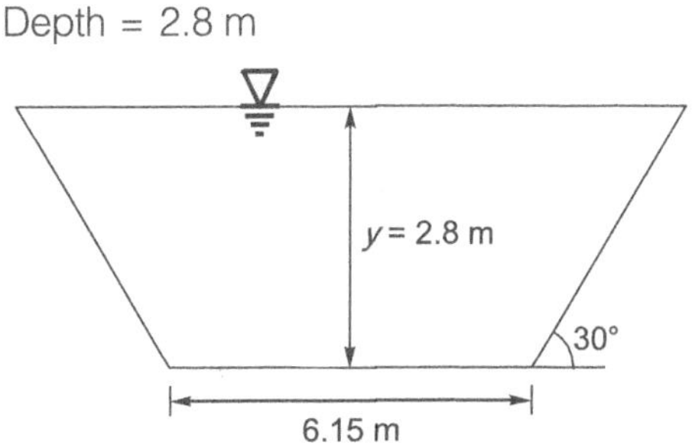

3. Design and Construction of Gravity Dam

Level-1

3.1 What are the factors to be considered for the selection of site for a reservoir?

[10 Marks]

Sol:

Factors to be considered are:

- Geological condition of the catchment area should be such that percolation losses are minimum and maximum runoff is obtained.
- Quantity of leakage through the reservoir site should be minimum. Reservoir site having the presence of highly permeable rocks reduce the water tightness of the reservoir. The dam should be founded on sound water tight rock base, and percolation below the dam be minimum.
- Cost of dam is often a controlling factor in selection of a site.
- The reservoir basin should have narrow opening in the valley so that length of the dam is less.
- The cost of real estate for the reservoir, including road, rail, dwelling re-location etc. must be as less as possible.
- Topography of reservoir site should be such that it has adequate capacity without submerging excessive land and other properties.
- The site should be such that a deep reservoir is formed. A deep reservoir is preferable to a shallow one because of
  - (i) low submergence area
  - (ii) less evaporation loss
  - (iii) less likelihood of weed growth.
- Reservoir site should be such that it avoids or excludes water from those tributaries which carry a high percentage of silt.
- Reservoir site should be such that water stored in it is suitable for the purpose for which the project is undertaken.
- The soil and rock mass at the reservoir site must not contain any objectionable minerals and salts.

3.2 Describe various types of dams on the basis of type and materials of construction and principle of function.

[10 marks]

Sol:

Based on type and materials of construction dams are classified as following:

1. **Rigid dam:** It is constructed with rigid material such as stone, masonry, concrete, steel, or timber. Steel dams (steel plates supported on inclined struts) and timber dams (wooden planks supported on a wooden framework) are constructed only for small heights.
2. **Non-rigid dam (embankment dams):** It is constructed with non-rigid material such as earth, tailings, rockfill etc. Following are types of non-rigid dam with materials used:
   - (a) Earthen dam - gravel, sand, silt, clay etc.
   - (b) Tailings dam - waste or refuse obtained from mines
   - (c) Rockfill dam - rock material supporting a water tight material on the u/s face
   - (d) Rockfill composite dam - rockfill on the d/s side and earth fill on the u/s side

Dams can be classified on the basis on principle of function are as following:

1. **Storage dam or impounding dam:** It is constructed to create a reservoir to store water during periods when there is huge flow in the river in excess of demand for utilisation later during periods of low flow when demand exceeds flow in the river, water stored in the reservoir is used for irrigation, power generation, water supply etc.
2. **Detention dam:** It is primarily constructed to temporarily detain all or part of the flood water in a river and to gradually release the stored water later at controlled rates so that the entire region on the downstream side of the dam is protected from possible damage due to floods.
3. **Diversion dam:** It is constructed to divert part of or all the water from a river into a conduit or a channel. For diverting water from a river into an irrigation canal, mostly a diversion weir is constructed across the river.
4. **Coffer dam:** It is a temporary dam constructed to exclude water from a specific area. It is constructed on the u/s side of the site where a dam is to be constructed so that the site is dry. In this case, it behaves like a diversion dam.
5. **Debris dam:** It is constructed to catch and retain debris flowing in a river.

3.3 **Calculate uplift force on a gravity dam of ht. 84 m due to seepage forces only. The U/S and D/S height of water level is 80 m and 6 m respectively. Base width of dam is 56 m and drainage gallery is provided at 8 m from heel of dam.**

[8 Marks]

Sol:

$$U_1 = \frac{1}{2}\left[\gamma_\omega H_1 + \gamma_\omega H_2 + \frac{1}{3}\gamma_\omega (H_1 - H_2)\right] \times 8$$

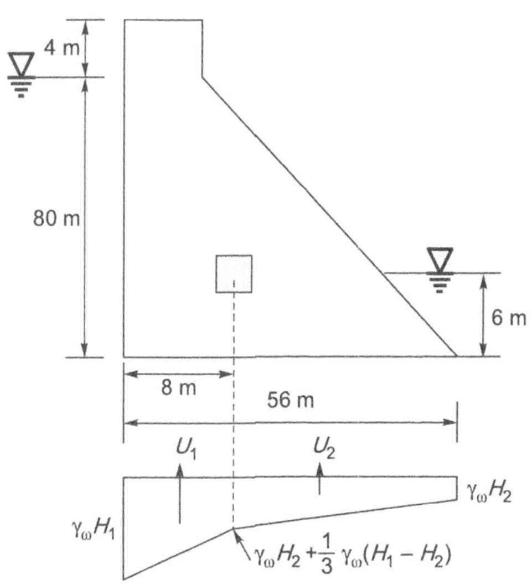

$$= \frac{1}{2} \times 9.81\left[80 + 6 + \frac{1}{3}(80 - 6)\right] \times 8 = 4342.56 \text{ kN}$$

$$U_2 = \frac{1}{2}\left[\gamma_\omega H_2 + \gamma_\omega H_2 + \frac{1}{3}\gamma_\omega (H_1 - H_2)\right] \times 48$$

$$= \frac{1}{2}\left[6+6+\frac{1}{3}(80-6)\right]\times 9.81\times 48 = 8632.8 \text{ kN}$$

$$U = U_1 + U_2 = 4342.56 + 8632.8 = 12975.36 \text{ kN}$$

3.4 The C/S of a trapezoidal gravity dam has top width of 5 m and bottom width 20 m, height 30 m, out of which 2 m is freeboard. Check the stability of the dam section neglecting uplift forces.

[10 Marks]

Sol:

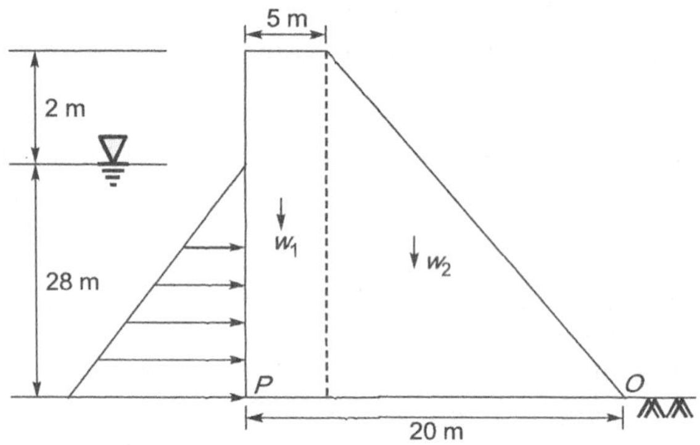

S/W of dam = 24 kN/m^3

Force	Calculation	Magnitude (kN/m)	LA (m) about O	Moment (kNm) about O
W_1	$24\times 30\times 5$	3600 (↓)	17.5	63000
W_2	$\frac{1}{2}\times 24\times 30\times 15$	5400 (↓)	10	54000
P(Water)	$\frac{1}{2}\times 9.81\times 28^2$	3845.52 (→)	9.33	35878.7

Resisting moment, M_R = 63000 + 54000 = 117000 kN-m/m

Overtaking moment, M_0 = 35878.7 kN-m/m

$$\text{FOS} = \frac{M_R}{M_0} = \frac{117000}{35878.7} = 3.26 > 1.5(\text{Safe})$$

For safety against sliding, (Take μ = 0.5)

$$(\text{FOS})_{\text{sliding}} = \frac{\mu\Sigma V}{H} = \frac{0.5\times(3600+5400)}{3845.52}$$

$$= 1.17 > 1 \text{ (Safe)}$$

So, dam is safe against sliding and overturning.

3.5 The following information is available regarding the relationship between trap efficiency and capacity, inflow ratio for a reservoir:

Capacity/inflow ratio	0.1	0.2	0.3	0.4	0.5	0.6	0.7	0.8	0.9	1.0
Trap efficiency (%)	87	93	95	95.5	96	96.5	97	97.2	97.3	97.5

Find the probable life of the reservoir of an initial reservoir capacity of 30 Mm^3, if the annual flood inflow is 60 Mm^3 and average annual sediment inflow is 360000 tonnes. Assume a specific weight of sediment equal to 1200 kg/m^3. The useful life of a reservoir will terminate when 80% of initial capacity is filled with sediment.

[12 Marks]

Sol:

Annual sediment inflow = 360000t = 3.6×10^8 kg

$$\therefore \quad \text{Volume of sediment inflow} = \frac{3.6\times10^8}{1200} = 0.3\,\text{Mm}^3$$

Initial reservoir capacity = 30 Mm^3

Annual flood inflow = 60 Mm^3

$$\therefore \quad \text{Initial capacity inflow ratio} = \frac{30}{60} = 0.5$$

This ratio goes on decreasing as the sedimentation takes place, and hence trap efficiency goes on decreasing. The computations for the life of reservoir are done below:
Assume the volume interval as 6 Mm^3. (i.e. 20% of the initial capacity)

Capacity		Capacity/inflow ratio	η	Avg. η for interval	Annual sediment trapped; S_t (Mm^3)	Yrs. to fill = $6/S_t$
%	Vol(Mm^3)					
100	30	0.5	0.96	0.9575	0.3 × 0.9575 = 0.2872	20.9
80	24	0.4	0.955	0.9525	0.3 × 9525 = 0.2857	21.0
60	18	0.3	0.95	0.9400	0.3 × 0.94 0.282	21.3
40	12	0.2	0.93	0.9000	0.3 × 0.9 = 0.27	22.2
20	6	0.1	0.87		Total	85.4 years

Hence, the useful life of the reservoir is ≈ 85 years.

3.6 **A reservoir has capacity of 50 Mm^3 and average annual inflow of 75 Mm^3, area of catchment is 1000 Km^2. Sediment yield and in place specific weight of sediment are estimated at 5400 kN/Km^2 and 13.5 kN/m^3 respectively. The trap efficiency may be approximated by**

$$y = 100\left[1-\frac{1}{65x+1}\right]^2$$

where y = Trap efficiency (in %)
x = Capacity - inflow ratio
Calculate how long it will take reservoir to fill up with sediments.

[10 Marks]

Sol:

Initial capacity of reservoir (C_0) = 50 Mm^3.

Average annual inflow, I = 75 Mm^2.

$$\text{Trap efficiency} = y = 100\left[1-\frac{1}{65x+1}\right]^2$$

Total inflow of sediment, W = 5400 × 1000 = 54×10^5 kN.

$$\text{Total volume of sediment, } V = \frac{54\times10^5}{13.5} = 4\times10^5\,\text{m}^3 = 0.4\ \text{Mm}^3$$

Taking 20% capacity loss as a volume interval for the reservoir.

Capacity (in Mm^3)	$x = CI$	y (Trap efficiency)(%)	$\bar{y}$ (%)	Life = $1000/V\bar{y}$ (in yrs.)
50	0.67	95.56	95	26.316
40	0.53	94.44	93.59	26.71
30	0.40	92.73	91.12	27.44
20	0.27	89.51	84.74	29.50
10	0.13	79.96	39.98	62.53
0	0	0		

$\Sigma = 172.5$ years

So, it will take 172.5 years for reservoir to be filled of sediments.

3.7 **The relative density of sediment particles is 2.6 and unit weight (dry) of 1 m³ of sediment is 980 kg, estimate the weight of 1 m³ of deposited sediment in the reservoir bed.**

[5 Marks]

Sol:

Volume of solid sediments, $V_S = \dfrac{980}{2.6 \times 1000} = 0.377\ m^3$

Volume of voids, $V_V = 1 - V_S = 1 - 0.377$

$= 0.623\ m^3$

Weight of sediments in reservoir bed (w) = Weight of solid + Weight of water

$\therefore \quad w = w_s + w_w$

$= 980 + 0.623 \times 1000 = 1603$ kg

Level-2

3.8 Figure shows the section (non-overflow portion) of a straight gravity dam built with concrete. Considerating water pressure and uplift pressure, and neglecting the other external forces acting on the dam, check whether the resultant passes through the middle third of the base for the reservoir full condition. In the figure, RL stands for Reduced Level in metres and MWL stands for Maximum Water Level. (Unit weight of water is 1000 kg/m³ and that of concrete is 2400 kg/m³)

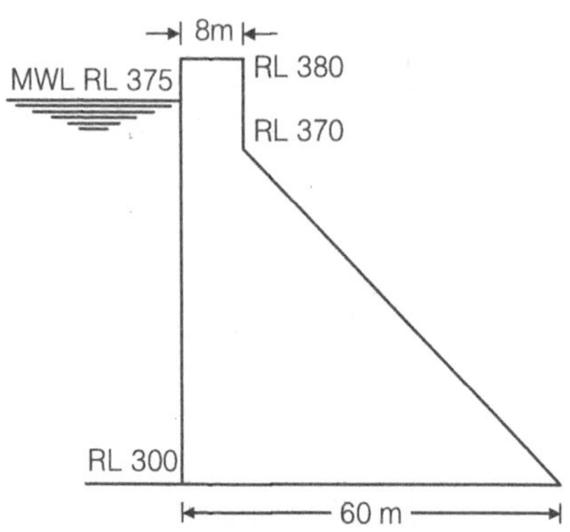

[20 Marks]

Sol:

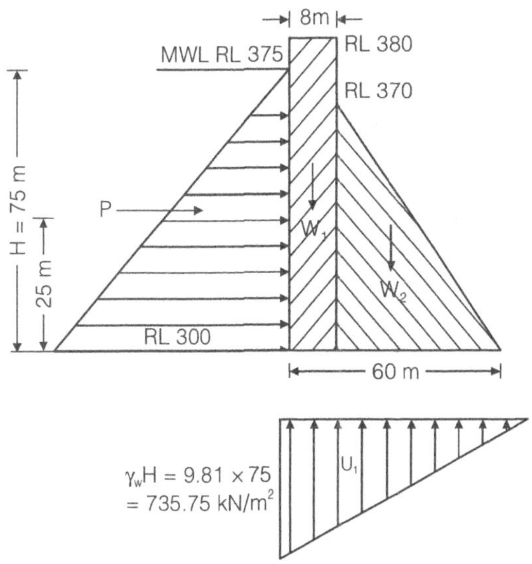

Unit weight of water

$$= 1000 \text{ kg/m}^3 = 1000 \times 9.81 \text{ N/m}^3 = 9810 \text{ N/m}^3 = 9.81 \text{ kN/m}^3$$

Unit weight of concrete

$$= 2400 \times 9.81 \text{ N/m}^3 = 23544 \text{ N/m}^3 = 23.544 \text{ kN/m}^3$$

Consider 1 m length of dam

The various forces and their moments about the toe are then calculated and tabulated in the table

Name of the force	Designation if given	Magnitude in kN	Lever arm in m	Moment about toe
Vertical Forces				
Downward weight of the dam	W_1	$8 \times 80 \times 1 \times 23.54$ $= 15068.16$	56	(+) 843816.96
	W_2	$\frac{1}{2} \times 52 \times 70 \times 23.54$ $= 42842.8$	34.67	(+) 1485359.88
		$\Sigma V_1 = 57910.96$		$\Sigma M_1 = 2329176.84$
Uplift pressure	U_1	$-\frac{1}{2} \times 60 \times 735.75$	40	882900
		$\Sigma V_2 = 22072.5$		$\Sigma V_2 = -882900$
Horizontal				
Water pressure on U/s face	P	$\frac{1}{2} \times 9.81 \times 75 \times 75 \times 1$ $= 27591$	25	(–)689775

Net moment,

$$\Sigma M = 2329176.84 - 882900 - 689775 = 756501.84 \text{ kNm}$$

$$\Sigma V = 57918 - 22072.5 = 35838.46 \text{ kN}$$

Distance of resultant from toe

$$\bar{x} = \frac{\Sigma M}{\Sigma V} = \frac{756501.84}{35838.46} = 21.10 \text{ m}$$

Eccentricity,

$$e = \frac{B}{2} - \bar{x}$$

$$= 30 - 21.11 = 8.89 \text{ m}$$

Since $$e < \frac{B}{6}$$

∴Resultant passes through the middle third of the base.

3.9 **For a given gravity dam so as to retain water with head of 60 m on upstream side and that of 12 m on downstream side,**

μ = 0.85 and concrete = 24 kN/m^3. Calculate

(i) FOS against overturning

(ii) FOS against sliding

(iii) Shear friction factor

Permissible shear stress of joint = 15 kg/cm^2.

Neglect EQ and wind forces, as well as wave pressure.

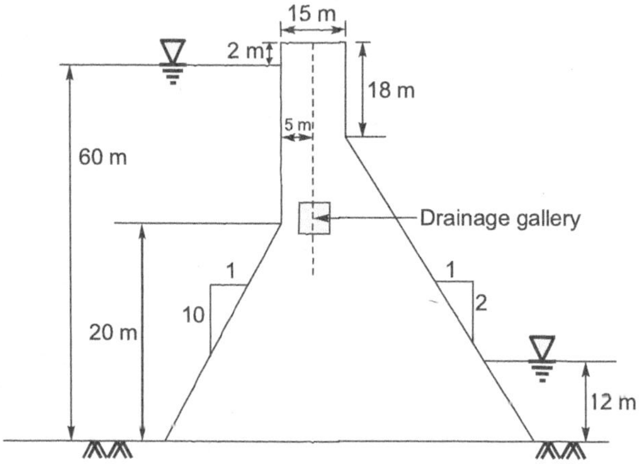

[20 Marks]

Sol:

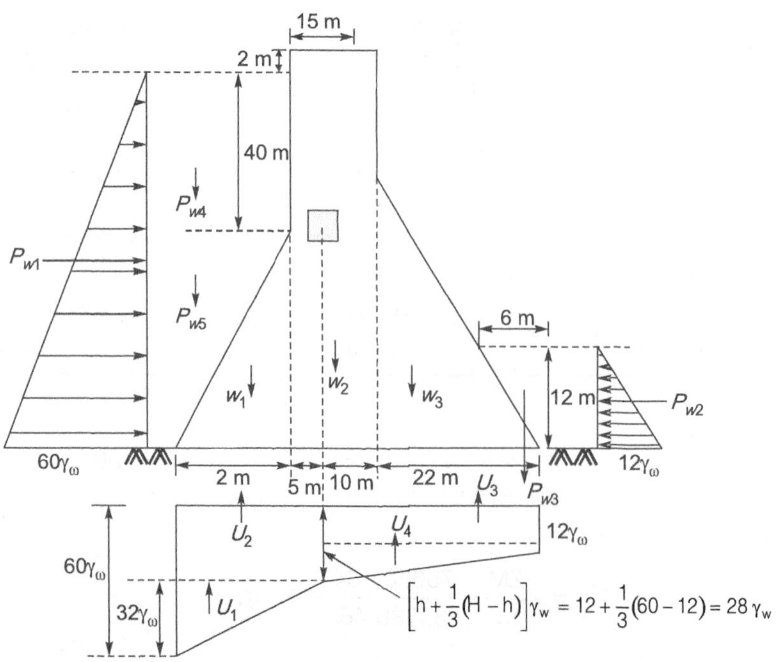

Force	f_H(kN)	f_V(kN)	Lever Arm (m)	M_{resist} (↑)	M_{over} (↓)
$W_1 = \frac{1}{2} \times 2 \times 20 \times 24$		480 (↓)	37.67	18081.6	
$W_2 = 15 \times 62 \times 24$		22320 (↓)	29.5	658440	
$W_3 = \frac{1}{2} \times 22 \times 44 \times 24$		11616 (↓)	14.67	170406.7	
$P_{w1} = \frac{1}{2}\, 60\, \gamma_\omega \times 60$	17658 (→)		20		353160
$P_{w2} = \frac{1}{2} \times 12 \times 12\, \gamma_\omega$	706.32 (←)		4	2825.28	
$P_{w3} = \frac{1}{2} \times 6 \times 12\gamma_\omega$		353.16 (↓)	2	706.32	
$P_{w4} = 40 \times 2\gamma_\omega$		784.8 (↓)	38	29822.4	
$P_{w5} = \frac{1}{2} \times 2 \times 20\gamma_\omega$		196.2 (↓)	38.33	7520.35	
$U_1 = \frac{1}{2} \times 7 \times 32\gamma_\omega$		1098.72 (↑)	36.67		–40290.06
$U_2 = 28 \times 7\gamma_\omega$		1922.76 (↑)	35.5		–68258
$U_3 = 12\gamma_\omega \times 32$		3767.04 (↑)	16		–60272.64
$U_4 = \frac{1}{2} \times 16 \times 32\gamma_\omega$		2511.36 (↑)	21.33		–53567.3
	$\Sigma f_H =$ 16951.7 kN	$\Sigma f_V =$ 26450.28 kN		$\Sigma M_R =$ 887802.65	$\Sigma M_0 =$ 575548

(i) Factor of safety against overturning,

$$(FOS)_D = \frac{\Sigma M_R}{\Sigma M_0} = \frac{887802.65}{575548} = 1.54 > 1.5 \qquad \text{(OK)}$$

(ii)
$$(FOS)_{sliding} = \frac{\mu \Sigma F_V}{\Sigma F_H} = \frac{0.85 \times 26450.28}{16951.7} = 1.32 > 1 \qquad \text{(OK)}$$

(iii)
$$\text{Shear friction factor} = \frac{\mu \Sigma F_V + (B \times 1) \times q}{\Sigma F_H}$$

$$q = 15 \text{ kg/cm}^2 = 15 \times 9.81 \times 10^4 \text{ kN/m}^2.$$
$$= 1471.5 \text{ kN/m}^2$$

$$= \frac{0.85 \times 26450.28 + 39 \times 1471.5}{16951.7} = 4.71 > 3 \qquad \text{(OK)}$$

3.10 **Design an elementary profile of gravity dam so as to retain water with head of 60 m., $\mu = 0.65$ and $G = 2.4$. Maximum compression strength of concrete is 2000 kN/m^2.**

[15 Marks]

Sol:

For uplift to be present

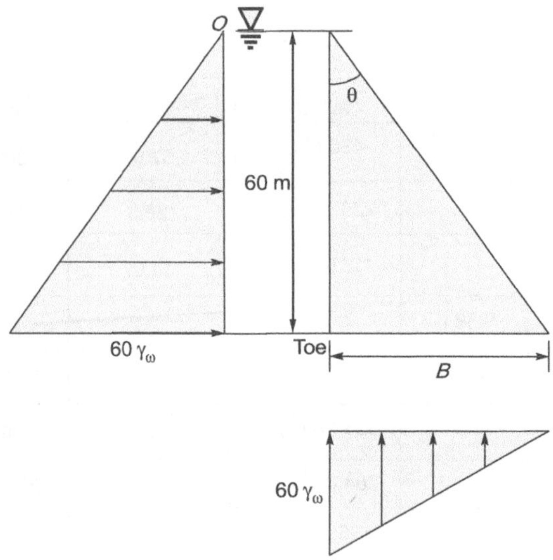

(i) Base width

For no tension failure, $$e = \frac{B}{6}$$

$$B_{min} = \frac{H}{\sqrt{G-C}} = \frac{60}{\sqrt{2.4-1}} = 50.71\text{m}$$

For safe to be in sliding

$$B_{min} = \frac{H}{\mu(G-C)} = \frac{60}{0.65(2.4-1)} = 65.93 \text{ m} \qquad (C = 1, \because \text{ uplift present})$$

$$B_{min} = \text{Maximum of } 65.93 \text{ m and } 50.71 \text{ m}$$

$\therefore$ Adopt, $B \geq 65.93$ m

$$B = 66 \text{ m}$$

For compression:

Maximum compression stress = $\sigma = \sigma_y [1 + \tan^2\alpha]$

$$\tan\alpha = \frac{66}{60} = 1.1$$

$$@ \text{ toe, } \sigma_y = \frac{\Sigma V}{B}\left[1 + \frac{6e}{B}\right]$$

For $$e = \frac{B}{6}$$

$\Rightarrow$ $$\sigma_y = 2 \times \frac{1}{2} \times \gamma_\omega H(G-C)$$

$$= 9.81 \times 60 \times (2.4 - 1) = 824.04 \text{ kN/m}^2.$$

$\therefore$ $$\sigma = \sigma_y [1 + \tan^2\alpha]$$

$$= 824.04 [1 + 1.1^2] = 1821.13 \text{ kN/m}^2 < 2000 \text{ kN/m}^2$$

$$H_{limiting} = \frac{f}{\gamma_\omega (S_C - C + 1)} = \frac{200}{9.81(2.4 - 1 + 1)}$$

$$= 84.94 \text{ m} > 60 \text{ m} \qquad \text{(OK)}$$

4. Water Logging, Theories of Seepage, Spillways and Miscellaneous

Level-1

4.1 Explain the term "exit gradient". Using Khosla's theory, estimate the value of exit gradient for a weir with a horizontal floor on a permeable foundation having width $b = 10$ m, and the depth of downstream sheet pile = 1.5 m, given the difference between upstream and downstream water levels is 4 m.

[10 marks]

Sol:

Exit gradient: The gradient i.e., head loss per unit length at the exit of seepage flow is called exit gradient.

According to Khosla's theory:

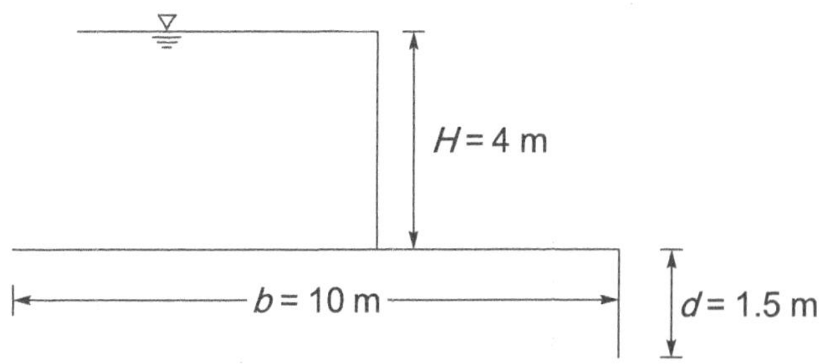

Exit gradient, $$G_e = \frac{H}{d} \times \frac{1}{\pi\sqrt{\lambda}}$$

$$\lambda = \frac{1+\sqrt{1+\alpha^2}}{2}$$

$$\alpha = \frac{b}{d} = \frac{10}{1.5} = 6.67$$

$$\therefore \quad \lambda = \frac{1+\sqrt{1+(6.67)^2}}{2} = 3.872$$

$$\therefore \quad G_e = \frac{4}{1.5} \times \frac{1}{\pi\sqrt{3.872}} = 0.431 = \frac{1}{2.32}$$

4.2 Explain the following terms:

(i) River bed aggradation

(ii) Armouring

(iii) Thalwegs

[12 marks]

Sol:

(i) **River bed aggradation :** When the sediment transporting capacity of a river at a point becomes less than the sediment load being carried, as a result of reduction in the velocity due to an increase in cross-section or reduction in slope of the river, the excess sediment gets deposited on the river bed. As a result, the riverbed rises and this phenomenon is being termed as aggradation. Often this phenomenon is noticed on the upstream of the dam.

(ii) **Armouring :** A phenomenon related to stream bed degradation is called armouring, where the coarsening of the bed material size results during degradation as finer particles get washed away. When the applied bed shear stress is sufficiently large to mobilize the large bed particles, degradation continues. When the bed shear stress cannot mobilize the coarse bed particles, an armour layer forms on the bed surface. The armour layer becomes coarser and thicker as the bed degrades until it is sufficiently thick to prevent any further degradation.

(iii) **Thalwegs :** The locus of the deepest point of the river along the length is called as thalweg. Most thalwegs pass through a succession of pools in the channel bed that are separated by riffles which might be sedimentary bed forms or bed rock ledger.

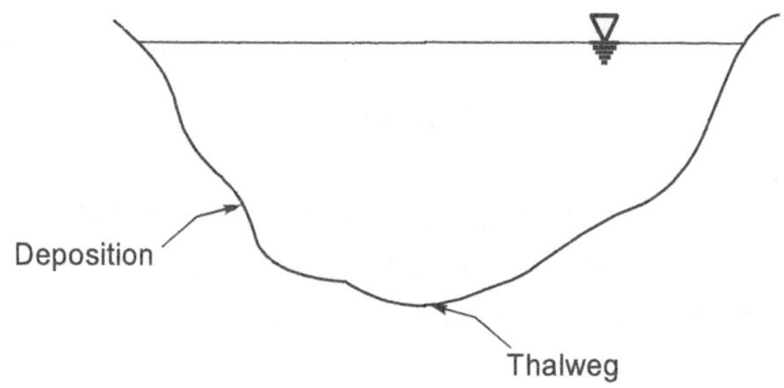

Cross sectional view of a river

4.3 **Name different river training works. What are various methods adopted for river training works?**

[8 Marks]

Sol:

On the basis of purpose required to be served, the river training works may be classified into the following types:

(i) High water training
(ii) Low water training
(iii) Medium water training

The various methods adopted for river training are as follows:

(i) Marginal embankments/levels.
(ii) Guide banks/guide bunds.
(iii) Gryones or spurs
(iv) Pitching of banks and provision of launching aprons.
(v) Pitched island
(vi) Miscellaneous methods

4.4 **What is meant by the terms 'Flexibility', 'Proportionality', 'Setting' and 'Sensitivity' as applied to modulus.**

[12 Marks]

Sol:

Flexibility (F): It is defined as the ratio of rate of the change of discharge of the outlet to the rate of change of discharge of the distributary channel.

Thus, $$F = \frac{dq / q}{dQ / Q}$$

where,

F = Flexibility of the outlet.
q = Discharge passing through the outlets.
Q = Discharge in the distributary channel.

$$q = C.H^m$$

(H = head acting on the outlet)
(C, m - Constant depending upon type of outlet)

$$Q = k.y^n$$

(y = Depth of water in the distributary)
k, n - Constant

So, $$F = \frac{m}{n} \times \frac{y}{H}$$

- For modular outlet, $F = 0$, $\Rightarrow$ Hence, it is known as a Rigid outlet or module.
- For semi-modular outlet, $F \neq 0 \Rightarrow$ Hence, it is known as a Flexible module.

Proportionality:

- The outlet is said to be proportional when rate of change of outlet discharge equal the rate of change of channel discharge, in other words, the outlet is 'Proportional' when 'Flexibility' equal unity.
- So, for a proportional outlet,

$$F = \frac{m}{n} \times \frac{y}{H} = 1$$

or

$$\frac{H}{y} = \frac{m}{n} = \frac{\text{Outlet index}}{\text{Channel index}}$$

- Ratio $\frac{H}{y}$ i.e. ratio of the depth of silt level of the outlet below the FSL of the distributary to the full supply depth of the distributary is known as setting.
- For a proportional outlet, $\frac{H}{y} = \frac{\text{Outlet index}}{\text{Channel index}}$
- For a wide trapezoidal channel, discharge is proportional to $y^{5/3}$. Therefore, for such channel, $n = 5/3$. Hence, the channel index is generally 5/3.

Sensitivity: Ratio of the rate of discharge through the outlet to the rate of change of water level of the distributary, referred to the normal depth of the channel. For rigid module, the distributary is referred to the normal depth of the channel. For rigid modules, the discharge is fixed and hence the sensitivity is zero.

- For flexible modulus, where the discharge through the outlet is independent of the water level of the water course and depends only upon the water level of the distributary a gauge can be fixed and calibrated so as to indicated its reading $G = 0$ when $q = 0$. Thus,

$$\text{Sensitivity, } S = \frac{dq/q}{dG/y}$$

Note: Sensitivity of a modular outlet is zero.

Sensitivity of an outlet for a wide trapezoidal or rectangular channel is equal to $\frac{5}{3}F$.

4.5 **Compute the discharge over an ogee weir coefficient of discharge equal to 2.4 at a head of 2 m. The length of the spillway is 100 m. The weir crest is 8 m above the bottom of the approach channel having the same width as that of the spillway.**

[8 Marks]

Sol:

Neglecting approach velocity and end contractions:

$$Q = CLH^{3/2}$$
$$= 2.4 \times 100 \times 2^{3/2} = 680 \text{ m}^3/\text{s}$$

The approach velocity is given by

$$V_a = \frac{Q}{\text{Height} \times \text{Width of channel}} = \frac{Q}{(8+2) \times 100} = \frac{680}{10 \times 100} = 0.68 \text{ m/s}$$

$$\therefore \quad h_a = \frac{V_a^2}{2g} = \frac{(0.68)^2}{2 \times 9.81} = 0.023 \text{ m}$$

$$H_e = H + h_a = 2 + 0.023 = 2.023 \text{ m}$$

$$\because \quad Q = 2.4 \times 100 \times H_e^{3/2} = 2.4 \times 100 \times 2.023^{3/2} = 691 \text{ m}^3/\text{s}$$

4.6 A dam has spillway whose cross-section is 1 m high and 4 m wide. The tail water elevation at design flow is 6 m below the summit of the siphon and the head water elevation is 1.5 m above the summit. Assuming a coefficient of discharge of 0.6, what is the capacity of the siphon? What head would be required on an ogee spillway (C = 2.25) 4 m long to discharge this flow? What length of ogee weir would be required to discharge the same flow as the siphon with a head of 1.5 m on the weir crest?

[6 Marks]

Sol:

For the siphon spillway:

$$\text{Net operative head} = 6 + 1.5 = 7.5 \text{ m}$$

$$Q = CA\sqrt{2gH} = 0.6\times1\times4\times\sqrt{2\times9.81\times7.5} = 29.1 \text{ m}^3/\text{s}$$

For the ogee spillway 4 m long,

$$Q = CLh^{3/2}$$

$$29.1 = 2.25\times4\times h^{3/2}$$

$$\Rightarrow \quad h = 2.18 \text{ m}$$

For the ogee spillway of h = 1.5 m

$$Q = CLh^{3/2}$$

$$29.1 = 2.25\times L\times(1.5)^{3/2}$$

$$\Rightarrow \quad L = 7.04 \text{ m}$$

4.7 A saddle siphon has the following data:

Full reservoir level = 435.00 m

Level of centre of siphon outlet = 429.60 m

High flood level = 435.85 m

High flood discharge = 600 m^3/s

If the dimensions of the throat of the siphon are:

Width = 4 m and height = 2 m.

Determine the number of siphon units required to pass the flood safely. The siphon discharges freely in air.

[5 Marks]

Sol:

$$\text{Maximum operative head, } H = \text{HFL} - \text{RL of outlet centre}$$
$$= 435.85 - 429.60 = 6.25 \text{ m}$$

The discharge is given by

$$Q = CA\sqrt{2gH}$$

Assume, $C = 0.65$

Now, $A = 4\times2 = 8 \text{ m}^2$

$$\therefore \quad Q = 0.65\times8\times\sqrt{2\times9.81\times6.25} = 57.58 \text{ m}^3/\text{s}$$

$$\therefore \quad \text{Number of units required} = \frac{600}{57.58} = 10.42 \approx 11.$$

4.8 The section of throat of saddle siphon spillway is width = 5 m and height = 2.5 m

Full reservoir level = +252.00 m

Crest level of spillway = + 248.00 m

Level of centre in the reservoir = +240.00 m

High flood level in the reservoir = + 254.00 m

If the high flood discharge is 650 m^3/s, determine the number of siphon units required to pass the high flood discharge safety. Coefficient of discharge = 0.85 and the siphon discharges freely into air.

[10 Marks]

Sol:

Throat section of saddle spillway = 5 × 2.5 = 12.5 m^2

Full reservoir level = 252.0 m

Crest level of spillway = 248 m

Level of centre of siphon outlet = 240 m.

High flood level = 254 m

High flood discharge = 650 m^3/s

Coefficient of discharge, C_d = 0.85

For each siphon of saddle spillway, operating head,

$$H = (\text{HFL} - \text{Level of centre of siphon outlet})$$

$$= 254 - 240 = 14 \text{ m}$$

$$a = 5 \times 2.5 = 12.5 \text{ m}^2$$

$\because$ Discharge through each siphon, $q = C_d \times a \times \sqrt{2gH}$ (For free discharge in the air)

$\therefore$ $q = 0.85 \times 12.5 \times \sqrt{2 \times 9.81 \times 14} = 176.093 \text{ m}^3/\text{s}$.

Let 'n' be the number of siphon for discharge of high flood discharge.

$\therefore$ $Q = nq$

$$n = \frac{Q}{q} = \frac{650}{176.093} = 3.69 \approx 4$$

≈ 4 number of saddle siphon to be installed

4.9 **Classify the irrigation water having following characteristics on analysis:**
$Ca = 2.32\ M_{eq}/l$, $M_{eq} = 1.44\ m_{eq}/l$, $Na = 17.73\ m_{eq}/l$, $EC = 480$ μmho/cm @ 25°C.
Also, indicate the suitability of the water for irrigation;

[5 Marks]

Sol:

$$\text{SAR} = \frac{Na^+}{\sqrt{\frac{Ca^{2+} + Mg^{2+}}{2}}} = \frac{17.73}{\sqrt{\frac{2.32 + 1.44}{2}}} = 12.93$$

$\because$ SAR in between 10 – 18 ⇒ Medium sodium water, S_2.

EC in between 250 – 750 μmho/cm @ 25°C

⇒ C_2 water (Medium salinity water)

So, given water is classified C_2S_2 group and is a moderate water.

This water can be used on coarse-textured or organic soils of good permeability. It can be used for irrigating almost all crops of moderate leaching. It may prone to be hazardous if used on fine-textured soils. So, addition of Gypsum ($CaSO_4$) either to soil or water can make its suitability for irrigating fine. textured soil as well.

4.10 Classify irrigation water based on its EC value and SAR criterion?

[5 Marks]

Sol:

(μ mhos/cm at 25°C)	Type of water	SAR value	Type of water
100 - 250	Low salinity water (C_1)	0 - 10	Low sodium water (S_1)
250 - 750	Medium salinity water (C_2)	10 - 18	Medium sodium water (S_2)
750 - 2250	High salinity water (C_3)	18 - 26	High sodium water (S_3)
> 2250	Very high salinity water (C_4)	> 26	Very high sodium water (S_4)

Level-2

4.11 An overflow ogee spillway has six bays each of 8 m clear length and five rounded nose piers (k_p = 0.01) each 2.5 m thick. Both the abutments are sharp edged (k_a = 0.2). The upstream face of the spillway is vertical. Compute the discharge that will pass over the spillway at the design head, H_d = 4 m. The discharge coefficient for design head C_0 = 2.1. Also compute the passing discharge over spillway when head over the crest is H = 4.45 m.

Neglect the effect due to approach velocity to the spillway. Discharge coefficients for head other than design head are given below.

$\frac{H}{H_d}$	0.6	0.85	1.11	1.5
$\frac{C}{C_0}$	0.94	0.98	1.01	1.06

C is discharge coefficient for head *H*.

[12 marks]

Sol:

$$\text{Design head, } H_d = 4 \text{ m}$$

Design head discharge coefficient, $C_0 = 2.1$

$$H_e = H_d + H_a$$

where, H_a = Head due to velocity of approach = $\frac{V^2}{2g}$

$$\text{Spillway discharge, } Q = CL_eH_e^{3/2}$$

where $$L_e = L - 2\left[Nk_p + k_a\right]H_e$$

$\because$ $$H_a = 0 \quad \text{(given)}$$

$\therefore$ $$H_e = H_d = 4 \text{ m}$$

$\therefore$ $$L_e = 6 \times 8 - 2\,[5 \times 0.01 + 0.2] \times 4 = 46 \text{ m}$$

Also $$\frac{H}{H_d} = \frac{4.45}{4} = 1.11$$

For $$\frac{H}{H_d} = 1.11,$$

$$\frac{C}{C_0} = 1.01$$

$$\Rightarrow \quad C = 1.01 \times 2.1 = 2.121$$

Discharge that will pass over the spillway at design head,

$$Q_1 = C_0 L_e H_e^{3/2}$$

$$\Rightarrow \quad Q_1 = 2.1 \times 46 \times 4^{3/2}$$

$$\Rightarrow \quad Q_1 = 772.8 \text{ m}^3/\text{s}$$

Discharge over spillway for head $H (= 4.45 \text{ m})$ over the crest

$$Q_2 = C L_e H^{3/2}$$

$$Q_2 = 2.121 \times 46 \times 4.45^{3/2}$$

$$Q_2 = 915.88 \text{ m}^3/\text{s}$$

4.12 What is called water logging? State the causes and effects of water lagging? Describe in detail reclamation methods of water logged area?

[15 Marks]

Sol:

A water lagging is a phenomena in which productivity of land gets affected due to high water table leading to the flowing of root zone of the plants and making the root zone of the plants aerated.

- The water lagging affects the productivity or fertility of the land and thus leads to a reduction in the crop yield.
- The water logging is usually called by rise of sub-soil water table.

Causes of water logging:

(i) Over and intensive irrigation of fields.
(ii) Seepage of water through the canals.
(iii) Inadequate surface drainage.
(iv) Heavy rains
(v) Obstruction of natural drainage will cause flooding in the area and consequent water logging of land.
(vi) Incorrect and defective methods of cultivation.
(vii) Inadequate natural drainage.
(viii) Seepage of water from adjoining high lands.
(ix) Low permeability of soil like black cotton etc.

Causes of water logging:

(i) Absence of aeration of soil in the RZ of plants.
(ii) Difficulty in cultivation operations.
(iii) Growth of waterweeds and wild aquatic plants.
(iv) Rise of salts in surface layers.
(v) Restricted root growth due to high water table.
(vi) Plants become vulnerable to attack by pests and diseases.

Reclamation methods of water logged area:

- Land reclamation is a process by which an uncultivable land is made fit for cultivation.
- Saline and water logged lands give very less crop yields and are therefore, almost unfit for cultivation unless they are reclaimed.
- All those measures which were suggested for preventing water logging holds good for preventing salinity of land also.
- After the high water table has been lowered by suitable drainages, the soil is made from existing salts by 'leaching' process.
- 'Leaching' is a process in which land is flooded of adequate depth of water. The alkali salts present in the soil gets dissolved in this water which percolate down to the water table.
- This process is repeated till the salts in the top layer of the land are reduced to such an extent that some salt resistance crops can be grown. Ex-Bajra, fodder etc.
- If the salinity is reduced to such an extent that ordinary crop like wheat, cotton etc. can be grown, the land is said to here been reclaimed.

4.13 **Figure below shows the section of a hydraulic structure founded on sand. Calculate the average hydraulic gradient. Also, find the uplift pressures at Points 6, 12 and 18 m from the U/S end of the floor and find the thickness of the floor at those points. Use Bligh's method.**

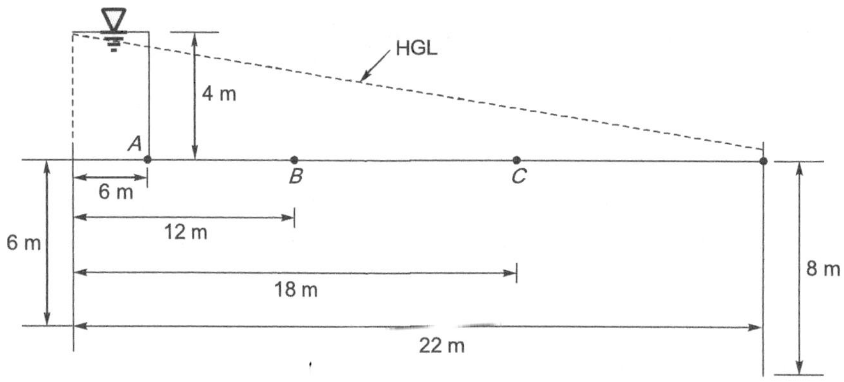

[15 Marks]

Sol:

$$\text{Total length of creep} = (2 \times 6) + 22 + (2 \times 8) = 50 \text{ m}$$

$$\text{Hydraulic gradient} = \frac{4}{50} = \frac{1}{12.5}.$$

(i) Uplift pressure at a point A, 6 m from U/S:

$$\text{Length of creep upto A} = 6 \times 2 + 6 = 18 \text{ m}$$

$$\therefore \quad \text{Unbalanced head, } (h_1) = 4\left(1 - \frac{18}{50}\right) = 2.56 \text{ m}$$

$$\therefore \quad \text{Uplift pressure} = \rho_w h_1 = 1 \times 2.56 = 2.56 \text{ t/m}^2$$

$$\text{Thickness, } t = \frac{4}{3} \times \frac{h_1}{G-1} = \frac{4}{3} \times \frac{2.56}{(2.24-1)} = 2.75 \text{ m} \qquad \text{(Assume } G = 2.24\text{)}$$

(ii) Uplift pressure at point B, 12 m from U/S.

$$\text{Length of creep upto } B = 6 \times 2 + 12 = 24 \text{ m}$$

$$\therefore \quad \text{Unbalanced head, } h_2 = 4\left(1 - \frac{24}{50}\right) = 2.08 \text{ m}$$

$\therefore$ Uplift pressure $= \rho_w h_2 = 1 \times 2.08 = 2.08$ t/m^2

$$\text{Thickness, } t = \frac{4}{3} \times \frac{h_2}{G-1} = \frac{4}{3} \times \frac{2.08}{(2.24-1)} = 2.24 \text{ m}$$

(iii) Uplift pressure at point C, 18 m from U/S.

Length of creep upto $C = 6 \times 2 + 18 = 30$ m

$$\therefore \quad \text{Unbalanced head, } h_3 = 4\left(1 - \frac{30}{50}\right) = 1.6 \text{ m}$$

$\therefore$ Uplift pressure $= \rho_w h_3 = 1 \times 1.6 = 1.6$ t/m^2

$$\text{Thickness, } t = \frac{4}{3} \times \frac{h_3}{G-1} = \frac{4}{3} \times \frac{1.6}{(2.24-1)} = 1.72 \text{ m}$$

4.14 **Samples of two waters yielded the following chemical analysis:**

Chemical	Sample - I	Sample - II
Na (mg/l)	114	410
Ca (mg/l)	102	106
Mg (mg/l)	32	30
EC (μ mho/cm)	1230	2860

Assuming these samples represent average conditions, compute the SAR, for each water and classify then. How much gypsum ($CaSO_4$) would be required to reduce the SAR of one hectare meter of sample -II water to 5?

[15 Marks]

Sol:

Convert Na, Ca, Mg salt contents in m_{eq}/ml from mg/l

$\therefore$ Salt content of two samples of water in m_{eq}/ml:

	Sample– I	Sample– II
Na	$\frac{114}{23} = 4.95$	$\frac{410}{23} = 17.8$
Ca	$\frac{102}{20} = 5.1$	$\frac{106}{20} = 5.3$
Mg	$\frac{32}{12} = 2.66$	$\frac{30}{12} = 2.50$

SAR of sample - I:

$$\text{SAR}_I = \frac{Na^+}{\sqrt{\dfrac{Ca^{+2} + Mg^{+2}}{2}}} = \frac{4.95}{\sqrt{\dfrac{5.1 + 2.66}{2}}} = 2.51 \Rightarrow S_1$$

EC value = 1230 μmho/cm $\Rightarrow C_3$.

So, water of sample I = C_3S_1 and is moderate water.

SAR of sample - II:

$$\text{SAR}_{II} = \frac{17.8}{\sqrt{\dfrac{5.3 + 2.5}{2}}} = \frac{17.8}{\sqrt{3.9}} = 9.01 \Rightarrow S_1$$

EC value = 2250 μmho/cm ⇒ C_4.

Classified as C_4S_1 water and is poor quality water.

Requirement of gypsum for sample - II:

$$SAR_{II} = 9.01$$

$$(SAR_{II})_{new\ reqd.} = 5$$

So,
$$SAR = \frac{Na^+}{\sqrt{\frac{Ca^{+2} + Mg^{+2}}{2}}} \Rightarrow 5 = \frac{17.8}{\sqrt{\frac{Ca^{+2} + 2.50}{2}}}$$

⇒
$$Ca^{+2} = 22.84\ m_{eq}/ml = 22.84 \times 20 = 456.8\ mg/l.$$

$$\%\text{ of Ca in } CaSO_4 \text{ (gypsum)} = \frac{40}{136} \times 100 = 29.41\%$$

∴
$$\text{Quantity of gypsum } (CaSO_4) = \frac{456.8}{0.294} mg/l = 1553.74\ mg/l = 1.55\ kg/m^3.$$

∴ Quantity of $CaSO_4$ required per hectare meter = 1.55 × 100 × 100 kg = 15.5 tonnes

4.15 **The annual rainfall in a certain area is 80 cm. Find the spacing of drains if 1% of average annual rainfall is to be drained in 24 hrs. Given depth of impervious stratum from top of soil surface = 10 m. Position of drains is 1.8 m below top soil surface and the depth of highest position of WT below top soil surface = 1.5 m. Take Permeability, $k = 1 \times 10^{-4}$ m/s.**

[15 Marks]

Sol:

1% of average rainfall in an area, $A = \frac{0.80A}{100} m^3$. This is to be removed in 24 hours.

So,
$$Q_D = \frac{0.80A}{100 \times 24 \times 3600} m^3/s$$

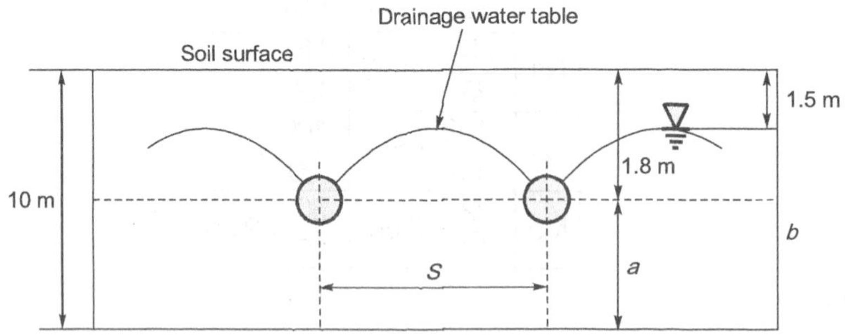

Now,
$$S = \frac{4k(b^2 - a^2)}{Q_D} = \frac{4 \times (1 \times 10^{-4})(b^2 - a^2)}{Q_D}$$

$a = 10 - 1.8 = 8.2$ m

$b = 10 - 1.5 = 8.5$ m,

δ = Spacing between the drains.

∴
$$S = \frac{4 \times 10^{-4}(8.5^2 - 8.2^2) \times 100 \times 24 \times 3600}{0.80A}$$

But the area to be catered by drain is $S \times B$.

Where B = Breadth of the drain or its longitudinal length.

Taking this to be 1 m.

$$A = S \times 1 = S$$

$$\therefore \quad S = \frac{21643.2}{S}$$

$$S = 147.12 \text{ m}$$

∴ Drains should be at 147 m c/c.

4.16 **A river discharge 1000 m^3/s of water at high flood level of RL = 103. A weir is constructed for flow diversion of a crest length of 255 m and total length of concrete floors as 40 m. The weir has to sustain the under seepage at a maximum static head of 2.4 m. The silt factor and the safe exit gradient for the river bed material are 1.1 and 1/6 respectively. Determine the depth of cut-off required at the D/S end of the concrete floor. Take the length of D/s concrete floor as RL = 100. Check for exit gradient.**

[15 Marks]

Sol:

Given: $Q = 1000$ m^3/s, $L = 255$ m

∴ Discharge per unit length of weir, $(q) = \frac{Q}{L} = \frac{1000}{255} = 3.92 \text{ m}^3/\text{s}$

Now, for a discharge intensity, (q) the normal depth of scour is given by Lacey's equation i.e.

$$R = 1.35\left(\frac{q^2}{f}\right)^{1/3} = 1.35\left(\frac{3.92^2}{1.1}\right)^{1/3} = 3.25 \text{ m}$$

Adopting D/S cutoff upto 1.5 R below the D/S water level.

$$\therefore \quad 1.5R = 1.5 \times 3.25 = 4.875 \text{ m}$$

U/S water level = U/S HFL = 103 m (Given)

H = Maximum static head causing seepage = 2.4 m (Given).

∴ D/S water level = U/S HFL − H = 103 − 2.4 = 100.6 m

Hence, RL of bottom of D/S cutoff = 100.6 − 4.875 = 95.725 m

Now, RL of D/S floor = 100 m (Given)

∴ Depth of D/S cutoff = 100 − 95.725 = 4.275 m

Hence, $d = 4.275$ m

Check for exit gradient, $$G_E = \frac{H}{d} \times \frac{1}{\pi\sqrt{\lambda}}$$

where, $$\lambda = \frac{1+\sqrt{1+\alpha^2}}{2}$$

where, $$\alpha = \frac{b}{d} = \frac{\text{Length of weir floor}}{\text{Depth of D/S cutoff}} = \frac{b}{d} = \frac{40}{4.275} = 9.357$$

$$\therefore \quad \lambda = \frac{1+\sqrt{1+9.357^2}}{2} = 5.2$$

$$\therefore \quad G_E = \frac{2.4}{4.275} \times \frac{1}{\pi\sqrt{5.2}} = \frac{1}{12.77} < \frac{1}{6} \text{ (Safe exit gradient).}$$

Hence, the weir is safe from exit gradient considerations of bottom of D/S cutoff at RL of 95.725 m.

■■■■

04 Environmental Engineering

1. Water Demand

Level-1

1.1 The population of a city in three consecutive years i.e., 1991, 2001 and 2011 is 80000, 250000 and 480000 respectively. Determine

(a) The saturated population

(b) The equation of logistic curve

(c) The expected population in 2021

[8 Marks]

Sol:

In this questions, we have $P_0 = 80000\ ;\ t_0 = 0$

$P_1 = 250000\ ;\ t_1 = 10$ years

$P_2 = 480000\ ;\ t_2 = 20$ years

(a) Saturation population:

$$P_s = \frac{2P_0 P_1 P_2 - P_1^2(P_0 + P_2)}{P_0 P_2 - P_1^2}$$

$$= \frac{2 \times 80000 \times 250000 \times 480000 - (250000)^2 \cdot (80000 + 480000)}{80000 \times 480000 - (250000)^2}$$

$$= 655601.6598 \approx 655602$$

(b) The equation of logistic curve

$$m = \frac{P_s - P_0}{P_0} = \frac{655602 - 80000}{80000} = 7.195$$

$$n = \frac{1}{t_1} \ln\left[\frac{P_0(P_s - P_1)}{P_1(P_s - P_0)}\right] = \frac{1}{10} \ln\left[\frac{80000(655602 - 250000)}{250000(655602 - 80000)}\right] = -0.1489$$

$\therefore$ Equation of logistic curve

$$P = \frac{P_s}{1 + m \cdot \log_e^{-1}(nt)} = \frac{655602}{1 + 7.195 \log_e^{-1}(-0.1489t)}$$

(c) The expected population in 2021 i.e., after 3-decades from the start.

i.e., $t = 30$ yrs

$$\therefore \quad P_{2021} = \frac{655602}{1 + 7.195 \log_e^{-1}(-0.1489 \times 30)} = 605436$$

1.2 Compute the population of the year 2000 and 2006 for a city whose population in the year 1930 was 25000, and in the year 1970 was 47000. Make use of geometric increase method.

[6 marks]

Sol:

In this question, the intermediate census data between 1930 to 1970 is not given, and hence geometric mean method of all known decades is not possible. The growth rate per decade (r) can, however, be computed by using as

$$r = \sqrt[t]{\frac{P_2}{P_1}} - 1 = \sqrt[4]{\frac{47000}{25000}} - 1$$

$$= 0.17095$$

$$= 17.095\% \text{ per decade}$$

Now, $$P_n = P_0\left(1+\frac{r}{100}\right)^n, \text{ we have}$$

Hence, $$P_{2000} = P_3 \text{ (after 3 decades from 1970 onward)}$$

$$= P_{1970}\left(1+\frac{r}{100}\right)^3$$

$$= 47000\,(1+0.17095)^3$$

$$= 47000(1.17095)^3$$

$$= 75459.9 \simeq 75460$$

Population for the year 2006, means that it is after 36 years (3.6 decades) from 1970 onward.

$$P_{2006} = P_{3.6} = P_{1970}\,(1+0.17095)^{3.6}$$

$$= 47000\,(1.17095)^{3.6} = 82954$$

1.3 **The population of 5 decades from 1930 to 1970 are given in table below. Find out the population after one, two and three decades beyond the last known decade, by using arithmetic increase method.**

Year	1930	1940	1950	1960	1970
Population	25000	28000	34000	42000	47000

[6 marks]

Sol:

(i) The given data in question table is extended in table below, so as to compute the increase in population (x) for each decade (col. 3), the total increase, and average increase per decade $(\bar{x})$, as shown

Year	Population	Increase in population (x)
(1)	(2)	(3)
1930	25000	
1940	28000	3000
1950	34000	6000
1960	42000	8000
1970	47000	5000
Total		22000
Average increase per decade $(\bar{x})$		$\bar{x} = \frac{22000}{4} = 5500$

The future populations are now computed by using equation as,

$$P_n = P_0 + n \cdot \bar{x}$$

∴ (a) Population after 1 decade beyond 1970

$$P_{1980} = P_1 = P_{1970} + 1 \cdot \bar{x}$$
$$= 47000 + 1 \times 5500$$
$$= 52500$$

(b) Population after 2 decades beyond 1970

$$P_{1990} = P_2 = P_{1970} + 2 \cdot \bar{x}$$
$$= 47000 + 2 \times 5500$$
$$= 58000$$

(c) Population after 3 decades beyond 1970

$$P_{2000} = P_3 = P_{1970} + 3 \cdot \bar{x}$$
$$= 47000 + 3 \times 5500 = 63500$$

Level-2

1.4 **A water supply scheme has to be designed for a city having a population of 100000. Estimate the important kinds of drafts which may be required to be recorded for an average water consumption of 250 lpcd. Also record the required capacities of the major components of the proposed water works system for the city using a river as the source of supply. Assume suitable figures and data where needed.**

[12 Marks]

Sol:

Given data : Population = 100000, Average water consumption – 250 lpcd

(i) Average daily draft, q = (per capita average consumption in litres/person/day) × population
= 250 × 100000 litres/day = 25 × 10^6 litres/day
= 25 millions litres/day = 25 MLD

(ii) Maximum daily draft : May be assumed as 180% of annual average daily draft i.e.
Maximum daily draft = 1.8 × Average daily consumption = 1.8 × 25 = 45 MLD

(iii) Maximum hourly draft of the maximum day i.e., (peak demand)
= 1.5 × Average hourly consumption of the maximum day
= $1.5 \times (1.8q) = 2.7q$ = 2.7 × 25 = 67.5 MLD

(iv) Fire demand: Since population = 100000 < (2 lakh)

∴ As per national board of fire under writers formula.

$$Q = 4637\sqrt{P}\left[1 - 0.01\sqrt{P}\right] L/\text{min}$$

where P = Population in thousands = 100000 = 100 thousands

$$Q = 4637\sqrt{100}\left[1 - 0.01\sqrt{100}\right] \text{ L/min}$$

$$= 41733 \text{ L/min} = \frac{41733 \times 60 \times 24}{10^6} \text{ MLD} = 60.09 \text{ MLD}$$

(v) Coincident draft:

Coincident draft = Maximum daily draft + Fire draft
= 45 + 60.09 = 105.09 MLD

- Capacities of various components of water works:

S.No.	Water distribution system	Designed for	Design Capacity
1.	Intake structure	Maximum daily draft	45 MLD
2.	Pipe mains (to take water from source to service reservoir), filter and other treatment units	Maximum daily draft	45 MLD
3.	Distribution system (to carry water from service reservoir to water taps)	Maximum of 1. Maximum hourly demand of the maximum day 2. Coincident draft	Maximum of {67.5, 105.09} = 105.09 MLD

1.5 Obtained from census record is as following table,

Year	1970	1980	1990	2000	2010
Population	15000	20000	24500	29500	32500

Estimate the population of the locality in 2040, using arithmetic increasing, geometrical progressions, increase incremental, and rate of decrease methods.

[12 Marks]

Sol:

Year	Population	Increment per decade	Incremental increase	Percentage increment per decade	Decrease in the % increase
1970	15000				
		5000		$\frac{5000}{15000}\times 100 = 33.33\%$	
1980	20000		–500		10.83%
		4500		$\frac{4500}{20000}\times 100 = 22.5\%$	
1990	24500		500		2.092%
		5000		$\frac{5000}{24500}\times 100 = 20.408\%$	
2000	29500		–2000		10.239%
		3000		$\frac{3000}{29500}\times 100 = 10.169\%$	
2010	32500				
		$\bar{x} = \frac{5000+4500+5000+3000}{4} = 4375$	$\bar{y} = \frac{-500+5000-2000}{3} = -666.67$	$r = (33.33\times 22.5\times 20.405\times 10.169)^{1/4} = 19.862$	$r' = \frac{(10.83+2.092+10.239)}{3} = 7.75$

Now, Arithmetic increase method:

Population after n-decades from the present, (i.e., last known census)

$$P_n = P_0 + n\bar{x}$$

Where, P_0 = Population in present , n = Number of decades between now and future

$\bar{x}$ = Average of population increase in the known decades

In 2040,

i.e., $n = 3$

$$P_{2040} = P_{2010} + 3\bar{x} = 32500 + 3\times 4375 = 45625$$

Geometric Progression Method:

$$P_n = P_0\left(1+\frac{r}{100}\right)^n \qquad r = \text{growth rate (in \%)}$$

$$P_{2040} = P_{2010}\left(1+\frac{19.862}{100}\right)^3 = 32500\left(1+\frac{19.862}{100}\right)^3 = 55966.47 \simeq 55967$$

Incremental Increase Method:

$$P_n = P_0 + n\bar{x} + \frac{n(n+1)}{2}\cdot\bar{y}$$

$\bar{y}$ = Average of incremental increase over the known decades.

$$P_{2040} = P_{2010} + 3x + \frac{3(3+1)}{2}\cdot\bar{y}$$

$$= 32500 + 3 \times 4375 + \frac{3 \times 4}{2} \times \left(-\frac{2000}{3}\right) = 41625$$

Decreasing rate of growth method:

$$P_n = P\left[1 + \frac{r_0 - r'}{100}\right]\left[1 + \frac{r_0 - 2r'}{100}\right]\left[1 + \frac{r_0 - 3r'}{100}\right] \ldots \left[1 + \frac{r_0 - nr'}{100}\right]$$

r_0 = growth rate of last decade

r' = Average decrease in percentage growth rate

$$\therefore \quad P_{2040} = P_{2010}\left[1 + \frac{10.169 - 7.72}{100}\right]\left[1 + \frac{10.169 - 2 \times 7.72}{100}\right]\left[1 + \frac{10.169 - 3 \times 7.72}{100}\right]$$

$$= 32500\left[1 + \frac{2.449}{100}\right]\left[1 + \left(-\frac{5.271}{100}\right)\right]\left[1 + \left(\frac{-12.991}{100}\right)\right]$$

$$= 32500 \times (1.02449) \times (0.94729) \times (0.87009) = 27443.418$$

$$\therefore \quad P_{2040} = 27444$$

1.6 **(a) Discuss variation in water demand and their effects on the design of various components of water supply scheme.**

(b) What is meant by coincident draft and how it is used in estimating design draft for design of distribution system.

[10 + 10 = 20 Marks]

Sol:

(a) The per capita consumption or demand varies not only from year to year and from season to season but more important from day to day and hour to hour.

Assessment of variation in demand:

- The smaller the town, the more variable is the demand.
- The maximum demands (monthly, daily, hourly) are generally expressed as ratios of their means. These ratio may vary considerably for different communities. The following figures are generally adopted:

  Maximum daily demand = 1.8 × Average daily demand

  Maximum hourly demand of maximum day i.e., peak demand

$$= 1.5 \times \text{Average hourly demand} = 1.5 \times \frac{\text{Maximum daily demand}}{24}$$

$$= 1.5 \times \left(\frac{1.8 \times \text{Avg. daily demand}}{24}\right) = 2.7 \times \text{Annual average hourly demand}$$

  Maximum weekly demand = 1.48 × Average weekly demand

  Maximum monthly consumption/demand = 1.28 × Average monthly demand

- A water supply system has several units and design of each unit should match with the hourly, daily and seasonal variation in the demand.

Seasonal Variation: The demand peaks during summer fire breakouts are generally more in summer, increasing demand. So, there is seasonal variation.

Daily Variation: Depends on the activity, people draw out more water on sunday and festival days, thus increasing demand on these days.

Hourly Variation: It is very important as they have a wide range during active household working hours i.e., from six to ten in the morning and four to eight in the evening, the bulk of the daily requirement is taken. So, an adequate quantity of water must be available to meet the peak demand. To meet all the fluctuations, the supply pipes, service reservoirs, distribution pipes must be properly proportioned.

- The water is supplied by pumping directly and the pumps and distribution system must be designed to meet peak demand.

- The effect of monthly variation influences the design of storage reservoirs and the hourly variations influences the design of pumps and service reservoirs.

(b) Coincident draft: The maximum daily demand (i.e., 1.8 times the average daily demand) when added to the fire demand is known as the coincident draft.

- For general community purposes, the total draft is not taken as the sum of maximum hourly demand and fire demand, but it is taken as sum of maximum daily demand and fire demand or maximum hourly demand, whichever is more.
- The distribution system should be designed for maximum hourly draft of maximum day or coincident draft with fire, whichever is more.

1.7 **(a) Describe the 'incremental method' by which the population of a locality on a future data can be estimated. Also state its advantages over arithmetical and geometrical methods.**

(b) In a town, it has been decided to provide 200 litres per head per day in the 21st century. Estimate the domestic water requirements of this town in the year 2000 by projecting the population of the town by incremental increase method.

Year	1940	1950	1960	1970	1980
Population	2,37,98,624	4, 69, 78, 325	5, 47, 86, 437	6, 34, 67, 823	6, 90, 77, 421

[15 Marks]

Sol:

(a) Incremental Increase Method (or method of varying increment):

- In this method, growth rate is not assumed to be constant as in the arithmetic or geometric progression methods, but is progressively increasing or decreasing, depending upon whether the average of the incremental increases in the past data is positive or negative.
- The population for the future decade is found out by adding the mean arithmetic increase ($\bar{x}$) to the last known population as in arithmetic increase method and this is added to the average of the incremental increase $\bar{y}$, once for the first decade, twice for second decade, and so on.
- The population after n decades from present (i.e. last known census) is given by:

$$P_n = P_0 + n\bar{x} + \frac{n(n+1)}{2}\bar{y}$$

where, $\bar{x}$ = average increase of population of known decades.

$\bar{y}$ = average of incremental increases of the known decades.

- This method will give results, somewhere between the results given by arithmetic increase method and geometric increase method and thus considered to be giving quite satisfactory results.
- The geometric progression method, gives highest values of forecasted population. This method is suitable for new and younger cities expanding at faster rate. For old cities, the arithmetic method may be better, although incremental method is considered to be the best for any city whether old or new.

(b)

Year	Population	Increase in population	Increment over the increase, i.e. (incremental increase)
1940	2,37, 98,624		
		2,31, 79,701	
1950	4, 69, 78, 325		–153, 71, 589
		78, 08, 112	
1960	5, 47, 86, 437		8, 73, 274
		86, 81, 386	
1970	6, 34, 67, 823		–30, 71, 788
		56, 09, 598	
1980	6, 90, 77, 421		
Total		4,52,78,797	–1,75,70,103
Average per decade		$\bar{x} = \frac{45278797}{4}$ = 1, 13, 19, 699	$\bar{y} = \frac{-17570103}{3}$ = –58, 56, 701

$\therefore$ Expected population at the end of year (2000)
i.e. after 2-decades from last known census

$$\therefore \quad P_{2000} = P_{1980} + 2\bar{x} + \frac{2(2+1)}{2} \cdot \bar{y}$$
$$= 69077421 + (2 \times 11319699) + 3 \times (-5856701)$$
$$= 74146716$$

$\therefore$ Water requirement in 2000@200 *l*/head/day

$$= \frac{200 \times 74146716}{10^6} \text{ ML/day} = 14829 \text{ MLD}$$

2. Conduits for Transporting Water & Distribution System

Level-1

2.1 For a town daily requirement of water for supply to the population is 2 lakh liters. The pattern of demand of is as follows:

6 a.m to 9 a.m - 30% of day; supply
9 a.m to 4 p.m - 35% of day; supply
4 p.m to 7 p.m - 25% of day; supply
7 p.m to 6 a.m - 10% of day; supply

The pumping is done for 10 hrs a day from 8 a.m to 6 p.m. Find out the storage capacity of the distribution reservoir for the water supply scheme.

[10 Marks]

Sol:

Total daily demand/requirement = 200000 litres

Now, the cumulative demand covered is tabulated in the table given below:

Period	Rate of demand	Demand (litres)	Cumulative demand
6 am to 9 am	30% of 200000	60000	60000
9 am to 4 pm	35% of 200000	70000	130000
4 pm to 7 pm	25% of 200000	50000	180000
7 pm to 6 am	10% of 200000	20000	200000

Total demand is met in (10 hrs)

$$\text{Rate of supply} = \frac{200000}{10} = 20000 \text{ liters/hr}$$

The mass curves of demand has been plotted in figure.

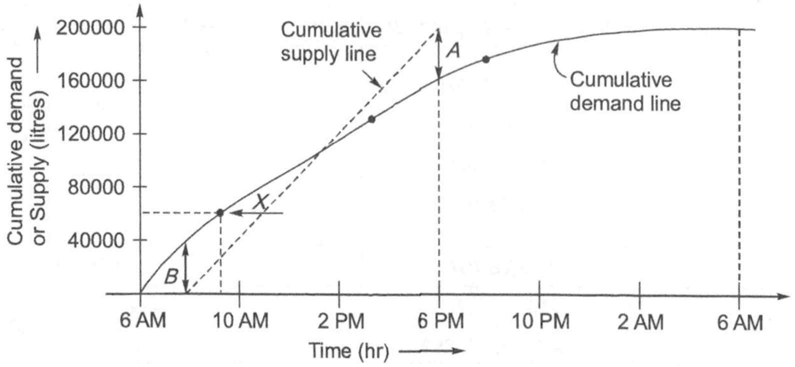

Two maximum ordinates between supply and demand lines are

$$A = 30000 \text{ liters}$$
$$B = 42000 \text{ liters}$$

$\therefore$ Total storage capacity = $A + B = 42000 + 30000 = 72000$ litres

2.2 **Discuss the functioning and applicability of the following:**

(i) Air value **(ii) Reflux value**

(iii) Flanged joint **(iv) Expansion joint**

Sol:

(i) Air value:

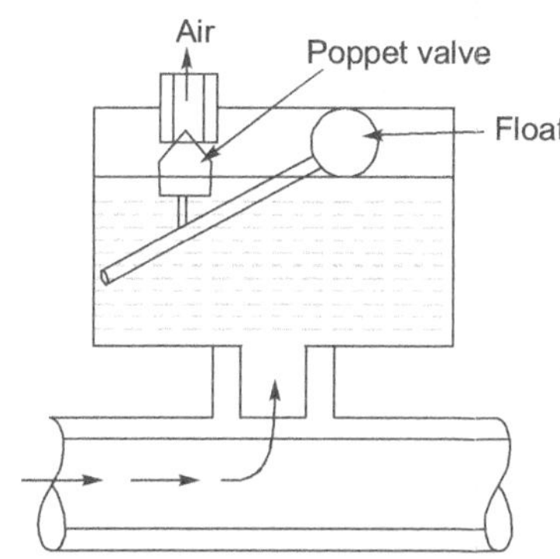

- These are the special kind of valves which are generally placed along the pipeline at "summits" on both sides of the sluice valves and also on the down streams side of all other sluice valves.
- When placed on summits which are very near the HGL or sometimes above HGL (especially during negative water hammers) they ensure the safety of the pipe against collapse.
- Similarly, when placed below the ordinary sluice valves, they will protect the pipe against the negative pressures.
- A valve which will open out automatically as soon as the pressure in the pipe falls below a certain fixed predetermined value, and thus allowing air to enter the pipe is known as air inlet valve.
- Sometimes, air gets accumulated at high points when the supply is restored and the pipe is refilled after repairs. The accumulated air obstructs the free flow of water and the pipe may get air locked. Air relief valves are therefore required to be provided at all summits to remove the accumulated air under such circumstances.
- The most commonly used air valve is single orifice air valve. However, sometimes double acting orifice air valve is also provided.

Reflux Valve:

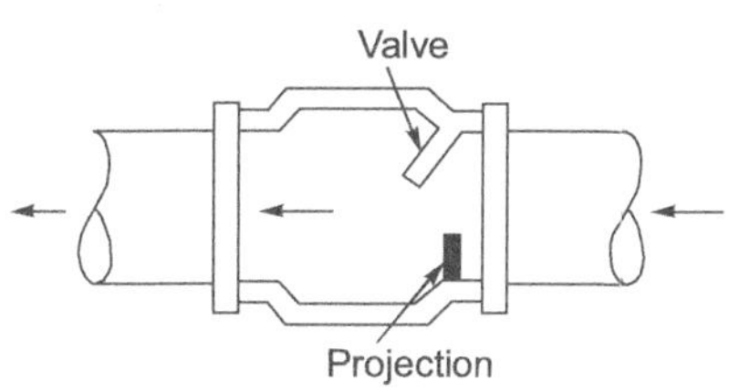

- These are sometimes, called non-return valves because they prevent water to flow back in the opposite direction.
- They may be installed on the delivery side of the pumping set, so as to prevent the back flow of stored or pumped water, when the pump is stopped.
- These are also installed on pump discharges to reduce water hammer forces on the pump.
- These are also required at interconnections between a polluted water system and a potable water system, so as to prevent the entry of pollution into the pure water.
- The simplest type of reflux valve or check valve is a flap shutter hinged at the outlet and of pipe, which opens out in the direction of flow but closes of its own weight when flow in the permissible direction ceases.

Flanged Joint:

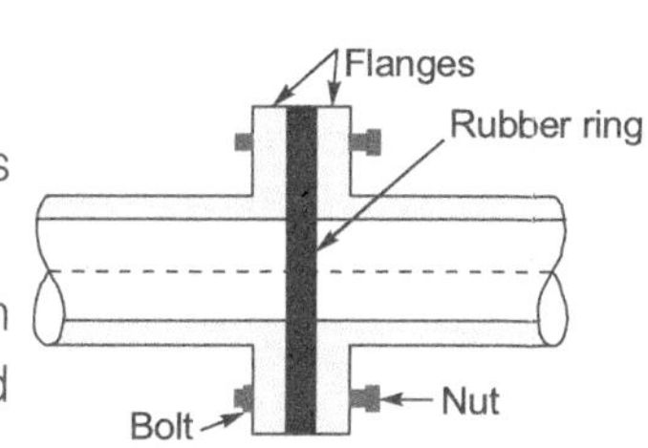

- These are used for pumping stations, filter plants, and at other locations where it may be necessary to occasionally disjoint the pipe.
- Two flanged pipes are brought together, keeping a rubber washer in between so as to make them water tight. They are fixed by means of nuts and bolts.
- These joints are strong but rigid and hence cannot be used where deflections or vibrations are expected.
- They are expensive and mostly used for indoor works such as pumping stations, filter plants etc.

Expansion Joint:

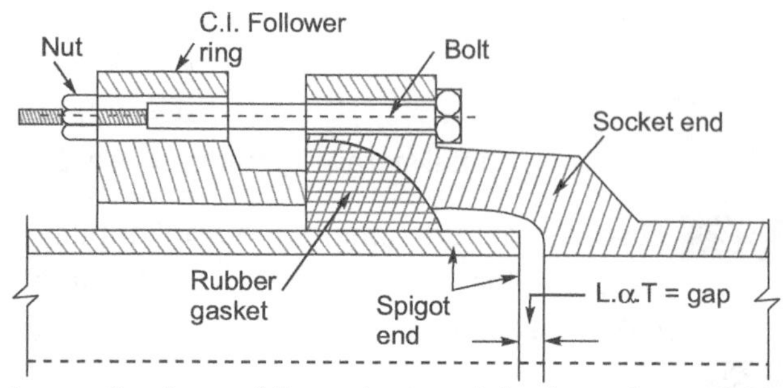

- These are provided at suitable intervals in the pipeline, so as to counteract the thermal stresses produced due to temperature variations
- The socket end is cast flanged and the spigot end is kept plain to provide an expansion joint in cast iron pipe.
- While making this joint, a small space is kept between the face of the spigot and the inner face of the socket and the spigot is filled up by means of a rubber gasket.
- The flanges are then tightend by means of nuts and bolts.
- When the pipe expand, the socket end moves forward and the gap left just closed.
- Similarly, when the pipes contracts the socket moves backward creating the gap.

2.3 **A centrifugal pump driven by an electric motor lifts water through a total height of 60 m from reservoir to the discharge end. The pump efficiency is 77% and the motor efficiency is 85%. The lift is through 300 m length, 10 cm diameter pipe and the pumping rate is 1500 lit/min. If the friction factor is 0.025, and the power of cost is 25 paise per kWh, what is the cost of power for pumping four millions litre of water.**

[12 Marks]

Sol:

Given data: $H = H_s + H_d = 60$ m {Total lift including suction and delivery pipe}

$$\eta_p = 77\%$$

$$\eta_m = 85\%$$

Diameter of pipe (d) = 10 cm = 0.1

$$Q = 1500 \text{ L/min}$$

$$= \frac{1500 \times 10^{-3}}{60} = 25 \times 10^{-3} = 0.025 \text{ m}^3/\text{sec}$$

$$f = 0.025$$

$$h_f = \frac{fLQ^2}{12.1d^5} = \frac{0.025 \times 300 \times (0.025)^2}{12.1 \times (0.1)^5} = 38.73 \text{ m}$$

∴ Head against which pump has to be work = 60 + 38.73 = 98.73 m

∴ Work horse power of the pump $= \frac{\gamma_W QH}{\eta_p(0.746)} = \frac{9.81 \times 0.025 \times 98.73}{0.77 \times 0.746} = 42.153$ HP {1 HP = 746 watt}

Horse power of motor $= \frac{42.153}{\eta_m} = \frac{42.153}{0.85} = 49.59$ Hp

∴ Power required = 49.59 × 0.746 = 36.995 kW

Time during which 4 million litres of water will be lifted

$$= \frac{\text{Volume of water to be pumped}}{\text{pumping rate}} = \frac{4 \times 10^6}{1500 \times 60} \text{ hr} = 44.44 \text{ hr}$$

∴ Energy required = 36.995 kW × 44.44 hr

= 1644.078 kWh

∴ Cost of the electrical energy at the rate of 25 paise per kWh

$$= 1644.078 \times \frac{25}{100} = 411 \text{ Rs}$$ **Ans.**

Level-2

2.4 Find out the balancing storage required in a distribution reservoir with the following data:
Population to be served = 50000
Average per capita demand = 2000 litres per day
Pumping into reservoir is done for 15 hours only
(i.e. from 5 A.M to 12:30 P.M; and from 1 : 30 P.M to 9.00 P.M)
Hourly variation factors between different hours are as under:

0 - 1	1 - 2	2 - 3	3 - 4	4 - 5	5 - 6	6 - 7	7 - 8	8 - 9	9 - 10	10 - 11	11 - 12
0.25	0.25	0.25	0.25	0.35	0.45	1.0	1.5	1.9	2.5	2.1	1.5
12 - 13	13 - 14	14 - 15	15 - 16	16 - 17	17 - 18	18 - 19	19 - 20	20 - 21	21 - 22	22 - 23	23 - 24
1.2	0.9	1.0	1.2	1.5	1.6	1.3	1.0	0.8	0.5	0.4	0.3

[20 Marks]

Sol:

Given data: Average per capita demand = 200 lpcd , population to be served = 50000

$\therefore$ Total demand = Population × per capita water demand
= 50000 × 200 = 10 MLD

Total supply during the day = Total demand volume of water to be supplied per day = 10 ML

By using hourly variation factors, demand can be calculated by

$$= \text{hourly variation factor} \times \frac{\text{Volume of water}}{\text{Duration}}$$

$$\therefore \quad \text{Rate of water supply} = \frac{\text{Volume of water to be supplied}}{\text{Duration of supply}} = \frac{10}{15} = \frac{2}{3} = 0.67 \text{ ML/hr}$$

1	2	3	4	5	6	7(Col.4 – Col.6)	8(Col.6 – Col.4)
Time (in hrs.)	Hourly Variation factors	Demand (ML/hr) Col. 2 × 10/24	Cumulative demand (ML)	Supply (ML)	Cumulative supply (ML)	Excess of demand (ML)	Excess supply (ML)
0 - 1	0.25	0.25 × 10/24 = 0.104	0.104	-	-	0.104	Nil
1 - 2	0.25	0.104	0.208	-	-	0.208	Nil
2 - 3	0.25	0.104	0.312	-	-	0.312	Nil
3 - 4	0.25	0.104	0.416	-	-	0.416	Nil
4 - 5	0.35	0.146	0.562	-	-	0.562	Nil
5 - 6	0.45	0.1875	0.7495	0.67	0.67	0.0795	Nil
6 - 7	1	0.417	1.665	0.67	1.34	Nil	0.1735
7 - 8	1.5	0.625	1.7915	0.67	2.01	Nil	0.2185
8 - 9	1.9	0.792	2.5835	0.67	2.68	Nil	0.0965
9 - 10	2.5	1.042	3.6255	0.67	3.35	0.2755	Nil
10 - 11	2.1	0.875	4.5005	0.67	4.02	0.4805	Nil
11 - 12	1.5	0.625	5.1255	0.67	4.69	0.4355	Nil
12 - 13	1.2	0.5	5.6255	0.335	5.025	0.6005	Nil
13 - 14	0.9	0.375	6.0005	0.335	5.36	0.6405	Nil
14 - 15	1	0.417	6.4175	0.67	6.03	0.3875	Nil
15 - 16	1.2	0.5	6.9175	0.67	6.7	0.2175	Nil
16 - 17	1.5	0.625	7.5425	0.67	7.37	0.1725	Nil
17 - 18	1.6	0.67	8.2125	0.67	8.04	0.1725	Nil
18 - 19	1.3	0.542	8.7545	0.67	8.71	0.445	Nil
19 - 20	1.0	0.417	9.1715	0.67	9.38	Nil	0.2085
20 - 21	0.8	0.33	9.5015	0.67	10	Nil	0.4985
21 - 22	0.5	0.208	9.7095	-	10	Nil	0.2905
22 - 23	0.4	0.167	9.8765	-	10	Nil	0.1235
23 - 24	0.3	0.125	10	-	10	Nil	Nil

∴ The maximum excess of demand = 0.6405 ML

The maximum excess of supply = 0.4985 ML

∴ Required balancing storage = 0.6405 + 0.4985 = 1.139 ML **Ans.**

2.5 **Explain the need of service reservoir and describe briefly the graphical method to evaluate the reservoir capacity of such reservoir?**

[10 Marks]

Sol:

Distribution reservoir, also called service reservoirs, are the storage, which store the treated water for supplying water during emergencies (such as fires, breakdowns repairs etc) and also to help in absorbing the hourly fluctuations in the normal water demand. The main functions served by the distribution reservoirs are enumerated below:

(i) They absorb the hourly variations in demand, and allow the water treatment units and pumps to operate at constant rate. This will reduce the RMO costs and improve efficiency.

(ii) They help in maintaining constant pressure in the distribution mains. In their absence, the pressure will fall down as the demand of water increases, and if the pumps don't immediately respond to it. it will reduce the available head at the service point, thus will cause non-availability of water.

(iii) The pumping of water in shifts is made possible by them without affecting the supply. Thus 8 to 16 hours of pumping can be carried out, so as to pump the whole days demand.

(iv) The water stored in these reservoirs can be supplied during emergencies, such as breakdown of pumps, heavy fire demand etc.

(v) They lead to an overall economy by reducing the size of pumps, pipelines and treatment units.

Graphical Method:

A mass diagram is plot of accumulated inflow (i.e., supply) or outflow (i.e. demand) versus time.

The mass curve of supply (i.e., supply lines) is, therefore, first of all drawn, and is superimposed by the demand curve. The amount of balancing storage can then be easily determined by adding the maximum ordinates between the demand and the supply lines to construct such diagrams for a particular water supply project, we have to proceed as follows:

1. From the past records, determine the hourly demand (or draft) for all 24 hrs for typical days (maximum, average and minimum).
2. Calculate and plot the cumulative demand against time, and thus plot the mass curve of demand.

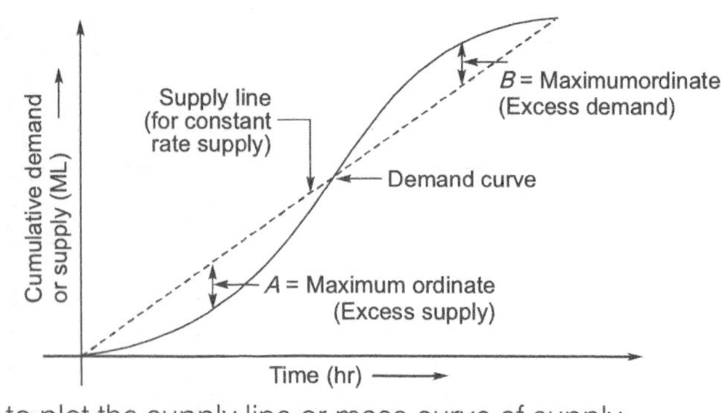

3. Draw the cumulative supply also against time to plot the supply line or mass curve of supply.
4. Read the storage required as the sum of the two maximum ordinates between demand and supply line.
5. Repeat the procedure for all the typical days (maximum, average and minimum) and determine the maximum storage required for the worst day.

2.6 **Evaluate the distribution of flow in the pipe network shown in figure. As the pipes are rough the flow may be assumed as turbulent and head loss h_L may be taken as Q^n. The value of r for each pipe is shown in figure (Take n = 2).**

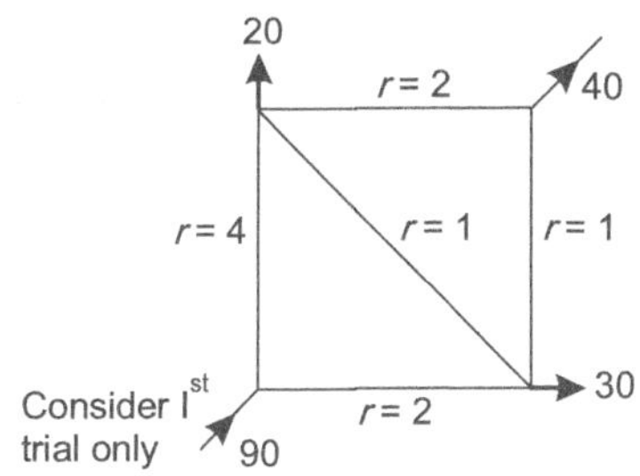

[20 Marks]

Sol:

1st trial:

Assumed discharge:

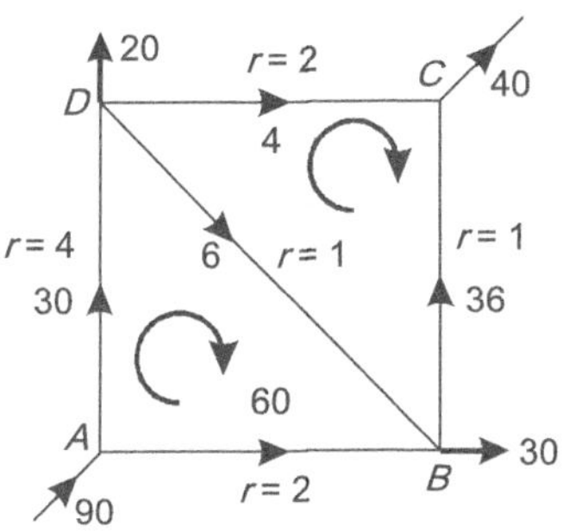

Loop ADB :

Pipe	r	Q_0	rQ_0^2	$\lvert 2rQ_0 \rvert$
AD	4	30	3600	240
DB	1	6	36	12
AB	2	–60	–7200	240
		Σ	**–3564**	**492**

$$\Delta Q = -\frac{\Sigma rQ_0^2}{2rQ_0} = \frac{-(-3564)}{492} = 7.243 \approx 7.2$$

Positive hence, adding to the flow in clockwise direction and substract to the flow in anticlockwise direction.

Now, discharge after 1st trial in loop ADB.

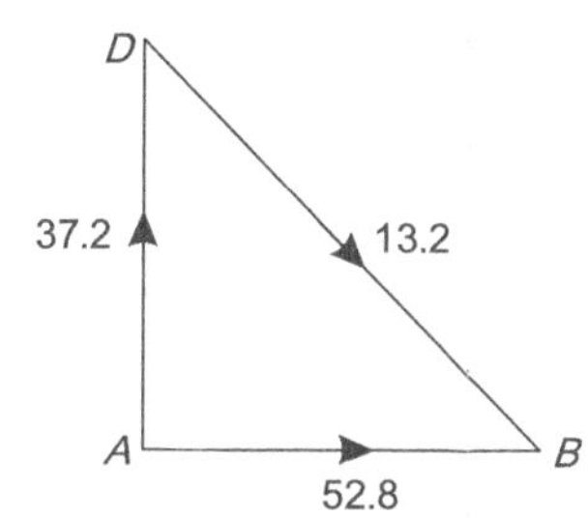

Loop DCB :

Pipe	r	Q_0	rQ_0^2	$\lvert 2rQ_0 \rvert$
DC	2	4	32	16
CB	1	–36	–1296	72
BD	1	–6	–36	12
		Σ	**–1300**	**100**

$$\Delta Q = -\frac{\Sigma rQ_0^2}{2rQ_0} = \frac{-(-1300)}{100} = 13$$

Positive hence, adding to the flow in clockwise direction and substract to the flow in anticlockwise direction.

Now, discharge after 1st trial in loop DCB.

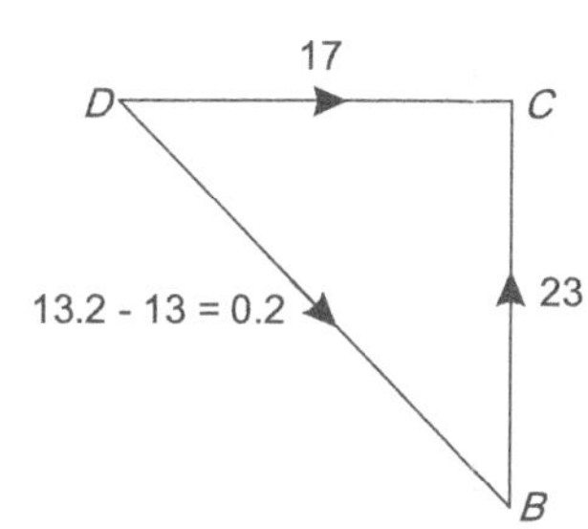

Hence, after 1st trial discharge in the pipe/network:

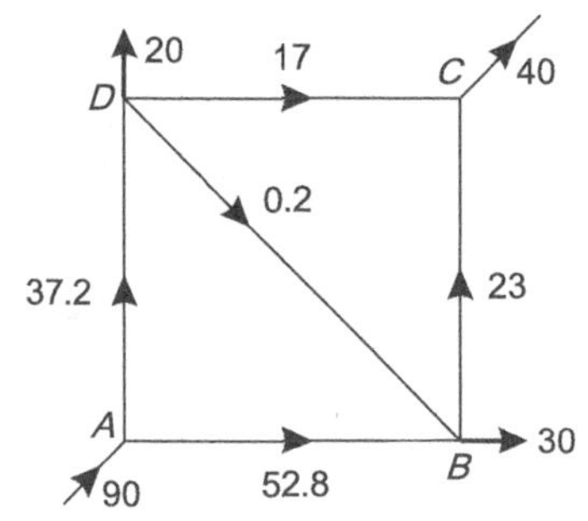

3. Development of Ground Water and Well Hydraulics

Level-1

3.1 Two tube wells each of 20 cm diameter are spaced at 100 m distance. Both the wells penetrate fully a confined aquifer of 12 m thickness

(i) Calculate the discharge if only one well is discharging under a head of 3.0 m.

(ii) What will be the percentage decrease in discharge of the well if both the wells are discharging under a depression head of 3.0 m.

Take, radius of influence for each well equal to 250 meters and coefficient of permeability of the aquifer as 60 m/day.

[12 Marks]

Sol:

Given, Diameter of tube wells = 20 cm

$$\text{Radius of tube well} = \frac{20}{2} = 10 \text{ cm} = 0.1 \text{ m}$$

Spacing between wells, B = 100 m

Thickness of confined aquifer, D = 12 m

Radius of influence, R = 250 m

Coefficient of permeability of aquifer, k = 60 m/day

(i) If only one well is discharged:

depression head, $H - h_w$ = 3.0 m

$$\therefore \quad Q = \frac{2\pi kD(H-h_w)}{\ln\left(\frac{R}{r_w}\right)} = \frac{2\pi \times 60 \times 12(3)}{\ln\left(\frac{250}{0.1}\right)}$$

$$Q = 1734.611 \text{ m}^3/\text{day}$$

(ii) If both the well discharges,

Depression head, $H - h_w$ = 3 m

then, discharge through each well is given by

$$\therefore \quad Q' = \frac{2\pi kD(H-h_w)}{\ln\left(\frac{R^2}{r_w B}\right)} = \frac{2\pi \times 60 \times 12 \times 3}{\ln\left(\frac{(250)^2}{0.1 \times 100}\right)} = 1552.764 \text{ m}^3/\text{day}$$

percentage decrease in discharge,

$$= \frac{Q - Q'}{Q} \times 100 = \frac{1734.611 - 1552.764}{1734.611} \times 100 = 10.48\%$$

3.2 600 m³/day of water is to be obtained from a proposed infiltration gallery, which is placed at 6 m depth from sub-surface water table. The coefficient of permeability of soil aquifer is 100 m/day. Find the length of the gallery if the drawdown in the gallery on pumping is not to exceed 4 m. The radius of influence may be assumed as 100 m.

[5 Marks]

Sol:

The discharge from infiltration gallery is given by

$$Q = kL\left(\frac{H^2 - h^2}{2R}\right)$$

Given, $Q = 600$ m³/day; $H = 6$ m, $k = 100$ m/day; $R = 100$ m, $h = 6 - 4 = 2$ m

$$\therefore \quad 600 = 100 \times L\left(\frac{6^2 - (2)^2}{2 \times 100}\right)$$

$$L = 37.5 \text{ m}$$

3.3 During recuperation test, the water level in an open well was depressed by pumping upto 2.5 m. The water level was raised by 1 m within 60 minutes just after stopping the pumping. Determine the yield from the well of 2 m diameter, when the depression head is 3 m. (Derive the formulae used)

[12 Marks]

Sol:

(Recuperation test) :

This test is done to calculate the specific capacity of the well so that we can calculate discharge corresponding to a particular drawdown.

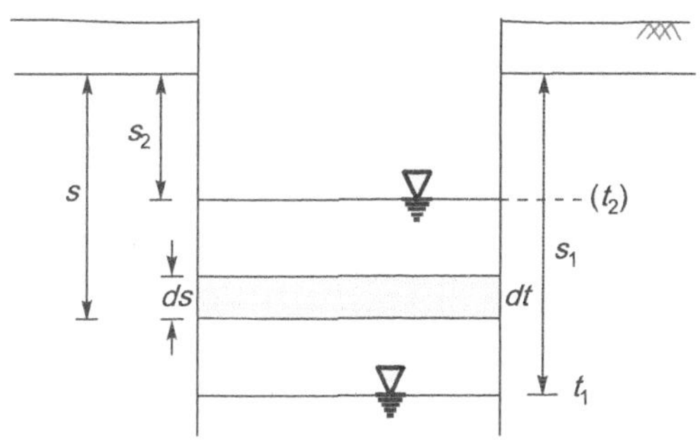

By pumping at a heavy rate a drawdown S_1 is created at time t_1 and then the pumping is stopped. Water level in the well recuperates and the drawdown at time t_2 becomes S_2. At any time 't' the drawdown was S and in time 'dt' the drawdown decreases by amount dS.

Hence, $$Qdt = -A.ds.$$

$$C = \text{Specific capacity}$$

$$= \frac{Q}{S}$$

$$CSdt = -\text{Ads}$$

$$C\int_{t_1}^{t_2} dt = -A\int_{S_1}^{S_2} \frac{ds}{s}$$

$$C(t_2 - t_1) = -A \cdot \ln\left(\frac{S_2}{S_1}\right)$$

$$c = \frac{1}{t} A \ln\left(\frac{S_1}{S_2}\right)$$

$$\frac{C}{A} = \frac{1}{t}\ln\left(\frac{S_1}{S_2}\right)$$

A = open area of well.

Now, given : $S_1 = 2.5$ m, $t = 60$ min; $S_2 = 2.5 - 1 = 1.5$ m, $d_w = 2$ m, Depression head, (S) = 3 m

$$C = \frac{1}{t}A\cdot\ln\left(\frac{S_1}{S_2}\right) = \frac{1}{60}\times\frac{\pi}{4}\times(2)^2\ln\left(\frac{2.5}{1.5}\right) = 0.02674 \text{ m}^2/\text{min}$$

Now, discharge or yield from well,

$$Q = C.S = 0.02674 \times 3 = 0.08022 \text{ m}^3/\text{min} = 0.08022 \times 60 \text{ m}^3/\text{hr}$$

$$Q = 4.813 \text{ m}^3/\text{hr}$$

3.4 Explain the reason for the functional end physical failure of tube wells. What are the methods for reducing these failures?

[10 Marks]

Sol:

(a) Naturally tube wells last about 15 to 20 years. They may fail due to incrustation and corrosion.

- **(Incrustation):** It occurs due to deposition of alkali salts on the inside walls of the pipe. The important alkali salts causing incrustation is calcium carbonate. The incrustation of well pipe reduces the effective diameter of the well pipe and hence reduces the discharge of the tube well.

The incrustation can be reduced by:

(i) Reducing the drawdown and hence pumping rate.
(ii) Using screens having larger area of openings so as to keep some allowance for the future incrustation.
(iii) Using such materials for the strainers (screens) that may easily permit the removal of incrustating material by chemical action without affecting that strainer material.
(iv) Properly maintaining and periodically cleaning the well screens.

- **Corrosion**: The well pipe is gradually destroyed by corrosion due to the action of acidic waters on the pipe material. When chlorides and sulphates or carbon dioxide are present in the water, the well pipe will definitely get corroded. The aquifer sand, surrounding the well pipe finds a way out into the corroded pipe through the worn out pipe walls, thus bringing sands along with water. Hence, corrosion results in excessive withdrawal of sand along with well water. Thicker pipe may be used to avoid corrosion. Other measures which help in reducing corrosion and thereby increasing the life of tubewell as are:

(i) By reducing the drawdown and pumping rate.
(ii) By using corrosion resistant materials for pipes.
(iii) By using corrosion resistant coating on the pipes.
(iv) By using thicker pipe.

3.5 Explain (i) Aquifer; (ii) Aquitard and (iii) Aquiclude

[10 Marks]

Sol:

(a) (Aquifer) : It is a saturated formation of earth material which not only stores water but yields in sufficient quantity. Thus, an aquifer transmits water relatively easily due to its high permeability. Unconsolidated deposits of sands and gravel form good aquifers. The availability of ground water from an aquifer at a place depends on the rates of withdrawal or replenishment. Aquifer play the roles of both a transmission conduit and a storage. Aquifers are classified as unconfined aquifers and confined aquifer on the basis of their occurance and field situations. An unconfined aquifer (water table aquifer) is one in which a free water surface i.e. a water table exists. A confined aquifer (artesian aquifer) in an aquifer which is confined between two impervious beds.

(Aquitard) : An aquitard is a formation through which only seepage is possible and thus the yield is insignificant compared to an aquifer. It is partly permeable but highly porous. A sandy clay unit or silty units are example of aquitard. Through an aquitard appreciable quantities of water may leak to an aquifer below it.

(Aquiclude): Aquiclude is a geological formation which is essentially impermeable to the flow of water. It may be considered as close to water movement even though it may contain large amount of water due to its high porosity.

Ex. Clay.

Level-2

3.6 **A town having a population of 50,000 is supplied with a per capita water supply of 130 litres per day. The sewage from this town enters the pumping station through a low levelled sewer at R.L. 195.0 m and leaves through the high levelled sewer at R.L. 205.0 m. The town has been provided with a separate sewerage system, and there are no chances of storm water getting infiltrated into sewers. Design (a) the size of the rising main, (b) size of the sump well, and (c) B.H.P. of the pump motor required. Assume the length of the rising main as 100 m also velocity of flow in the rising main as 1.2 m/s. Consider losses due to bends = 0.3 m; Efficiency of pump = 60%; Friction factor = 0.01. Assume 80% of water supplied gets converted into sewage and the maximum flow is three times the average flow.**

[15 marks]

Sol:

Given that 80% of the water supplied appears as sewage, we have

Average quantity of sewage produced

$$= 0.8\left[\frac{50{,}000 \times 130}{1000 \times 24 \times 60 \times 60}\right] = 0.06 \text{ cumecs}$$

The maximum flow is three times the average flow, we have

$$\text{Peak sewage discharge} = 3 \times 0.06 = 0.18 \text{ cumecs}$$

(a) **Design of the rising main.** Assuming the velocity of flow in the rising main as 1.2 m/sec, we have

The required cross-sectional area of the rising-main

$$= \frac{Q}{V} = \frac{0.18}{1.2} \text{m}^2 = 0.15 \text{ m}^2$$

$$\text{Required dia. of the rising main} = \sqrt{\frac{0.15 \times 4}{\pi}} = 0.437 \text{ m} \simeq 0.45 \text{ m (say)}$$

Hence, use 0.45 m dia. rising main.

$$\therefore \quad \text{Velocity in the rising main} = \frac{0.18}{\frac{\pi}{4}(0.45)^2} = \frac{0.18}{0.159} = 1.13 \text{ m/sec}$$

(b) **Design of the sump well.** The sump well can be designed for the peak flow of 15 minutes (i.e., 15 minutes detention period with peak flow). The capacity of the rising main may also be added to the above quantity, in case the rising main has to be emptied into the sump at the time of cleaning the sump.

$$\text{Peak flow rate} = 0.18 \text{ m}^3\text{/sec}$$

$\therefore$ Peak flow for 15 minutes

$$= 0.18 \times (15 \times 60) = 162 \text{ cu-m}$$

$$\text{Capacity of the rising main} = \text{Area} \times \text{Length}$$
$$= \frac{\pi}{4} \times (0.45)^2 \times 100$$
$$= 15.9 \text{ cu-m} \simeq 16 \text{ cu-m (say)}$$

∴ Total capacity of the sump well

$$= 162 + 16 = 178 \text{ cu-m.}$$

Let us now provide 3 number of sumps, any two taking the maximum flow and the third as stand-by unit, and thus we have the capacity of each sump well

$$= \frac{178}{2} = 89 \text{ cu-m}$$

Assuming the depth of the sump well as 3 m, we have the required cross-sectional area of the sump well

$$= \frac{\text{Capacity}}{\text{Depth}} = \frac{89}{3} m^2 = 29.67 \text{ m}^2$$

∴ The required dia. of the sump well

$$= \sqrt{\frac{29.67 \times 4}{\pi}} = 6.15 \text{ m}$$

(c) **Design of pump.** Each pump has to lift 89 cu-m of waste water in 15 minutes.

$$\therefore \quad \text{Capacity of each pump} = \frac{89}{15 \times 60} = 0.0989 \text{ cumecs}$$

The lift of pump required

= R.L. of high levelled sewer – R.L. of low levelled sewer + Depth of sump well

$$= 205.0 - 195.0 + 3 = 13 \text{ m}$$

The pump has to work against this lift as well as the losses. Let us now evaluate the losses.

Frictional loss of head in the rising main

$$= \frac{fLV^2}{2gd} = \frac{0.01 \times 100 \times (1.13)^2}{2 \times 9.81 \times 0.45} = 0.145 \text{ m}$$

Given the loss in bends, etc. = 0.3 m

$$\therefore \quad \text{Total losses} = 0.145 + 0.3 = 0.445 \text{ m}$$

∴ Total lift against which pumps have to work

$$= 13 + 0.445 = 13.445 \text{ m}$$

∴ H.P. of the pump

$$= \frac{w \cdot QH}{75} = \frac{1000 \times 0.0989 \times 13.445}{75} = 17.7 \text{ H.P.} \simeq 18 \text{ (say)}$$

Given the pump efficiency is 60%, we have

The Brake Horse Power (BHP) of the motor required

$$= \frac{18}{0.6} = 30 \text{ BHP}$$

3.7 A 40 cm well penetrates an aquifer of 30 m thickness, and the length of strainer (screened portion) is 10 m. The yield is 2000 rpm with a drawdown of 3 m in the well. If the length of screen is increased to 20 m, what will be the drawdown in the well and the increase in the specific capacity? **[16 Marks]**

Sol:

Correction factor(F) for confined aquifers for obtaining discharge or specific capacity values of partially penetrating well is given by

$$F = \left[1 + 7\sqrt{\frac{r_w}{2H_1}} \times \cos\left(\frac{\pi\alpha}{2}\right)\right] \times \alpha$$

where $\alpha = \dfrac{H_1}{H}$

H_1 = Depth upto which well penetrates

H = Full depth of confined aquifer

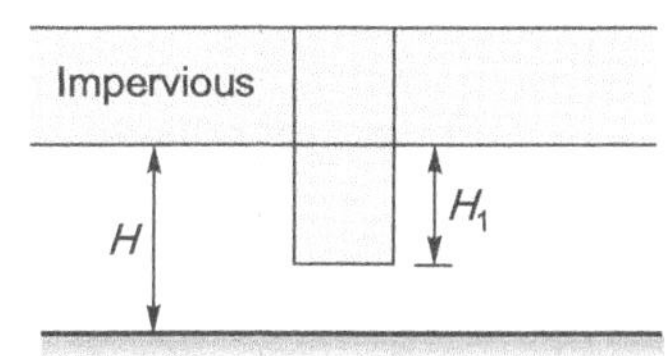

Case 1: Given, $H = 30$ m; $H_1 = 10$ m; $r_w = 20$ cm; $Q = 2$ m³/min

$$\therefore \quad \alpha = \frac{10}{30} = \frac{1}{3}$$

$$F = \left[1 + 7\sqrt{\frac{0.2}{2 \times 10}} \cos\frac{\pi}{2} \times \frac{10}{30}\right] \times \frac{10}{30} = 0.567$$

Case 2: Similarly, when $H_1 = 20$ m

$$F' = \left[1 + 7\sqrt{\frac{0.2}{2 \times 10}} \cos\frac{\pi}{2} \times \frac{20}{30}\right] \times \frac{20}{30} = 0.996$$

Let C represents the specific capacity of fully penetrating well, end c_p be the specific of partially penetrating well, we have

$$c_p = \text{C.F.}$$

where $c_p = \dfrac{Q_p}{S_p} = \dfrac{2 \text{ m}^3/\text{min}}{3\text{m}} = 0.667 \text{ m}^2/\text{min}$

Now, $0.667 = c \times 0.567$

$$c = 1.176 \text{ m}^2/\text{min}$$

For the 2nd case, of 20 m penetration, we have

$$c_p' = c.F'$$

$$c_p' = 1.176 \times 0.996 = 1.171 \text{ m}^2/\text{min}$$

$$c_p' = \frac{Q_p'}{S_p'} = \frac{2 \text{ m}^3/\text{min}}{S_p'}$$

$$1.171 = \frac{2 \text{ m}^3/\text{min}}{S_p'}$$

$$s_p' = 1.706 \text{ m}$$

Drawdown in 2nd case = 1.706 m

Percentage inc. in specific capacity of well

$$= \frac{c_p' - c_p}{c_p} \times 100 = \frac{1.171 - 0.667}{0.667} \times 100 = 75.56\%$$

3.8 A 20 cm dia. well penetrates 30 m below static water level water is pumped at the rate of 1800 rpm and drawdown in the well is observed as 1.2 m end 0.6 m. The observational wells are 12 m and 36 m from well. Determine:

(i) Transmissibility of aquifer.

(ii) Drawdown in the pumped well (Taking, R = 300 m).

(iii) Specific capacity of well.

[15 Marks]

Sol:

Given, Q = 1800 l/min, d_w = dia. of well = 20 cm, $r_w = \frac{20}{2}$ = 10 cm = 0.1 m

(i)

$$h_1 = 30 - 1.2 = 28.8 \text{ m}$$

$$h_2 = 30 - 0.6 = 29.4 \text{ m}$$

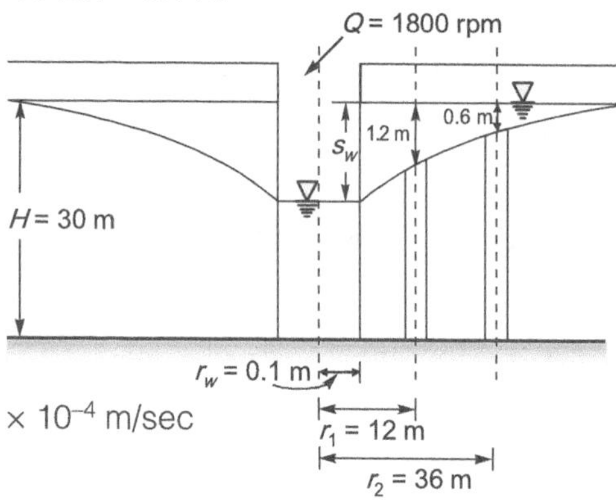

For an unconfined aquifer;

Now,
$$Q = \frac{\pi k\left(h_2^2 - h_1^2\right)}{\ln\left(\frac{r_2}{r_1}\right)}$$

$$\frac{1800 \times 10^{-3}}{60} = \frac{\pi k\left(29.4^2 - 28.8^2\right)}{\ln\left(\frac{36}{12}\right)} = 3 \times 10^{-4} \text{ m/sec}$$

Now, transmissibility of aquifer

$$T = kH = 3 \times 10^{-4} \times 30$$

$$= 90 \times 10^{-4} \text{ m}^2/\text{sec}$$

$$= 90 \times 10^{-4} \times 86400 \text{ m}^2/\text{day}$$

$$T = 778.7 \text{ m}^2/\text{day}$$

(ii) Drawdown in the pumped well (S_w) :

$$Q = \frac{\pi k\left(H^2 - h_w^2\right)}{\ln\left(\frac{R}{r_w}\right)}$$

$$\frac{1800 \times 10^{-3}}{60} = \frac{\pi \times 3 \times 10^{-4}\left(30^2 - h_w^2\right)}{\ln\left(\frac{300}{0.1}\right)}$$

$$h_w = 25.40 \text{ m}$$

∴ Drawdown in the well,

$$S_w = H - h_w = 30 - 25.4$$

$$S_w = 4.6 \text{ m}$$

(iii) Specific capacity of well,

$$c = \frac{\text{Discharge}}{\text{Drawdown of well}} = \frac{1800 \times 10^{-3} \times 60}{4.6} = 23.48 \text{ m}^3/\text{hr}$$

3.9 A 30 cm diameter well completely penetrates a confined aquifer of permeability 45 m/day. The length of strainer is 20 m under steady state of pumping the drawdown at the well was found to be 3.0 m end the radius of influence was 300 m. Calculate the discharge with increase in diameter of well by 50%. What will be the increase in percentage of discharge, derive the equation used.

[20 Marks]

Sol:

(Assumption):

i. The soil is assumed to be homogenous end isotropic.
ii. The soil medium is semi-infinite.
iii. Flow is laminar.
iv. The horizontal flow velocity is independent of depth.
v. The GWT (Ground water table) is assumed as horizontal in all directions.
vi. The hydraulic gradient at any point on the drawdown is equal to the slope of tangent at that point.

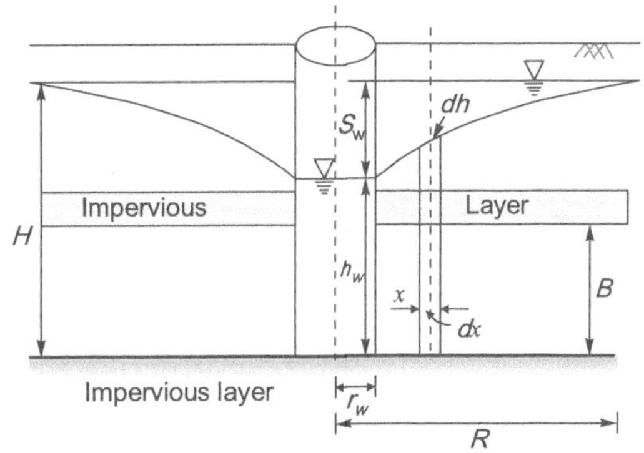

By Dercy's law,

$$q_x = ki_xA_x = k\frac{dh}{dx}(2\pi x)B$$

$$Q = \int q_x = \int(2\pi x)Bk\frac{dh}{dx}$$

$$Q\int_{r_w}^{R}\frac{dx}{x} = 2\pi kB\int_{h_w}^{H} dh$$

$$Q[\ln x]_{r_w}^{R} = 2\pi kB[h]_{h_w}^{H}$$

$$Q = \frac{2\pi kB(h-h_w)}{\ln\left(\frac{R}{r_w}\right)}$$

Now, given : $d_w = 30$ cm, $r_w = 15$ cm $= 0.15$ m, $k = 45$ m/day, $B = 20$ m; $S_w = 3$ m; $R = 300$ m

$$S_w + h_w = H$$

$$S_w = H - h_w = 3 \text{ m}$$

$$Q = \frac{2\pi kB(h-h_w)}{\ln\left(\frac{R}{r_w}\right)} = \frac{2\pi\times45\times20\times3}{\ln\left(\frac{300}{0.15}\right)} = 2231.92 \text{ m}^3\text{/day}$$

When diameter of well (increases) by 50%.

i.e. $d_w' = 1.5\, d_w = 1.5 \times 30 = 45$ cm

$r_w' = 22.5$ cm $= 0.225$ m

$$Q' = \frac{2\pi kB(H-h_w)}{\ln\left(\frac{R}{r_w'}\right)} = \frac{2\pi\times45\times20\times3}{\ln\left(\frac{300}{0.225}\right)} = 2257.68 \text{ m}^3\text{/day}$$

$$\text{Percentage increase in discharge} = \left(\frac{Q'-Q}{Q}\right)\times100 = \left(\frac{2357.68-2231.92}{2231.92}\right)\times100 = 5.63\%$$

3.10 (a) What is well function?

(b) A well is located in a 25 m confined aquifer of permeability 30 m/day and storage coefficient of 0.005. If the well is being pumped at the rate of 1750 lpm. Calculate the drawdown at a distance of (a) 100 m and (b) 50 m from the well after 20 hr. Using following expression for well function $w(u) = -0.5772 - \ln(u) - u$.

[5 + 15 = 20 Marks]

Sol:

(a) Well function is a special integral which is used to calculate the drawdown from a well foran unsteady flow in a confined aquifer.

The flow in a confined aquifer in unsteady state can be given as

$$\frac{\partial^2 h}{\partial x^2}+\frac{\partial^2 h}{\partial y^2}+\frac{\partial^2 h}{\partial z^2} = \frac{S}{T}\times\frac{\partial h}{\partial t}$$

$$\nabla^2 h = \frac{S}{T}\times\frac{\partial h}{\partial t}$$

The equation above is basic differential equation governing unsteady ground water flow in a homogeneous isotropic confined aquifer. This is known as diffusion equation. In polar coordinates, to represent the radial flow in a well, this equation takes the form,

$$\frac{\partial^2 h}{2r^2}+\frac{1}{r}\times\frac{\partial h}{\partial r} = \frac{S}{T}\times\frac{\partial h}{\partial t}$$

the solution to this equation is given as:

$$S = (H-h) = \frac{Q}{4\pi T}\int_u^\infty \frac{e^{-u}}{u}du$$

here $\int_u^\infty \frac{e^{-u}}{u}du$ is called well function

end it is denoted by $w(u)$.

$$S = (H-h) = \frac{Q}{4\pi T}w(u)$$

where, u = a parameter = $\frac{r^2 s}{4Tt}$, S = storage coefficient, t = time from start of pumping, T = transmissibility of the aquifer

(b) Given, K = 30 m/day; H = 25 m

$$\therefore \quad T = kH = 30 \times 25 \text{ m}^2/\text{day}$$

$$= \frac{750}{86400}\text{ m}^3/\text{sec} = 8.68\times10^{-3}\text{ m}^2/\text{sec}$$

$$S = 0.005$$

(a) when $r = 100$ m

$$u = \frac{r^2 s}{4Tt} = \frac{(100)^2\times0.005}{4\times8.68\times10^{-3}\times(20\times60\times60)} = 0.02$$

Now, $w(u) = -0.5772 - \ln(0.02) - 0.02 = 3.3148$

Now, drawdown at a distance of 100 m from well

$$S_{100} = \frac{Q}{4\pi T}w(u) = \frac{1.750}{60\times4\pi\times(8.68\times10^{-3})}\times3.3148$$

$$S_{100} = 0.886\text{ m}$$

(b) when $r = 50$ m

$$u = \frac{r^2 s}{4Tt} = \frac{(50)^2 \times 0.005}{4 \times 8.68 \times 10^{-3} \times (20 \times 60 \times 60)} = 0.005$$

Now, $w(u) = -0.5772 - \ln(0.005) - 0.005 = 4.7161$

Now, drawdown at a distance 50 m from the well.

$$S_{50} = \frac{Q}{4\pi T} w(u) = \frac{1.750}{60 \times 4\pi \times 8.68 \times 10^{-3}} \times 4.7161$$

$$S_{50} = 1.261 \text{ m}$$

4. Water Quality

Level-1

4.1 The analysis of a sample of a sample of water with pH 7.5 has produced the following concentrations (in mg/L)

Cations	Anions
Ca^{2+} = 80	Cl^- = 100
Mg^{2+} = 30	SO_4^{2-} = 100
Na^+ = 72	HCO_3^- = 165
K^+ = 6	

Find total hardness, carbonate hardness, non-carbonate hardness and the alkalinity, all expressed as $CaCO_3$.

[10 Marks]

Sol:

Total hardness → Hardness will be due to multivalent cations only

So, only Ca^{2+} and Mg^{2+} will produce hardness,

$$Ca^{2+} \text{ in equivalent of } CaCO_3 = \frac{Ca^{2+}(\text{in mg/L})}{\text{Equivalent weight of } Ca^{2+}} \times \text{Equivalent weight of } CaCO_3$$

$$= \frac{80}{20} \times 50 = 200 \text{ mg/L as } CaCO_3$$

$$Mg^{2+} \text{ in equivalent of } CaCO_3 = \frac{Mg^{2+}(\text{in mg/L})}{\text{Equivalent weight of } Mg^{2+}} \times \text{Equivalent weight of } CaCO_3$$

$$= \frac{30}{(24/2)} \times 50 = 125 \text{ mg/L as } CaCO_3$$

∴ TH (Total hardness) in mg/L as $CaCO_3$ = 200 + 125 = 325 mg/L as $CaCO_3$

Alkalinity will be due to HCO_3^- only

$$\text{Alkalinity} = \frac{\text{Concentration } HCO_3^- (\text{in mg/L})}{\text{Equivalent weight of } HCO_3^-} \times \text{Equivalent weight of } CaCO_3$$

$$= \frac{165}{61} \times 50 = 135.25 \text{ mg/L as } CaCO_3$$

Carbonate hardness (CH) = Minimum of {TH, Alkalinity}

= Minimum of {325, 135.25} = 135.25 mg/L as $CaCO_3$ **Ans.**

Now, Non-carbonate hardness (NCH) = Total Hardness – CH

= 325 – 135.25 = 189.75 mg/L as $CaCO_3$ **Ans.**

4.2 **Determine the resultant pH of two electrolytic solutions, one with a pH of 6.0 and other with a pH of 8.0.**

[5 Marks]

Sol:

$$pH_1 = 6 \qquad pH_2 = 8$$

$$\left[H_1^+\right] = 10^{-6}\,M \qquad \left[H_2^+\right] = 10^{-8}\,M$$

When two solutions are mixed equally, the hydrogen 10 y concentration, will become

$$[H^+] = \frac{\left[H_1^+\right]+\left[H_2^+\right]}{2} = \frac{10^{-6}+10^{-8}}{2} = 5.05\times10^{-7}\text{ M}$$

How, pH of resultant solution,

$$pH = -\log[H^+] = -\log[5.05\times10^{-7}] = 7-\log_{10}(5.05)$$

$$pH = 6.296$$

4.3 **Consider a glucose solution ($C_6H_{12}O_6$) of molarity 1.75×10^{-3} that is completely oxidized to CO_2 and H_2O. Find the amount of oxygen required for this reaction. The chemical mass balance equation for the above reaction is given as**

$$C_6H_{12}O_6 + 6O_2 \rightarrow 6CO_2 + 6H_2O$$

(Atomic weights are: C = 12; H = 1, O = 16)

[10 Marks]

Solution:

Molecular weight of

$$C_6H_{12}O_6 = 6\times12 + 12\times1 + 6\times16 = 180$$

$$\text{Molarity} = \frac{\text{Wt. of } C_6H_{12}O_6 \text{in gm}}{\text{Molecular wt. of } C_6H_{12}O_6}\times1000$$

Molecular weight of $O_2 = 2\times16 = 32$ or $1.75\times10^{-3} = \frac{W_1}{180}\times1000$

$\Rightarrow$ $W_1 = 3.15\times10^{-4}$ gm

Now 1 mole $C_6H_{12}O_6$ require $O_2 = 6\times32$ gm

1 gm $C_6H_{12}O_6$ require

$$O_2 = \frac{6\times32}{180}\text{gm}$$

$\Rightarrow 3.15\times10^{-1}$ gm $C_6H_{12}O_6$ require

$$O_2 = \frac{6\times32}{180}\times3.15\times10^{-4}$$

$$= 3.36\times10^{-4}\text{ gm}$$

Hence amount of O_2 required to oxidize 1 lit. of glucose is 3.36×10^{-4} gm.

4.4 Write short notes on water quality.

[10 Marks]

Sol:

Water quality refers to the chemical, physical, biological and radiological characteristics of water. It is a measure of the condition of water relative to the requirements of one or more biotic species and or to any human need or purpose. It is the most frequency used reference to a set of standards against which compliance can be assessed. The most common standards used to assess water quality relate to health of ecosystem, safety of human contact.

Standards: In the setting of standards, agencies make political and technical/scientific decisions about how the water will be used. In the case of natural water bodies, they also make some reasonable estimate of present condition natural water bodies will vary in response to environmental conditions. Environmental scientists work to understand how these system functions, which in turn helps to identify the sources and fates of contaminants. Environmental lawyers and policy makers work to define legislation with the intention that water is maintained at an appropriate quality for its identified used.

The vast majority of surface water on the earth is neither potable nor toxic. This remains true when sea water in the oceans (which is too salty to drink) is not counted.

Another general perception of water quality is that of a simple property that tells whether water is polluted or not. In fact, water quality is a complex subject, in part because water is a complex medium intrinsically tied to the ecology of the earth. Industrial and commercial activities (e.g. manufacturing, mining, construction, transport) are a major cause of water pollution as are run off from agricultural areas, urban runoff and discharge of treated and untreated sewage.

Sampling and Measurement: The complexicity of a water quality as a subject is reflected in the many types of measurements of water quality indicators. The most accurate measurements of water quality are made on site and in direct contact with the water source question include temperature, pH, dissolved oxygen, conductivity, oxygen reduction potential (ORP), turbidity and secchi disk depth.

Drinking water indicators: An electrical conductivity meter is used to measure total dissolved solids. The following is a list of indicators often measured by situational category.

- Alkalinity
- Color of water
- pH
- Taste and odour
- Dissolved metals and salts
- Microorganism such as fecal Coliform Bacteria (Escherichia coil) Cryptosporidium and Giardia Lambila
- Dissolved metals and metalloids (lead, mercury, arsenic etc.)
- Redon
- Heavy metals
- Dissolved organics.

4.5 **Write short term and long term implications on human health of concentrations of following parameters if significantly increased than permissible levels.**

1. Coliform 2. Chlorides 3. Cadmium

[10 Marks]

Sol:

(1) **Coliform :** Coliforms are bacteria that are always present in digestive tracts of animal, humans and are found in the their wastes. They are also found in plant and soil material presence of coliform bacterial in drinking water is taken as evidence of recent pollution with human and animal faces most bacteria do not cause disease, but some rare strain of E.Coli can cause serious illness. These can remain in water for long time. These serves as indicator organisms for more harmful pathogenic bacteria.

(2) **Chlorides:** These are present in water in form of sodium chloride or may be due to leaching of marine sedimentary deposits. The presence of high concentrations of chloride in river or stream water may indicate pollution from sea water, industrial water or domestic waste. The high concentration of chlorides in water produces bitter taste in mouth. It also causes airway irrigation, wheezing, difficulty breathing, sore throat, cough, chest tightness, eye irritation, skin irritation. It can also create dangerous carcinogenic compound due to reaction with some organic matter.

(3) **Cadmium:** Cadmium in water supply is usually due to e-waste, metal industries. Short term effect of high level cadmium can cause nausea, vomiting, diarrhea, muscle cramp, salivation sensory disturbances, liver injury and renal failure. While long term effect of high concentrations of cadmium include kidney, liver bone and blood damage.

Level-2

4.6 **A 100 mL sample require 10.2 mL of 0.02 NH_2SO_4 solution to reach the pH of 8.3 and a total of 25.6 mL of 0.02 NH_2SO_4 to reach the pH of 4.5.**
(a) Calculate the phenopthalien and total alkalinity of the given sample (as $CaCO_3$).
(b) Determine the concentrations of HCO_3^- and CO_3^{2-} (in mg/L), if the initial pH was 11.

[15 Marks]

Sol:

Given, 100 ml sample requires,

(100 m*l*) (1000 m*l*)

pH = 11 → 10.2 m*l* of 0.02 NH_2SO_4 102 m*l*
pH = 8.3
pH = 4.5 → 15.4 m*l* 154 m*l*
Total = 25.6 m*l* 0.02 NH_2SO_4

Given, pH = 11, pOH = 3.

$$[OH^-] = 10^{-3}\text{ M} = 10^{-3} \times 1 \text{ gm.eq/liter}$$
$$= 10^{-3} \times 50 \times 10^3 \text{ mg/L as } CaCO_3 = 50 \text{ mg/L as } CaCO_3$$
$$OH^- \text{ alkalinity} = 50 \text{ mg/L as } CaCO_3$$

In 1000 ml sample,

102 ml of 0.02 NH_2SO_4 required for 1 liter of water to bring the sample to the pH to 8.3.

$$\frac{1}{2}\left[CO_3^{2-}\right] + [OH^-] = 102 \text{ mg/L as } CaCO_3$$
$$\left[CO_3^{2-}\right] = (102 - 50) \times 2 = 104 \text{ mg/L as } CaCO_3$$

∴ Carbonate alkalinity = 104 mg/L as $CaCO_3$

Also, $$\frac{1}{2}CO_3^{2-} + HCO_3^- = 104 \text{ mg/L as } CaCO_3$$

∴ $$HCO_3^- = (154 - 52) = 102 \text{ mg/L as } CaCO_3$$

∴ Bicarbonate alkalinity = 102 mg/L as $CaCO_3$

How, total alkalinity (in mg/L as $CaCO_3$) = 104 + 102 + 50 = 256 mg/L as $CaCO_3$

Phenopthalien alkalinity (up to pH 8.3) = 102 mg/L as $CaCO_3$

Amount of acid required to reduce the pH to 8.3 corresponds to phenoptheuen alkalinity and the acid required to reduce pH 4.5 corresponds to total alkalinity.

4.7 A sample of water at pH 10 has 32 mg/L of CO_3^{2-}, find the alkalinity as $CaCO_3$, the reaction rate constant for reaction is 4.68×10^{-11}.

$$HCO_3^{2-} \rightleftharpoons H^+ + CO_3^{2-}$$

[15 Marks]

Sol:

Given data: pH = 10 ; CO_3^{2-} (in mg/L) = 32 mg/L

Total alkalinity in water sample is mainly due to CO_3^{2-}, HCO_3^-, OH^-

CO_3^{2-} alkalinity 'or' carbonate alkalinity (in mg/L as $CaCO_3$)

$$= \frac{\text{Concentration of } CO_3^{2-} \text{ (in mg/ L)}}{\text{Equivalent Weight of } CO_3^{2-}} \times \text{Equivalent Weight of } CaCO_3$$

$$= \frac{32}{(60/2)} \times 50 = 53.33 \text{ mg/L as } CaCO_3$$

Now, (bicarbonate alkalinity):

Given reaction,
$$HCO_3^- \rightleftharpoons H^+ + CO_3^{2-}$$

$$K = \frac{[H^+]\cdot\left[CO_3^{2-}\right]}{[HCO_3^-]}$$

$$[H^+] = 10^{-10} \text{ M}$$

$$\left[CO_3^{2-}\right] = \left(\frac{32 \times 10^{-3}}{60}\right) \text{M}$$

$$\therefore \quad 4.60 \times 10^{-11} = \frac{10^{-10} \times \left(\frac{32 \times 10^{-3}}{60}\right)}{\left[HCO_3^-\right]}$$

$$\left[HCO_3^-\right] = 1.139 \times 10^{-3} \text{ M}$$

Now, $$\left[HCO_3^-\right] \text{in mg/L} = 1.139 \times 10^{-3} \times 61 \text{ gm/L} = 1.139 \times 10^{-3} \times 61 \times 10^3 \text{ mg/L}$$

$$= 69.516 \text{ mg/L}$$

$\therefore$ Bicarbonate alkalinity (in mg/L as $CaCO_3$)

$$= \frac{\text{Concentration of } HCO_3^- \text{ (in mg/ L)}}{\text{Equivalent Weight of } HCO_3^-} \times \text{Equivalent Weight of } CaCO_3$$

$$= \frac{69.516}{61} \times 50 = 56.98 \text{ mg/L as } CaCO_3$$

Caustic alkalinity (i.e., alkalinity due to OH^-)

$$[OH^-] = 10^{-4} \text{ M}$$

$$[OH^-] \text{ in mg/L} = (10^{-4} \times 17) \text{ gm/L} = 10^{-4} \times 17 \times 10^3 \text{ mg/L} = 1.7 \text{ mg/L}$$

$\therefore$ OH^- alkalinity (in mg/L as $CaCO_3$)

$$= \frac{\text{Concentration of } OH^{-1} \text{(in mg/ L)}}{\text{Equivalent Weight of } OH^{-1}} \text{ Equivalent Weight of } CaCO_3$$

$$= \frac{1.7}{1.7} \times 50 = 5 \text{ mg/L as } CaCO_3$$

$\therefore$ Total alkalinity = Carbonate alkalinity + Bicarbonate alkalinity + Caustic alkalinity

$$= 53.33 + 56.98 + 5 = 115.31 \text{ mg/L as } CaCO_3 \quad \textbf{Ans.}$$

5. Water Treatment

Level-1

5.1 **In a continuous flow settling tank 3 m deep, 70 m long. What flow velocity of water would you recommend for effective removal of 0.02 mm particles at 25°C. The specific gravity of particles is 2.65; and kinematic viscosity of water may be taken as 0.01 cm²/sec.**

[10 Marks]

Sol:

Given data : Height of tank = 3 m, Length of tank = 70 m, Diameter of particles = 0.02 mm, Specific gravity of particle (G_S) = 2.65, Kinematic viscosity (ν) = 0.01 cm²/sec

Since $d < 0.1$ mm; Hence Stoke's equation is applicable to find out the settling velocity.

$$\therefore \quad V_S = \frac{g}{18}(G_S - 1)\frac{d^2}{\nu}$$

$$= \frac{9.81 \text{ m/s}^2}{18} \times (2.65 - 1) \times \frac{(0.02 \times 10^{-3})^2 \text{ m}^2}{0.01 \times 10^{-4} \text{m}^2/\text{sec}}$$

$$= 3.597 \times 10^{-4} \text{ m/s}$$

Calculation of flow velocity of water:

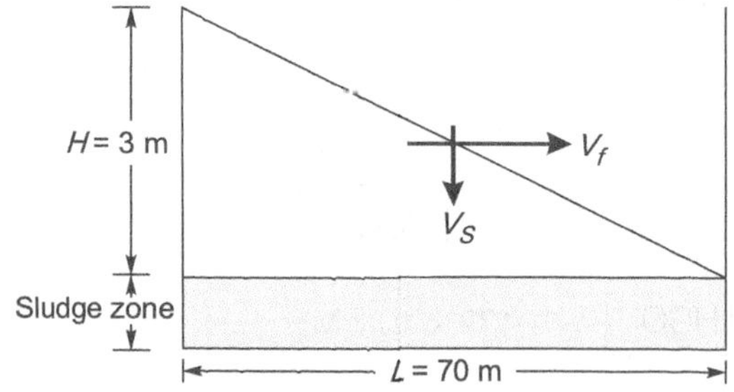

$$\text{Detention time, } D_t = \frac{L}{V_f} = \frac{H}{V_S}$$

$$\therefore \quad V_f = \frac{L}{H} \times V_S = \frac{70}{3} \times 3.597 \times 10^{-4} \text{m/sec}$$

$$V_f = 8.393 \times 10^{-3} \text{ m/sec}$$

Hence, in order to ensure effective removal of particles upto 0.02 mm the flow velocity in the settling tank should not be more than 8.393 mm/sec. **Ans**

Check against the scour of deposited particles:

$$V_d = \sqrt{\frac{8\beta}{f} g(G_s - 1)d} = \sqrt{\frac{8 \times 0.04}{0.03} \times 9.81(2.65 - 1) \times 0.02 \times 10^{-3}}$$

$$= 0.0587 \text{ m/sec} \simeq 5.87 \text{ cm/sec}$$

$\therefore$ ($V_f < V_d$); Hence design flow velocity is OK.

5.2 A rectangular settling tank without mechanical equipment is to treat 1.8 million litres per day of raw water. The sedimentation period is to be 4 hrs, the velocity of flow is 8 cm/min, and the depth of the water and sediment 4.2 m. If an allowance of 1.2 m for sediment is made, what should be

(a) the length of the basin

(b) the width of the basin

[8 Marks]

Sol:

Given data : Quantity of water to be treated (Q_0) = 1.8 MLD, Detention period (D_t) = 4 hrs., Velocity of flow (V_f) = 8 cm/min, Depth of water and sediment = 4.2 m, Depth of sediment = 1.2 m

$\therefore$ Effective water depth = 4.2 – 1.2 = 3 m

$$\text{Capacity of tank } (V) = Q_0 \cdot D_t$$

$$= \frac{1.8 \times 10^6 \times 10^{-3}\, m^3}{24 \text{ hrs}} \times 4 \text{ hrs} = 300\, m^3$$

(a)
$$\text{Length of tank } (L) = V_f \times D_t = \frac{8 \times 10^{-2} m}{60 \text{ sec}} \times 4 \times 3600 \text{ sec} = 19.2 \text{ m}$$

(b) Width of tank (B)

$$L \times B \times H = 300$$

$$B = \frac{300}{19.2 \times 3} = 5.21 \text{ m}$$

5.3 Design the dimensions of a suitable sedimentation tank for the raw water supplied for a town of population 100000 with a daily per capita water supply of 120 litres. Assume D_t = 6 hrs, velocity of flow as 20 cm/min.

[12 Marks]

Sol:

Given data : Population = 100000, Per capita supply = 120 lpcd, Detention time (D_t) = 6 hrs., Velocity of flow (V_f) = 20 cm/min.

$\therefore$ Average daily demand = 120 × 100000 L/day = 12 MLD

Since, sedimentation tank is designed for maximum daily demand.

$$\therefore \quad Q_{design} = 1.8 \times 12 = 21.6 \text{ MLD}$$

$$\therefore \quad \text{Capacity of tank } (V) = Q_{design} \times D_t = \frac{21.6 \times 10^3 m^3}{24 \text{ hrs}} \times 6 \text{ hrs} = 5400\, m^3$$

$$\text{Length of tank } (L) = V_f \times D_t$$

$$= \frac{20 \times 10^{-2} m}{60 \text{ sec}} \times 6 \times 3600 \text{ sec} = 72 \text{ m}$$

Assuming,
$$\frac{L}{B} = 4$$

$$B = \frac{72}{4} = 18 \text{ m}$$

$$\therefore \quad LBH = 5400$$

$$H = \frac{5400}{72 \times 18} = 4.167 \text{ m} \simeq 4.2 \text{ m}$$

Assuming free board (FB) = 0.5 m

$\therefore$ Depth of tank = 4.2 + 0.5 = 4.7 m

Dimension of tank = **72 m × 18 m × 4.7 m**

5.4 It is required to supply water to a population of 20000 at a per capita demand of 150 litres per day. The disinfectant used for chlorination is bleaching powder which contains 30% of available chlorine. Determine how much of bleaching powder is required annually at the water works, if 0.3 ppm of chlorine dose is required for disinfection.

[10 Marks]

Sol:

Given data : Population = 20000, Per capita demand = 15 *l*pcd

$$\therefore \quad \text{Average daily demand} = \frac{20000 \times 150}{10^6} = 3\,\text{MLD}$$

$$\text{Chlorine dosage} = 0.3 \text{ ppm} = 0.3 \text{ mg}/l$$

$$\therefore \quad \text{Total chlorine required} = 3 \times 10^6 l/\text{day} \times 0.3 \times 10^6 \text{kg}/l$$

$$= 0.9 \text{ kg/day}$$

Since, bleaching powder contains 30% of chlorine

$$30 \text{ kg of } Cl_2 \simeq 100 \text{ kg of bleaching powder}$$

$$\therefore \quad 1 \text{ kg of } Cl_2 \simeq \frac{100}{30} \text{ kg of bleaching powder}$$

$$\therefore \quad 0.9 \text{ kg of } Cl_2 \simeq \frac{100}{30} \times 0.9 = 3 \text{ kg bleaching powder}$$

$$\therefore \quad \text{Bleaching powder required} = 3 \text{ kg/day}$$

$$\therefore \quad \text{Annual bleaching powder requirement} = 3 \times 365 = 1095 \text{ kg}$$

5.5 The following is a chlorination curve for a water sample

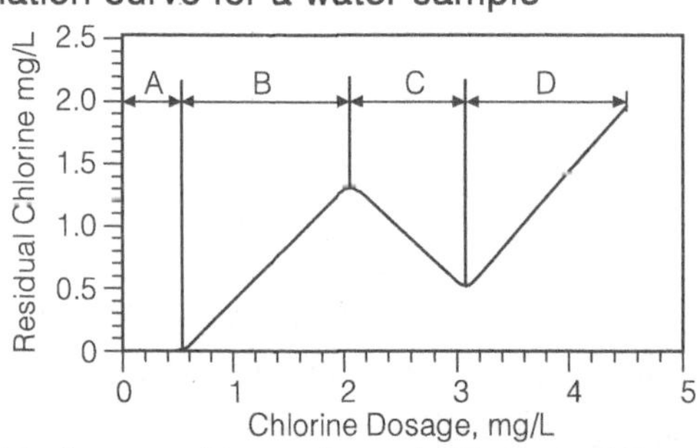

(a) Calculate what is the chlorine used in regions A, B, C and D.

(b) If you are to design a chlorination facility for this water, what minimum chlorine dose will you choose?

[10 Marks]

Sol:

(a) In region A

Applied chlorine dosage is 0.5 mg/*l* and residual chlorine dose is zero.

Hence chlorine used = 0.5 mg/*l* – 0 = 0.5 mg/*l*

In region B

Applied chlorine dosage is 2 mg/*l* and residual chlorine is 1.3 mg/*l*

∴ Chlorine used = 2 – 1.3 = 0.7 mg/*l*

In region C

Applied chlorine dosage is 3 mg/*l* and residual chlorine is 0.5 mg/*l*

∴ Chlorine used = 3 – 0.5 = 2.5 mg/*l*

In region D

Any addition of chlorine will appear equally as free residual chlorine.

(b) The minimum chlorine dose will be 2.5 mg/*l*

5.6 Briefly explain different methods of Defluoridation. Also give their respective advantages and disadvantages.

[12 marks]

Sol:

Comparative Table of Available Technologies for Defluoridation

Name of technology	Developer and process	Advantages	Disadvantages
1. Absorption by activated Alumina (AA). Commonly known as Prashanti technology.	• Satya Sai University for Higher Learning • Andhra Pradesh : Cost Rs. 35,000 at community level and Rs. 1,300 - 1700 at household level, Recurring cost of AA replenishment. • In this process, raw water is passed through AA which absorbs fluoride, passing out defluoridated water.	• Useful at community and household level. • Filters can be attached with handpumps or stand-posts. • Low sludge formation	• Cost is inhibiting for vilagers • Poor after sales service in villages • Regeneration of AA poses problem
2. Ion Exchange Absorption Method.	• Ion Exchange (India) Ltd. • In this process, resin is used to absorb fluoride from water. • Capacity ranges from 500 litres per hour to 5000 litres per hour.	• Useful at community and household level. • Filters can be attached with handpumps or standposts. • Low sludge formation	• Regular replacement of resin • Large amount of salts are involved in regeneration • Poor after sales service in villages
3. Nalgonda Technique	• National Environmental Engineering and Research Institute, Nagpur in 1975. • In this process, alum and lime is added to raw water and stirred for 10 minutes. Fluoride gets adsorbed by alum, and flocs are formed. These flocs are allowed to settle down for 90 minutes, after which is filtered by addition of bleaching powder.	• Useful at community and household level. • High fluoride removal efficiency.	• Technical expertise required • Time consuming process • Disposal of flocs is a problem • Alum dose needs regular calculation
4. Reverse Osmosis	• In this process, water is passed through a membrane which blocks fluoride flow, allowing only defluoridated water to pass • 60-litre per day filter costs Rs. 20,000 • Annual maintenance costs about Rs. 3000	• Different kind of membranes available. • High fluoride removal efficiency.	• Cost inhibitive for villagers • Poor after sales service in villages

5.7 Enumerate the chemicals which are used for coagulation. Discuss their comparative merits and demerits.

[10 Marks]

Sol:

Important coagulant used for coagulation process are as follows:

(i) Alum (ii) Copperas (iii) Chlorinated copperas (iv) Sodium aluminate (v) Lime

S.No.	Coagulants	Remarks
1.	Alum	• Chemical formula : $Al_2(SO_4)_3 \cdot 18\,H_2O$ • Alum requires alkalinity in water, (its normal dose 10-30 mg/*l*). • It is cheap and flocs formed are stable. • Disadvantages is that it is difficult to dewater the sludge formed
2.	Copperas	• Chemical formula : $FeSO_4 \cdot 7H_2O$ • It is added to water in conjugation with lime. • It is used for water which is not coloured. • It works in the pH range of 8.5 and above. • It forms heavy flocs. • Quantity used in same as that of alum (10 - 30) mg/*l*.
3.	Chlorinated Copperas	• Chemical formula : $[Fe_2(SO_4)_3 + Fe(Cl)_3]$ • It is effective in removing colour also. • It works in large pH range.
4.	Sodium Aluminate	• Chemical formula : $[Na_2Al_2O_4]$ • It reduces the temporary and permanent hardness. • It is costlier than alum. • It is useful for water which does not require alkalinity. • It is used for biolerfeed water which permits very low value of hardness

5.8 **What do you understand by the term "Optimum dosage of coagulant"? Explain the procedure to determine the optimum coagulant dose in laboratory. Duly taking into consideration the necessary optimum pH condition?**

[10 Marks]

Sol:

Optimum dosage of coagulant : It is defined as a minimum quantity of given coagulant (in mg) which produces a good floc when mixed with the given raw water.

- The quantity of coagulant greater or lesser then this optimum dose will give unsatisfactory floc formation.
- Optimum coagulant dose in decided by jar test.

Jar Test

The jar test involves the use of a stirring device. The stirrer consists of six paddles capable of rotation with variable speed between 0 to 100 rpm.

- In Jar test, 1 lit of water is placed in each of the jar or breakers, which are dosed with different amount of coagulant (say Alum).
- After rapid mixing to disperse the chemicals, the samples are stirred slowly for floc formation and then allowed to settle under quiescent conditions. For this purpose the jars are initially mixed at a speed of 60 to 80 rpm for 1 min. and then stirred at a speed of 30 rpm for 15 min.
- After the stirrer is stopped, the nature and setting characteristics of the flocs are observed and recorded in qualitative form such as poor, fair, good, very good, excellent.
- The lowest alum dosage that provides good turbidity removal during the jar test is considered as the first trial dosage in plant operation, and the final optimum quantity is adjusted by actual observations at the WTP. (Water treatment plant).

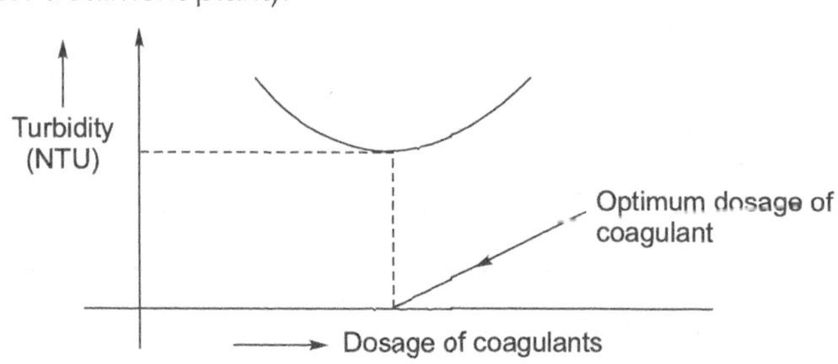

5.9 **Design six slow sand filters beds from the following data:**

1. **Population to be served = 50000 persons.**
2. **Per capita demand = 150 lpcd.**
3. **Rate of filtration = 180 *l*/hr/m^2.**
4. **Length of each bed = 2 × breadth.**

Assume maximum demand as 1.8 times the average daily demand and one unit out of six will be kept as stand by

[10 Marks]

Sol:

Given data : Population = 50000, Per capita demand = 150 *l*pcd, Rate of filtration (f_r) = 180 *l*/hr/m^2,

Average daily demand = 150 × 50000 *l*/day = 7.5 MLD

Maximum daily demand = 1.8 × 7.5 = 13.5 MLD

Since, slow sand filter is designed for maximum daily demand

$$\therefore \quad Q_{design} = 13.5 \text{ MLD}$$

$$\therefore \quad \text{Plan area} = \frac{Q_{design}}{f_r} = \frac{\frac{13.5\times10^6}{24}\ l/\text{hr}}{180\ l/\text{hr/m}^2} = 3125\ \text{m}^2$$

Number of operational unit = 6 – 1 = 5

$$\therefore \quad \text{Area of each filter} = \frac{3125}{5} = 625\ \text{m}^2$$

$$\therefore \quad L \times B = 625$$

$$2B \times B = 625$$

$$B = \left(\frac{625}{2}\right)^{1/2} = 17.68\ \text{m} \simeq 18\ \text{m}$$

$$L = 2B = 2 \times 18 = 36\ \text{m}$$

Adopt, $H = 3$ m

∴ Dimension of each filter 36 m × 18 m × 3 m

5.10 **What do you understand by residual chlorine and breakpoint chlorination?**

[10 Marks]

Sol:

- Breakpoint chlorination is a term which gives us an idea of the extent of chlorine added to the water. In fact, it represents that how much dose of chlorination, beyond which any further addition of chlorine will equally appear as free residual chlorine.
- When chlorine is added to the water, it first of all, generally reacts with ammonia present in the water, so as to form chloramines. These chloramines respond to the DPD test in the same manner as does free chlorine. Therefore, the DPD, test will indicate the quantum of total residual chlorine, combined as well as free chlorine. Hence, if chlorine is slowly added to the water, and the residual is tested, it will be found that the residual will go on increasing with addition of chlorine. However, some chlorine is consumed for killing bacteria, and thus the amount of residual chlorine shall be slightly less than that added.

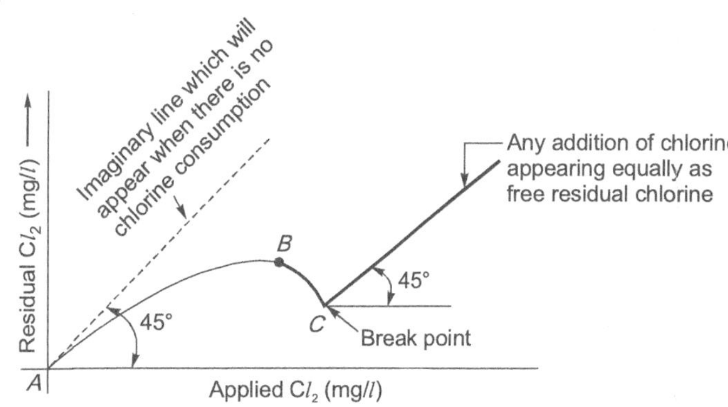

The addition of chlorine is continued beyond the point B, the organic matter present in water starts getting oxidised and therefore, the residual chlorine content suddenly falls down, as shown by the curve BC. The point C is the point beyond which any further addition of chlorine will equally appear as free chlorine, since nothing of it shall be utilized. This point C is called the breakpoint, as any chlorine that is added to water beyond this point, breaks through the water, and totally appears as residual chlorine. The addition of chlorine beyond breakpoint is called breakpoint chlorination.

5.11 **Write short notes on chlorination and desalination.**

[10 Marks]

Sol:

Chlorination

Chlorine in its various forms is invariably and almost universally used for disinfecting public water supplies. It is cheap, reliable, easy to handle, easily measurable and above all capable of providing residual disinfecting effects for long periods. Thus, affording complete protection against future contamination of water in distribution system. It's only disadvantage is that when used in great amount, it imparts bitter taste to the water.

Desalination

It is the method of converting salty water into fresh water. It is costly process. The water obtained by desalination proves much costlier then naturally available tracted water. Following methods can be adopted for desalination forecasting.

(i) Desalination by evaporation and distillation.
(ii) Electrodialysis method
(iii) Reverse osmosis method
(iv) Freezing process
(v) Solar distillation method

5.12 **A dual media rapid sand filter plant is to be constructed for treatment of 72 million litres of water per day. A pilot plant study indicated that a filtration rate of 15 m/h would be acceptable. Allowing one unit out of service for backwashing, how many 5 m x 8 m filter units will be required? Determine the net production in million litres per day of each filter unit if backwashing is done at 36 m/h for 20 minutes and the water is wasted for the first 10 minutes of each filter run.**

[12 Marks]

Sol:

Quantity of water to be treated = 72 MLD

Filtration rate = 15 m/hour

Design discharge = 72×10^6 lit/day

$$= \frac{72 \times 10^3}{24} = 3000 \text{ m}^3\text{/hour}$$

∴ Area of filter required,

$$= \frac{3000}{15} = 200 \text{ m}^2$$

Area of one unit = $5 \times 8 = 40 \text{ m}^2$

∴ Total number of filter units required

$$= \frac{200}{40} = 5 \text{ units}$$

As one unit required for back washing

Therefore, total number of filter unit

= 5 + 1 = 6 unit

Water treated by one unit = $40 \text{ m}^2 \times 15$ m/hour

= 600 m^3/hour

$$\text{Amount of water used in cleaning} = 36 \times 40 \times \frac{20}{60} = 480 \text{ m}^3\text{/hour}$$

$$\therefore \quad \text{Amount of water wasted} = 36 \times 40 \times \frac{10}{60} = 240 \text{ m}^3\text{/hour}$$

∴ Total water to be treated by one filter unit

(10 min = 0.166 hour water in each filter for cleaning)

$$= \frac{600 \times 24}{24.166} = 595.86 \text{ m}^3\text{/hour}$$

∴ Net production of water in each filter unit

$$= 595.86 - 480 = 115.86 \text{ m}^3/\text{hour}$$
$$= 115.86 \times 10^3 \times 24$$
$$= 2.78 \text{ MLD}$$

5.13 A rectangular sedimentation tank is designed for a surface overflow rate (surface loading) of 12,000 litre/hour/m^2. What percentage of the suspended particles of diameter (a) 0.03 mm and (b) 1.2 mm will be removed in the tank. Appropriate expressions for settling velocity (V_s, mm/s) may be selected from the formulate given below. Kinematic viscosity (n) of water = 0.897 mm^2/s. Specific gravity of particles = 2.65.

Stoke's : $V_S = \dfrac{g}{18}\dfrac{d^2}{\nu}(G_S - 1)$

Hazen's : $V_s = \left[\dfrac{4}{3}\dfrac{gd}{C_d}(G_s - 1)\right]^{0.5}$

$$C_D = \frac{24}{R_e} + \frac{3}{\sqrt{R_e}} + 0.34$$

Where R_e is the Reynolds number

Newton's : $V_s = [3.33\, gd\, (G_s - 1)]^{0.5}$

[12 Marks]

Sol:

$$V_0 = 12000 \text{ litre/hour/m}^2$$
$$\nu = 0.897 \text{ mm}^2/\text{s} = 0.897 \times 10^{-6} \text{ m}^2/\text{sec}$$
$$G_s = 2.65$$
$$d = 0.03 \text{ mm} = 0.03 \times 10^{-3} \text{ m} < 0.1 \text{ mm}$$

For $d < 0.1$ mm

Use stoke's law

$$V_s = \frac{g}{18}\cdot\frac{d^2}{\nu}(G_s - 1)$$
$$= \frac{9.81}{18} \times \frac{(0.03 \times 10^{-3})^2}{0.897 \times 10^{-6}} \times (2.65 - 1) = 9.022 \times 10^{-4} \text{ m/s}$$

Surface loading

$$V_0 = 12000 \text{ litre/hr/m}^2$$
$$= \frac{12000 \times 10^{-3}}{60 \times 60} \text{ m}^3/\text{m}^2/\text{sec}$$
$$= 3.33 \times 10^{-3} \text{ m/s}$$

∴ Percentage removal of particle of size = 0.03 mm

$$= \frac{V_s}{V_0} \times 100 = \frac{9.022 \times 10^{-4}}{3.33 \times 10^{-3}} \times 100 = 27\%$$

(b) For $d = 1.2$ mm

Use Newton's equation

$$V_s = [3.33\, gd(G - 1)]^{0.5}$$
$$= [3.33 \times 9.81 \times 1.2 \times 10^{-3} \times 1.65]^{0.5} = 0.25 \text{ m/s}$$

which is greater than the surface overflow rate.

Hence all particles will settle.

Level-2

5.14 **(a) Clearly distinguish between different settling process also giving examples of treatment units in water or wastewater treatment, where these process are utilized.**

(b) Show that the settling velocity of a spherical particle in a liquid under condition when Reynold's number is less than 0.5 may be given by expression

$$V_s = \frac{g}{18}(G_s - 1)\frac{d^2}{v}; \text{ where } G_s \text{ is the specific gravity of particle}$$

[20 Marks]

Sol:

(a) Settling/sedimentation of raw water and wastewater has generally been divided into the following four types:

(i) Type 1 or Discrete Settling:

- This type of settling refers to the settling of discrete particles, such as the removal of grit and sand from raw waters, containing low concentration of solids.
- It occurs when particles do not change their size, shape or mass during settling.
- For calculating settling velocity either Stoke's law or transition law is applied.

(ii) Type 2 or Flocculent Settling:

- This type of settling refers to settling of flocculent particles in rather dilute suspensions.
- The degree of flocculation depends on the contact opportunities which in turn is affected by the surface overflow rate, the depth of basin, the concentration of particles, the range of particle size and the velocity gradient of the system.
- No adequate mathematical equation exist to describe flocculent settling and therefore overflow rate to achieve a given removal efficiency are determined using settling column analysis.
- Removal of raw sewage, organic suspended solids in PST, settling of chemical flocs in settling tank and bioflocs in upper portion of SST are example of flocculent settling.

(iii) Type 3 or Zone or Hindered Settling :

- This type of settling involves the settling of the flocculent particles in concentrated suspensions, in which interparticle forces are sufficient to hinder the settling of the neighbouring particle, thereby tending the particle to remain in a fixed positions with respect to each other, causing settling of the mass of particle as a unit, developing a solid liquid interface at the top of the settling mass.
- The settling that occurs in the secondary settling tank of a biological sewage treatment unit is an example of this type of settling.

(iv) Type 4 or Compression Settling :

- When solids are present in excessive concentration, such as at the bottom of SST or in a sludge thickener unit, the settling occurs only by compression caused by weight of particles over it.
- Such settling of particles in form of sludge blanket in wastewater treatment is usually represented as type 4 settling.

(b) **Derivation of Stoke's law :** When a solid particles settles down in water, its downward settlement is opposed by the drag force offered by the water. The effective weight of the particle (i.e. actual weight-buoyancy) causes the particles to accelerate in the beginning, till it attains a sufficient velocity (V_S) at which the drag force becomes equal to the effective weight of the particle. After attaining that velocity, the particle falls down with that constant velocity (V_S). Now, the drag force offered by the fluid is given by Newton's law as,

$$\text{Drag force, } F_D = \frac{1}{2}C_D\rho_w AV^2 \quad \text{...(i)}$$

where, C_D = Coefficient of drag, A = Area of particle, ρ_w = Density of water, V = Velocity of fall

$\therefore$ The effective weight of the particle = Total weight – Buoyancy

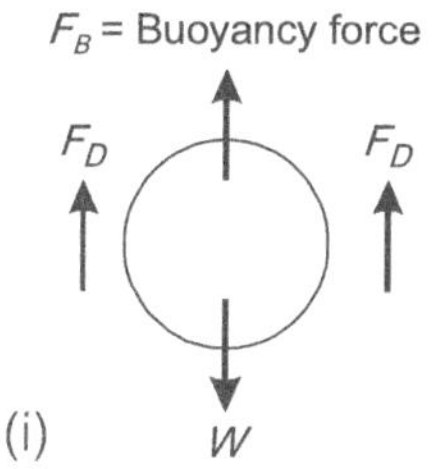

$$= \frac{4}{3}\pi r^3 \gamma_s - \frac{4}{3}\pi r^3 \gamma_w$$

$$= \frac{4}{3}\pi r^3 (\gamma_s - \gamma_w) \quad \text{...(ii)}$$

Equation (i) and (ii) will become equal when V becomes equal to V_S in equation (i)

$$\therefore \quad \frac{1}{2}C_D \rho_w A V_S^2 = \frac{4}{3}\pi r^3 (\gamma_S - \gamma_w)$$

$$V_S^2 = \frac{\frac{4}{3}\pi r^3 (\gamma_s - \gamma_w)}{\frac{1}{2}C_D \rho_w (\pi r^2)} = \frac{8}{3}\frac{r}{C_D}(G_s - 1)g$$

$$V_S^2 = \frac{8}{3}\times\frac{d}{2}\times g \times \frac{1}{C_D}(G_s - 1)$$

$$V_S^2 = \left[\frac{4}{3}\frac{gd(G_s - 1)}{C_D}\right]^{1/2} \quad \text{...(iii)}$$

- The coefficient of drag (C_D) changes and depends upon the flow regimes surrounding the particle. The flow regime may be characterized qualitatively as laminar, turbulent and transitional.
- However, for laminar flow region ($Re < 0.5$), the relation become simple as

$$C_D = \frac{24}{R_e}; \; R_e = \frac{V_s d}{\nu}$$

$\therefore$ From equation (iii),

$$V_S = \left(\frac{\frac{4}{3}gd(G_s - 1)}{C_D}\right)^{1/2}$$

$$V_S^2 = \frac{4}{3}gd\frac{(G_S - 1)}{24} \times \frac{V_S d}{\nu}$$

$$V_S = \frac{g}{18}(G_S - 1)\frac{d^2}{\nu} \quad \text{[Proved]}$$

5.15 The population of a town is 100000 and the average per capita demand is 135 litres/day/capita. Design the coagulation-cum-sedimentation tank for supplying water to the town. The maximum demand may be taken as 1.5 times the average demand. Detention period is 5 hrs. for settling tank and 30 minutes for flocculent chamber. Flow rate is 900 *l*/hr/m² of plan area.

[20 Marks]

Sol:

Given data : Population = 100000, Average per capita demand = 135 *l*pcd, Detention time for settling tank (D_t) = 5 hrs., Overflow rate (V_0) = 900 l/hr/m^2

Average daily discharge/demand = (135 × 100000 *l*/day) = 13.5 MLD

$\therefore$ Maximum daily demand = 1.5 × 13.5 = 20.25 MLD

Design of settling tank:

$$\text{Capacity of tank } (V) = Q_{design} \times D_t$$

$$= \frac{20.25 \times 10^3 m^3}{24 \text{ hrs}} \times 5 = 4218.75 \text{ m}^3$$

$$V_0 = 900 \text{ } l/\text{hr/m}^2$$

$$\therefore \quad \text{Plan area} = \frac{Q}{V_0} = \frac{20.25 \times 10^6}{24 \times 900} = 937.5 \text{ m}^2$$

Using width of tank (B) = 12 m

$$\therefore \quad L = \frac{937.5}{12} = 78.125 \text{ m} \simeq 78 \text{ m}$$

$$\therefore \quad \text{Height of tank} = \frac{(4218.75)}{78 \times 12} = 4.5 \text{ m}$$

Hence, use a tank of 78 m x 12 m x 4.5, tank with extra depth of sludge storage, say use 5.0 m depth at starting end and $5 + \frac{78}{50} = 6.56$ m at the downstream end (using 1 in 50 slope). Use a free board of 0.5 m above the water level.

Design of the floc chamber:

In addition to 78 m length of the settling tank, the floc chamber at the entry has to be provided. Assume the effective depth in the floc chamber as half of the depth in the tank near the floc chamber i.e. $\left(\frac{5}{2}\right) = 2.5$ m

Given Period of flocculation = 30 minutes

$$\therefore \quad \text{Capacity of the chamber} = \frac{20.25 \times 10^3 m^3}{24 \text{ hrs}} \times \frac{30}{60} \text{hrs} = 421.875 \text{ m}^3$$

$$\text{Plan area} = \frac{421.85}{2.5} = 168.75 \text{ m}^2$$

Using the same width as 12 m,

$$\therefore \text{ Length of flocculation chamber } (L) = \frac{168.75}{12} = 14 \text{ m}$$

$\therefore$ Use (14 m x 12 m x 2.5 m) floc chamber.

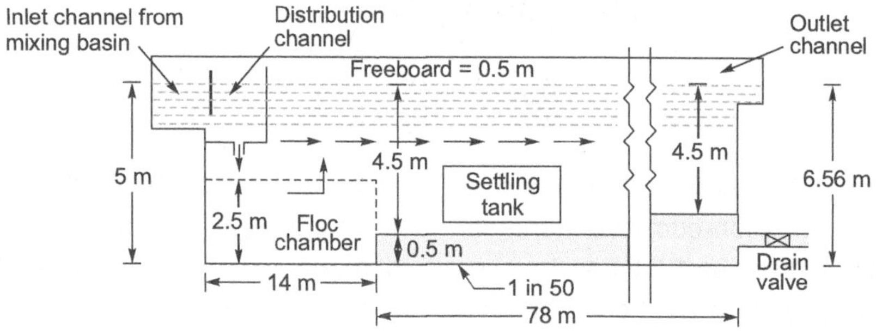

5.16 **Design a primary circular sedimentation tank for a town of population 50000 making all the necessary but reasonable assumptions.**

[15 Marks]

Sol:

Given data : Population = 50000

Assuming average per capita demand as 325 $l$pcd.

$\therefore$ Average daily demand = (325 × 50000 L/day) = 16.25 MLD

As we know that,

Sedimentation tank is designed for maximum daily demand

$\therefore$ $Q_{design} = 1.8 \times$ Average daily demand $= 1.8 \times 16.25 = 29.25$ MLD

Assuming detention time $(D_t) = 6.5$ hrs

$\therefore$ Volume of tank $(V) = Q_{design} \times D_t$

$$= \frac{29.25 \times 10^3 m^3}{24 \text{ hrs}} \times 6.5 \text{ hrs} = 7921.875 \text{ m}^3$$

Assuming surface overflow rate,

$$(V_0) = 15 \text{ m/day} = \frac{15}{24 \times 3600} = 1.736 \times 10^{-4} \text{m/s}$$

$$\therefore \quad \text{Plan area of tank} = \frac{Q_{design}}{V_0} = \frac{29.25 \times 10^3 / 24 \times 3600}{1.736 \times 10^{-4}} = 1956.79 \text{ m}^2$$

Let D be the diameter of the tank

$$\therefore \quad \frac{\pi}{4} D^2 = 1956.79$$

$$D = 49.91 \text{ m} \simeq 50 \text{ m}$$

Also, volume of circular sedimentation tank,

$$V = D^2 (0.785H + 0.011D)$$

$$7921.875 = (50)^2 (0.785H + 0.011 \times 50)$$

$$H = 3.33 \text{ m} \simeq 3.4 \text{ m}$$

Providing a free board of 0.3 m

$\therefore$ Size of circular primary sedimentation tank should be 50 m (diameter) × 3.7 m (height)

5.17 For the given particle distribution, compute the overall efficiency of the sedimentation tank, Overflow rate = 32.6 m/day, Specific gravity of solids = 1.2, Dynamic viscosity = 1.027 centipoises, $\rho_{water} = 0.997$ gm/cm^3

Particle (mm)	0.1	0.08	0.07	0.06	0.04	0.02	0.01
Weight fraction greater in size	10	15	40	70	93	99	100

[20 Marks]

Sol:

Given data : $V_0 = 32.6$ m/day, $G_S = 1.2$,

$$\mu = 1.027 \text{ poise} = \frac{1.027}{100} \text{poise}$$

$$= \frac{1.027}{100 \times 10} \left(\frac{\text{N-s}}{\text{m}^2}\right) = 1.027 \times 10^{-3} \left(\frac{\text{N-s}}{\text{m}^2}\right)$$

$$\rho_w = 0.997 \text{ gm/cm}^3 = \frac{0.997 \times 10^{-3} \text{kg}}{10^{-6} \text{m}^3} = 997 \text{ kg/m}^3$$

Since, particle present in water are of all possible size. Hence to find the efficiency of sedimentation tank we use cumulative frequency distribution curve as:

Settling velocity:

$$V_S = \frac{g}{18}(G_s - 1)\frac{d^2}{\nu} \qquad \text{Assuming } d \text{ (in mm)}$$

$$V_S = \frac{9.81 \times 1000}{18} \frac{\text{mm}}{\text{sec}^2} \times (1.2 - 1) \times \frac{d^2 \text{mm}^2}{\left(\dfrac{1.027 \times 10^{-3} \text{kg} \dfrac{\text{m}}{\text{sec}^2} \times \dfrac{\text{sec}}{\text{m}^2}}{997 \text{ kg/m}^3} \times 10^6\right) \dfrac{\text{mm}^2}{\text{sec}}}$$

$$V_S\text{(mm/sec)} = 105.8\, d^2 \text{ (d in mm)}$$

$$V_0\text{ (SOR)} = 32.6 \text{ m/day} = \frac{32.6 \times 1000}{86400} = 0.377 \text{ mm/sec}$$

Particle size (mm)	0.1	0.08	0.07	0.06	0.04	0.02	0.01
Weight fraction greater in size	10	15	40	70	93	99	100
Weight fraction smaller in size	90	85	60	30	7	1	0
$V_S = 105.8\, d^2$ (mm/sec)	1.058	0.677	0.518	0.381	0.169	0.042	0.010

0.377 mm/sec

From table, x_s = Fraction of particle having settling velocity less than SoR

$$x_s = 7 + \frac{30-7}{0.381-0.169}(0.377-0.169) = 29.56\%$$

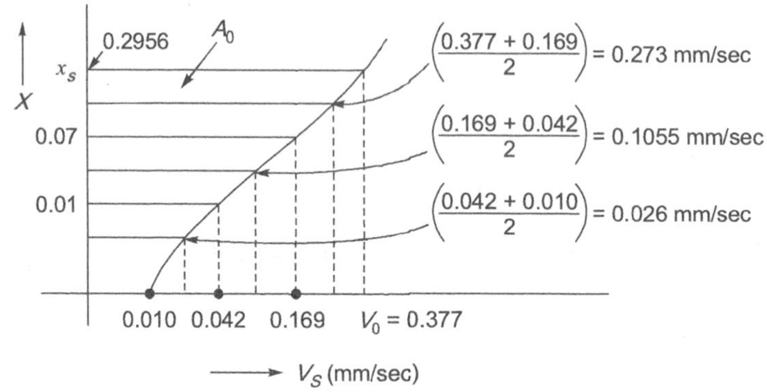

X = Fraction of particles having settling velocity less than stated velocity.

$$A_0 = \Sigma \bar{V}_S \Delta x$$

$$= 0.273 \times (0.2956 - 0.07) + 0.1055\,(0.07 - 0.01) + 0.026\,(0.01 - 0)$$

$$= 0.0682$$

∴ Efficiency of sedimentation tank,

$$\eta = \left[(1 - x_S) + \frac{A_0}{V_0}\right] \times 100 = \left[(1 - 0.2956) + \frac{0.0682}{0.377}\right] \times 100$$

$$\eta = 88.53\%$$ **Ans.**

5.18 **A coagulation sedimentation plant clarifies 40 MLD of water everyday. The quantity of filter alum required at the plant is 18 mg/*l*. If the raw water is having an alkalinity equivalent to 5 mg/*l* as $CaCO_3$, determine the quantity of filter alum and quick time (containing 85% of CaO) required per year.**

[15 Marks]

Sol:

Quantity of water required to be treated = 40 MLD

Alum required = 18 mg/*l*

∴ Total required of alum required per year = $40 \times 10^6\, l\text{/day} \times 18 \times 10^{-6}$ kg/lit $\times 365$ day/yr

= 262800 kg/year = 262.8 tonne/year

As we know,

1 gm of alum required 0.45 gm of alkalinity as $CaCO_3$.

∴ Total alkalinity requirement = 18 × 0.45 = 8.1 mg/*l* as $CaCO_3$

Alkalinity present = 5 mg/*l* as $CaCO_3$

$\therefore$ Additional alkalinity requirement = (8.1 – 5) = 3.1 mg/L as $CaCO_3$

$$= \frac{3.1\times10^{-3}}{50}\text{gm. eq/lit as CaO} = \frac{3.1\times10^{-3}}{50}\times 28\text{gm. eq/lit as CaO}$$

$$= \frac{3.1\times10^{-3}}{50}\times 28\times10^{3}\text{ mg/lit as CaO} = 1.736\text{ mg/L as CaO}$$

$\therefore$ Quantity of quickline (CaO) required = 40×10^{6} lit/day $\times 1.736\times10^{-6}$ kg/lit

$$= (40\times1.736)\text{ kg/day} = \left(\frac{40\times1.736\times365}{1000}\right)\text{ tonnes/yr}$$

$$= 25.3456\text{ t/yr}$$

$\therefore$ Quantity of pure 85% CaO required = $\left(\frac{25.3456}{0.85}\right) = 29.88$ t/years

5.19 **A water treatment plant is to treat a design discharge of 60 MLD. The plant has a paddle flocculator 20 m long, 12 m wide and 3.5 m deep. It is fitted with four horizontal shafts to which revolving paddles are attached which rotates at 150 rpm. Each shaft supports four paddles which are 10 m long, 20 cm wide and centered 1.5 cm from the shaft.**

- **Assume mean velocity of flow of water as 20% of the paddle velocity.**
- **C_D = 1.8**

Determine the difference in velocity between paddle and water, mean velocity gradient and detention time. Take dynamic viscosity of water = 1.31 × 10^{-4} N-s/m^2.

[20 Marks]

Sol:

Given data : Quantity of water to be treated, Q_0 = 60 MLD

Cross-section/Dimension of flocculation chamber (20 m × 12 m × 3.5 m)

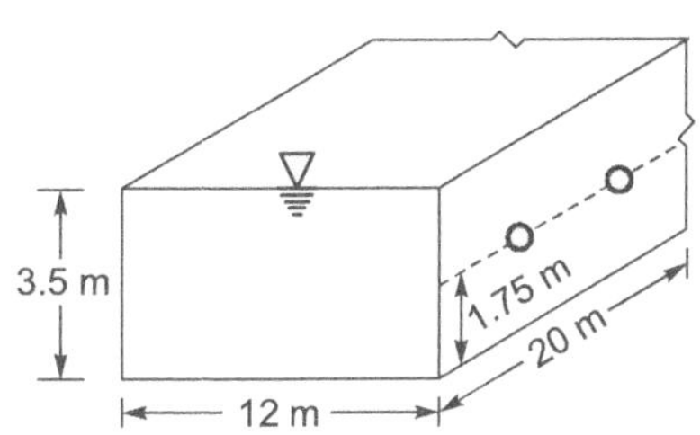

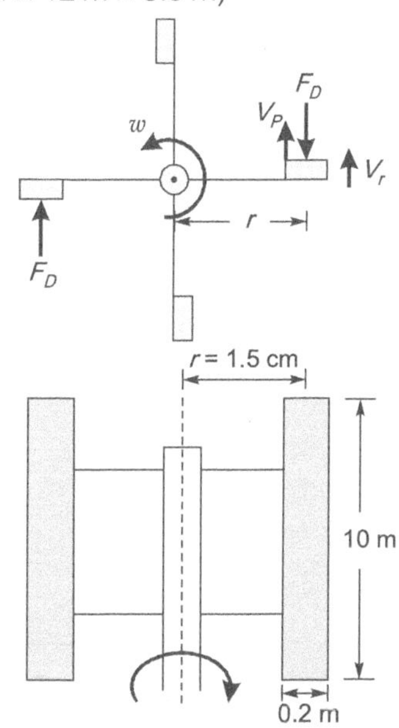

ω = 150 rpm

V_P(Velocity of paddles) = ωr

$$V_P = \frac{2\pi N}{60}r = \frac{2\pi\times150}{60}\times0.015 = 0.235\text{ m/sec}$$

(a) Difference of velocity between paddle and water

$$V_r = V_P - V_W = V_P - 0.2V_P = 0.8\ V_P = 0.8 \times 0.235 = 0.188 \text{ m/sec } \textbf{Ans}$$

(b) Power required, $P = (F_D \times V_r)$

$$= \frac{1}{2} C_D \rho A V_r^2 \times V_r = \frac{1}{2} \times 1.8 \times 1000 \times (4 \times 10 \times 0.2 \times 4) \times (0.188)^3$$

$$= 191.367 \text{ W}$$

Mean velocity gradient, $G = \left(\frac{P}{\mu V}\right)^{1/2} = \left(\frac{191.367}{1.31 \times 10^{-4} \times 20 \times 12 \times 3.5}\right)^{1/2} = 41.70 \text{ S}^{-1}$ **Ans.**

(c) Detention time,

$$D_t = \frac{V}{Q} = \frac{(20 \times 12 \times 3.5) \times 24 \times 60}{60 \times 10^3} = 20.16 \text{ min}$$ **Ans.**

5.20 **A town has requested proposals for a new coagulation water treatment plant. The design flow for the plant is 0.1065 m³/sec.**

The following design assumptions for a rapid mix tank have been made.

1. **Number of tank = 1.**
2. **Tank configuration; circular with depth = 2 × diameter.**
3. **Detention time, (D_t) = 10 sec.**
4. **Velocity gradient = 800 s^{-1}**
5. **6 flat blades [Area of each plates = 0.4 m²]**
6. **Distance of centre of plates from centre of tank = 0.22 m.**

Determine the following

(a) Power input (in kW) **(b) Tank dimension (in m)**

(c) Rotational speed of central shaft (in rpm)

Take $C_D = 1.6$; $\mu = 1.5 \times 10^{-3}$ N-s/m²; $V_r = 0.8V_P$

[20 Marks]

Sol:

Given data : Design flow rate, Q_0 = 0.1065 m³/sec; Detention time, D_t = 10 sec ; Velocity gradient, G = 800 S^{-1}; Ap (Area of each plate) = 0.4 m²; $\mu = 1.5 \times 10^{-3} \frac{\text{N-s}}{\text{m}^2}$

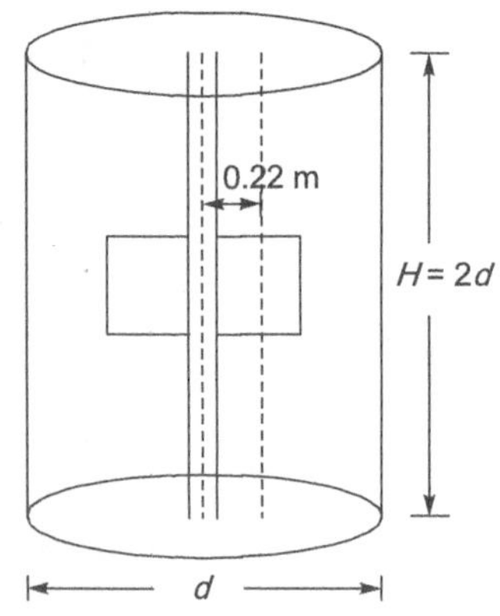

Volume/capacity of tank, $V = Q_0 D_t = 0.1065 \times 10 = 1.065 \text{ m}^3$

(i) $P = \mu V G^2 = 1.5 \times 10^{-3} \times 1.065 \times (800)^2 = 1022.4$ Watt

(ii)

$$\frac{\pi}{4} d^2 \times H = V$$

$$\frac{\pi}{4} d^2 \times 2d = 1.065$$

$$d = \left(\frac{1.065 \times 4}{\pi \times 2}\right)^{1/3} = 0.878 \text{ m}$$

$$\text{Height of tank, } H = 2d = 2 \times 0.878 = 1.756 \text{ m}$$

(iii)
$$P = F_D V_r = \frac{1}{2} C_D P A_P V_r^3$$

$$1022.4 = \frac{1}{2} \times 1.6 \times 1000 \times (6 \times 0.4) \times (0.8 \times V_P)^3$$

$$V_P = 1.013 \text{ m/sec}$$

$$\omega r = 1.013 \text{ m/sec}$$

$$\omega = \frac{1.013}{0.22} = 4.605 \text{ rad/sec}$$

$$N(\text{in rpm}) = \frac{60\omega}{2\pi} = \frac{60 \times 4.605}{2 \times 3.14} = 43.97 \text{ rpm} \simeq 44 \text{ rpm}$$

5.21 **A rapid sand filter is to be provided in a water treatment plant, to process the water for town with a population of 400000. The water demand is 250 litres/capita/day. The rate of filtration is 25 m³/m²/hr. Allow 3% of filtered water for backwash requirement. Each backwash period is of 30 min. Determine the number of filter required. The available surface area of filter unit is 10 m x 5 m. Also compute the upflow velocity and headloss to expand the bed to 0.75 m. The porosity of the bed is 0.60. Specific gravity is 2.6. The average particle size is 0.8 mm, drag coefficient is 26 and kinematic viscosity coefficient is 1.02 x 10^{-6} m²/sec. Assume unexpanded length of bed as 0.6 m.**

[20 Marks]

Sol:

Given data : Population = 400000, Per capita demand = 250 *l*pcd, Rate of filtration (f_r) = 25000 *l*/hr/m²

Rapid sand filter is designed to meet the requirement of maximum daily demand plus backwashing demand.

∴ Total water to be filtered per day = Maximum daily demand + Backwashing demand

$$= 1.8 \times \text{Average daily demand}\left(1 + \frac{3}{100}\right)$$

$$= 1.8 \times \frac{4 \times 10^5 \times 250}{10^6} \times 1.03 = 185.4 \text{ MLD}$$

Since, duration of backwashing = 30 min

$$\text{This total amount of water to be filtered in} = \left(24 - \frac{30}{60}\right) = 23.5 \text{ hr}$$

$$\therefore \quad Q_{\text{design}} = \frac{185.4 \text{ m}l\text{/day}}{23.5 \text{ hr/day}} = 7.8893 \text{ ML/hr}$$

$$\therefore \quad \text{Total plan area required, } A = \frac{Q_{\text{design}}}{f_r} = \frac{7.8893 \times 10^6 \; l\text{/hr}}{25000 \; l\text{/hr/m}^2} = 315.57 \text{ m}^2$$

$$\therefore \quad \text{Number of filter required} = \frac{A}{(\text{Area of each filter unit})} = \frac{315.57}{50} = 6.31$$

∴ Adopt 7 operational filters and kept one unit as standy

∴ Total number of filters required = 7 + 1 = 8 unit

Calculation of upflow velocity/backwash velocity:

$$n' = \left(\frac{V_b}{V_s}\right)^{0.22} \; ; V_b = \text{backwash velocity, } V_s = \text{Settling velocity}$$

$$V_s = \sqrt{\frac{4}{3} gd \frac{(G_s - 1)}{C_D}}$$

$$= \sqrt{\frac{4}{3} \times \frac{9.81 \times 0.8 \times 10^{-3}(2.6-1)}{26}} = 0.02537 \text{m/sec}$$

$$\frac{D'}{D} = \frac{(1-n)}{1-n'}$$

$$\frac{0.75}{0.6} = \frac{1-0.6}{1-n'}$$

$$n' = 0.68$$

$$\therefore \quad 0.68 = \left(\frac{V_b}{0.02537}\right)^{0.22}$$

$$V_b = 4.395 \times 10^{-3} \text{ m/sec}$$ **Ans.**

Head loss to expand the bed, $h_L = D(1-n)(G_s - 1)$

$$= 0.6\,(1-0.6)\,(2.6-1)$$

$$= 0.384 \text{ m}$$ **Ans.**

5.22 **A rapid sand filter is proposed for water supply treatment plant for a town with a population of 75000. The average rate of water supply is 150 lpcd and rate of filtration is to be 100 *l*/min/m². Design:**

(i) Size and number of filter beds.

(ii) Manifold lateral under drainage system.

(iii) Wash water discharge required, if rate of washing is 45 cm/min.

[25 Marks]

Sol:

(i) Size and number of filter beds :

Given data : Population 75000, Per capita water demand = 150 lpcd, $f_r = 100 \times 60$ *l*/hr/m² = 6000 *l*/hr/m²

Assuming 5% of daily water required to be kept as wash water and each backwashing period is 30 min. Since, rapid sand filter (RSF) is designed to meet the demand of maximum daily demand plus backwashing demand.

∴ Total amount of water required to be filtered per day = 1.8 × Average daily demand $\left[1 + \frac{5}{100}\right]$

$$= 1.8 \times \frac{75000 \times 150}{10^6} \times 1.05 = 21.2625 \text{ MLD}$$

This total amount of water to be filtered in $\left(24 - \frac{30}{60}\right) = 23.5$ hr

Now, $$Q_{design} = \frac{21.2625 \text{ mL/day}}{23.5 \text{ hr/day}} = 0.904787 \text{ ML/hr}$$

$$\therefore \quad Q_{design} = 904787 \text{ } l\text{/hr}$$

∴ Total plan area required, $A = \frac{Q_{design}}{f_r} = \frac{904787}{6000} = 150.79 \text{ m}^2$

Number of operational unit, $N = 1.22\sqrt{Q} = 1.22\sqrt{21.2625} = 5.625$

Adopt 6 number of operational unit and one unit kept as standby

$$\therefore \quad \text{Size of one unit} = \frac{150.79}{6} = 25.13 \text{ m}^2$$

$$\text{Adopt } \frac{L}{B} = 1.25$$

$$\therefore \quad 1.25\,B^2 = 25.13$$

$$B = 4.48 \text{ m}$$

$$L = 1.25 \times 4.48 = 5.6 \text{ m}$$

$$\therefore \quad \text{Adopt, } B = 4.5 \text{ m}, L = 5.6 \text{ m}$$

(ii) Design of manifold under drainage system

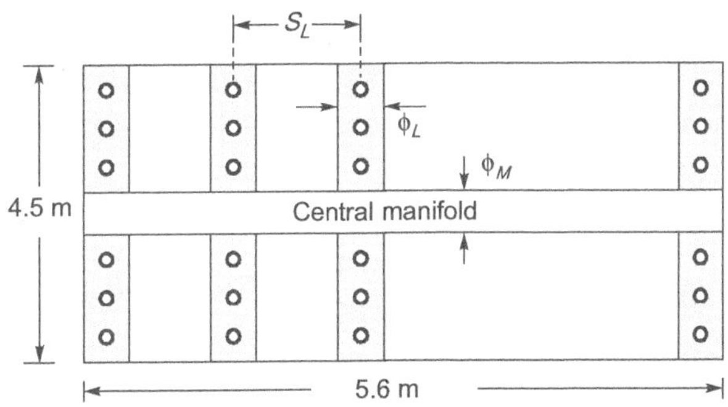

Adopt, S_L (Spacing between laterals) = 0.3 m

$$\therefore \quad \text{Number of laterals} = 2\left[\frac{L}{S_L}+1\right] = 2\left[\frac{5.6}{0.3}+1\right] = 39.33 \simeq 40$$

Assuming, total area of perforations on all laterals taken together = 0.3% of filter area

$$= \frac{0.3}{100} \times (5.6 \times 4.5) = 0.0756 \text{ m}^2$$

$$\therefore \quad \text{Area of perforations on one lateral} = \frac{0.0756}{40} = 1.89 \times 10^{-3} \text{m}^2$$

Adopting 12 mm perforations,

$$\text{Assuming, area of one lateral} = 2 \times \text{Area of perforations in it}$$

$$= 2 \times 1.89 \times 10^{-3} = 3.78 \times 10^{-3} \text{ m}^2$$

$$\frac{\pi}{4}\phi_L^2 = 3.78 \times 10^{-3}$$

$$\phi_L = 0.06937 \text{ m}$$

$$\phi_L = 0.07 \text{ m}$$

$$\text{Assuming, area of manifold pipe} = 2 \times \text{Cross section area of all laterals}$$

$$\frac{\pi}{4}\phi_M^2 = 2 \times \left[40 \times \frac{\pi}{4}\phi_L^2\right]$$

$$\frac{\pi}{4}\phi_M^2 = 80 \times \frac{\pi}{4} \times (0.07)^2$$

$$\phi_M = 0.626 \text{ m}$$

$$\phi_M \simeq 0.63 \text{ m}$$

$$\therefore \quad \text{Projected length of lateral} = \left(\frac{B-\phi_M}{2}\right) = \frac{4.5-0.63}{2} = 1.935 \text{ m}$$

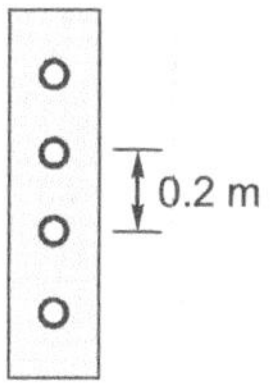

$$\text{Area of one perforation} = \frac{\pi}{4} \times (0.012)^2 = 1.131 \times 10^{-4} \text{m}^2$$

$$\therefore \quad \text{Number of perforations on one lateral} = \frac{1.89 \times 10^{-3}}{1.131 \times 10^{-4}} = 16.71 \simeq 18$$

For 18 perforations, length of lateral pipe should be greater than $(9 \times 0.2) = 1.8$ m, the projected length itself is greater than 1.8 m, and also the inclined length will be greater than even projected length. Hence, the holes can easily be accommodated on the lateral.

Check:

(1) $$\frac{\text{Length of lateral}}{\text{Diameter of lateral}} = \frac{\left(\frac{B-\phi_M}{2}\right)}{\phi_L} = \frac{1.935}{0.07} = 27.64 < 60 \text{ (OK)}$$

(2) Maximum velocity through manifold pipe $= \dfrac{21.2625 \times 10^3 (m^3)}{86400 \times 6} \times \dfrac{1}{\frac{\pi}{4} \times (0.63)^2}$

$= 0.1315$ m/sec < 2.4 m/sec (OK)

According to GOI manual, velocity of flow in manifold be 1.8 – 2.4 m/sec

(3) Wash water discharge required

Rate of washing of filter = 45 cm/min

$\therefore$ Wash water discharge required per unit $= \dfrac{0.45}{60} \times 4.5 \times 5.6 = 0.189\ m^3/sec$ **Ans.**

5.23 **A raw water to be treated by the lime-soda process to the minimum hardness possible has the following characteristics:**

$Ca^{2+} = 80$ mg/l; $CO_2 = 22.2$ mg/l; $Mg^{2+} = 12$ mg/l;

$Na^{+} = 46$ mg/l; $HCO_3^{-} = 152.5$ mg/l; $SO_4^{2-} = 216$ mg/l

(a) Check if the number of equivalent of positive and negative ions are balanced.

(b) For a flow of 25000 m^3/day, calculate the chemical requirement and the mass of solids produced. Assume that the lime used is 90% pure and the soda used is 85% pure. Also assume that the specific gravity of the sludge is 1.04. Assume, 2% solids in sludge produced.

[25 Marks]

Sol:

(a) Bar chart for water analysis

- It is the graphical representation of the water analysis, and is generally obtained by plotting milliequivalent per litre values of various radicals, to a scale. In order to plot such a chart. It is therefore necessary to complete milliequivalent/litre (meq/l) values for cations (positive radicals) as well as anions (negative radicals). The sum of the meq/l values for cation must be equal to the sum of the meq/l values for anions. In perfect evaluation, they would exactly be the same, because water in equilibrium is electrically balanced.
- The values of different radicals in mg/l as given in question are therefore converted to meq/l values.

Impurities	Number of milli.eq/l
$Ca^{2+} = 80$ mg/l	$\frac{80}{20} = 4$
$CO_2 = 22.2$ mg/l	$\frac{22.2}{(44/2)} = 1.009$
$Mg^{2+} = 12$ mg/l	$\frac{12}{12} = 1$
$Na^{2+} = 46$ mg/l	$\frac{46}{23} = 2$
$HCO_3^{-} = 152.5$ mg/l	$\frac{152.5}{61} = 2.5$
$SO_4^{2-} = 216$ mg/l	$\frac{216}{48} = 4.5$

Bar chart:

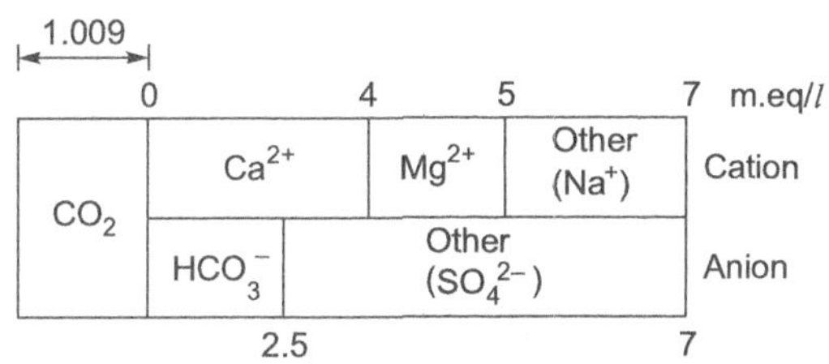

From this bar chart, it is clear that number of equivalents of +ve (cations) and –ve (anions) are balanced.

(b) Calculation of lime and soda requirement :

- Lime is required for CO_2, Mg^{2+}, HCO_3^-.
- Soda ash is required for all NCH (non carbonate hardness).

$\therefore$ Total lime required = 1.009 + 1 + 2.5 = 4.509 milli.eq/l

$\therefore$ Total pure CaO required = 4.509 × 28 = 126.25 mg/l as CaO

$$\text{Actual CaO required} = \frac{126.25}{0.9} = 140.28 \text{ mg}/l$$

Total water to be treated, Q = 25000 m³/day = 25 MLD

$\therefore$ Actual CaO required = 140.28 × 25 = 3507 kg/day **Ans.**

$$\text{NCH} = \text{TH} - \text{Alkalinity}$$

$$= 5 \times 50 - 2.5 \times 50 = 125 \text{ mg}/l \text{ as } CaCO_3$$

$$\therefore \quad \text{Pure soda ash required} = \frac{125}{50} \times 53 = 132.5 \text{ mg}/l \text{ as } Na_2CO_3$$

$$\text{Actual soda ash required} = \frac{132.5}{0.85} = 155.88 \text{ mg}/l$$

Actual soda ash required (in kg/day) = 25 × 155.88

= 3897.058 kg/day **Ans.**

The sludge formed in the lime soda process is corresponding to

(i) precipitate of Ca^{2+} in the lime as $CaCO_3$.

(ii) precipitate of Ca^{2+} in the calcium hardness as $CaCO_3$.

(iii) precipitate of Mg^{2+} in the magnesium hardness as $Mg(OH)_2$.

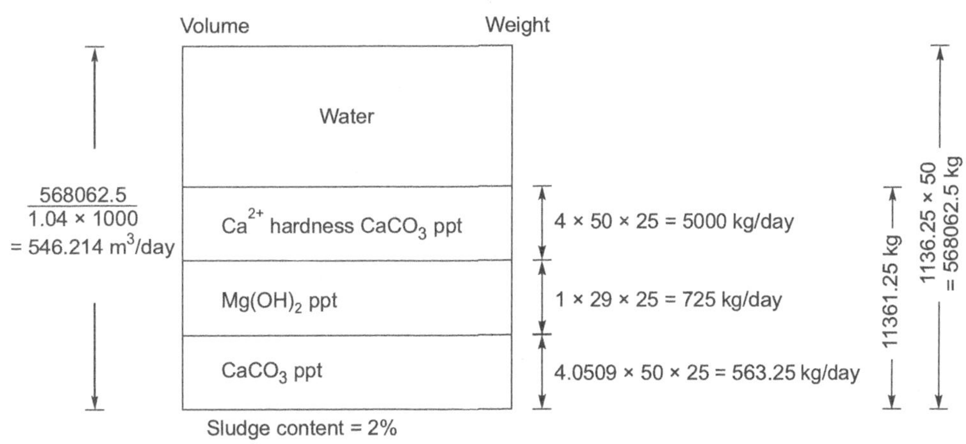

2 kg of solids ≃ 100 kg of sludge

1 kg of solids ≃ 50 kg of sludge

$\therefore$ Total sludge produced = 50 × 11361.25

= 568062.5 kg/day **Ans.**

5.24 Design a rectangular sedimentation tank to remove spherical particles of size equal to and greater than 50 mm (micrometers) with a specific gravity of 2.3 from 100 MLD (Million Litres per Day) of turbid water. Also determine the detention time of the tank.

Assume:

(a) Stoke's law of sedimentation is applicable.

(b) length to width ratio is 3 : 1 for the tank

(c) depth of tank is 3 m, and

(d) kinematic viscosity of water is 1.01×10^{-6} m²/s.

[15 Marks]

Sol:

Quantity of water required = 100×10^6 lit/day = 10^5 m³/day

Given,

Diameter of particle, $d = 50 \times 10^{-6}$ m $= 0.05$ mm

Specific gravity of particle, $G = 2.30$

Kinematic viscosity of water, $\nu = 1.01 \times 10^{-6}$ m²/sec

$\therefore$ Setting velocity,

$$V_s = \frac{g}{18}(G-1)\frac{d^2}{\nu} = \frac{9.81}{18}(2.30-1)\frac{(50\times10^{-6})^2}{1.01\times10^{-6}}$$

$$= 1.753 \times 10^{-3} \text{ m/s}$$

Given, $\frac{L}{B} = \frac{3}{1}$

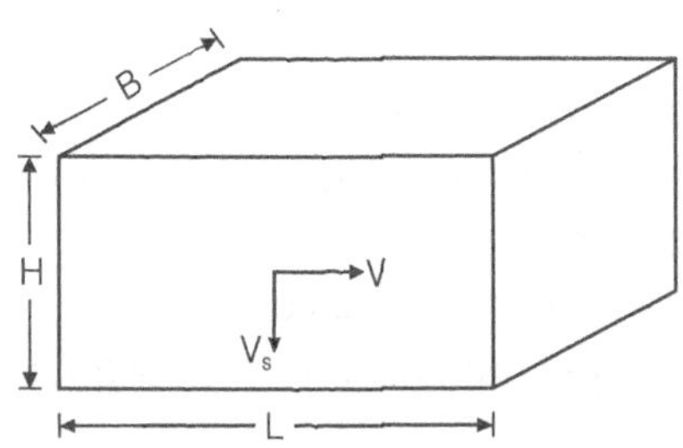

and $H = 3$ m

$\because$ $\frac{V}{V_s} = \frac{L}{H}$

$\Rightarrow$ $V = V_s \times \frac{L}{H} = 1.753\times10^{-3} \times \frac{L}{3} = 0.584 \times 10^{-3} L$

Assume detention time of the tank is T days

$\therefore$ capacity of the tank

$$V = Q \times T = 10^5 \times T \text{ m}^3$$

$\Rightarrow$ $LBH = 10^5 T$ m³

$\Rightarrow$ $3 B.B.3 = 10^5 T$

$\Rightarrow$ $9 B^2 = 10^5 T$ ...(i)

$\because$ $V_s = \frac{Q}{BL}$

$$\Rightarrow \qquad 1.753 \times 10^{-3} = \frac{10^5 m^3/day}{3B^2} = \frac{\frac{10^5}{86400} m^3/sec}{3B^2}$$

$$\Rightarrow \qquad B^2 = \frac{10^5}{86400 \times 3 \times 1.753 \times 10^{-3}}$$

$$B = 14.835 \text{ m}$$

$\therefore$ Length of the tank, $L = 3B = 3 \times 14.835 = 44.50$ m

$\therefore$ from equation (i) detention time,

$$T = \frac{9 \times (14.835)^2}{10^5} = 0.0198 \text{ days} = 0.475 \text{ hour}$$

5.25 In order to test a filtration process, clear water is made to pass through a bed of uniform sand at a filtering velocity of 3.0 m/hour. Sand bed has the following properties.

Depth of bed : 0.6 m
Sand grain size : 0.5 mm
Sand grain specific gravity : 2.65
Sand grain shape factor used to calculate filtration Reynold's number
: $\phi = 0.80$
Porosity of sand bed : n = 0.4
Kinematic viscosity of water : $\nu = 1.0 \times 10^{-6}$ m²/s
Calculate the loss of head in filtration

[20 Marks]

Sol:

The head loss through a clean stratified sand filter with uniform porosity and having representative particle size d, is given by Rose's equation as

$$h_L = \frac{1.067 (V)^2 D}{\phi g n^4} \left[\frac{C_D}{d}\right]$$

where,

V = approach velocity or filtration velocity in m/s
D = depth of filter in m
ϕ = shape factor
g = acceleration due to gravity in m/s²
n = porosity
C_D = drag coefficient

The equation to be used for computing C_D depends upon the value of Reynold's number, which is given by

$$\therefore \qquad R_e = \frac{Vd}{\nu}$$

where,

V = velocity of flow in m/s
ν = kinematic viscosity of water in m²/s
d = dia. of spherical sand particles
= ϕd for non spherical particles where ϕ is shape factor

$$R_e = \frac{V\phi d}{\nu}$$

Now, V = filtration velocity in m/s

$= 3$ m/hr (given)

$$= \frac{3}{(60 \times 60)} \frac{\text{m}}{\text{sec}} = 8.33 \times 10^{-4} \text{ m/s}$$

d = sand grain size

$= 0.5 \text{ mm} = 0.5 \times 10^{-3}$ m

$\phi d = 0.8 \times 0.5 \times 10^{-3}$ m

$= 0.4 \times 10^{-3}$ m

$\nu = 1 \times 10^{-6}$ m²/s

$$\therefore \quad R_e = \frac{8.33 \times 10^{-4} \times 0.4 \times 10^{-3}}{1 \times 10^{-6}}$$

$= 0.33 < 0.5$

Since $R_e < 0.5$, the coefficient of drag force (C_D) can be calculated by

$$C_D = \left[\frac{24}{R_e}\right] = 72$$

$$\therefore \quad h_L = \frac{1.067 \times \left(8.333 \times 10^{-4}\right)^2 \times 0.6}{0.8 \times 9.81 \times (0.4)^4} \left[\frac{72}{0.5 \times 10^{-3}}\right]$$

$$\Rightarrow \quad h_L = 0.3186 \text{ m}$$

6. Design of Sewer

Level-1

6.1 **A 450 mm sewer pipe, $n = 0.013$, is placed on a slope of 0.0025. At what depth of flow, the velocity is equal to 0.60 m/sec. Table of hydraulic elements of sewers flowing full, $n = 0.013$**

D(mm)	Slope (m / m)	Velocity (m / sec)
500	0.0025	1.18
450	0.0025	0.90
400	0.004	1.40
300	0.0025	1.10

Table of hydraulic element of sewers partially full:

d / D	0.20	0.23	0.28	0.41
v / V	0.60	0.67	0.70	0.80

[10 Marks]

Sol:

For 450 mm sewer,

$n = 0.013$, slope $= 0.0025$

Velocity at full depth of flow (V) $= 0.90$ m/sec (from table)

Also, $v = 0.6$ m/sec (for partial flow)

$$\frac{v}{V} = \frac{0.6}{0.9} = \frac{6}{9} = 0.67$$

Hence, for $\frac{v}{V} = 0.67;$

$$\frac{d}{D} = 0.23 \quad \text{(from table of sewers partially full)}$$

$\therefore$ $d = 0.23 \times 450 = 103.5 \text{ mm} = 10.35 \text{ cm}$ **Ans.**

Hence, in 450 mmϕ sewer depth of flow at which velocity of flow is equal to 0.6 m/sec is equal to 10.35 cm.

6.2 **Find the relation between the sides of a square section of one sewer and the diameter of a circular sewer when both are hydraulically equivalent.**

[8 Marks]

Sol:

Let D be the diameter of the circular sewer, and B be the side of equivalent square sewer. The discharge carrying capacity of circular sewer while running full, at a gradient of 1 in S,

$$= \frac{1}{N}\left(\frac{\pi}{4}D^2\right)\left(\frac{D}{4}\right)^{2/3} S^{1/2} \quad ...(1)$$

The discharge carrying capacity of square section while running full at a gradient of 1 in S,

$$= \frac{1}{N}(B^2)\times\left(\frac{B^2}{4B}\right)^{2/3} \times S^{1/2} \quad ...(2)$$

equating eq. (1) and eq. (2), we get

$$\frac{1}{N}\left(\frac{\pi}{4}D^2\right)\times\left(\frac{D}{4}\right)^{2/3} \times S^{1/2} = \frac{1}{N}(B)^2\times\left(\frac{B}{4}\right)^{2/3} \times S^{1/2}$$

$$\frac{\pi}{4}\times\frac{D^{8/3}}{(4)^{2/3}} = \frac{B^{8/3}}{(4)^{2/3}}$$

$$D^{8/3} = \frac{4}{\pi}\times B^{8/3}$$

$$D = \left(\frac{4}{\pi}\right)^{3/8} \times B$$

$$D = 1.094 \text{ B} \quad \textbf{Ans.}$$

(This is the required relation)

6.3 **A 450 mm sewer pipe, $n = 0.013$, is placed on a slope of 0.0025. At what depth of flow, the velocity is equal to 0.60 m/sec. Table of hydraulic elements of sewers flowing full, $n = 0.013$**

D(mm)	Slope(m / m)	Velocity(m / sec)
500	0.0025	1.18
450	0.0025	0.90
400	0.004	1.40
300	0.0025	1.10

Table of hydraulic element of sewers partially full:

d / D	0.20	0.23	0.28	0.41
v / V	0.60	0.67	0.70	0.80

[10 Marks]

Sol:

For 450 mm sewer,

$$n = 0.013, \text{ slope} = 0.0025$$

Velocity at full depth of flow (V) = 0.90 m/sec (from table)

Also, $v = 0.6$ m/sec (for partial flow)

$$\frac{v}{V} = \frac{0.6}{0.9} = \frac{6}{9} = 0.67$$

Hence, for $\frac{v}{V} = 0.67$;

$$\frac{d}{D} = 0.23 \quad \text{(from table of sewers partially full)}$$

$\therefore$ $d = 0.23 \times 450$

$= 103.5 \text{ mm} = 10.35 \text{ cm}$ **Ans.**

Hence, in 450 mmϕ sewer depth of flow at which velocity of flow is equal to 0.6 m/sec is equal to 10.35 cm.

Level-2

6.4 Determine the capacity of a combined sewer and the minimum velocity of flow required, given the following data:

Area of locality served	:	50 sq. km
Population of the locality	:	200 heads/ha
Average rate of sewage flow	:	300 lpcd
Maximum rate of flow	:	1.5 times of the average flow
Runoff from rainfall	:	15 mm/24 hr
Diameter of sand particles carried by the flow	:	1 mm
Specific gravity of sand	:	2.65
For solids in suspension	:	$k = 0.04$
Friction factor	:	0.012.

[10 Marks]

Sol:

Population = 200 head/ha = 200 × (5000) = 10^6

Average sewage flow = 300 l/cd = 300 × (200 × 5000) l/d

$$= \frac{300 \times 10^6}{10^3 \times 24 \times 60 \times 60} \text{m}^3/\text{s} = 3.47 \text{ m}^3/\text{s}$$

Max. sewage flow = 1.5 × 3.47 = 5.21 m^3/s

$$\text{Storm water flow} = (50 \times 10^6)\text{m}^2 \left(\frac{15}{1000}\right) \frac{1}{24 \times 60 \times 60} \left(\frac{m}{s}\right) = 8.68 \text{ m}^3/\text{s}$$

Total max. flow of the combined sewer

$= 5.21 + 8.68 = 13.89 \text{ m}^3/\text{s}$

Hence, the capacity of the sewer

$= 13.89 \text{ m}^3/\text{s}$

Minimum velocity of flow reqd. to transport the given size and type of sand; i.e., self-cleansing velocity is given by

$$V_s = \sqrt{\frac{8k}{f'}(G-1)\cdot gd'}$$

Here, $f' = 0.012$; $k = 0.04$; $G = 2.65$;

$$d' = \text{dia. of grain} = 1\text{ mm} = \frac{1}{1000}\text{m}$$

$$\therefore \quad V_s = \sqrt{\frac{8\times 0.04}{0.012}(2.65-1)\times 9.81\times\frac{1}{1000}}$$

$$= 0.657\text{ m/s} \simeq 0.7\text{ m/s (say)}$$

7. Quality and Characteristics of Sewage

Level-1

7.1 **Explain the importance of BOD test in waste water treatment. Differentiate BOD and COD. Also explain the use of population equivalent in municipal sewage treatment.**

[10 Marks]

Sol:

Importance of BOD test:

- BOD (Biochemical oxygen demand) is defined as the amount of oxygen required to carryout the decomposition of biodegradable organic matter.
- It is a chemical procedure for determining how fast biological organisms use up oxygen in a body of water.
- It is used in water quality management, and assessment, ecology and environmental science.
- There are 2 stages of decomposition of BOD test :

(a) Carbonaceous stage

(b) Nitrogenous stage

Difference between COD and BOD :

- COD is a measure of oxygen required for complete oxidation of both biodegradable and non-biodegradable organic matter, while BOD is a measure of oxygen required for biodegradable organic matter only.
- Hence COD is always greater than BOD.

Population equivalent :

Population equivalent is an important parameter for characterizing industrial wastewater. Population equivalent reflects the equivalence between the polluting potential of an industry (commonly in terms of biodegradable organic matter) and a certain population which produces the same polluting load.

The formula of calculation of population equivalent based on BOD test is

$$PE\text{(Population equivalent)} = \frac{\text{BOD load from industry (kg/day)}}{\text{Per capita BOD load (kg/person/day)}}$$

7.2 **15 m*l* of the waste sample has been added into a 300 m*l* BOD incubation bottle. The initial D.O. of the diluted sample was 7.8 mg/*l* and final D.O. after 5 days was 1.7 mg/*l*. The corresponding initial and final D.O. of the seeded dilution water was 8.1 and 6.9 respectively. What is the 5-day BOD of the wastewater sample?**

[10 Marks]

Sol:

Nutrients and buffer solutions are added to the dilution water to provided nutrient for bacterial growth and maintain pH near neutral. Sufficient amount of seed is added to the bottle to ensure adequate concentration of bacterial population to carry out the biodegradation.

$$BOD_5 = \frac{(DO_i - DO_f) - (B_i - B_f)(1-P)}{P}$$

where, DO_i = DO, of mixture, initially, DO_f = DO of mixture, after incubation, B_i = DO of seeded solution, initially, B_f = DO of seeded solution, after incubation

$$BOD_5 = \frac{(7.8-1.7)-(8.1-6.9)\left(1-\frac{15}{300}\right)}{\left(\frac{15}{300}\right)}$$

$$BOD_5 = 99.2 \text{ mg}/l$$

7.3 **Determine the ultimate BOD of a waste water sample which was subjected to the BOD determination as follows:**

6 mL of waste water containing no dissolved oxygen (DO) was mixed with 294 mL of dilution water containing 8.6 mg/L of DO. After incubation at 20° C for 5 days, the DO of the mixture was 5.4 mg/L. The BOD rate constant, k to the base e is 0.25 /d.

[12 Marks]

Sol:

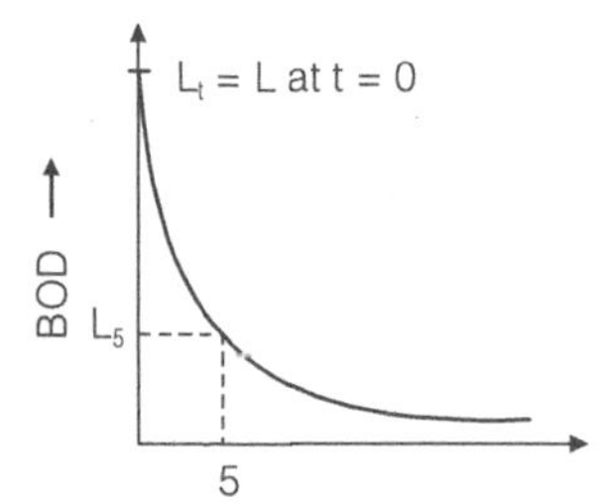

Volume of sewage sample = 6 ml

D.O of sewage sample = 0

Volume of dilution water = 294 ml

D.O of dilution water = 8.6 mg/lit

5-day D.O after incubation= 5.4 mg/lit

Volume of mixture = 6 + 294 = 300 ml

$$\text{D.O of mix} = \frac{6\times 0 + 294 \times 8.6}{300}$$

$$= 8.428 \text{ mg/lit}$$

It is given that after incubation at 20°C for 5-days, the DO of the mixture = 5.4 mg/lit

∴ DO consumed = 8.428 − 5.40 = 3.028 mg/lit

∴ 5-day BOD at 20°C, L_5

= DO consumed × DF

$$\Rightarrow \quad L_5 = 3.028 \times \frac{300}{6}$$

$$= 151.4 \text{ mg/lit}$$

$$L_t = L[1-(10)^{-k_D t}]$$

$$\Rightarrow \quad L_5 = L[1-(10)^{-k_D \cdot 5}]$$

$$k_D = 0.434\,k = 0.434 \times 0.25$$
$$= 0.1085$$

$$\therefore \quad L_5 = L\,[1-(10)^{-0.1085 \times 5}]$$

$$\Rightarrow \quad 151.4 = L\,[1-(10)^{-0.5425}]$$

$$\Rightarrow \quad L = 212.267 \text{ mg/lit}$$

$\therefore$ ultimate BOD = 212.267 mg/lit.

Alternate method:

$$\text{D.O of mix} = \frac{6\times 0+294\times 8.6}{6+294}$$
$$= 8.428 \text{ mg/lit}$$

i.e. $(D.O)_{initial} = 8.428$ mg/lit

Now, after 5 day,

$$(D.O)_{final} = 5.40 \text{ mg/lit}$$

$$\therefore \quad (BOD)_5 \text{ or } L_5 = [(D.O)_i-(D.O)_f]\,D.F$$
$$= (8.428-5.40)\times\frac{300}{6}$$
$$= 151.4 \text{ mg/lit}$$

$$\because \quad L_t = L[1-(10)^{-k_D t}]$$

and $$k_D = 0.434\,k = 0.434\times 0.25$$
$$= 0.1085$$

$$\Rightarrow \quad L_5 = L[1-(10)^{-0.1085\times 5}]$$

$$\Rightarrow \quad 151.4 = L[1-(10)^{-1085\times 5}]$$

$$\Rightarrow \quad L = 212.267 \text{ mg/lit}$$

$\therefore$ ultimate BOD = 212.267 mg/lit

7.4 **A dairy processing 113000 kg of milk daily produces an average of 246 m³/day of wastewater with a BOD of 1400 mg/*l*.**

Compute:

(i) The wastewater flow per 1000 kg of milk received.

(ii) BOD per 1000 kg of milk received.

(iii) The BOD equivalent population.

(iv) Hydraulic equivalent population.

Assuming 80 gm of BOD per person per day, and the wastewater at 400 litres per person per day. **[12 Marks]**

Sol:

Average flow of wastewater = 246 m³/day = 0.246 MLD

BOD of wastewater = 1400 mg/*l*

Daily milk processed = 113000 kg

$\therefore$ Total BOD produced by diary (in kg/day) = $0.246 \times 1400 = 344.6$ kg/day

(i) Wastewater produced per 1000 kg of milk $= \dfrac{246}{113000}\times 1000 = 2.177 \text{ m}^3$ **Ans.**

(ii) BOD produced per 1000 kg of milk $= \dfrac{1400 \times 10^{-6}\text{kg}}{10^{-3}} \times 2.177$

$= 1.4 \times 2.177 = 3.0478$ kg **Ans.**

(iii) BOD population equivalent $= \dfrac{\text{Total BOD}}{\text{Per capita BOD}} = \dfrac{344.6}{80 \times 10^{-3}}$

$= 4307.5 = 4605$ kg **Ans.**

(iv) Hydraulic equivalent population $= \dfrac{245 \times 1000\ l/\text{day}}{400\ l/\text{day/person}} = 615$ **Ans.**

Level-2

7.5 **(a) Explain the procedure for estimating BOD of a given sample of sewage.**

(b) 5% diluted waste water was subjected to BOD test and the following observations were recorded,
DO of water used for dilution = 3.6 mg/l
DO of waste water sample = 0.8 mg/l
DO of sample after 5-days incubation = 0.7 mg/l

Determine,
(i) BOD of waste water
(ii) Ultimate BOD

[10 + 10 = 20 Marks]

Sol:

(a) Standard lab test: In the lab, BOD is determined by diluting a known volume of sample of wastewater with a known volume of Aerated water. The $(DO)_{mix}$ is found out and is represented as $(DO)_L$. After incubation of 5-days at 20°C, DO is again found out and represented as DO_f. The total oxygen consumed for the diluted sample $\{DO_i - DO_f\}$;

BOD of the wastewater is found out as follows:

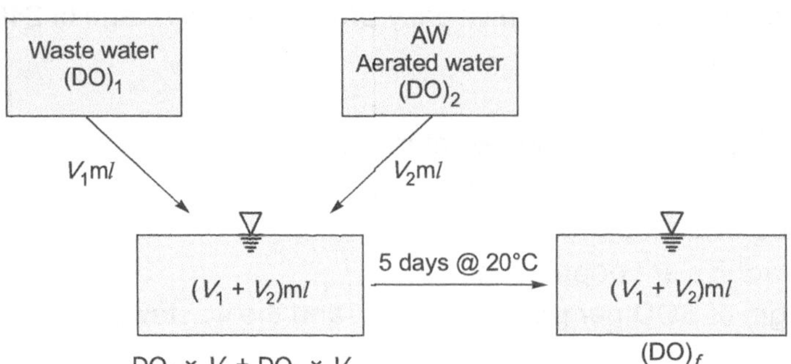

$$(DO)_{mix} = (DO)_i = \frac{DO_1 \times V_1 + DO_2 \times V_2}{V_1 + V_2}$$

$\therefore$ Total quantity of dissolved oxygen consumed (mg/l)

$$= DO_i - DO_f \text{ for } (V_1 + V_2)\text{m}l$$

$\therefore$
$$\text{BOD5 (at 20°C)} = [DO_i - DO_f] \times \left(\frac{V_1 + V_2}{V_1}\right)$$

$$= [DO_i - DO_f] \times \text{Dilution factor (DF)}$$

(b)

Given data : 5% of diluted wastewater

It implies that in 100 ml sample, for 5 ml of wastewater, there is 95 ml Aerated water

$$\therefore \quad D.F. = \frac{100}{5} = 20$$

$V_1 = 5$ ml; $V_2 = 95$ ml

$(DO)_{W.W} = 0.8$ mg/l $\quad (DO)_{A.W.} = 3.6$ mg/l $\quad (DO)_f = 0.7$ mg/l

$(DO)_{W.W}$ = waste water DO

$(DO)_{A.W}$ = Aerated water DO

$$(DO)_{mix} = \frac{5 \times 0.8 + 95 \times 3.6}{100} = 3.46 \text{ mg}/l$$

$$\therefore \quad (BOD_5) = [DO_i - DO_f] \times D.F. = [3.46 - 0.7] \times 20 = 55.2 \text{ mg}/l \quad \text{Ans.}$$

Now, $\quad BOD_5 = (BOD)_u\left[1 - 10^{-k_D \times t}\right] \quad$ At 20°C, $k_D = 0.1\ d^{-1}$

$$55.2 = (BOD)_u\left[1 - 10^{-0.1 \times 5}\right]$$

$$(BOD)_u = 80.73 \text{ mg}/l \quad \text{Ans.}$$

7.6 A wastewater sample is tested for BOD determination. The following observations curve made when 15 ml of sample is diluted to 250 ml in a BOD bottle:

DO of aerated water = 3 mg/l

DO diluted sample after 5 days of incubation = 0.8 mg/l

DO of original sample = 0.4 mg/l

Calculate the BOD_5 and ultimate BOD of the sample assuming $k_d = 0.1d^{-1}$. Also calculate the population equivalent for the 1 MLD flow the given wastewater, if the per capita per day BOD_5 contribution is 75 gm.

[12 Marks]

Sol:

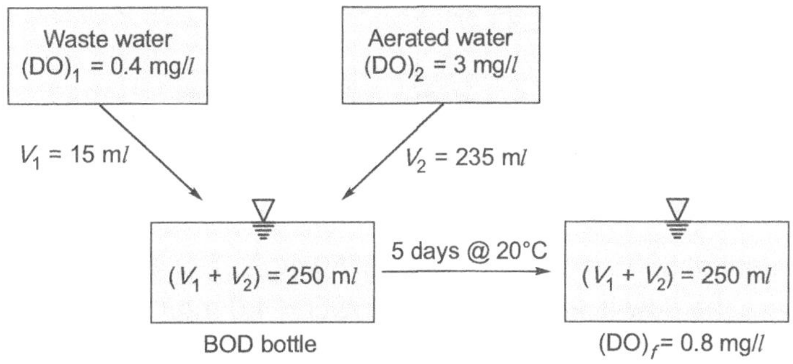

$$(DO)_i = (DO)_{mix} = \frac{0.4 \times 15 + 3 \times 235}{250} = 2.844 \text{ mg}/l$$

(i) $\quad BOD_5 = \left[(DO)_i - (DO)_f\right] \times D.F.$

$$= [2.844 - 0.8] \times \frac{250}{15} = 34.067 \text{ mg}/l \quad \text{Ans.}$$

Also,
$$BOD_5 = BOD_u\left[1-10^{-k_D t}\right]$$
$$34.067 = BOD_u\left[1-10^{-0.1\times5}\right]$$
$$BOD_u = 49.82 \text{ mg}/l$$

(ii) $$\text{Population equivalent} = \frac{\text{Total BOD}}{75 \text{ gm/capita/day}} = \frac{1\times10^6 \, l/\text{day} \times 34.067 \times 10^{-6} \text{ kg/lit}}{0.075 \text{ kg/capita/day}}$$
$$\simeq 454.22 \simeq 455$$ Ans.

7.7 A sample of water has the 5-day 20°C BOD = 200 mg/*l* with k_D (base 10) = 0.1d^{-1}

Determine
1. 5-day 30°C BOD
2. 3-day 40° BOD

[12 Marks]

Sol:

Given data : $Y_5 = (BOD_5) = 200$ mg/*l* ; $k_D = 0.1 \text{ d}^{-1}$,
$$Y_5 = L_0\left(1-10^{-k_D\times t}\right) \qquad L_0 = \text{Ultimate BOD}$$
$$200 = L_0\left(1-10^{-0.1\times5}\right)$$
$$L_0 = \frac{200}{1-10^{-0.5}} = 292.495 \text{ mg}/l$$

Since, ultimate BOD is independent of temperature

(i) At 30°C,
$$k'_{D(T=30°C)} = k_{D(T=20°C)}[1.047]^{T-20} = 0.1[1.047]^{30-20} = 0.158 d^{-1}$$

Now,
$$(Y_5)_{\text{at } T-30°C} = L_0\left(1-10^{-k_D\times5}\right) = 292.495\left(1-10^{-0.158\times5}\right) = 245.057 \text{ mg}/l$$

(ii) At 40°C,
$$k''_{D(T=40°C)} = k_{D(T=20°C)}(1.047)^{T-20} = 0.1(1.047)^{40-20} = 0.2505 d^{-1}$$

Now, 3-day BOD,
$$(Y_3)_{T=40°C} = L_0\left(1-10^{-k''_D\times3}\right)$$
$$= 292.495\,(1-10^{-0.2505\times3}) = 240.66 \text{ mg}/l$$ Ans.

8. Disposal of Sewage Effluents

Level-1

8.1 A wastewater effluent of 560 L/sec with a BOD of 50 mg/L, DO = 3.0 mg/L and temperature of 23°C enters a river where the flow is 28 m³/sec, and BOD = 4.0 mg/L, DO = 8.2 mg/L, and temperature of 17°C, K_1 of the waste is 0.1 d^{-1} at 20°C. The velocity of water in the river downstream is 0.18 m/s and depth of 1.2 m. Determine the following after mixing of wastewater with the river water.

(i) Combined discharge
(ii) BOD
(iii) DO
(iv) Temperature

[12 Marks]

Sol:

(Particulars of sewage thrown) (Particulars of River)

Q_S = 560 lit/sec Q_R = 28 m^3/sec

(Concentrations)

Sewage	River
BOD = 50 mg/L	BOD = 4.0 mg/L
DO = 3 mg/L	DO = 4.0 mg/L
Temperature = 23°C	Temperature = 17°C
k_D at 20°C = 0.1 d^{-1}	

(i) Combined discharge = $Q_S + Q_R = 0.56 + 28 = 28.56$ m^3/sec

(ii)
$$(\text{BOD})_{mix} = \frac{(\text{BOD})_5 \times Q_S + (\text{BOD})_R \times Q_R}{Q_R + Q_S}$$

$$= \frac{50 \times 0.56 + 28 \times 4}{28.56} = 4.9 \text{ mg/L} \qquad \textbf{Ans.}$$

(iii)
$$(\text{DO})_{mix} = \frac{(\text{DO})_5 \times Q_s + (\text{DO})_R \times Q_R}{Q_R + Q_S}$$

$$= \frac{3 \times 0.56 + 28 \times 8.2}{28.56} = 8.098 \text{ mg/L} \qquad \textbf{Ans.}$$

(iv)
$$\text{Temperature of mix} = \frac{T_S \times Q_S + T_R \times Q_R}{Q_R + Q_S} = \frac{23 \times 0.56 + 2178 \times 28}{28.56} = 17.12°\text{C} \qquad \textbf{Ans.}$$

8.2 What is sewage sickness? State various methods to prevent sewage sickness.

[6 Marks]

Sol:

In sewage farming, when sewage is applied continuously on the land, the voids of the soil go on clogging gradually. A time comes, when the soil voids get completely clogged, air circulating through the soil gets totally ceased and sediment gets deposited on the surface. An anaerobic condition is developed. At this state, the soil is unable to absorb further sewage load. This condition is associated with liberation of foul small. This condition is termed as sewage sickness.

[Measure to prevent sewage sickness]

(i) Primary treatment should be given to sewage to eliminate the suspended solids.
(ii) Intermittent supply of sewage should be adopted considering the nature of soil.
(iii) Sub-soil drainage system should be provided to drain out the subsoil effluent.
(iv) Deep ploughing should be adopted to increase the soaking capacity of soil.

8.3 Discuss the phenomenon of the self purification of streams. How can it be evaluated and respected? What are the factors that affect the self-purification abilities of a river or streams?

[10 Marks]

Sol:

- When sewage is discharged into a natural body of water, the receiving water gets polluted due to waste products, present in sewage effluents. But the conditions donot remain so forever, because the natural forces of purification such as dilution, sedimentation, oxidation-reduction, sunlight etc. go on acting upon the pollution elements and bring back the water into original condition. This automatic purification of polluted water, in due course, is called the self purification of river.
- However, if the self-purification is not achieved successfully either due to too much of pollution discharged into it or due to other causes, the river water itself will get polluted, which, in turn, may also pollute the sea where the river out falls.

The various natural forces of purification which helps in effecting self purification are discussed below:

(i) Physical forces are:

(a) Dilution and dispersion:

- When the sewage is discharged into the river it gets diluted in the river i.e. the concentration of various species diminishes.
- When sewage of concentration C_S flows at a rate Q_S into a river stream with concentration C_R flowing at a rate Q_R, the concentration of the resulting mixture is given by

$$C = \frac{Q_S C_S + Q_R C_R}{Q_S + Q_R}$$

This equation is applicable separately to concentrations of different impurities, such as oxygen content, BOD; suspended sediments and other characteristics content of sewage.

(b) Sedimentation : The settleable solids, if present in sewage effluents, will settle down into the bed of the river, near the outfall of sewage, thus helping in the self purification process.

(c) Sunlight : The sunlight has a bleaching and stabilizing effect of bacteria. It also helps certain micro-organisms to derive energy from it, and convert themselves into food for other forms of life, thus absorbing carbon dioxide and releasing oxygen by a process known as photosynthesis. The evolution of oxygen in river water due to sunlight will help in achieving self-purification through oxidation.

(ii) Chemical forces are:

(a) Oxidation : The oxidation of the organic matter present in sewage effluents will start as soon as sewage outfalls into the river water containing dissolved oxygen. The deficiency of oxygen so created, will be filled by the atmospheric oxygen. The process of oxidation will continue till the organic matter has been completely oxidized.

This is the most important action responsible for affecting self purification of rivers.

(b) Reduction : Reduction occurs due to hydrolysis of organic matter settled at the bottom either chemically or biologically. Anaerobic bacteria, will help in splitting the complex organic constituents of sewage into liquids and gases and thus paving the way for their ultimate stabilization by oxidation.

The various factors on which these natural forces of purification depends are:

(a) Temperature
(b) Turbulence
(c) Hydrography such as the velocity and surface expanse of the river stream.
(d) Available dissolved oxygen and the amount and type of organic matter.
(e) Rate of reaeration etc.

8.4 Derive the formula for reaeration coefficient (k_R) of a stream deficient in DO but without BOD load. The DO deficits at upstream and downstream stations are 3.55 mg/L and 2.77 mg/L respectively. The time of travel in this stream reach is 0.67 days. The mean water temperature is 26.5°C. What is value of k_R at 20°C.

[12 Marks]

Sol:

For a stream, deficit in DO without organic load

$$\frac{dD_t}{dt} = k_D \overset{0}{\cancel{L_t}} - k_R \cdot D_t$$

$$\frac{dD_t}{dt} = -k_R \cdot D_t \qquad ...(i)$$

Here, $\frac{dD_t}{dt}$ represents the absolute change of DO deficit (in mg/L) over an increment of time '*dt*'

k_R = Reaeration constant (d^{-1})

Integrating above equation (i) from t_1 to t_2

$$\int_{D_1}^{D_2} \frac{dD_t}{dt} = \int_{t_1}^{t_2} -k_R dt$$

$$[\ln D_t]_{D_1}^{D_2} = -k_R (t_2 - t_1)$$

$$[\ln D_2 - \ln D_1] = -k_R (t_2 - t_1)$$

$$\ln\left(\frac{D_2}{D_1}\right) = -k_R (t_2 - t_1)$$

$$\frac{D_2}{D_1} = e^{-k_R(t_2-t_1)}$$

or,

$$k_R = -\frac{\ln\left(\frac{D_2}{D_1}\right)}{\Delta t}$$

Given, At temperature = 26.5°C

D_1 = DO at U/S = 3.55 mg/L

D_2 = DO at D/S = 2.77 mg/L

Δt = 0.67 days

$$k_R = -\frac{\ln\left(\frac{2.77}{3.55}\right)}{0.67} = 0.37 d^{-1}$$

As we know that

$$k_{R(T°C)} = k_{R(T=20°C)} [1.016]^{T-20}$$

$$0.37 = k_{R(T=20°C)} [1.016]^{26.5-20}$$

$$k_{R(T = 20°)} = \frac{0.37}{(1.016)^{6.5}} = 0.33 d^{-1} \qquad \text{Ans.}$$

8.5 A wastewater treatment plant discharges 1.5 m^3/sec of effluent having an ultimate BOD of 40.0 mg/lit into a stream flowing at 10 m^3/sec. Just upstream of the constant to the base 'e' is estimated as 0.32 /day.

(a) Assuming complete mixing, find the ultimate BOD of the mixture of waste and stream just downstream of the outfall.

(b) Assuming a constant cross-section area for the stream equal of 50 m^2, estimate the BOD of the stream at a point 2.5 km downstream from the outfall.

[12 Marks]

Sol:

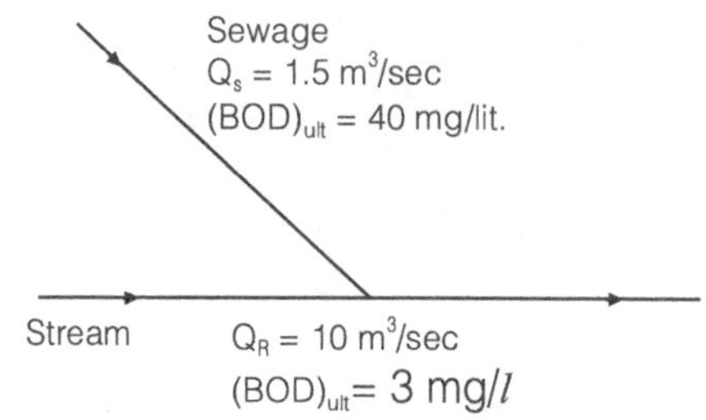

$$Q_R = 10\ m^3/sec$$
$$(BOD)_{ult} = 3\ mg/l$$
$$k_D = \frac{0.32}{2.303} = 0.139/day$$

(a)
$$(Y_5)_{sewage} = 40(1 - 10^{-0.139 \times 5})$$
$$= 40 \times 0.798$$
$$= 31.92\ mg/l$$
$$(Y_5)_{river} = 3(1 - 10^{-0.139 \times 5})$$
$$3 \times 0.798 = 2.394\ mg/l$$
$$(Y_5)_{mixture} = \frac{31.92 \times 1.5 + 2.394 \times 10}{11.5}$$
$$= 6.24\ mg/l$$

Ultimate BOD mixture

$$= \frac{6.24}{0.798} = 7.82\,mg/l$$

(b) Velocity of flow

$$= \frac{11.5}{50} = 0.23\,m/sec$$

$$\text{Time taken} = \frac{2.5 \times 1000}{0.23}$$
$$= 10870\ sec = 0.126\ days$$
$$Y_t = L\left[1 - 10^{-k_D t}\right]$$
$$= 7.82[1 - 10^{-0.139 \times 0.126}]$$
$$= 0.3086\ mg/l$$

Level-2

8.6 **The population of a town is 60000 and the domestic sewage is 160 litres/capita/day, having per capita BOD of 70 gm/day. The dairy waste of town is 2.5 x 10^6 litres/day with a BOD of 1600 mg/l and waste from other industries is 1.7 x 10^6 litres/day with BOD of 2000 mg/L. An overall expansion factor of 15% is to be provided. The wastewater is discharged in the natural stream having a minimum discharge of 7500 litres/sec and a saturation dissolved oxygen of 9 mg/L. It is necessary to maintain a DO of 4 mg/L in the stream. Determine the degree of treatment required to be given to the wastewater prior to its discharge into stream. Assume k_D = 0.1 d^{-1} and k_R = 0.3 d^{-1}.**

[20 Marks]

Sol:

Given data : Population = 60000, Domestic sewage production rate = 160 lpcd, BOD of domestic sewage = 70 gm/capita/day, dairy waste = 2.5 MLD, BOD of diary waste = 1600 mg/L, Industrial waste = 1.7 MLD, BOD of industrial waste = 2000 mg/L, expansion factor = 15%

Discharge flowing in river, Q_R = 7500 litres/sec

$$= \frac{7500 \times 10^{-3} \times 1000 \times 86400}{10^6} \text{MLD}$$

$$= 648 \text{ MLD}$$

$(DO)_{sat} = 9$ mg/L; $(DO)_{min} = 4$ mg/L; $k'_D = 0.1\ d^{-1}$

$k'_R = 0.3\ d^{-1}$ [Assuming on base 10]

For domestic sewage

Q_{ds} = 60000 × 160 L/day = 9.6 MLD

BOD = $60000 \times 70 \times 10^{-3}$ = 4200 kg/day

BOD (in mg/L) = $\frac{4200}{9.6}$ = 437.5 mg/L

Diary waste

Q_{dw} = 2.5 MLD

BOD = 1600 mg/L

Industrial waste

Q_{is} = 1.7 MLD

BOD = 2000 mg/L

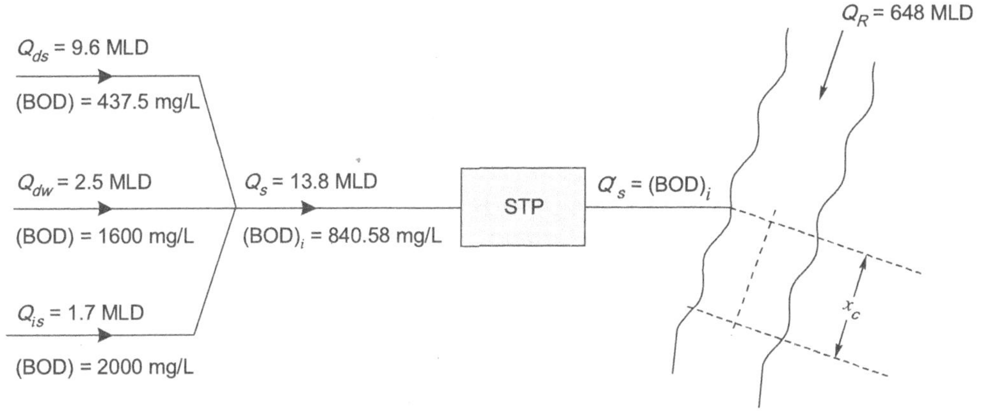

$(BOD)_i$ = 2000 mg/L

$Q'_S = 1.15 \times 13.8 = 15.87$ MLD

$$(BOD)_i = \frac{Q_{ds} \times (BOD)_{ds} + Q_{dw} \times (BOD)_{dw} + Q_{is} + (BOD)_{is}}{Q_{ds} + Q_{dw} + Q_{is}}$$

$$= \frac{9.6 \times 437.5 + 2.5 \times 1600 + 1.7 \times 2000}{13.8} = 840.58 \text{ mg/L}$$

$$f = \frac{k'_R}{k'_D} = \frac{0.3}{0.1} = 3$$

Critical deficit, $D_C = (DO)_{sat} - (DO)_{min} = 9 - 4 = 5$ mg/L

D_0 (initial deficit) $= (DO)_{sat} - (DO)_{\text{mix at t}} = 0$

$$(D_0)_{\text{mix at } t=0} = \frac{648 \times 9 + 15.87 \times 0}{648 + 15.87} = 8.78 \text{ mg/L}$$

$$D_0 = 9 - 8.78 = 0.22 \text{ mg/L}$$

Now,

$$\left(\frac{l_0}{D_c f}\right)^{f-1} = f\left(1 - (f-1)\frac{D_0}{l_0}\right)$$

$$\left(\frac{l_0}{5 \times 3}\right)^{3-1} = 3\left[1 - (3-1) \times \frac{0.22}{l_0}\right]$$

$$\frac{l_0^2}{(15)^2} = 3\left[1-\frac{0.44}{l_0}\right]$$

$$l_0^2 = 225\times 3 - \frac{225\times 3\times 0.44}{l_0}$$

$$l_0^3 + 297 - 675 l_0 = 0$$

$$l_0^3 - 675 l_0 + 297 = 0$$

$$l_0 = 25.76; 0.44$$

We will adopt higher value, because it leads to lesser degree of treatment,

$$\therefore \quad l_0 = 25.76 \text{ mg/L}$$

$$(BOD_5)_{mix} = l_0\left(1-10^{-k_D\times 5}\right)$$

$$= 25.76\left(1-10^{-0.1\times 5}\right) = 17.61 \text{ mg/L}$$

$$\therefore \quad 17.61 = \frac{(BOD_5)_e \times Q'_S + (BOD_5)_R \times Q_R}{Q_R + Q'_S}$$

$$17.61 = \frac{(BOD_5)_e \times 15.87 + 0}{648 + 15.87}$$

$$(BOD_5)_e = 736.657 \text{ mg/L}$$

Now, degree of treatment required $= \dfrac{(BOD)_i - (BOD)_e}{(BOD)_i} \times 100$

$$= \left(\frac{840.58 - 736.657}{840.58}\right)\times 100 = 12.36\% \qquad \text{Ans.}$$

8.7 125 m^3/sec of sewage of a city is discharged in a perennial river which is fully saturated with oxygen and flows at a minimum rate of 1600 cumec. With a minimum velocity of 0.12 m/sec. If the 5 day BOD of the sewage is 300 mg/L, find out where the critical DO will occur in the river. Assume

(i) The coefficient of purification of the river as 4.0.

(ii) The deoxygenation rate as 0.11 d^{-1}.

(iii) The ultimate BOD is 125% of the 5 day BOD of mixture of sewage and river water $(DO)_{sat}$ = 9.2 mg/L.

[20 Marks]

Sol:

Given data : Q_R = 1600 m^3/sec, Velocity of flow in river = 0.12 m/sec,

f = 4; K'_D = 0.11d^{-1} (Assuming on base 10)

$$(BOD)_u = l_0 = 125\% \text{ of } (BOD)_{mix}$$

$$(DO)_{sat} = 9.2 \text{ mg/L}$$

$$D_0 = (DO)_{sat} - (DO)_{mix \text{ at } t=0}$$

$$(D_0)_{mix\,(t=0)} = \frac{Q_S \times (DO)_S + Q_R \times (DO)_R}{Q_R + Q_S}$$

$$= \frac{0 + 1600\times 9.2}{1600 + 125} = 8.53 \text{ mg/L}$$

$$\therefore \quad D_0 = 9.2 - 8.53 = 0.67 \text{ mg/L}$$

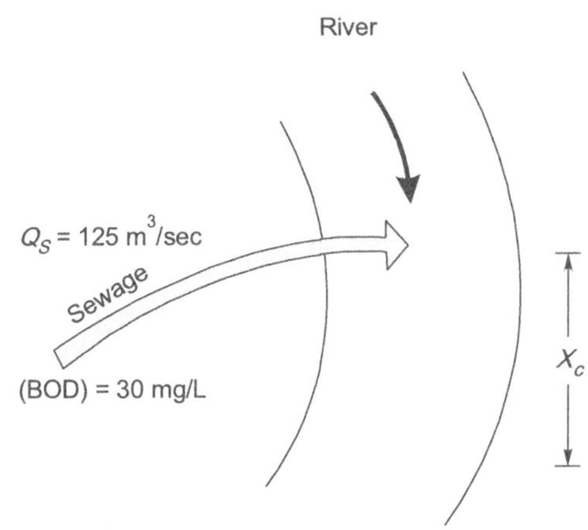

$$(BOD_5)_{mix} = \frac{(BOD_5)_S \times Q_S + (BOD)_R \times Q_R}{Q_R + Q_S}$$

$$= \frac{300 \times 125 + 0}{1600 + 125} = 21.74 \text{ mg/L}$$

$\therefore$ $$l_0 = 1.25 \times 21.74 = 27.17 \text{ mg/L}$$

Now, critical time after which critical deficit will occur

$$t_c = \frac{1}{k'_D(f-1)} \log_{10}\left[f\left\{1-(f-1)\cdot\frac{D_0}{l_0}\right\}\right]$$

$$= \frac{1}{0.11(4-1)} \log_{10}\left[4\left\{1-(4-1)\times\frac{0.67}{27.17}\right\}\right]$$

$$= 1.723 \text{ days}$$

$\therefore$ Distance after which critical deficity will occur

$$X_c = Vt_c$$

$$= 0.12 \times \frac{1.723 \times 86400}{1000} = 17.87 \text{ km}$$ **Ans.**

8.8 A city discharges 20000 m^3/day of sewage into a river whose rate of flow is 0.7 m^3/sec. Determine the DO deficit profile for 100 km from the following data

River	Sewage effluent from STP
5 day BOD at 20°C = 3.4 mg/L	5 day BOD at 20°C = 45 mg/L
Temperature = 23°	Temperature = 26°
DO = 8.2 mg/L	DO = 2 mg/L

Assume Velocity of mix = 0.25 m/sec

$k_R = 0.4\ d^{-1}$; $k_D = 0.23d^{-1}$ **[25 Marks]**

Sol:

Given data: River discharge, Q_R = 0.7 m^3/sec;

Sewage discharge (Q_S) = 20000 m^3/day = $\frac{20000}{86400} = 0.231 \text{ m}^3\text{/sec}$

$$(BOD_5)_{mix} = \frac{(BOD_5) \times Q_S + (BOD_5)_R \times Q_R}{Q_R + Q_S}$$

$$= \frac{45 \times 0.231 + 3.4 \times 0.7}{0.7 + 0.231} = 13.72 \text{ mg/L}$$

$$(DO)_{mix} = \frac{(DO)_S \times Q_s + (DO)_R \times Q_R}{Q_R + Q_S}$$

$$= \frac{8.2 \times 0.231 + 2 \times 0.7}{0.7 + 0.231} = 6.66 \text{ mg/L}$$

$$\text{Temperature of mix} = \frac{T_S \times Q_S + T_R \times Q_R}{Q_R + Q_S} = \frac{23 \times 0.7 + 26 \times 0.231}{0.7 + 0.231} = 23.74°C$$

As we know that,

$$(DO)_{sat} = 9.2 \text{ mg/L at temperature} = 20°C$$

$$(DO)_{sat} = 7.6 \text{ mg/L at temperature} = 30°C$$

$\therefore$ $(DO)_{sat}$ at temperature = 23.74°C

$$(DO)_{sat} = 9.2 + \frac{7.6 - 9.2}{30 - 20}(23.74 - 20) = 8.6 \text{ mg/L}$$

$$k_D = 0.23 d^{-1};\ k_R = 0.4 d^{-1}$$ (Assuming as base e)

Now,

$$Y_5 = l_0(1 - e^{-kt})$$

$$13.72 = l_0(1 - e^{-0.23 \times 5})$$

$\therefore$

$$\text{Ultimate BOD, } l_0 = \frac{13.72}{1 - e^{-0.23 \times 5}} = 20.08 \text{ mg/L}$$

$$\text{Initial deficit, } D_0 = (DO)_{sat} - (DO)_{mix}$$

$$= 8.6 - 6.66 = 1.94 \text{ mg/L}$$

Now, deoxygenation and reoxygenation constant at mix temperature = 23.74°C

$$k_{D(T = 23.74°C)} = k_{D(T = 20°C)}(1.047)^{(T-20)}$$

$$= 0.23(1.047)^{23.74 - 20} = 0.273 \text{ d}^{-1}$$

$$k_{R(T = 23.74°C)} = k_{R(T = 23.74°C)}(1.016)^{(T-20)}$$

$$= 0.4(1.016)^{23.74 - 20} = 0.424 \ d^{-1}$$

$\therefore$

$$f = \frac{k_R}{k_D} = \frac{0.424}{0.273} = 1.55$$

$\therefore$

$$\text{Critical time, } t_c = \frac{1}{k_D(f-1)} \cdot \ln\left[\left\{1 - (f-1) \cdot \frac{D_0}{l_0}\right\} f\right]$$

$$= \frac{1}{0.273(1.55 - 1)} \cdot \ln\left[\left\{1 - (1.55 - 1) \cdot \frac{1.94}{20.08}\right\} \times 1.55\right]$$

$$= 2.555 \text{ days}$$

$$\text{Critical deficit, } D_C = \frac{k_D l_0}{k_R} \cdot e^{-k_D t_c}$$

$$D_C = \frac{l_0}{f} e^{-k_D t_c} = \frac{20.08}{1.55} e^{-0.273 \times 2.555}$$

$$D_C = 6.45 \text{ mg/L}$$

Critical distance after which critical deficit will occur,

$$x_c = Vt_c = 0.25 \times \frac{2.555 \times 86400}{1000}$$

$$= 55.19 \text{ km}$$

(Deficit profile for 100 km)

Time required for mix to reach at 20 km distance,

$$t_{20} = \frac{20 \times 1000}{0.25 \times 86400} = 0.926 \text{ days}$$

Using Streeter Phelps equation,

$$D_t = \frac{k_D L_0}{k'_R - k'_D}\left[e^{-k_D t} - e^{-k_R t}\right] + D_0 e^{-k_R t}$$

$$= \frac{0.273 \times 20.08}{0.424 - 0.273}\left[e^{-0.273 \times 0.926} - e^{-0.424 \times 0.926}\right] + 1.94 \times e^{-0.424 \times 0.926}$$

$$= 4.99 \text{ mg/L}$$

$\therefore \quad (DO)_{sat} - (DO)_t = 4.99$

$$(DO)_t = 20 \text{ km} = 8.6 - 4.99 = 3.61 \text{ mg/L}$$

Similar for $x = 40$ km

$$t = \frac{40 \times 1000}{0.25 \times 86400} = 1.852 \text{ days}$$

$$D_t = \frac{0.273 \times 20.08}{0.424 - 0.273}\left[e^{-0.273 \times 1.852} - e^{-0.424 \times 1.852}\right] + 1.94 e^{-0.424 \times 1.852}$$

$$= 6.22 \text{ mg/L}$$

$\therefore \quad (DO)_{t = 40 \text{ km}} = 8.6 - 6.22 = 2.38$ mg/L

For $x = 80$ km

$$t = \frac{80000}{0.25 \times 86400} = 3.704 \text{ days}$$

$$D_t = \frac{0.273 \times 20.08}{0.424 - 0.273}\left[e^{-0.273 \times 3.704} - e^{-0.424 \times 3.704}\right] + 1.94 e^{-0.424 \times 3.704}$$

$$= 6.06 \text{ mg/L}$$

Now, $\quad (DO)_{t = 80 \text{ km}} = 8.6 - 6.06 = 2.54$ mg/L

For $x = 100$ km

$$t = \frac{100000}{0.25 \times 86400} = 4.63 \text{ days}$$

$$D_t = \frac{0.273 \times 20.08}{0.424 - 0.273}\left[e^{-0.273 \times 4.63} - e^{-0.424 \times 4.63}\right] + 1.94 \times e^{-0.424 \times 4.63}$$

$$= 5.43 \text{ mg/L}$$

$\therefore \quad (DO)_{t = 100 \text{ km}} = 8.6 - 5.43 = 3.17$ mg/L

$\therefore$ The DO deficit at different points along the length of river is as below:

Distance (km)	Time (in days)	DO deficit (mg/L)	DO (in mg/L)
0	0	1.94	6.66
20	0.926	4.99	3.61
40	1.852	6.22	2.38
55.18	2.555	6.45	2.15
80	3.704	6.06	2.54
100	4.63	5.43	3.17

8.9 A city discharges 20 millions litres per day of wastewater at temperature 25°C into a stream, whose discharge is 2.0 cumecs and whose temperature is 25°C. The 5-day BOD of wastewater is 150 mg/L and k_D (Deoxygenation constant) value as 0.1d^{-1} and k_R (Reoxygenation constant) value as 0.2d^{-1}. Determine the critical oxygen deficit and its occurrence. Assume stream is 90% saturated with oxygen before the waste is added. Take solubility of oxygen in water at 25°C as 9 mg/L.

[20 Marks]

Sol:

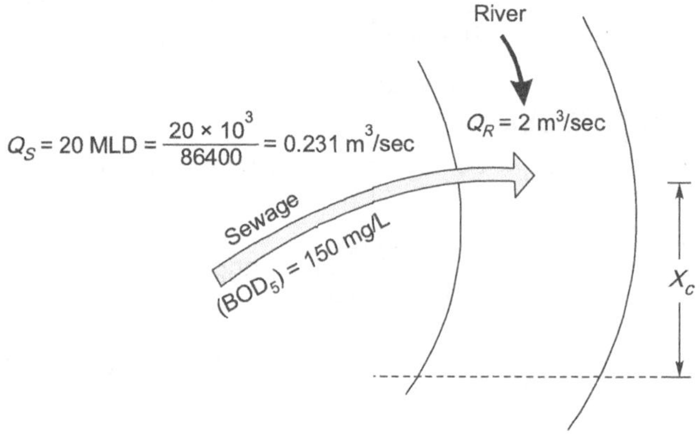

$$(DO)_{sat} = 9 \text{ mg/L}$$

$$(DO)_{River} = 9 \times 0.9 = 8.1 \text{ mg/L}$$

$k_D = 0.1\text{d}^{-1}$; $k_R = 0.2\text{d}^{-1}$; (Assuming as base 10)

$$\text{DO(initial deficit)} = (DO)_{sat} - (DO)_{mix\ at\ t=0}$$

$$(DO)_{mix\ at\ t=0} = \frac{Q_S \times (DO)_S + Q_R \times (DO)_R}{Q_R + Q_S} = \frac{0.231 \times 0 + 2 \times 8.1}{2 + 0.231} = 7.26 \text{ mg/L}$$

$$D_0 = 9 - 7.26 \approx 1.74 \text{ mg/L}$$

$$(BOD_5)_{mix} = \frac{(BOD_5) \times Q_S + (BOD_5)_R \times Q_R}{Q_R + Q_S} = \frac{150 \times 0.231 + 0}{2 \times 0.231} = 15.53 \text{ mg/L}$$

$$BOD_5 = l_0\left(1 - 10^{-k_D \times 5}\right)$$

$$15.53 = l_0\left(1 - 10^{-0.1 \times 5}\right)$$

$$l_0 = 22.71 \text{ mg/L}$$

$\therefore$ f = Self purification factor

$$= \frac{k_R}{k_D} = \frac{0.2}{0.1} = 2$$

Now, critical time, after which critical deficit will occur

$$t_c = \frac{1}{k_D(f-1)} \cdot \log_{10}\left[f\left\{1 - (f-1)\frac{D_0}{l_0}\right\}\right]$$

$$= \frac{1}{0.1(2-1)} \log_{10}\left[\left\{1 - (2-1) \times \frac{1.74}{22.71}\right\}\right] = 2.664 \text{ days}$$

$$\text{Critical deficit, } D_C = \frac{l_0}{f} \cdot 10^{-k_D t_c} = \frac{22.71}{2} \times 10^{-0.1 \times 2.664} = 6.15 \text{ mg/L}$$ **Ans.**

8.10 The minimum flow of a river is 50 m^3/s having a dissolved oxygen (DO) content of 7.0 mg/L (80% saturation) and BOD_5 of 8.0 mg/L. It receives a wastewater discharge of 5 m^3/s with BOD_5 of 200 mg/L and no DO. If the rate constants for deoxygenation and reaeration (both base e) are 0.5/d and 0.1/d respectively and the velocity of river flow is 0.8 m/s, calculate the distance in kilometer downstream from the point of wastewater discharge where the minimum DO occurs.

[15 Marks]

Sol:

$$Q_S = 5\ m^3/s$$
$$Q_R = 50\ m^3/s$$

BOD_5 of sewage = 200 mg/l

BOD_5 of river = 8 mg/l

k_d = 0.5/day, k_R = 1/day

$$f = \frac{k_R}{k_D} = 2$$

$$BOD_{mix} = \frac{C_SQ_S + C_RQ_R}{Q_S + Q_R}$$

$$\therefore \quad BOD_{mix} = \frac{200 \times 5 + 8 \times 50}{55} = 25.45\ mg/l$$

Dissolved oxygen

$$\therefore \quad C_{mix} = \frac{C_SQ_S + C_RQ_R}{Q_S + Q_R}$$

Assume dissolved oxygen of sewage = 0

$$\therefore (DO)_{mix} \text{ initially} = \frac{0 + 7 \times 50}{55} = 6.36\ mg/l$$

$\therefore$ Initial O_2 deficit = 7 – 6.36 = 0.636 mg/l

Assume ultimate BOD = 125% BOD of sewage and river water

$$= 1.25 \times 25.45 = 31.81\ mg/l$$

$$\therefore \left(\frac{L}{D_c f}\right)^{f-1} = f\left[1-(f-1)\frac{D_0}{L}\right]$$

$$\Rightarrow \quad \left(\frac{31.81}{D_c \times 2}\right) = 2\left[1-(2-1)\frac{0.636}{31.81}\right]$$

$$\Rightarrow \quad \frac{15.905}{D_c} = 2\left[1-\frac{0.636}{31.81}\right]$$

$$\Rightarrow \quad D_c = 8.115\ mg/l$$

$$\therefore \quad D_c = \frac{k_D}{k_R} \times 10^{-k_D t_c}$$

$$\Rightarrow \quad 8.115 = \frac{0.5}{1} \times 31.81 \times 10^{-0.5 \times t_c}$$

$$0.510 = 10^{-0.5 \times t_c}$$

$$\Rightarrow \quad t_c = 0.585\ \text{days}$$

$\therefore$ Distance at which maximum deficit will occur

$$= V \times t_C$$

$$= [0.8 \times 0.585 \times 24 \times 60 \times 60]\ m = 40.43\ kms$$

9. Treatment of Sewage

Level-1

9.1 **What is grit? Why should it (grit) be removed from waste water? Basic principle behind the design of grit chamber? Reason to have the content velocity of flow in grit chamber (conventional horizontal flow) and how is it achieved?**

[15 Marks]

Sol:

Grits are small loose stones or sand particle with size > 2 μ.

Why should it be removed:

- If there is grit in waste water then in sedimentation tank overflow velocity have to be kept higher. If velocity is low then so much grits will be collected in it and we have to clean it in short interval.
- Self cleansing velocity will be high.
- If velocity high then removal efficiency will less.

Principles in design of grit chamber:

- Grit chambers are designed such that no organic material should settle and also does not create nuisance.
- They separate non-decaying heavy inorganic material from sewage so that inorganic material may be disposed off without causing nuisance.
- Hence, the grit chamber must be designed to scour the lighter organic particles, while the heavier grit particles remain settled.

Reason behind keeping horizontal velocity constant:

- We keep velocity to be constant in grit chamber to ensure that only a certain range of particles settle.
- A low change in horizontal velocity can cause to wash out the grits.
- Horizontal flow velocity is artificially controlled by a proportioning weir on the effluent end of the tank or a parabolic section is often used to maintain steady flow.

9.2 **Design an aerated grit chamber, for treatment of average flow of 60 MLD, assume the peak flow rate to be 3 times the average.**

[10 Marks]

Sol:

$$\text{Average flow rate} = 60\text{ MLD} = \frac{60 \times 10^3}{86400} = 0.694\text{ m}^3/\text{sec}$$

$$\text{Peak flow rate} = 0.694 \times 3 = 2.082\text{ m}^3/\text{sec}$$

Assume,average liquid detention time = 3 min = 180 sec

$$\text{Aerator volume} = 2.082 \times 180 = 374.76\text{ m}^3$$

In order to drain the channel periodically for routine cleaning and maintenance, use two chambers.

$$\text{Volume of one aerated channel} = \frac{374.76}{2} = 187.38\text{ m}^3$$

To determine the dimensions of aerated channel,

Assume depth of 3 m

and $$\frac{B}{H} = 2:1$$

$$\text{Width of channel} = 6\text{ m}$$

$$\text{Length of channel} = \frac{187.38}{6 \times 3} = 10.41 \text{ m}$$

Increase the length by about 20% to account for the inlet and outlet conditions

$$\text{provided length} = 1.2 \times 10.41 = 12.492 \text{ m}$$

Hence, use 2 chambers,

each of size (12.5 m × 6 m × 3 m)

Air supply requirement:

Assume that 0.03 m^3/min per m length may be adequate

$$\text{air required} = 0.03 \frac{m^3}{\text{min} \times m} \times 12.5\text{m} = 0.375 \text{ m}^3/\text{min}$$

Volume of grit produced daily:

Assume that 50 $m^3/10^6 m^3$ of sewage of grit is produced by incoming sewage

$$\text{daily grit volume produced} = 2.08 \text{ m}^3/\text{sec} \times \frac{50\text{m}^3}{10^6 \text{m}^3} \times 86400 \text{ sec/day} = 8.98 \text{ m}^3/\text{day}$$

9.3 Mention the operational troubles of a standard rate trickling filter and their remedies.

[8 Marks]

Sol:

The major operational troubles of a standard rate trickling filter are:

(i) **Fly Nuisance:**

- The filter fly psychoda which develops in the filter particles, may prove to be nuisance, as the same may be carried away into the habitation.
- This problem may be controlled by flooding the filter with sewage for 24 hours or more. The flooding will destroy the larvae, and usually interferes slightly with the results of operation.
- Another method of controlling fly nuisance is by using insecticides such as DDT, chlordane and BHC in the filter plant.

(ii) **Odour Nuisance:**

- Odours generally do not prevail in trickling filter using rotary distributions, but however when fixed nozzles are used, H_2S and other odourous gases are frequently released from the sprays into the atmosphere.
- The usual remedy is to chlorinate the sewage to prevent formation of H_2S or to neutralize that already formed.

(iii) **Ponding trouble:**

- Sometimes, the voids in the filter media gets clogged due to heavy growth of fungi and algae. This may result in ponding of the sewage over the filter bed.
- This trouble can be controlled by chlorinating the sewage which kills the algae, thus causing unloading of the accumulated material.
- Another method of controlling algae is to add $CuSO_4$ to the sewage and resting the bed for sometime.

9.4 Determine the BOD of the effluent from a standard rate trickling filter of diameter 25 m and depth of 1.8 m, the hydraulic loading is 2000 m^3/day and influent BOD is 150 mg/L. Take k = 2d^{-1} and n = 0.67. Use Ecken felder equation.

[10 Marks]

Sol:

Given data: Diameter of filter, (d) = 25 m; As per Eckenfelder trickling filter equation,

$$\frac{y_e}{y_1} = e^{-kD/Q_L^n}$$

y_1 = influent BOD = 150 mg/L

Q_L = Hydraulic loading rate ($m^3/min/m^2$) = $\frac{Q}{A}$

$$= \frac{2000}{\frac{\pi}{4} \times (25)^2} = 4.074 \text{ m}^3/\text{day/m}^2$$

$k = 2d^{-1}$

D = Depth of filter = 1.8 m

Now,
$$\frac{y_e}{150} = e^{-\left(\frac{2 \times 1.8}{(4.074)^{0.67}}\right)}$$

y_e = 36.81 mg/L **Ans.**

Effluent BOD = 36.81 mg/L **Ans.**

9.5 **What are oxidation ponds? What are their merits and demerits?**

[10 Marks]

Sol:

Oxidation pond:

- It is an open flow through low cost earthern basin, which provides a longer detention time.
- The organic matter present in the waste water gets stabilized by the action of microorganisms the oxidation pond may be classified as aerobic, anaerobic and facultative depending upon the nature of purification.
- Oxidation ponds are designed to operate aerobically in which there exists a mutual relationship between algae and microorganism called as algal symbiosis.
- The algae while growing in the presence of sunlight produces oxygen by the action of photosynthesis and this O_2 is utilized by microorganism for oxidizing the organic matter, the end product of decomposition serves as nutrients for the growth of algae.

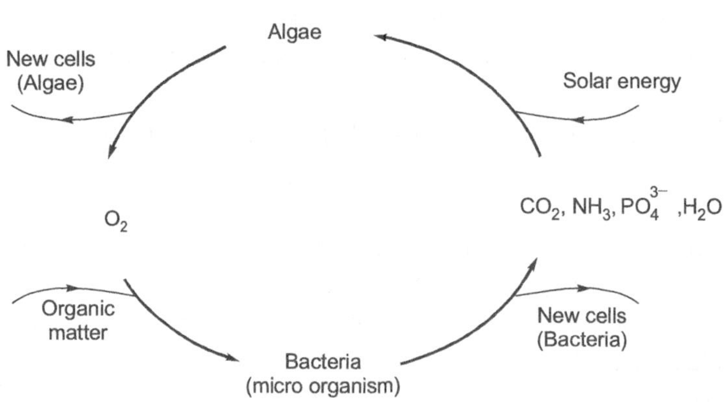

(Merits):

- Pathogen bacteria removal efficiency = 98%.
- BOD removal efficiency = 95%
- No electricity supply or power source is required and are generally constructed in villages.
- Effluents are sufficiently clarified that they are used in sewage farming.

(Demerits):

- Difficult to construct in urban area due to large area requirement.
- Due to overloading of oxidation pond, odour problem may persist.

9.6 **How do imhoff tank differ from septic tank in principle of operation? Mention the advantages and disadvantages of imhoff tanks.**

[10 Marks]

Sol:

An imhoff tank is an improvement over septic tank, in which the incoming sewage is not allowed to get mixed up with the sludge produced and the outgoing effluent is not allowed to carry with it large amount of organic load as in the case of septic tank. An imhoff tank is infact a two storeyed tank, which is sometimes also known as two storey digestion tank.

Advantages of Imhoff tank:

- Detention period of imhoff tank is very less compressed to septic tanks.
- Imhoff tank combines the advantages and disadvantage of both septic as well as sedimentation tanks and as such they find use in case of small treatment plants requiring only primary treatment.
- They are quite economical and don't require skilled attention during operations.
- The results obtained are quite good, with 60 to 65% removal of solids and 30 to 40% removal of BOD.
- There is no problem of sludge disposal, as in the case of sedimentation tank.

Disadvantages of Imhoff tanks:

- Depth of tank is more which may make the constructions costlier, especially in hard rocks or quick sands. At such places, these tanks may become uneconomical.
- They may give out offensive odour, when improperly operated.
- They are unsuitable and do not function properly where sewage is highly acidic in nature.
- These tank have tendency to foam or boil. This may cause the scum to go up to the top of tank and it may also force the sludge particles to enter the sedimentation chamber through the slot. The foaming may thus, adversely affect the quality of the effluent.

9.7 **A municipal waste water treatment plant is to work with average and peak loading rates of 4000 and 8000 m³/day respectively. Design a primary clarifier to remove 65% suspended matter at average flow. An average overflow rate of 35 m³/m² day is expected to correspond to 65% suspended matter removal efficiency. Obtain the diameter, side wall depth, detention time and calculate the overflow rate at peak condition.**

[10 Marks]

Sol:

$$\text{Average loading rate} = 4000 \text{ m}^3\text{/day}$$

$$\text{Peak loading rate} = 800 \text{ m}^3\text{/day}$$

$$\text{Average overflow rate} = 35 \text{ m}^3\text{/m}^2 \text{ day}$$

$$\text{Peak overflow rate} = \frac{35}{4000} \times 8000 = 70 \text{ m}^3\text{/m}^2\text{day}$$

Let us assume detention period = 6 hours

∴ Quantity of water to be treated during the detention period of 6 hours

$$= \frac{8000}{24} \times 6 \text{ m}^3 = 2000 \text{ m}^3$$

$$\text{Now, surface area of} = \frac{\text{Peak loading rate}}{\text{Peak overflow rate}} = \frac{8000}{70} \text{m}^2 = 114.28 \text{ m}^2 \cong 115 \text{ m}^2$$

$$\therefore \quad \text{Depth of water} = \frac{\text{Volume}}{\text{Surface area of clarifier}} = \frac{2000}{115} \text{m} \cong 17.5 \text{ m}$$

Now, we know for a circular tank

$$V = d^2(0.011\,d + 0.785\,H)$$

$$\Rightarrow \quad 2000 = d^2(0.011\,d + 0.785 \times 17.5)$$

$$\Rightarrow \quad 2000 = 0.011\,d^3 + 13.74\,d^2$$

Solving, we get

$$d = 12.00 \text{ m}$$

$$\therefore \text{Diameter} = d = 12 \text{ m}$$

Side wall depth = 17.5 m

Detention period = 6 hours

Overflow rate at peak condition = 70 m^3/m^2 day

9.8 **For a waste water biological treatment system, the growth phase of microorganism is very important draw a biomass growth curve and show and describe various phases on it.**

[10 Marks]

Sol:

The entire process of biological decomposition takes place in four phases:

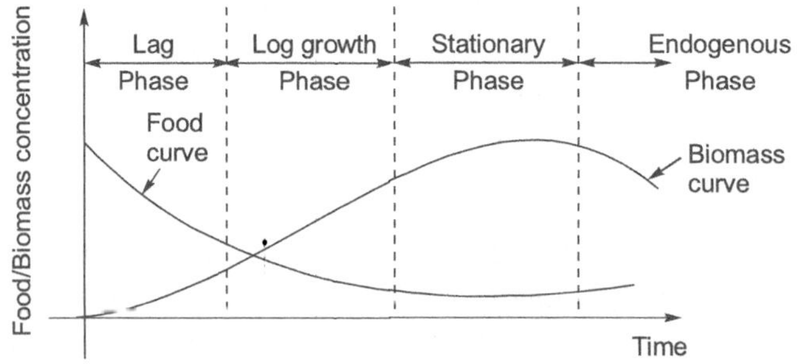

(i) **Lag phase**: The microorganism first get acclimatized to their environment and to the food provided. During this phase, the growth of microorganism is negligible, this acclimatization period is called lag phase, this lag phase becomes negligible if the microorganism are acclimatized to a similar environment and to a similar food. i.e. If the microorganism is already familiar to the food end environment given to them duration of lag phase reduces.

(ii) **Log growth phase**: In this phase, bacterial cell reproduce by cell division and the rate of division is very high due to abundant supply of food.

- Microorganism gets used to food and environment given to them hence their growth rate increases.

(iii) **Stationary phase**: In this phase, the production of new bacteria is roughly offset and the food supply becomes limited.

(iv) **Endogenous phase**: The endogenous respiration starts/predominates in the final stage in which food availability is very scare.

9.9 **Design a septic tank for a colony of 400 people. The colony is supplied water at the rate of 135 lpcd. Assume a detention period of 24 hrs and 70% of water becomes wastewater. The tank is cleaned once in a year. The rate of deposition of sludge is 60 litres/person/year. Depth of tank is to be kept as 2.0 m and free board is 0.3 m. Length to breadth ratio may be kept at 3 : 1.**

[12 Marks]

Sol:

$$\text{Number of people} = 400$$
$$\text{Water supplied} = 135 \text{ L/person/day}$$
$$\text{Detention period} = 24 \text{ hours}$$
$$\text{Rate of deposition of sludge} = 60 \text{ L/person/year}$$
$$\text{Quantity of water supplied} = 400 \times 135 \text{ L/day} = 54000 \text{ L/day}$$
$$\text{Quantity of sewage produced} = 0.7 \times 54000 = 37800 \text{ L/day}$$
$$\text{Detention period} = 24 \text{ hrs}$$

Quantity of sewage produced during detention period

$$= 378005 \times \frac{24}{24} = 37800 \text{ L}$$

$$\text{Period of cleaning} = 1 \text{ year}$$
$$\text{Volume of sludge deposited} = 60 \times 400 \times 1 = 24000 \text{ lit.}$$
$$\text{Total capacity of tank required} = (\text{capacity for sewage}) + (\text{capacity of sludge})$$
$$= 37800 + 24000 = 61800 \text{ lit} = 61.8 \text{ m}^3$$
$$\text{Depth of tank} = 2 \text{ m}$$
$$\text{Surface area of tank} = \left(\frac{61.8}{2}\right) = 30.9 \text{ m}^2$$
$$\frac{L}{B} = 3$$
$$3B \times B = 30.9$$
$$B = 3.209 \text{ m}$$
$$L = 3.209 \times 3 = 9.627 \text{ m}$$
$$\text{Total depth} = 2 + 0.3 = 2.3 \text{ m}$$
$$\text{Size of septic tank} = 9.627 \times 3.209 \times 2.3 \text{ m}$$

Ans.

9.10 **Design a sludge digestion tank for 40,000 people. The sludge content per capita per day is 0.068 kg. The moisture content of the sludge is 94%. The specific gravity of the wet sludge is 1.02 and 3.5 per cent of the digester volume is daily filled with the fresh sludge, which is mixed with the digested sludge.**

[10 marks]

Sol:

(ii) Dry sludge content produced by 40,000 persons

$$= 0.068 \times 40{,}000 \text{ kg} = 2{,}720 \text{ kg/day}$$

94% moisture content means that 6 kg of dry sludge will produce 100 kg of wet sludge.

$\therefore$ 6 kg of dry sludge produces wet sludge = 100 kg

2720 kg of dry sludge produces wet sludge

$$= \frac{100}{6} \times 2720 = 45333 \text{ kg} = 45.3 \text{ t/day}$$

Volume of wet sludge produced per day

$$= \frac{\text{Mass of wet sludge}}{\text{Density of wet sludge}}$$

$$= \frac{45.3}{1.02} \text{ m}^3/\text{day} = 44.4 \text{ m}^3/\text{day}$$

[∵ Density of sludge in t/m^3 = Sp. gr. × Density of water (which is 1 t/m^3) = 1.02 × 1 = 1.02 t/m^3]

∴ 44.4 m^3 of fresh sludge is added to the tank daily, to fill 3.5% of the digester capacity.

$$\therefore \frac{3.5}{100} \times \text{Capacity of digester} = \text{Volume of fresh sludge produced daily} = 44.4 \text{ m}^3$$

$$\Rightarrow \text{Capacity of digester required} = \frac{44.4 \times 100}{3.5} = 1268.6 \text{ cu-m}$$

Providing 30% additional capacity for fluctuations, we have

The required digester capacity

$$= 1268.6 \times 1.3 \simeq 1650 \text{ cu-m (say)}$$

Now, providing 6 m depth of the cylindrical digestion tank, we have the cross-sectional area of the tank

$$= \frac{1650}{6} = 275 \text{ m}^2$$

$$\therefore \quad \text{Dia. of tank} = \sqrt{\frac{275}{\pi/4}} = \sqrt{\frac{275}{0.785}} = 18.7 \text{ m}$$

Hence, provide a cylindrical sludge digestion tank 6 m deep, 18.7 m in diameter, with an additional hoopered bottom of 1 : 1 slope for collection of digested sludge.

Level-2

9.11 Describes the operational principles and specific features of

(a) Simple grit chamber

(b) Aerated grit chamber.

[10 + 10 = 20 Marks]

Sol:

(i) **Simple grit chamber:**

- Also known as horizontal flow type grit chamber.
- Grit chambers are provided in the form of channels of longer length and smaller cross-section area.
- These chambers are designed in the form of long narrow channels which may either have rectangular or parabolic cross-section.
- Velocity control devices like proportional or sutro weir and parshell flume are provided at the end of these chambers, if rectangular and parabolic sections are adopted respectively.
- Grit chambers should not allow the settlement of organic materials.
- Velocity in the grit chamber should be such that the organic matter do not settle, or if it is settled the velocity should be sufficient to scour the organic matter.
- Such a critical scouring velocity is given by modified shields formula.

$$V_H = 3 \text{ to} 4.5\sqrt{gd(G_s - 1)}$$

- These are used for removal of inorganic particles of sizes greater than 0.2 mm and here Stoke's law is not valid, therefore SOR is calculated using transition flow.

$$C_D = \frac{18.5}{R_e^{0.6}}$$

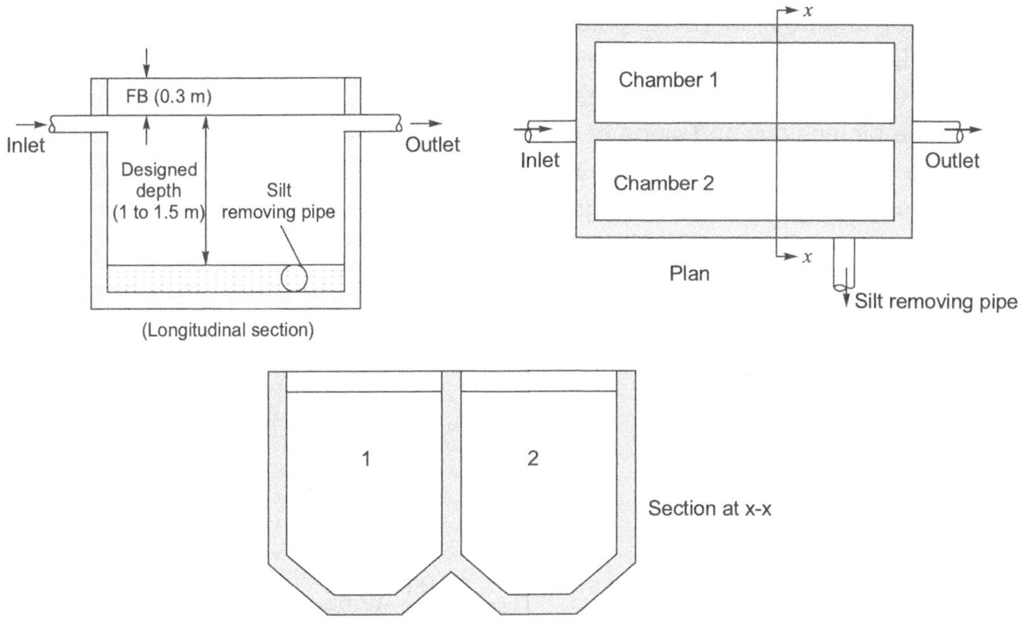

Sectional view of a rectangular grit chamber

(ii) **Aerated grit chamber**:

- These are like spiral flow aeration tanks, usually designed to provide detention periods of about 3 minutes at the maximum rate of flow.
- The spiral roll or helical flow of waste water in the grit chamber drives the grit in the hopper.
- The shearing action of air bubbles strips the inert grit of much of the organic material that adheres to its surface.
- The performance of an aerated grit chamber is a function of the roll velocity and detention time.
- If velocity of roll is too high, grit will be carried through the chamber, and if it is too low organic matter may settle down (along with grit).
- The roll velocity can however be easily controlled by adjusting the air feed rate which may range between 0.15 to 0.45 m^3/min of air per metre length of tank.
- In case of aerated grit chamber, the blowing air and the corresponding spiral motion of water keeps the lighter organic matter in suspension and allow settling of only grit (heavy particle).
- Anaerobic condition in this case can be avoided.
- Grit washing can be avoided and almost 100% removal of grit can be achieved.

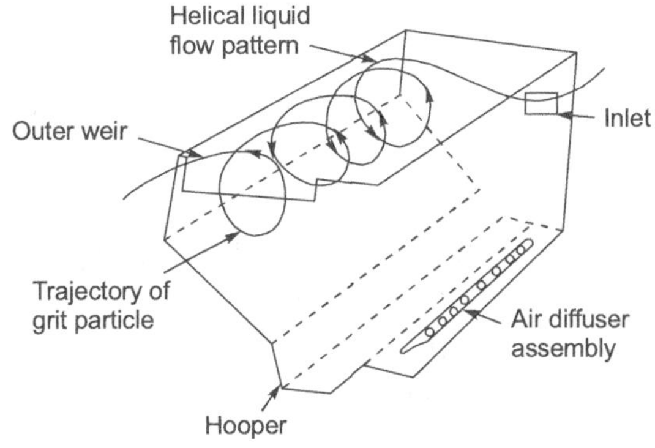

9.12 Design a grit chamber of rectangular cross-section following information is provided, flow (Q) = 50 MLD, minimum particle diameter to be completely removed (d) = 0.2 mm, specific gravity of particles (G_s) = 2.65, horizontal mean flow velocity (v) = 0.3 m/sec. Theoretical depth = 1 m. Calculate the dimensions and slope of the grit chamber. While designing the actual grit chamber, add 25% to depth for grit collection and 0.25 m freeboard. Also add 50% to the theoretically calculated length. Dynamic viscosity (μ) = 1.002 × 10^{-3} $N\text{-}s/m^2$. Density of water (ρ) = 1000 kg/m^3. g = 9.81 m/sec^2. S is the slope of chamber expressed in (m/m); value of manning; n = 0.2; Assume Stoke's law to be valid.

[20 Marks]

Sol:

Given data: Q = 50 MLD; d = 0.2 mm; G_s = 2.65; V_H = 0.3 m/sec; H = 1 m

Since, stoke's law is valid,

Settling velocity

$$V_S = \frac{g}{18}(G_s - 1)\frac{d^2}{\nu} = \frac{9.81}{18}(2.65-1)\times\frac{(0.2\times10^{-3})^2}{\left(\frac{1.002\times10^{-3}}{1000}\right)}$$

$$V_s = 0.0359 \text{ m/sec}$$

$$\text{Plan area, (A)} = \frac{Q}{V_s} = \frac{\left(\frac{50\times10^3}{86400}\right)}{0.0359} = 16.12 \text{ m}^2$$

Assume detention time, D_t = 50 sec

$$L = V_H \cdot D_t = 0.3 \times 50 = 15 \text{ m}$$

$$B = \frac{16.12}{15} = 1.075 \text{ m}$$

Dimensions to be provided,

$$H = 1 + 0.25 \text{ m} + 0.25 \text{ m} = 1.5 \text{ m}$$

$$L = 1.5 \times 15 = 22.5 \text{ m}$$

$$B = 1.075 \text{ m}$$

Required dimension of grit chamber ($L \times B \times H$) = 22.5 m × 1.075 m × 1.5 m **Ans.**

As per manning's

$$Q = \frac{1}{n}AR^{2/3} \times S_0^{1/2}$$

$$\frac{50\times10^3}{86400} = \frac{1}{0.2}\times(1.075\times1)\times\left(\frac{1.075\times1}{2+1.075}\right)^{2/3}(S_0)^{1/2}$$

So, $$S = \frac{1}{21.24}$$ **Ans.**

9.13 Population of a town is 50000. Domestic sewage produced 120 lpcd having BOD of 200 mg/litre. Industrial sewage produced is 4×10^5 L/day having BOD of 500 mg/lit. Design a high rate single stage trickling filter with the following data:
BOD removed in PST is 35%
Organic loading rate excluding recirculation = 10000 kg/ha-m/day
Hydraulic loading rate including recirculation = 170×10^6 L/hr/day
Recirculation ratio is 1.
Find the efficiency of trickling filter and BOD effluent?

[20 Marks]

Sol:

Given: Population = 50000
Domestic sewage produced per capita = 120 lpcd.
Total domestic sewage produced

$$Q_{d_S} = 50000 \times 120 = 6 \times 10^6 \text{ L/day} = 6 \text{ MLD}$$

$$\text{Industrial sewage produced} = 4 \times 10^5 \text{ L/day} = 0.4 \text{ MLD}$$

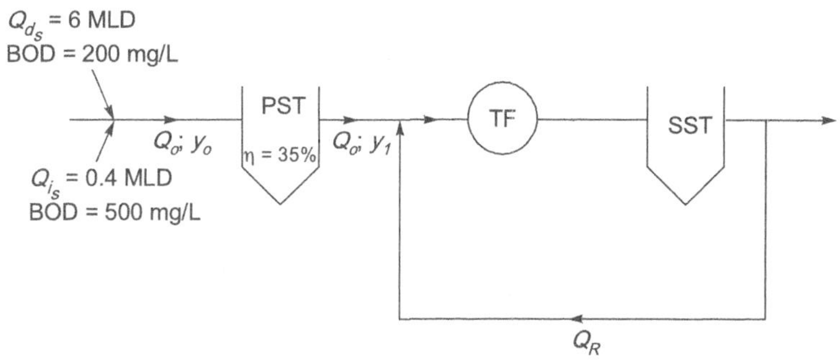

$$y_o = \frac{Q_{dS} \times (\text{BOD})_{ds} + Q_{is} \times (\text{BOD})_{is}}{Q_{dS} + Q_{is}} = \frac{6 \times 200 + 0.4 \times 500}{6.4} = 218.75 \text{ mg/L}$$

$$Q_0 = Q_{dS} + Q_{is} = 6 + 0.4 = 6.4 \text{ MLD}$$

$$\text{Recirculation ratio, } R = 1$$

$$(Q_R = Q_0)$$

Now, Efficiency of PST w.r.t. BOD removal = 35%

$$\text{BOD applied to the filter } (y_1) = 0.65 \times 218.75 = 142.1875 \text{ mg/L}$$

$$(\text{HLR})_{\text{IR}} = 170 \text{ ML/ha/day}$$

$$\text{SA(Surface area of filter)} = \frac{Q_0 + Q_R}{(HLR)_{IR}} = \frac{2 \times 6.4}{170} \times 10^4 = \text{m}^2 = 752.94 \text{ m}^2$$

$$\frac{\pi}{4} D^2 = 752.94$$

$$D = 30.96 \text{ m} < 60 \text{ m} \qquad \text{(OK)}$$

$$\text{Diameter of trickling filter} = 31 \text{ m}$$

$$(\text{OLR})_{\text{ER}} = 10000 \text{ kg/ha-m/day}$$

Volume of trickling filter

$$V = \frac{Q_0 y_1}{(OLR)_{ER}} = \frac{6.4 \times 142.1875}{10000} = 0.091 \text{ ha-m} = 910 \text{ m}^3$$

Height of trickling filter

$$H = \frac{V}{sA} = \frac{910}{752.94} = 1.208 \text{ m}$$

$$H \simeq 1.2 \text{ m}$$

Now, efficiency of single stage high rate trickling filter

$$\eta = \frac{100}{1+0.0044\sqrt{\frac{W}{VF}}}$$

$$F = \frac{1+R}{(1+0.1R)^2} = \frac{1+1}{(1+0.1\times1)^2} = 1.653$$

$$= \frac{100}{1+0.0044\sqrt{\frac{10000}{1.653}}} = 74.5\%$$

$$\eta = \frac{(\text{BOD})_i - (\text{BOD})_e}{(\text{BOD})_i}\times100$$

$$74.5 = \frac{142.1875 - y_e}{142.1875}\times100$$

$$y_e = 36.25 \text{ mg/L}$$

Effluent BOD = 36.25 mg/L **Ans.**

9.14 Determine the dimension of a high rate trickling filter for the following data:

(i) Sewage flow = 3 MLD
(ii) Recirculation ratio = 1.5
(iii) BOD of raw sewage = 250 mg/L
(iv) BOD removed in primary tank = 25%
(v) Final effluent BOD desired = 30 mg/L

By what percentage the diameter of the filter will have to be modified if it is to be designed as a standard rate trickling filter for the above requirement?

[15 Marks]

Sol:

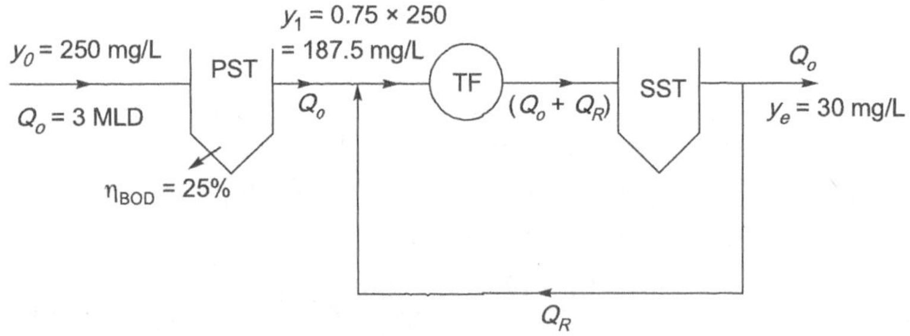

BOD, applied to filter (y_1) = 0.75 × 250 = 187.5 mg/L

Recirculation ratio, R = 1.5

$$\frac{Q_R}{Q_0} = 1.5$$

$$\eta = \frac{(\text{BOD})_i - (\text{BOD})_e}{(\text{BOD})_i}\times100 = \frac{187.5-30}{187.5}\times100 = 84\%$$

$$W_1 = Q_0 y_1 = 3 \times 187.5 = 562.5 \text{ kg/day}$$

$$F_1 = \frac{1+R}{(1+0.1R)^2} = \frac{1+1.5}{(1+0.1\times1.5)^2} = 1.89$$

$$\eta = \frac{100}{1+0.0044\sqrt{\frac{W_1}{V_1F_1}}}$$

$$84 = \frac{100}{1+0.0044\sqrt{\frac{562.5}{V_1(ha\text{-}m)\times 1.89}}}$$

$$V_1 = 0.1588 \text{ ha-m}$$

$$V_1 = 1588 \text{ m}^3$$

Assuming depth of filter as 2 m.

$$\text{Area of filter} = \frac{1588}{2} = 794 \text{ m}^2$$

$$\frac{\pi}{4}D_1^2 = 794$$

$$D_1 = 31.79 \text{ m} \qquad < 60 \text{ m (O.K.)}$$

(b) A STRF (Standard rate trickling filter) is simply a HRTF (High rate trickling filter) with zero recirculation (F = 1).

For similar efficiency

$$\frac{100}{1+0.0044\sqrt{\frac{W_1}{V_1F_1}}} = \frac{100}{1+0.0044\sqrt{\frac{W}{VF}}}$$

$$V_1F_1 = VF \qquad (F = 1)$$

$$\Rightarrow \quad V = V_1F_1 = 1588 \times 1.89$$

$$= 3001.32 \text{ m}^3$$

$$\frac{\pi}{4}D^2H = 3001.32$$

$$\frac{\pi}{4}D^2 \times 2 = 3001.32$$

$$D = 43.71 \text{ m}$$

Diameter of standard rate trickling filter = 43.71 m **Ans.**

$$\text{percentage change in dia.} = \left(\frac{D-D_1}{D_1}\right)\times 100$$

$$= \left(\frac{43.71-31.79}{31.79}\right)\times 100 = 37.5\%$$ **Ans.**

9.15 **What do you mean by activated sludge? Describe biological process reaction in activated sludge process with schematic diagram. Also, list basic components of activated sludge process.**

[25 Marks]

Sol:

Activated sludge process:

- It is an aerobic suspended growth system biological process that uses active microorganisms kept in suspension in the reactor/aerator.
- In the ASP a part of the settled sludge is returned to the reactor and oxygen is supplied externally to maintain aerobic condition.

- It increases the speed of the reaction between the organic matter and microorganism. The returned sludge from SST (Secondary sedimentation tank) is biologically very active and hence this sludge is called activated sludge and the process is called activated sludge process.
- ASP is a suspended culture process with sludge return, the process can be:
  (i) Completely mix process.
  (ii) Plug flow process.
- These processes are aerobic and for this oxygen is supplied from outside.

(Working of activated sludge process/completely mix process):

Assumptions:

(i) All reaction takes place in aerator only (it implies that BOD changes only in the aeration tank.

(ii) The influent and effluent biomass concentrations is negligible.

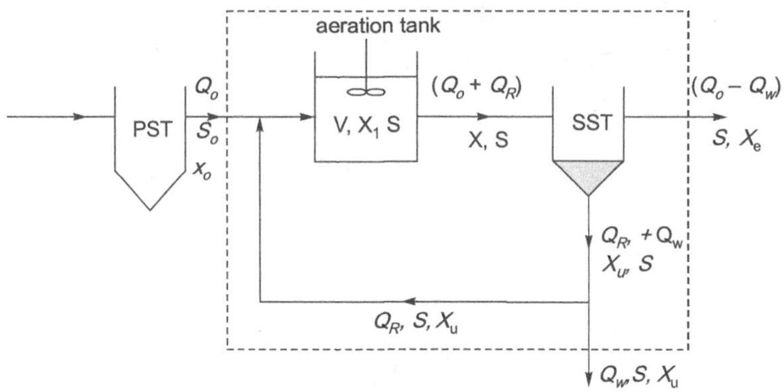

$X \rightarrow$ Suspended matter (concentration changes)

$S \rightarrow$ dissolved matter (concentration does not change).

Biomass in + Biomass growth = Biomass out

$$Q_0X_0 + (k - k_d)XV = (Q_0 - Q_w)X_e + Q_w X u$$

(Assuming influent and effluent biomass concentration is negligible).

$$(k - k_d) \times V = Q_w X_u \quad ...(1)$$

Also, Food in – Food consumed = Food out

$$Q_0S_0 - V\left(\frac{dS}{dt}\right) = (Q_0 - Q_w)S + Q_w S$$

$$Q_0S_0 - V\left(\frac{kx}{y}\right) = Q_0 S$$

$$Q_0(S_0 - S) = V\left(\frac{kx}{y}\right) \quad ...(2)$$

From eq. (1) and eq. (2)

$$\frac{Q_0(S_0 - S)y}{VX} = \frac{Q_W X_u}{VX} + k_d$$

$$\underbrace{\theta_0(S_0 - S)y}_{\text{(Total biomass produced)}} = \underbrace{\theta_W X_U}_{\text{(Biomass wasted)}} + \underbrace{K_d XV}_{\text{(Biomass billed)}}$$

Hence, ASP has three basic components:

(a) A reactor in which the microorganism responsible for treatment are kept in suspension.

(b) Liquid solid separation usually occurs in sedimentation tank.

(c) Recycle system for returning sludge solids removed from the liquid solid separation unit back to reactor.

(Operation):

- The most important aspect in the operation of activated sludge plant is to maintain proper F/M ratio i.e. food to biomass ratio, which is achieved by increasing or decreasing MLSS level in the aeration tank to suit the influent BOD.
- The MLSS in the aeration tank is regulated by controlling the rate of sludge return based on SVI determined experimentally.
- The excess sludge is wasted i.e. taken out of the system. The excess sludge is taken to the sludge digestion tank and then to sand drying bed.
- The excess sludge is wasted to maintain desired MLSS in aeration tank and also to reduce sludge age.

9.16 **An activated sludge tank is 30 m long, 9 m wide and has liquid depth of 4 m. The influent sewage flow rate is 3.5 MLD. The primary effluent has BOD of 130 mg/L and suspended load of 150 mg/L. The MLSS concentration in the aeration tank is 1800 mg/L. Calculate:**

(i) BOD loading in kg/ha-m
(ii) F/M ratio
(iii) Aeration period
(iv) Sludge age

[20 Marks]

Sol:

Given data: Q = 3.5 MLD, L = 30 m; B = 9 m; H = 4 m

$$X(\text{MLSS concentration}) = 1800 \text{ mg/L}$$
$$y_o = 130 \text{ mg/L}$$
$$S_o = 150 \text{ mg/L}$$

(i) BOD loading or organic loading $= \dfrac{Q_0 y_0}{V}$

$$\text{OCR} = \frac{\dfrac{3.5\times10^6 L}{\text{day}}\times130\times10^{-6}\text{kg/L}}{\left(\dfrac{30\times9\times4}{10^4}\right)\text{ha-m}} = 4212.96 \text{ kg/ha-m/day}$$ **Ans.**

(ii) F/M ratio

$$\text{F/M} = \frac{Q_0 y_0}{VX} = \frac{3.5\times10^6\text{L/day}\times130\times10^{-6}\text{kg/L}}{(30\times9\times4)\text{m}^3\times\dfrac{1800\times10^{-6}\text{kg}}{10^{-3}\text{m}^3}}$$

$$\text{F/M} = 0.234\text{d}^{-1}$$ **Ans.**

(iii) Aeration period

$$\theta = \frac{V}{Q_0} = \frac{(30\times9\times4)}{3.5\times10^3}\times24\text{hr}$$

$$\theta = 7.4 \text{ hr}$$ **Ans.**

(iv) Sludge age:

$$\theta_C = \frac{VX}{Q_0 S_0} = \frac{1080\text{m}^3\times1800\text{mg/L}}{(3.5\times10^3)\dfrac{\text{m}^3}{\text{day}}\times150\text{mg/L}}$$

$$\theta_C = 3.7 \text{ days}$$ **Ans.**

9.17 A wastewater plant produces 1500 kg of dry solids per day at a moisture content of 94%. The solids are 60% volatile with a specific gravity of 1.03 and the specific gravity of the non-volatile portion is 2.4. Determine the sludge volume.

(a) As produced

(b) After digestion which reduces the volatile solids by 60% and decreases the moisture content by 85%.

(c) After dewatering to 80% moisture.

(d) After drying to 8% moisture.

(e) After incineration

[25 Marks]

Sol:

Given data : Total solids production = 1500 kg/day

m.c. = 94%

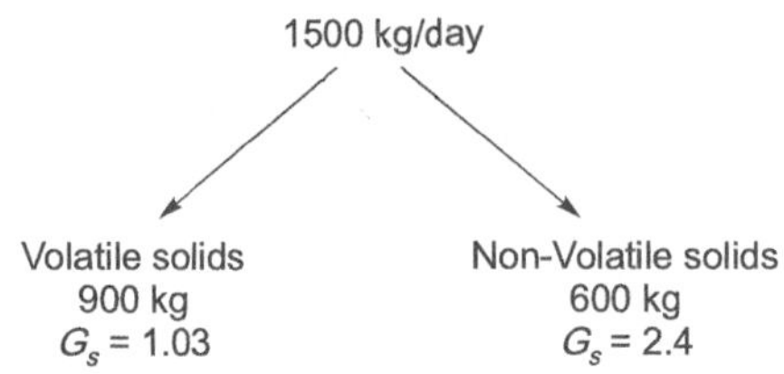

(a)

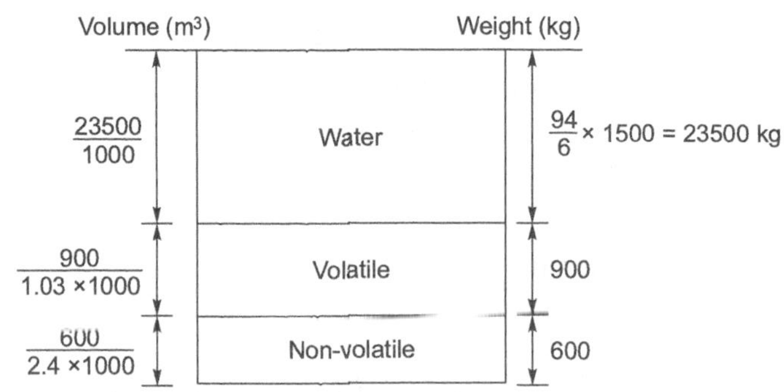

$$6 \text{ kg of solids} = 94 \text{ kg of water}$$

$$1 \text{ kg of solids} = \frac{94}{6} \text{ kg of water}$$

$$\text{Sludge volume produced} = 23.5 + \frac{900}{1.03 \times 1000} + \frac{600}{2.4 \times 1000}$$

$$= 24.62 \text{ m}^3$$ **Ans.**

(b) After digestion:

Volatile solids reduces by 60% and m.c. to 85%.

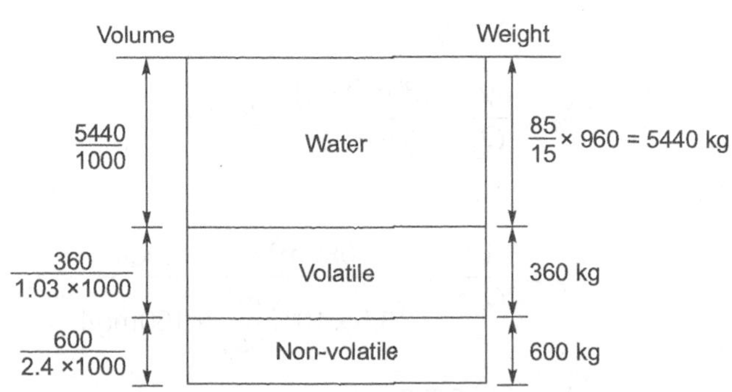

$$15 \text{ kg of solid} = 85 \text{ kg of water}$$

$$1 \text{ kg of solid} = \frac{85}{15} \text{ kg of water}$$

$$\text{Sludge volume after digestion} = 5.44 + \frac{360}{1.03 \times 1000} + \frac{600}{2.4 \times 1000} = 6.039 \text{ m}^3$$

(c) After dewatering to 80% moisture:

Dewatering will not change solid content, it will change only m.c.

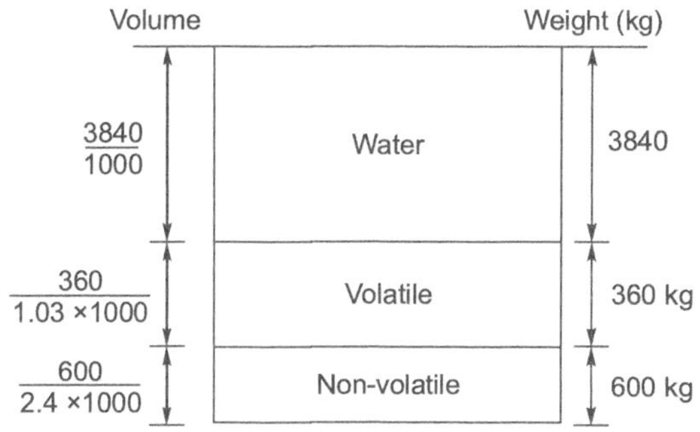

$$20 \text{ kg of solid} = 80 \text{ kg of water}$$

$$1 \text{ kg of solid} = 4 \text{ kg of water}$$

$$\text{Volume of dewatered sludge} = 3.84 + \frac{360}{1.03 \times 1000} + \frac{600}{2.4 \times 100} = 4.439 \text{ m}^3$$

(d) After drying to 10% moisture only.

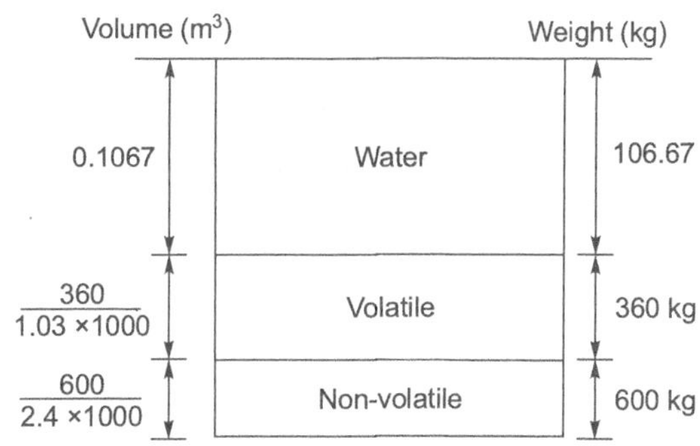

$$90 \text{ kg of solids} = 10 \text{ kg of water}$$

$$1 \text{ kg of solids} = \frac{1}{9} \text{ kg of water}$$

Now, $\text{volume of drying sludge} = 0.1067 + \frac{360}{1.03 \times 1000} + \frac{600}{2.4 \times 1000} = 0.706 \text{ m}^3$ **Ans.**

(e) After inclneration only non-volatile solid remain:

$$\text{Volume} = \frac{600}{2.4 \times 1000} = 0.25 \text{ m}^3$$ **Ans.**

9.18 **Calculate the dimensions of an oxidation pond and determine the detention time for treating sewage from a residential colony in southern state, with a population of 10000 persons. Assume that the sewage to be treated at the rate of 135 lpcd. BOD_5 of sewage = 300 ppm.**

[20 Marks]

Sol:

Given data : Population = 10000, Sewage produced per capita = 135 lpcd

Total sewage produced = (135 × 10000) L/day

$$= \frac{135 \times 10000}{10^6} = 1.35 \text{ MLD}$$

BOD_5 of sewage = 300 mg/L

BOD applied = 300 × 1.35 = 405 kg/day

It is given that pond is located in southern state

Latitude is between 24°N and 8°N approximately.

Let us assume OLR = 270 kg of BOD_5/day/ha

$$\text{Surface area/plan area required} = \frac{405}{270} = 1.5 \text{ ha} > (1 \text{ ha})$$

as surface of each oxidation pond should be between (0.5 – 1 ha)

Let us provide two oxidation pond of 0.75 ha each

Assuming $\frac{L}{B} = 2$, and depth of 1.5 m with FB of 0.3 m

$$\text{S.A.} = 0.75\text{ha}$$

$$2B \times B = 0.75 \times 10^4$$

$$B = 61.24 \text{m} = 62 \text{ m}$$

$$L = 2B = 2 \times 62 = 124 \text{m}$$

Dimension of tank = (124m × 62m × 1.8m (including FB))

Detention time

Now,
$$D_t = \frac{V}{Q} = \frac{(124 \times 62 \times 1.5)}{\left(\frac{1.35 \times 10^3}{2}\right) \text{m}^3/\text{day}} = 17.08 \text{ day} \simeq 2.44 \text{ week}$$

between (2 – 6 week)

Hence design is OK.

Dimension of oxidation pond

124 m × 62 m × 1.8 m **Ans.**

(Design of inlet and outlet of oxidation pond):

$$Q = 1.35 \text{ MCL} = \frac{1.35 \times 10^3}{86400} = 0.015625 \text{ m}^3/\text{sec of sewage}$$

Assume velocity of flow = 0.9 m/sec

$$\text{area of pond, } A = \frac{0.015625}{0.9} = 0.017 \text{ m}^2$$

$$\frac{\pi}{4} d_i^2 = 0.017$$

$$d_i = 0.1486 \text{ m}$$

$$= 14.86 \text{ cm}$$

for oxidation pond, the diameter of outlet pipe should be 1.5 times the dia. of inlet pipe.

$$d_{outlet} = 1.5 \times 14.86 \simeq 22.30 \text{ cm}$$ **Ans.**

9.19 Design an oxidation pond to treat waste from a community of 10,000 people with a per capita BOD contribution of 40 grams/day. Water is supplied at a rate of 100 *l*pcd and 80% of this is converted to sewage. Determine the detention time and also whether the effluent is suitable for irrigation if 80% BOD removal is achieved in the pond.

Assume:

(a) pond's length to width ratio 4 : 1
(b) depth of the pond is 2 m
(c) permissible BOD ≤ 100 mg/L for irrigation water, and
(d) the permissible BOD loading rate to oxidation pond is 200 kg BOD/hectare/d

[15 Marks]

Sol:

Contribution of BOD per capita = 40 gm/day

$\Rightarrow$ Total contribution of BOD

$$= 40 \times 10000 \text{ gm/day} = 4 \times 10^5 \text{ gm/day} = 400 \text{ kg/day}$$

Total water supply

$$= 100 \times 10000 \text{ lit/day} = 10^6 \text{ lit/day} = 1000 \text{ m}^3\text{/day}$$

Total sewage produced

$$= 1000 \times \frac{80}{100} = 800 \text{ m}^3\text{/day}$$

Given, the permissible BOD loading rate to oxidation pond is 200 kg BOD/Ha/day

$\therefore$ Surface area required

$$= \frac{\text{Total contribution of BOD}}{\text{Permissible loading rate}} = \frac{400}{200} \text{Ha} = 2 \text{ Ha}$$

also, as $\frac{L}{B} = \frac{4}{1}$

$\Rightarrow$ $4B \times B = 2 \times 10^4$

$\Rightarrow$ $B = 70.70$ m

$\therefore$ Provide $B = 75$ m

$\therefore$ $L = 4 \times B = 4 \times 75 = 300$ m

$\therefore$ size of the oxidation pond is 300 m × 75 m × 2 m.

Given, depth of the pond = 2 m

$\therefore$ Capacity of the tank

$$= L \times B \times H = 300 \times 75 \times 2 = 45000 \text{ m}^3$$

Now, detention time,

$$T = \frac{\text{Capacity}}{\text{Sewage flow/day}} = \frac{45000}{800} = 56.25 \text{ days}$$

$\therefore$ **Detention time = 56.25 days**

$$\text{BOD} = \frac{\text{BOD contribution per day}}{\text{Sewage flow per day}} = \frac{400 \times 1000 \text{ gm}}{800 \times 1000 \text{ lit}}$$

= 0.5 gm/lit = 500 mg/lit

BOD removed = 80% of BOD = 0.8×500 = 400 mg/lit.

Residue BOD = 500 – 400 = 100 mg/lit

$\therefore$ Permissible BOD for irrigation $\leq$ 100 mg/lit

9.20 **Design a septic tank for a colony of 200 people. The colony is supplied water at a rate of 135 litres/person/day. Assume a detention period of 24 hour and 75% of the water becomes waste water. The tank is cleaned once in a year. The rate of deposition of sludge is 40 litres/person/year. Depth of tank is to be kept as 2.0 m. Provided a free board of 0.3 m. Length to breadth ratio may be kept at 3 : 1.**

[1999 : 15 Marks]

Sol:

Number of people = 200

Water supplied = 135 litres/person/day

Detention period = 24 hrs

Tank is cleaned once in a year.

Rate of deposition of sludge = 40 litres/person/year

Depth of tank = 2.0 m

Free board = 0.3 m

Length of breadth ratio = 3 : 1

Design of Septic Tank

Quantity of water supplied = 200×135 litre/day

= 27000 litre/day

Given 75% of water becomes waste water

$\therefore$ Quantity of sewage produced

= 0.75×27000 = 20250 litre/day

Detention period = 24 hrs

$\therefore$ Quantity of sweage produced during detention period

$$= 20250 \times \frac{24}{24} = 20250 \text{ litres}$$

Rate of deposition of sludge = 40 litre/person/year

Period of cleaning = 1 year

$\therefore$ Volume of sludge deposited = $40 \times 200 \times 1$ = 8000 litres

$\therefore$ Total capacity of tank required = Capacity of sewage + Capacity for sludge

= 20250 + 8000

= 28250 litres $\simeq$ 28.25 m^3

Depth of tank = 2 m

$$\therefore \quad \text{Surface area of tank} = \frac{28.25}{2} = 14.125 \text{ m}^2$$

Ratio of length to width is 3 : 1

$$\frac{L}{B} = 3$$

$\Rightarrow$ $L = 3B$

Also, $L \times B = 14.125$

$\Rightarrow$ $3B^2 = 14.125$

$\Rightarrow$ $B = 2.17 \approx 2.2 \text{ m}$

$\therefore$ Provide width, $B = 2.2 \text{ m}$

and length $= 3B = 6.6 \text{ m}$

Thus the dimensions of septic tank will be 6.6 m × 2.2 m × (2 + 0.3) m overall depth [0.3 m free board]
Hence use a tank of size
6.6 m × 2.2 m × 2.3 m

9.21 **A town having a population of 30000 persons is producing the following sewage:**

(i) Domestic sewage @ 120 lpcd having 200 mg/*l* of BOD.

(ii) Industrial sewage @ 300000 *l* having 800 mg/*l* of BOD.

Design high rate single stage trickling filters for treating the above sewage.

Assuming that the primary sedimentation removes 35% of BOD. Allow an organic loading of 10000 kg/ha-m/day (excluding recirculated sewage). The recirculation ratio is 1.0, and the surface loading should not exceed 170 Ml/ha/day (including recirculated sewage). Also determine the efficiency of the filter and the BOD of the effluent.

[20 marks]

Sol:

Quantity of domestic sewage reduced per day

$= 120 \times 30000 \text{ litres/day} = 3.6 \text{ M } l\text{/day}$

BOD for domestic sewage $= 200 \text{ mg}/l$

$\therefore$ Total BOD of domestic sewage per day

$= 3.6 \times 200 \text{ kg/day} = 720 \text{ kg/day}$ ...(i)

Quantity of industrial sewage produced per day

$= 300000 \text{ litres}$

BOD of industrial sewage $= 800 \text{ mg}/l$

$\therefore$ Total BOD of industrial sewage

$$= \frac{30000 \times 800}{10^6} \text{ kg/day} = 240 \text{ kg/day} \quad \text{...(ii)}$$

Total BOD of domestic as well as industrial sewage per day

$= 720 + 240 = 960 \text{ kg/day}$

Out of this BOD, 35% is already removed in primary clarifier.

$\therefore$ BOD to be removed by filter unit

$= 960 \times (0.65) = 624 \text{ kg/day}$

Volume of filter media required

$$= \frac{\text{Total BOD removed}}{\text{Organic loadig}} = \frac{624}{10000} \text{ha-m}$$

$$= \frac{624}{10000} \times 10^4 \text{ m}^3 = 624 \text{ m}^3$$

Now, the total volume of sewage flow

$= (3.6 \times 10^6 + 300000) \text{ litres/day}$

$= (3.9 \times 10^6) \text{ litres/day} = 3.9 \text{ M.}l\text{/day}$

A recirculation ratio of 1 means that the volume of recirculated sewage

$$R = \text{Original volume} = 3.9 \text{ M.}l\text{/day} \quad \text{...(iii)}$$

Total volume (i.e., original + recirculated)

$$= 2\times 3.9 \text{ M.}l\text{/day} = 7.8 \text{ M.}l\text{/day}$$

$$\therefore \quad \text{Filter area required} = \frac{\text{Total flow volume}}{\text{Surface loading}} = \frac{7.8 \text{ M}l\text{/d}}{170 \text{ M}l\text{/ha-d}}$$

$$= \frac{7.8}{170}\text{ ha} = \frac{7.8}{170}\times 10^4 \text{ m}^2 = 458.8 \text{ m}^2$$

∴ Dia of the filter tank required

$$= \sqrt{\frac{458.8\times 4}{\pi}} = 24.17 \text{ m}$$

Hence, use, say, 24 m dia. tank with area = **452.4 m^2**

∴ Depth of filter media required

$$= \frac{\text{Volume of filter media}}{\text{Surface area}} = \frac{624 \text{ m}^3}{452.4 \text{ m}^2} = 1.38 \text{ m}$$

Efficiency of this filter is given as

$$\eta = \frac{100}{1+0.0044\sqrt{\dfrac{Y}{V.F.}}}$$

where, Y = Total organic load, i.e., total BOD applied to filter in kg/day = 624 kg per day

V = Volume of filter in ha-m

$$= \frac{624}{10000} = 0.0624 \text{ ha-m}$$

F = Recirculation factor as given as

$$= \frac{1+\dfrac{R}{I}}{\left[1+0.1\dfrac{R}{I}\right]} \quad \text{where } \frac{R}{I} = 1 \text{ (given)}$$

$$= \frac{1+1}{(1+0.1)^2} = \frac{2}{(1.1)^2} = \frac{2}{1.21} = 1.65$$

$$\therefore \quad \eta = \frac{100}{1+0.0044\sqrt{\dfrac{624}{0.0624\times 1.65}}} = \frac{100}{1+0.0044\sqrt{\dfrac{10^4}{1.65}}} = 74.5\%$$

$$\therefore \quad \text{BOD left in the effluent} = \frac{(100-74.5)}{100}\times 624 \text{ kg/day}$$

$$= \frac{25.5}{100}\times 624 = 159.12 \text{ kg/day}$$

$$\text{Total volume of effluent} = 3.9 \text{ M.}l\text{/day}$$

∴ BOD concentration in the effluent

$$= \frac{159.12\times 10^6}{3.9\times 10^6} \text{ mg/}l = 40.8 \text{ mg/}l$$

9.22 Raw waste water is entering a treatment plant and contains 250 mg/*l* suspended solids. 55% of these solids are removed in sedimentation.

(i) Find the volume of raw sludge produced per million litre of waste water. Assume that the sludge has a moisture content of 96% and specific gravity of solids is 1.2.

(ii) Find the unit weight of raw sludge.

(iii) If 45% solids of raw sludge is changed to liquid and gas in the digestion tank, find the volume of digested sludge per million litre of waste water. Assume that the moisture content of the digested sludge is 90%.

[20 marks]

Sol:

(i) Suspended solids in waste water = 250 mg/*l*

Since 55% of these solids are removed in sedimentation, we have

The solids removed in sedimentation as sludge

$$= 55\% \times 250 \text{ mg}/l = 137.5 \text{ mg}/l$$

Let volume of waste water is 1 million litre, then solids removed as sludge

$$= \frac{137.5}{10^6} \times 10^6 \text{ kg} = 137.5 \text{ kg}$$

Sludge produced will, thus, have 137.5 kg solids and the rest will be water. Now, since the moisture content of sludge is 96%, we have,

4 kg of solids will produce 100 kg of wet sludge, by joining with 96 kg of water.

∴ Water contained in 4 kg of solids = 96 kg

∴ Water contained in 137.5 kg of solids = $\frac{96}{4} \times 137.5$ kg = 3300 kg

Hence, volume of sludge produced per million litre of waste water

$$= \frac{\text{Mass of solids}}{\text{Density of solids}} + \frac{\text{Mass of water}}{\text{Density of water}}$$

$$= \left[\frac{137.5}{1.2 \times 1000} + \frac{3300}{1000}\right] \text{cu-m}$$

[∵ Density of solids = sp. gravity of solids × Density of water = 1.2 × 1000 kg/m^3]

$$= 0.115 + 3.3 = 3.415 \text{ cu m.}$$

Hence, volume of sludge produced per million litre of waste water = 3.415 cu m.

(ii)

$$\text{Density of raw sludge} = \frac{\text{Mass of solids} + \text{Mass of water}}{\text{Volume of sludge}}$$

$$= \frac{137.5 + 3300}{3.415} \text{kg/m}^3$$

$$= \frac{3437.5}{3.415} \text{kg/m}^3 = 1006.59 \simeq 1007 \text{ kg/m}^3$$

$$\therefore \quad \text{Unit weightof raw sludge} = \text{Density of raw sludge} \times g$$

$$= 1007 \times 9.81 = 9878.7 \text{ N/m}^3 = 9.88 \text{ kN/m}^3$$

(iii) 45% of raw sludge is changed to liquid and gas, means that 45% of solids are consumed (i.e., digested).

∴ Mass of dry solids left in the digested sludge = (100 – 45)% of total solid

$$= \frac{55}{100} \times 137.5 \text{ kg} = 75.625 \text{ kg}$$

Since digested sludge contains 90% moisture content we have,

Total volume of digested sludge

$$= \left[\frac{\text{Mass of solids left in digested sludge}}{\text{Density of solids}}\right] + \left[\frac{\text{Mass of water}}{\text{Density of water}}\right]$$

$$= \left[\frac{75.625}{1.2 \times 1000} + \left(\frac{75.625}{0.1 \times 1000}\right) \times 0.9\right] \text{m}^3 = 0.063 + 0.681 = 0.744 \text{ cu m}$$

Hence, the volume of digested sludge per million litre of waste water = 0.744 cu-m

10. Air, Sound and Land Pollution and EIA + Solid Waste Management

Level-1

10.1 **While recording a weight sound levels, 4 readings have been taken at a site at different times of a day. These readings are 20, 56, 66 and 42 dB(A). What is the average sound level?**

[10 Marks]

Sol:

Average sound level, $\overline{L}_p = 20\log_{10}\left[\frac{1}{N}\sum_{i=1}^{N} 10^{Li/20}\right] dB$

where, $\overline{L}_p$ = Average sound pressure level in dB

N = Number of measurement readings

l_i = i^{th} sound pressure level (in dB)

$$\overline{L}_p = 20\log_{10}\left[\frac{1}{4}(10^{20/20} + 10^{56/20} + 10^{66/20} + 10^{42/20})\right]$$

$= 56.78$ dB

10.2 **Discuss the composing method for disposal of solid wastes. Also explain methods adopted for compositing in rural areas in India.**

[10 Marks]

Sol:

The municipal solid waste, is collected in India from houses, roads, streets, markets, and from the commercial office buildings by the private or municipal sweepers and scavengers in hand corries and carried and dumped into the municipal refuse chambers (popularly called dhalos). During the time refuse is discarded and thrown away and till the time it is loaded in trucks for transportation, a lot of recyclable materials like plastics, papers and glasses are removed by the rag pickers.

After the wastes are loaded in trucks, it is taken for its disposal either for dumping on low leveled land, or for sanitary land filling or for compositing or for incineration etc. As far as India, is concerned, the MSW is normally brought and dumped into designated landfill site, a low lying area on the outskirts of the city. About 90% of MSW is thus, disposed off by land dumping, about 8.5% is disposed off by compositing while the remaining about 1.5% is disposed off by other method likes pulverization, vermin compositing.

Methods:

(Shredding and pulverization): The size and volume reduction of MSW is accomplished by the physical processes of shredding and pulverization shredding refers to the actions of cutting and tearing, whereas, pulverization refers to the actions of crushing and grinding. Shredding and pulverizing may help in reducing the overall volume of the original MSW, by as much as 40%. The shredding and pulverizing not only helps in reducing the volume of MSW, but also helps in changing the physical character of the waste, which becomes practically odourless and unattractive to the insects.

Compositing: Compositing of refuse refers to the biological methods of decomposing solid wastes. This decomposition can be affected either under aerobic conditions, or under anaerobic conditions or both. The final end product is menure, called the compost or humus, which is in great demand in European countries as fertilizer for forms.

(Incineration): Burning of refuse at high temperatures in furnaces, called incineration, is quite a sanitary method of refuse disposal, end is widely adopted in developed countries like USA, where the collected waste/refuse is of high calorific value, and is, hence quite suitable to burning.

(Barging it out into the sea): This method had been used in the past to dispose off refuse by throwing it away into the sea, after carrying it at reasonable distance from the coast (say 16 to 20 kilometers inside the sea) on barges. The sea depth at such disposal point should not be less than 30 m or so, and the direction of current should be such as not be bring it back towards the shore. This method may, however, have a limited use and that too only in a few coasted towns.

10.3 Based on type of air pollution and climate differentiate between the 'Brown air cities' and the gray air cities with examples.

[10 Marks]

Sol:

Large cities generally fall into one of two basic air pollution classes the 'Brown air cities' and the 'Gray air cities'.

Brown air cities like Los Angeles, are relatively young cities where the main source of air pollution is the internal combustion engine in the automobile. Every automobile sprays several hundred different compounds into the air. One of these compounds is gaseous nitric oxide, which can react with oxygen, O_2 in the air to form nitrogen dioxide, NO_2 a yellow brown Poissonus gases that produces brown haze over car-oriented cities. NO_2 in the presence of sunlight reacts with hydrocarbons to form photochemical smog.

The second type of city, the gray air city, is an older city such as London, New york and Chicago, that depends heavily on the burning of coal or oil for heating, manufacturing and electric power generations. This burning produces, two major air pollutants:

(i) Particulates (dust and soot) which give the air over such cities its gray cast, and
(ii) Sulphur dioxide, one of the most dangerous pollutants in our air coal and soil are usually contaminated with sulphur.

10.4 An industry utilizes 0.3 ML of oil fuel per month. It has also been estimated for every 1 ML of fuel oil burst in the factory per year the quantities of various pollutants emitted are given as

PM = 2.9 t/yr | **SO_2 = 60 t/yr**
NO_x = 8 t/yr | **HC = 0.4 t/yr**
CO = 0.5 t/yr

Calculate the height of the chimney required to be provided for safe dispersion of pollutants.

[12 Marks]

Sol:

Given, fuel consumption per month = 0.3 ML

Fuel consumed in a year = 0.3 × 12 = 3.6 ML

PM produced per year = 2.9 × 3.6 = 10.41 t

SO_2 produced per year = 60 × 3.6 = 216 t

Let us assume number of working days in a year = 300 days

$$\text{PM produced per hr} = \frac{3.6\times2.9}{300\times24} = 1.45\times10^{-3}\ \text{t/hr}$$

Height of chimney $h_p = 74(Q_P)^{0.27} = 74(1.45\times10^{-3})^{0.27} = 12.67$ m

$$SO_2 \text{ produced per hr} = \frac{216\times1000}{300\times24} = 30\ \text{kg/hr}$$

Height of chimney, $h_s = 14(Q_s)^{1/3}$

$h_s = 14(30)^{1/3} = 43.50$ m

minimum height of chimney = 30 m

required height of chimney is maximum of all the above criteria.

height of chimney = 43.5 m **Ans.**

10.5 Define primary and secondary air pollutants with appropriate examples.

[10 Marks]

Sol:

The atmospheric air may contain hundreds of air pollutants from the natural or the anthropogenic (manmade) sources. All these pollutants, which are emitted directly from the identifiable sources, either from the natural hazardous events like dust storms, volcanoes etc or from human activities like burning of wood, coal, oil in homes or industries or automobiles are called primary pollutants. The following five primary pollutants contribute to about 90% of the global air pollution.

Example:

1. Oxides of sulphur, particularly the sulphur dioxide (SO_2).
2. Oxides of carbon like carbon monoxide (CO), and carbon dioxide (CO_2), particularly the carbon monoxide (CO).
3. Oxides of nitrogen, like NO, NO2, NO_3 (expressed as NO_x).
4. Volatile organic compounds, mostly hydrocarbons.
5. Suspended particulate matter (SPM).

Secondary pollutants:

These primary pollutants often react with one another or with water vapour, aided and abetted by the sunlight, to form entirely a new set of pollutants called the secondary pollutants.

Example:

1. Sulphuric acid (H_2SO_4)
2. Ozone (O_3)
3. Formaldehyde
4. Peroxy acetyl nitrate (PAN); etc

10.6 An electrostatic precipitator must be designed to process 5 m³/s of stack gas. The drift velocity of the flash particles has been determined to be

$$V_d = 1.5\times10^5\ d_p$$

Determine the plate area required to remove particles with diameters of 0.7 μm with 95% efficiency.

[12 Marks]

Sol:

Given, flow rate, $Q = 5\ m^3/sec$

The drift velocity $(V_d) = 1.5 \times 10^5\ d_p$

d_p = diameter of particle

$= 0.7\ \mu m = 0.7 \times 10^{-6}\ m$

$\therefore \quad V_d = 1.5 \times 10^5 \times 0.7 \times 10^{-6} = 0.105\ m/sec$

Efficiency of electrostatic precipitators,

$$\eta = 1 - e^{-\left(\frac{AV_d}{Q}\right)}$$

$$0.95 = 1 - e^{-\left(\frac{A\times 0.105}{5}\right)}$$

whrere, A = Area of collector plates

Q = flow rate

$$e^{-\left(\frac{A\times 0.105}{5}\right)} = 0.05$$

Taking log both sides

$$-\frac{A\times 0.105}{5} = \ln(0.05)$$

$$A = 142.65\ m^2$$

Ans.

Level-2

10.7 From the following sound observations 69, 66, 63, 66 dB find

(a) total sound produced, if they are emitted by different sources.

(b) average sound level, if they are produced by the same source.

(c) equivalent continuous noise, if weightage factor for each sound is same.

[20 Marks]

Sol:

Different sounds 69, 66, 63, 66 dB.

(a) (Total sound produced) (69 + 66 + 63 + 66 dB)

As we know,

$$L = 20\log_{10}\left[\frac{p_{rms\ 1}}{(p_{rms})_0}\right];\ p_{rms\ 0} = 20\ \mu pa$$

$$69 = 20\log_{10}\left[\frac{p_{rms\ 2}}{20}\right] dB$$

$$p_{rms\ 1} = 10^{69/20} \times 20 = 56367.658\ \mu pa$$

Similarly,

$$66 = 20\log_{10}\left[\frac{p_{rms\ 2}}{20}\right] dB$$

$$p_{rms\ 2} = 10^{66/20} \times 20 = 39905.246\ \mu pa$$

$$63 = 20\log_{10}\left[\frac{p_{rms\ 3}}{20}\right] dB$$

$$p_{rms\ 3} = 20 \times 10^{63/20} = 28250.75\ \mu pa$$

$$p_{rms\,4} = 39905.246\,\mu Pa$$

$$p_{rms} \text{ of resultant sound} = \sqrt{(p_{rms_1})^2 + (p_{rms_2})^2 + (p_{rms_3})^2 + (p_{rms_4})^2}$$

$$= \sqrt{(56367.658)^2 + (39905.246)^2 + (28250.751)^2 + (39905.246)^2}$$

$$= 84618.408\,\mu pa$$

$$\text{Now, total sound (in dB)} = 20\log_{10}\left(\frac{P_{rms}}{20}\right) = 20\log_{10}\left(\frac{84618.408}{20}\right)$$

$$= 72.53 \text{ dB}$$ **Ans.**

(b) Avg. sound level,

$$\overline{\tau}_p = 20\log_{10}\left(\frac{1}{N}\times\sum 10^{4/20}\right)$$

$$= 20\log_{10}\left[\frac{1}{4}\times\left(10^{69/20} + 10^{66/20} + 10^{63/20} + 10^{66/20}\right)\right]$$

$$\overline{\tau}_p = 66.25 \text{ dB}$$

(c) Equivalent noise level (L_{eq})

$$L_{eq} = 10\log_{10}\left[\sum_{i=1}^{N} t_i \times 10^{Li/10}\right]$$

N = Total no. of sounds

L_i = The noise level of any i[th] sample

t_i = time duration of ith sample expressed as fraction of total sample time

since, weighted factor of each sound is same this means the time period of each sound is same.

Now,
$$L_{eq} = 10\log_{10}\left[\frac{1}{4}\times 10^{69/10} + \frac{1}{4}\times 10^{66/10} + \frac{1}{4}\times 10^{63/10} + \frac{1}{4}\times 10^{66/10}\right]$$

$$L_{eq} = 66.508 \text{ dB}$$ **Ans.**

10.8 **Illustrates with diagrams the influence of environmental lapse rates on shapes and dispersion of plume emitted from stacks of a thermal power plant.**

[20 Marks]

Sol:

Typical types of environmental conditions, characterized by different relative positions of environmental lapse rate and adiabatic lapse rate lines, which are generally encountered in the lower atmosphere. The manner in which the emitted plume behaves under each of these seven conditions discussed below:

(a) Looping Plume : Looping has a wavy character and occur in super-adiabatic environment; which produces highly unstable atmosphere, because of rapid mixing, during the high degree of turbulence, the dispersion of plume would be rapid, yet higher concentrations near the ground may occur due to turbulence, before the dispersion is finally completed.

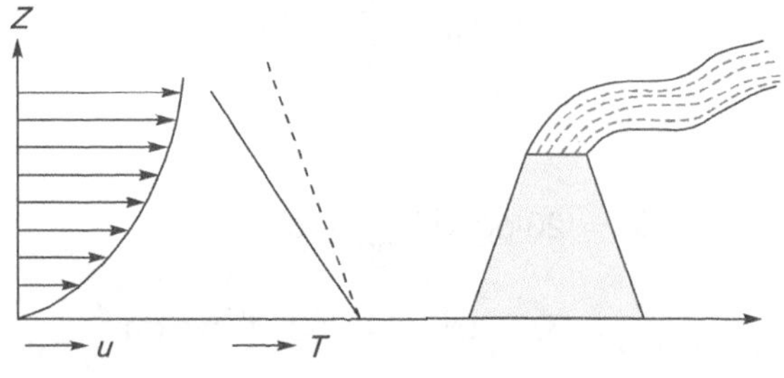

(b) Neutral Plume

- Neutral plume is the upward vertical rise of the plume from the stack, which occurs when the environmental lapse rate is equal to or very near to the adiabatic lapse rate. The upward lifting of plume will continue till it reaches an air of density similar to that of the plume itself.

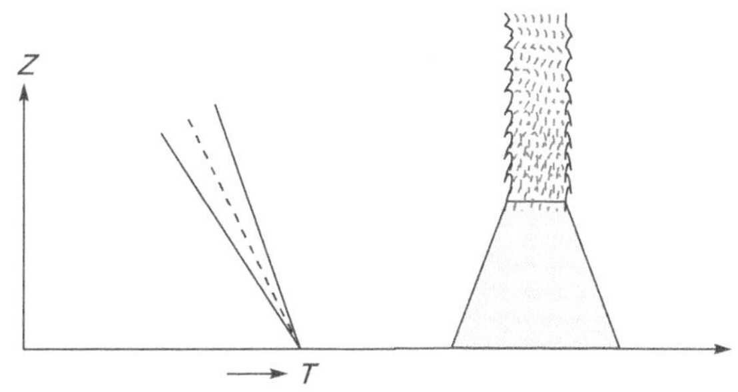

(c) Coning Plume

- Coning plume occurs on cloudy day or nights with strong wind velocity (> 32 km/hr) when the lapse rate is near neutral.
- The plume shape is vertically symmetrical about the plume line.
- However, the plume reaches the ground at greater distance than with looping plume.

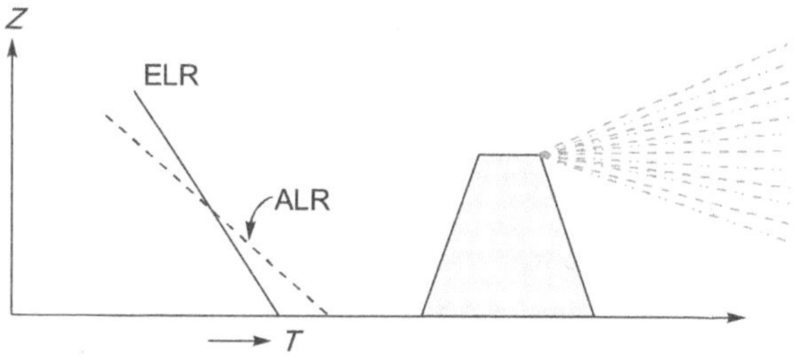

(d) Fanning Plume

- This occurs under extreme inversion conditions, in the presence of light wind.
- Most of the vertical dispersion is suppressed by extremely stable condition, and the plume fans out in the horizontal direction.

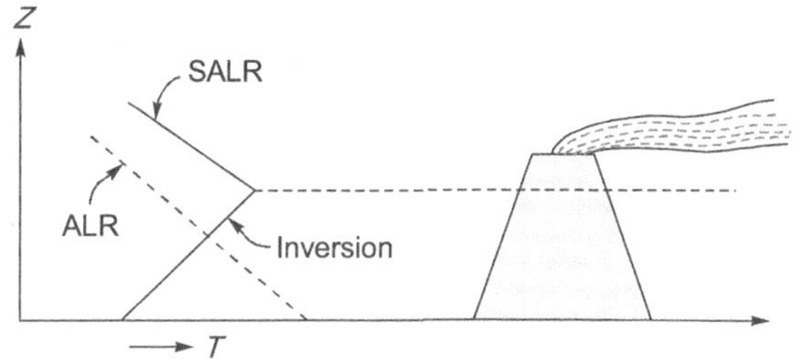

(e) Lofting Plume

- Plume is said to be lofting when there exists a strong super adiabatic lapse rate (SALR) above inversion.
- Fumigation takes place when an inversion layer occurs at a short distance above the top of the stack and super adiabatic conditions, prevail below it. Hence the pollutants cannot escape above the top of stack and they are brought down rapidly near ground due to turbulence in the region above the ground and below the inversion, caused by strong lapse rate.
- Fumigation represent quite a bad case of atmospheric dispersion.

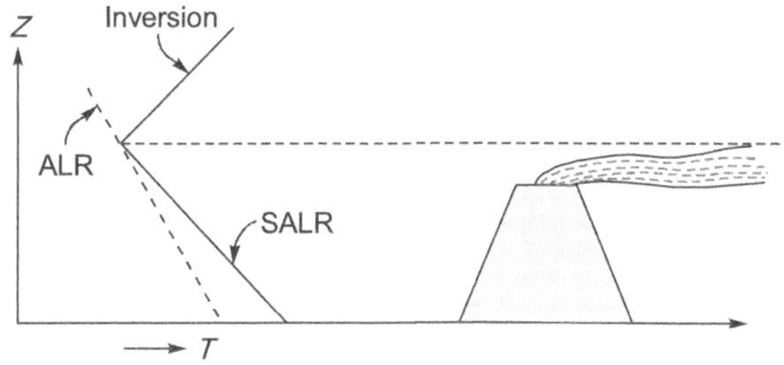

(f) Trapping Plume

This condition is achieved when the plume caught between two inversion layers. Hence, the emitted plume can neither go up nor down and will be trapped between two levels.

- A trapping plume is considered to be a bad condition for dispersion.

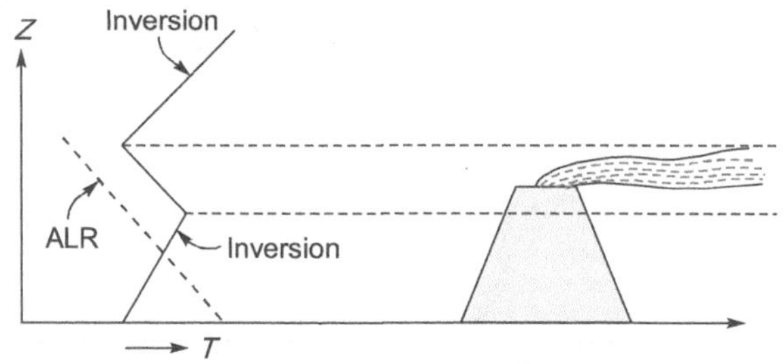

10.9 **A bag house is to design to handle 1000 m³/min of air. The filtration takes place at constant pressure so that air velocity through each bag decreases during the time between clearing according to the relation.**

$u = \dfrac{1}{0.267 + 0.08t}$; **where u is in $m^3/min/m^2$ of cloth and t is time in (min)**

The bags are shaken in a sequence row by row on a 30 minute cycle. Each bag is 20 cm in diameter and 3 m height. The bag house is to be square in cross-section with 30 cm spacing between the bags and 30 cm clearance from the wells. Calculate the number of bag required.

[15 Marks]

Sol:

Given data : Ratio of flow rate to cloth area, $u = \dfrac{1}{0.267+0.08t} m^3/min/m^2$, t = 30 min, diameter of bag (d) = 0.2 m, length of bag, L = 3 m and flow rate, Q = 1000 m³/min.

V_{avg} = Average velocity (in m/min), A_t = Total area of filter (m²), A_b = Area of single bag, (m²), N = Number of bags

$$V_{avg.} = \frac{1}{t}\int_0^t u\,dt = \frac{1}{t}\int_0^t \frac{dt}{0.267 + 0.08t}$$

$$= \frac{1}{30}\int_0^{30} \frac{dt}{0.267 + 0.08t}$$

$$= \frac{1}{30}\left[\frac{\ln(0.267 + 0.08t)}{0.08}\right]_0^{30} = 0.9589 \text{ m/min}$$

$$A_t = \frac{Q}{V_{avg}} = \frac{1000}{0.9589} = 1042.82 \text{ m}^2$$

$$A_b = \pi dL = \pi \times 0.2 \times 3 = 1042.82 \text{ m}^2$$

$$\therefore \quad N = \frac{A_t}{A_b} = \frac{1042.82}{1.8849} = 553.247$$

$$N \simeq 553$$ Ans.

The number of bags required for continuous removal of particulate matter are 553. Ans.

10.10 (a) For a population of 10000 with per capita waste generation of 1.0 kg/day, estimate the annual area requirements (excluding the buffer zone) for a normally compacted land fill with density of 450 kg/m^3 and having a refuse depth of 4 m excluding the cover material.

(b) A residential area of about 40 ha contains 400 single family residences and 8 ha with apartments housing 400 people. Assuming five residents per single family unit and per capita waste generation at 1.0 kg/day. Determine the number of trips on each collection day with two curbside pickups per week, one packer truck (4 to 5 tonne capacity) needs to make serve this area.

[12 Marks]

Sol:

(a) Given total waste generations in a year = $10000 \times 1.0 \times 365 = 365 \times 10^4$ kg

$$\text{volume of waste, } V = \frac{365\times10^4}{450} = 0.811 \times 10^4 \text{ m}^3$$

$$\text{area requirement} = \frac{0.811\times10^4}{4} = 0.2027 \times 10^4 \text{ m}^2 = 0.2027 \text{ ha} \quad \textbf{Ans.}$$

(b) Given, Total population = $400 \times 5 + 400 = 2400$

Total waste generation per week = $2400 \times 1 \times 7 = 16800$ kg

$$\text{Waste to be picked up per collection day} = \frac{16800}{2} = 8400 \text{ kg}$$

Taking truck capacity as 4.5 tones.

$$\text{Number of trips per collection day} = \frac{8400}{4.5\times1000} = 1.87 \simeq 2 \text{ trips} \quad \textbf{Ans.}$$

10.11 A residential area consisting of 1500 houses has an average of four residents per house. For estimating the quantity of solid waste generated, the following observations were made at disposal, site for a period of one week.

Type of vehicle	Number of trips	Volume (m^3)	Specific weight (kg/m^3)
I	10	15	300
II	8	2	150
II	25	0.50	100

Determine the unit rate of solid waste generation, also find density of MSW.

[15 Marks]

Sol:

Type of vehicle	Number of trips	Volume (m^3)	Specific weight (kg/m^3)
I	10	15	300
II	8	2	150
III	25	0.50	100

Total volume generated = $15 \times 10 + 8 \times 2 + 25 \times 0.5 = 178.5$ m^3

(Vehicle-I)

$$\text{Total solid waste transported} = \text{No. of trips} \times \frac{\text{Volume}}{\text{trip}} \times \rho_{sw} = 10 \times 15 \times 300 = 45000 \text{ kg}$$

(Vehicle-II)

Total solid waste transported = $8 \times 2 \times 150 = 2400$ kg

(Vehicle-III)

Total solid waste transported = $25 \times 0.5 \times 100 = 1250$ kg

Total solid waste generated = $45000 + 2400 + 1250 = 48650$ kg

$$\text{Total population} = \text{No. of houses} \times \frac{\text{No. of person}}{\text{house}} = 4 \times 1500 = 6000$$

$$\text{per capita MSW generation} = \frac{48650}{600} = 8.108 \simeq 8.11 \text{ kg/person/week}$$ **Ans.**

$$\frac{100}{\rho_{MSW}} = \frac{\%truck1}{(\rho_{sw})_1} + \frac{\%truck2}{(\rho_{sw})_2} + \frac{\%truck3}{(\rho_{sw})_3}$$

$$\frac{100}{P_{MSW}} = \frac{15 \times \frac{10}{178.5} \times 100}{300} + \frac{\frac{16}{178.5} \times 100}{150} + \frac{\frac{12.5}{178.5} \times 100}{100}$$

$$\rho_{MSW} = 243.96 \text{ kg/m}^3$$

10.12 **There is a thermal power plant of total capacity 915 MW with a load factor of 72.5 percent and an efficiency of 40 percent. Determine the amount of particulates, CO_2 and SO_2 that are generated annually if oil is the fuel source. The ultimate analysis of fuel are give below:**

Moisture	Ash	Carbon	Hydrogen	Nitrogen + Oxygen	Sulphur
0.3%	0.04%	85.2%	11.3%	0.36%	2.8%

[Assume 80% of ash is particulate, Calorific value of fuel = 40.5 MJ/kg]

[20 marks]

Sol:

$$\text{Power} = 915 \text{ MW}$$

$$= 915 \times 3600 \text{ MJ/hr} \qquad [\because \; 1W = 1 \text{ J/s} = 3600 \text{ J/hr}]$$

$$= 3294 \times 10^3 \text{ MJ/hr}$$

$$\text{Oil required} = \frac{3294 \times 10^3 \times 0.725}{0.4 \times 40.5} = 147417 \text{ kg/hr}$$

$$= 147.417 \text{ tonne/hr}$$

Given 80% of ash is particulates,

So, $$\text{Particulates} = 0.8 \times \frac{0.04}{100} \times 147.417 = 0.047 \text{ tonne/hr}$$

$$S + O_2 = SO_2$$

$\therefore$ 32 gm sulphur generates 64 gm of SO_2

$$SO_2 \text{ generated by 2.8\% of sulphur} = \frac{64}{32} \times \frac{2.8}{100} \times 147.417 = 8.255 \text{ t/hr}$$

$$C + O_2 = CO_2$$

$\therefore$ 12 gm carbon generates 44 gm of CO_2

$$\therefore \; CO_2 \text{ generated by 85.2\% of carbon} = \frac{44}{12} \times \frac{85.2}{100} \times 147.417 = 460.53 \text{ tonne/hr}$$

$\therefore$ Total annual production is as follows:

$$\text{Particulates} = 0.047 \times 24 \times 365 = 411.72 \text{ tonnes}$$

$$SO_2 = 8.255 \times 24 \times 365 = 72313.8 \text{ tonnes}$$

$$CO_2 = 460.53 \times 24 \times 365 = 4034242.8 \text{ tonnes}$$

■■■■

05 Soil Mechanics

1. Properties of Soil

Level-1

1.1 A sample of clay was coated with paraffin wax and its mass, including the mass of wax was found to be 697.5 gm. The sample was immersed in water and the volume of water displaced was found to be 350 ml. The mass of the sample without wax was 688.0 gm and the water content of the representative specimen was 16%.

Determine the bulk density, dry density, void ratio and the degree of saturation. The specific gravity of the solids was 2.71 and that of the wax was 0.88.

[10 Marks]

Sol:

$$\text{Mass of wax} = 697.5 - 688 = 9.5 \text{ gm}$$

$$\text{Volume of wax} = \left(\frac{9.5}{0.88 \times 1.0}\right) = 10.79 \text{ m}l$$

$$\text{Volume of soil} = 350 - 10.79 = 339.21 \text{ m}l$$

$$\text{Bulk density} = \left(\frac{688}{339.21}\right) = 2.03 \text{ gm/m}l$$

$$\text{Dry density} = \left(\frac{2.03}{1+\frac{16}{100}}\right) = 1.75 \text{ gm/m}l$$

$$\text{Dry density} = \left(\frac{G\gamma_w}{1+e}\right)$$

$$1.75 = \left(\frac{2.71 \times 1}{1+e}\right)$$

$$\Rightarrow \quad e = 0.548 \; ; \; e = \left(\frac{Gw}{s}\right)$$

$$S = \frac{Gw}{e} = \frac{2.71 \times 0.16}{0.548} = 0.7912 = 79.12\%$$

1.2 The following data were obtained during an in-in-situ weight sand replacement method:

Volume of calibrating container = 1000 ml; Weight of empty container = 900 gm

Weight of container + sand = 2550 gm

Weight of sand filling the conical portion of sand pouring cylinder = 460 gm

Initial weight of sand pouring cylinder + sand = 5450 gm

Weight of cylinder + sand after filling the excavation hole = 4180 gm

Weight of excavated soil = 954 gm; Insitu water content = 8%

Determine the in-situ unit weight and in-situ dry unit weight.

[10 Marks]

Sol:

Step 1: Calculation of unit weight of sand

Weight of sand filling the calibrated container of volume 100 ml = (2550 – 900) = 1650 gm

$$\text{Unit weight of sand} = \left(\frac{1650}{1000}\right) = 1.65 \text{ gm/cc}$$

Weight of sand filling the excavated hole and the conical portion of sand pouring cylinder = 5450 – 4180 = 1270 gm

Weight of sand used for filling the excavated hole = 1270 – 460 = 810 gm

$$\text{Volume of excavated hole} = \frac{\text{Weight of sand}}{\text{Density of sand}}$$

$$= \frac{810}{1.65} = 490.91 \text{ cm}^3$$

$$\text{Weight of excavated soil} = 954 \text{ gm}$$

$$\text{In-situ weight} = \frac{954}{490.91} = 1.94 \text{ gm/cm}^3$$

$$= 1.94 \times 9.81 = 19.03 \text{ kN/m}^3$$

$$\text{In-situ dry unit weight} = \left(\frac{19.03}{1+\frac{8}{100}}\right) = 17.62 \text{ kN/m}^3$$

1.3 **The following are the data obtained in shrinkage limit test:**

Initial weight of saturated soil = 96.5 gm

Initial volume of saturated soil = 70.0 cm^3

Final dry volume = 25.0 cm^3

Final dry weight = 45.5 gm

Determine the shrinkage limit, the specific gravity of grains, the initial and final dry unit weight, bulk unit weight and void ratio.

[10 Marks]

Sol:

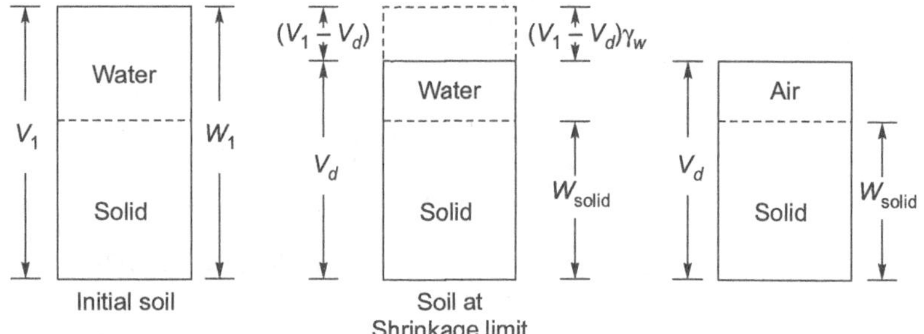

Above shrinkage limit ***Below shrinkage limit***

W_1 = 96.5 gm, V_1 = 70.0 cm^3, W_{solid} = 45.5 gm, V_d = 25.0 cm^3

Let the water content at shrinkage limit be w_s

$$w_s = \frac{(W_1 - W_{solid}) - (V_1 - V_d)\gamma_w}{W_{solid}}$$

$$= \frac{(96.50 - 45.5) - (70 - 25) \times 1}{45.5} = 0.1319$$

$$w_s = 13.19\%$$

Determination of specific gravity (G_s) :

$$w_s = \frac{W_w}{W_{solid}} = \frac{(V_d - V_s)\gamma_w}{W_{solid}} = \frac{\left(V_d - \frac{W_{solid}}{G_s\gamma_w}\right)\gamma_w}{W_{solid}}$$

$$0.1319 = \frac{\left(25 - \frac{45.5}{G_s \times 1}\right)\times 1}{45.5}$$

$$G_s = 2.395$$

$$\text{Initial bulk unit weight} = \left(\frac{96.50}{70}\right) = 1.38 \text{ gm/cc} = 13.54 \text{ kN/m}^3$$

$$\text{Water content} = \left(\frac{96.5 - 45.5}{45.5}\right) = 1.1209$$

$$\text{Initial dry unit weight} = \left(\frac{13.54}{1+1.1209}\right) = 6.384 \text{ kN/m}^3$$

$$\text{Initial void ratio} = w_i G_s = 1.1209 \times 2.395 = 2.684$$

$$\text{Final void ratio} = w_s G_s = 0.1319 \times 2.395 = 0.316$$

$$\text{Final dry unit weight} = \left(\frac{45.50}{25}\right) = 1.82 \text{ gm/cc} = 17.85 \text{ kN/m}^3$$

1.4 **A one litre capacity core-cutter of mass 1 kg was pushed into embankment under construction and mass of core cutter with soil was found to be 2.865 kg. If the soil had water content of 11%, specific gravity of soil is 2.67. Determine the bulk unit weight, dry unit weight, void ratio and degree of saturation of soil sample. The unit weight of water is 9.81 kN/m^3.**

[10 Marks]

Sol:

For given sample,

Mass of core cutter, $M_1 = 1$ kg

Mass of soil + Core cutter, $M_2 = 2.865$ kg

Volume of core cutter, $V = 1000$ cm^3

Mass of soil, $M = 2.865 - 1 = 1.865$ kg

Bulk unit weight of soil,
$$\gamma_{bulk} = \frac{Mg}{V} = \frac{1.865 \times 9.81}{1000} \times 1000 = 18.296 \text{ kN/m}^3$$

Dry unit weight,
$$\gamma_d = \frac{\gamma_{bulk}}{1+w} = \frac{18.296}{1+0.11} = 16.48 \text{ kN/m}^3$$

Using,
$$\gamma_d = \left(\frac{G}{1+e}\right)\gamma_w$$

$$16.48 = \left(\frac{2.67}{1+e}\right)\times 9.81$$

Void ratio, $e = 0.59$

Degree of saturation,
$$S = \frac{Gw}{e} = \frac{2.67 \times 0.11}{0.59} = 0.5 = 50\%$$

1.5 Two soils A and B are tested in a laboratory for the consistency tests and the following results were obtained:

Properties	Soil A	Soil B
Liquid limit	35%	65%
Plastic limit	20%	40%
Flow index	6	3
Water content	30%	36%

Which of two soils is
(a) more plastic
(b) better foundation material
(c) of better strength as a function of water
(d) of better strength at the plastic limit
(e) more compressible

[10 Marks]

Sol:

(a) Plasticity index, $I_P = W_L - W_P$

$$(I_P)_A = 35 - 20 = 15\%$$
$$(I_P)_B = 65 - 40 = 25\%$$
$$(I_P)_B > (I_P)_A$$

⇒ Soil B is more plastic than A

(b) Consistency index, $I_C = \left(\dfrac{W_L - W}{W_L - W_P}\right)$

$$(I_C)_A = \left(\frac{35-30}{35-20}\right) = 0.333 \;;\; (I_C)_B = \left(\frac{65-36}{65-40}\right) = 1.16$$
$$(I_C)_B > (I_C)_A$$

⇒ Soil *B* Is better foundation material.

(c) Flow index (I_F)

$$(I_F)_A > (I_F)_B$$

⇒ Soil *B* has better strength as function of water.

(d) Toughness index, $I_t = \left(\dfrac{I_P}{I_F}\right)$

$$(I_t)_A = \left(\frac{15}{6}\right) = 2.5 \qquad ; \; (I_t)_B = \left(\frac{25}{3}\right) = 8.33$$
$$(I_t)_B > (I_t)_A$$

⇒ Soil *B* has better strength as plastic limit.

(e) $(W_L)_B > (W_L)_A$

⇒ Soil *B* has more compressible than Soil *A*.

1.6 What are important clay minerals? Why do clays exihibit plasticity?

[12 marks]

Sol:

The clay minerals are basically composed of tiny crystalline substances of one or more members of a small group of minerals-commonly known as clay minerals. These clay minerals mainly evolved from the chemical weathering of certain rock forming minerals.

Important clay minerals are
1. Kaolinite
2. Montmorillonite
3. Illite

1. **Kaolinite :** The kaolinite structural unit consists of alternating layers of silica tetrahedral units with the tips embedded in alumina. The bonding is through hydrogen bond which results in considerable strength.

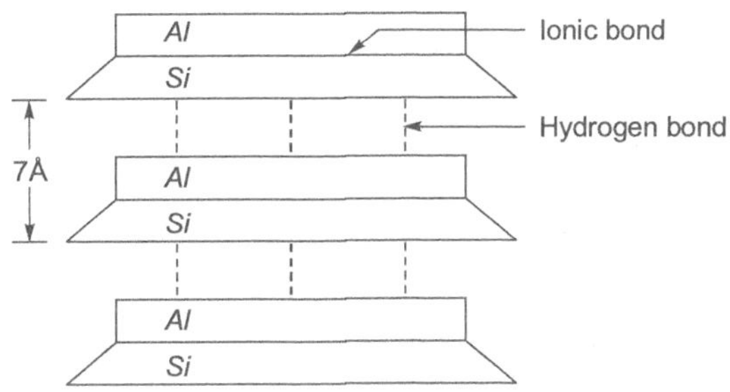

2. **Montmorillonite :** It is also called smectite. The structural unit of the mineral is composed of two silica sheets and one alumina sheet. The interlayer bonding between the top of silica sheets is mainly due to van der wall's forces which is very weak.

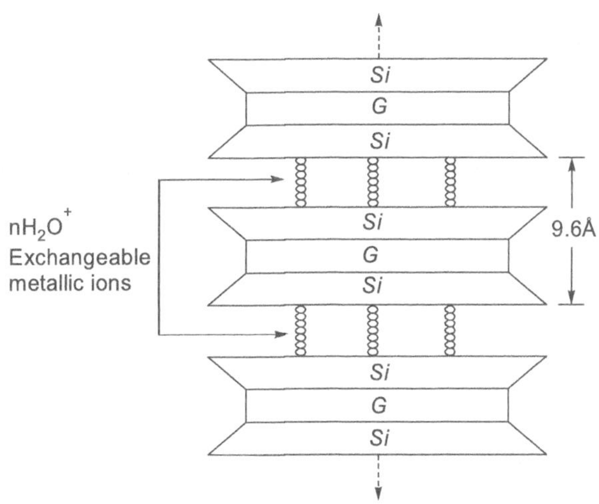

3. **Illite :** The Illite mineral is also a 2 : 1 mineral similar to montmorillonite. In Illite, potassium ion bonds are present and therefore Illite does not swell as much in the presence of water as montmorillonite.

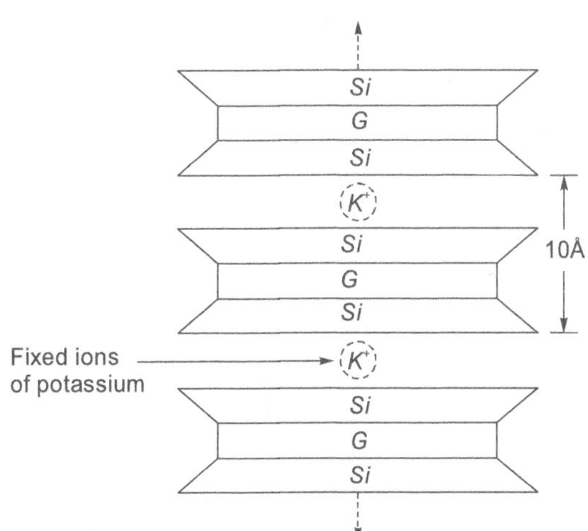

Clayminerals are electrically charged with negative ions. They have large specific surface area and are very flaky in shape. Clay attracts dipolar water towards its surface by adsorption. This induces plasticity in clay. Therefore plasticity increases with specific surface area. Water in clays exhibits negative pressure due to which two particles are held close to each other due to this apparent cohesion developed in clays.

1.7 Derive an expression for determination of water content in soil by pycnometer method.

[12 marks]

Sol:

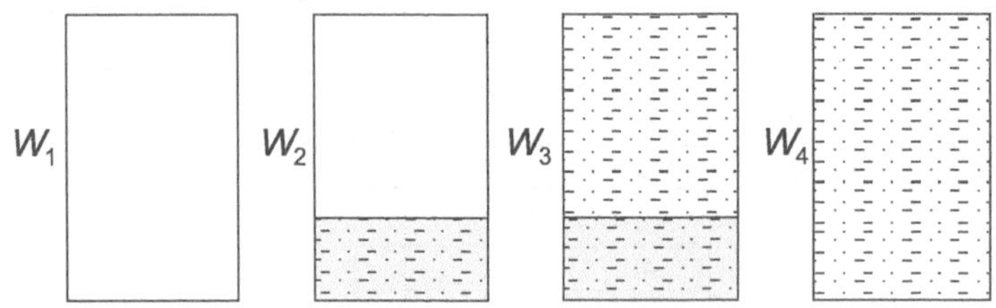

(1) Pycnometer (empty) (2) Pycnometer + Wet soil (3) Pycnometer + Wet soil + Water (4) Pycnometer + Water

$$\text{Water content will be, } w = \frac{W_w}{W_d}$$

and weight of water in wet soil will be, $W_w = (W_2 - W_1) - W_d$

As W_1 and W_2 are known in the pycnometer test, the weight of dry soil (W_d) is determined as follows:

$$G = \frac{\gamma_s}{\gamma_w} = \frac{W_d}{V_s\gamma_w}$$

$$\Rightarrow \quad V_s\gamma_w = \frac{W_d}{G}$$

Now,

$$W_4 = W_3 - W_d + \gamma_w V_s$$

$$\Rightarrow \quad W_4 = W_3 - W_d + \frac{W_d}{G}$$

$$\Rightarrow \quad W_d - \frac{W_d}{G} = W_3 - W_4$$

$$\Rightarrow \quad W_d\left(1 - \frac{1}{G}\right) = W_3 - W_4$$

$$\Rightarrow \quad W_d\left(\frac{G-1}{G}\right) = W_3 - W_4$$

$$\Rightarrow \quad W_d = (W_3 - W_4)\frac{G}{(G-1)}$$

$$\therefore \quad \text{Water content, } w = \frac{W_w}{W_d} = \frac{W_2 - W_1 - W_d}{W_d}$$

$$w = \frac{W_2 - W_1}{W_d} - 1$$

$$w = \frac{(W_2 - W_1)}{(W_3 - W_4)}\frac{(G-1)}{G} - 1$$

Level-2

1.8 Earth is required to be excavated from borrow pits for building an embankment of height 6 m, top width 2 m and side slopes 1 : 1. The unit weight of undisturbed soil in wet condition is 18 kN/m^3 and its natural water content is 8 percent. The dry density required in the embankment is 20 kN/m^3 with a water content of 10%. The specific gravity of soil solids is 2.70. Estimate the quantity of earth required to be excavated in the borrow area to construct one meter length of the embankment. If each truck has a capacity to carry 80 kN per trip, what is the number of truck loads required per meter length of embankment? What are the values of porosity and degree of saturation on the embankment?

[15 Marks]

Sol:

Given data: r_1 = 18 kN/m^3, w_1 = 8%, γ_{d_2} = 20 kN/m^3 w_2 = 10%, G = 2.70, V_1 = ?

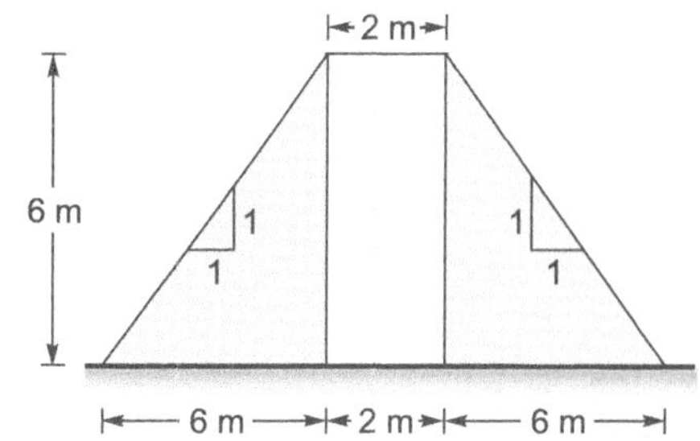

Area of embankment = $\left(\dfrac{2+14}{2}\right)\times 6$ = 48 m^2

Volume of embankment = Area × Length = 48 × 1 = 48 m^3 = V_2

Using subscript 1 and 2 for borrow pit and embankment respectively. All the symbols have their usual meanings

$$\therefore \quad \gamma_{d_2} = \frac{G\gamma_w}{1+e_2}$$

$$\Rightarrow \quad 20 = \frac{2.7\times 9.81}{1+e_2}$$

$$\Rightarrow \quad e_2 = \frac{2.7\times 9.81}{20} - 1 = 0.32$$

Now, porosity of the embankment soil,

$$n_2 = \frac{e_2}{1+e_2} = \frac{0.32}{1+0.32} = 0.2424$$

Degree of saturation of embankment soil can be given as

$$S_2 e_2 = w_2 G$$

$$\Rightarrow \quad S_2 = \frac{w_2 G}{e_2}$$

$$\Rightarrow \quad S_2 = \frac{0.10\times 2.70}{0.32}\times 100$$

$$\Rightarrow \quad S_2 = 84.37\%$$

Now, if V_1 is the volume of soil to be excavated from the borrow pit,

then $$\frac{V_1}{1+e_1} = \frac{V_2}{1+e_2} = V_s$$

But to calculate V_1, we require 'e_1' which can be calculated as

$$\gamma_{d_1} = \frac{G\gamma_w}{1+e_1} \quad \text{... (i)}$$

$$\gamma_{d_1} = \frac{\gamma_1}{1+w_1} \quad \text{... (ii)}$$

From (i) and (ii), we get $\dfrac{G\gamma_w}{1+e_1} = \dfrac{\gamma_1}{1+w_1}$

$$\Rightarrow \quad \frac{2.70\times 9.81}{1+e_1} = \frac{18}{1+0.08}$$

$$\Rightarrow \quad e_1 = \frac{2.70\times 9.81\times 1.08}{18} - 1 = 0.589$$

Now, we have

$$\frac{V_1}{1+e_1} = \frac{V_2}{1+e_2}$$

$$\Rightarrow \quad V_1 = \left(\frac{1+e_1}{1+e_2}\right)\times V_2 = \left(\frac{1+0.589}{1+0.32}\right)\times 48$$

$$= 57.78\ m^3$$

Hence, 57.78 m^3 of earth is to be excavated in the borrow area.

$$\text{Total weight of Borrow pit soil} = V_1 \times \gamma_2 = 57.78 \times 18 = 1036.8\ kN$$

$$\textbf{Number of truck loads required} = \frac{\text{Total weight of embankment soil}}{\text{Capacity of truck per trip}}$$

$$= \frac{1036.8}{80} = 12.96 \approx 13 \text{ number of truck loads, for construction}$$

of per meter length of embankment.

1.9 **Following are the results from the liquid and plastic limit tests for a soil:**

Liquid limit test:

Number of blows (N)	Moisture content (%)
16	42
20	40.8
28	39.1

Plastic limit test : PL : 28.5%

(i) Obtain the liquid limit and the plasticity index of the soil.

(ii) What is the shrinkage limit of this soil and explain the process of its determination.

[20 marks]

Sol:

(i) We know that liquid limit is defined as the water content at 25 nos. of blows.

Interpolating on semilog scale,

Hence liquid limit, $w_L = 40.8 - \frac{40.8 - 39.1}{\log 28 - \log 20} \times (\log 25 - \log 20) = 39.67\%$

Given plastic limit, $w_P = 28.5\%$

Plasticity index, $I_P = w_L - w_P$

$= 39.67 - 28.5 = 11.17\%$

(ii) For the shrinkage limit of the soil, we use the following procedure.

There is another use of the A-line and U-line in the plasticity chart. Casagrande has suggested that the shrinkage limit of a soil can be approximately determined if its plasticity index and liquid limit are known. This can be done in the following manner, with reference to the figure below:

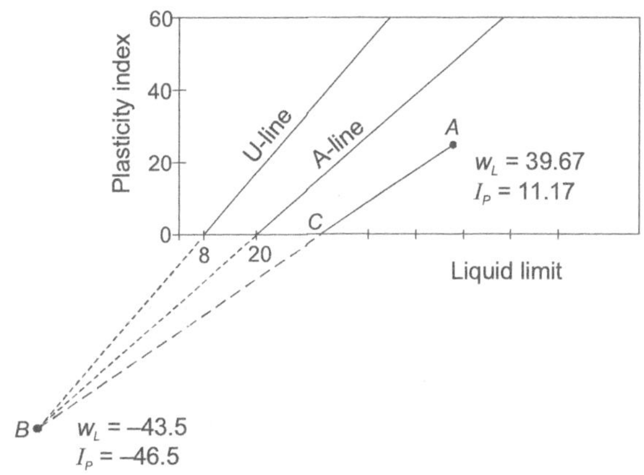

(a) Plot the plasticity index against the liquid limit of a given soil such as point A in figure.

(b) Project the A-line and U-line downward to meet at point B. Point B will have coordinates of $w_L = -43.5$ and $I_P = -46.5$.

(c) Join points B and A with a straight line. This will intersect the liquid limit axis at point *C*. The abscissa of point *C* is the estimated shrinkage limit.

Here,

Coordinates of point *A* = (39.67, 11.17)

Coordinates of point *B* = (–43.5, –46.5)

Equation of straight line *AB* between two points *A* and *B* is given by,

$$(y - y_1) = \left(\frac{y_2 - y_1}{x_2 - x_1}\right)(x - x_1)$$

$$(y - 11.17) = \left(\frac{-46.5 - 11.17}{-43.5 - 39.67}\right) \times (x - 39.67)$$

At point *C*, $y = 0$, $x = w_s$ (shrinkage limit)

On solving equation, $x = 23.56\%$

∴ Shrinkage limit, $w_s = 23.56\%$

1.10 **The following observations were taken during a pipette analysis for the determination of particle size distribution of a soil sample:**

(a) Depth below the water surface at which sample was taken = 120 mm

(b) Capacity of pipette = 10 m*l*

(c) Mass of sample when dried = 0.3 gm

(d) Time of taking sample = 7.5 minutes after start

(e) Volume of soil suspension in the sedimentation tube = 500 m*l*

(f) Dry mass of soil used in making suspension = 25 gm

Determine the coordinate of the point on the particle size distribution curve corresponding to above observations.

Take $G = 2.71$ and $\mu = 10.05$ millipoise, $\rho_w = 1$ gm/m*l*

[10 Marks]

Sol:

$$\text{Settling velocity} = \frac{H_e}{t} = \frac{(\gamma_s - \gamma_w)d^2}{18\mu}$$

H_e = Depth at which sample was taken
t = time of settling

$$\frac{120\times10^{-3}}{7.5\times60} = \frac{(2.71-1)\times9.81\times10^{3}\times d^{2}}{18\times10.05\times10^{-4}}$$

$$d = 1.696\times10^{-5}\text{ m} = 0.01696\text{ mm}$$

$$\text{Percentage finer} = \frac{\text{Concentration of suspension at anytime } t\times100}{\text{Concentration of original suspension}}$$

$$= \left(\frac{0.3\text{ gm}/10\text{ m}l}{25\text{ gm}/500\text{ m}l}\right)\times100 = 0.6\times100$$

$$\text{Percentage finer} = 60\%$$

Co-ordinate of the point on the particle size distribution curve is (0.01696, 60%).

2. Soil Classification

Level-1

2.1 **Classify the soil according to ISC system.**

- **Soil A : Liquid limit, W_L = 38%** **Plastic limit, W_P = 20%**
  Percentage passing 4.75 mm sieve = 78% **Percentage passing 0.075 mm sieve = 56%**
- **Soil B : Liquid limit, W_L = 18%** **Plastic limit, W_P = 12%**
  Percentage passing 4.75 mm sieve = 72% **Percentage passing 0.075 mm sieve = 53.50%**

[10 Marks]

Sol:

Soil A: $w_L = 38\%$
$w_P = 20\%$

Percentage passing 0.075 mm sieve = 56% > 50%

$\Rightarrow$ Soil is silt/clay

$$I_P = 38 - 20 = 18\%$$

As per A-line, $I_P = 0.73\,(w_L - 20)$
$= 0.73\,(38-20) = 13.14\% < 18\%$

Soil lies above A-line in plasticity chart

$\Rightarrow$ Soil A is clay

W_L = 38% lies between range 35% to 50%

Soil A can be classified as CI i.e. clay of intermediate compressibility.

Soil B:

Given, percentage 0.075 mm sieve > 50%

$\Rightarrow$ Soil silt/clay

$$w_L = 18\%,\ w_P = 12\%$$

Plasticity index, $I_P = 18 - 12 = 6\%$

As per A-line, $I_P = 0.73\,(w_L - 20)$
$I_P = 0.73\,(18-20) < 0$

Soil B lies above A-line, i.e. soil B is clay

$4 < I_P < 7$

⇒ Dual symbol must be given to the soil B.

Also, $w_L = 18\% < 35\%$

⇒ Soil B has low compressibility.

Best possible symbol given to soil B is CL-ML.

2.2 **Laboratory sieve analysis was carried out on a soil sample using a complete set of standard IS sieves. Out of 750 gm of soil used in the test, 300 gm was retained on IS 600 μ sieve, 375 gm was retained on IS 500 μ sieve and the remaining 75 gm was retained on IS 425 μ sieve. Find the coefficient of uniformity of the soil and also classify the soil.**

[8 Marks]

Sol:

Total quantity of soil = 750 gm

Soil retained on IS 600 μ sieve = 300 gm

$$\% \text{ retained on IS 600 } \mu \text{ sieve} = \left(\frac{300}{750}\right) \times 100 = 40\%$$

Soil passing through or finer than 600 μ = 100 – 40 = 60%

Hence, $D_{60} = 600\ \mu$

Soil retained on IS 500 μ sieve = 375 gm

$$\% \text{ retained on IS 500 } \mu \text{ sieve} = \left(\frac{375}{750} \times 100\right) = 50\%$$

% finer than IS 500 μ sieve = 100 – (50 + 40) = 10

Hence, $D_{10} = 500\ \mu$

$$\text{Coefficient of uniformly, } c_u = \left(\frac{D_{60}}{D_{10}}\right) = \left(\frac{600}{500}\right) = 1.20$$

Soil classification : Soil passing through 75 μ sieve = 0%.

Less than 50% is passing through 75 μ sieve so soil is coarse grained soil (either sand or gravel).

More than 50% of coarser fraction is passing through 4.75 mm sieve ⇒ sand.

$C_u < 6 \Rightarrow$ Poorly graded

Hence, soil is classified as SP.

2.3 **Draw plasticity chart and explain its use.**

[10 Marks]

Sol:

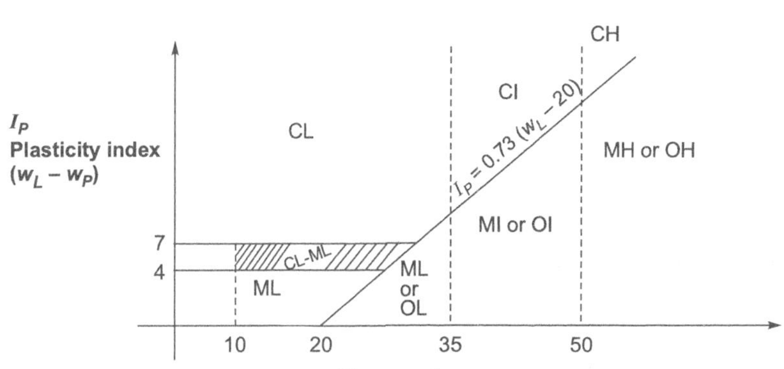

Plasticity Chart

Equations of A-line and U-line for plasticity chart:

A - line → $I_P = 0.73\,(w_L - 20)$

U - line → $I_P = 0.9\,(w_L - 8)$

- Plasticity chart is used to classify fine grained soils.
  L → Low compressible/Low plastic → $w_L < 35\%$
  I → Medium compressible/Medium plastic → $w_L = 35$ to 50%
  H → High compressible/High plastic → $w_L > 50\%$
  C → clay M → silt O → Organic soil.
- Plasticity index of soil is calculated and let it be $(I_P)_{soil}$.
- $(I_P)_{soil} > (I_P)_{A\text{-line}}$, then soil is clay else it is silt or organic soil.
- U-line represents upper boundary above which no result should lie. If results are found to be above U-line then test should be repeated.

2.4 Discuss the I.S. classification based on grain size and classify the given soil.
Percentage passing 4.75 mm IS sieve = 80%
Percentage passing 0.075 mm IS sieve = 10%
Uniformity coefficient, $c_u = 8$
Coefficient of curvature, $c_c = 2.5$
Plasticity index, $I_P = 3$

[10 Marks]

Sol:

I.S. classification based on grain size : This classification is based on particle size, coefficient of curvature and coefficient of uniformity, fineness modulus, plasticity index and liquid limit. It was given by A casagrande.

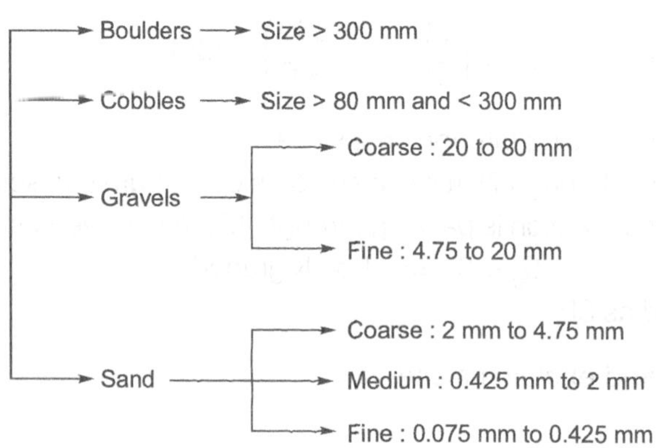

For given soil, more than 50% of particles passes through 4.75 mm sieve and only 10% passes through 0.075 mm sieve.

⇒ Soil is essentially sand.

$$c_c = 2.5\ (1 < c_c < 3)$$
$$c_c = 8\ (c_u > 6)$$

⇒ Soil must be well graded sand i.e. SW.

But percentage fines (10%) is between 5% and 12%.

Thus dual symbol must be adopted.

$$I_P = 3\ \text{(Given)} < 4$$

It satisfies the requirement of SM group.

Hence, soil is classified as SW – SM.

2.5 50 g of minus 75 micron, oven dried soil was used in a hydrometer analysis. The corrected hydrometer reading after 2 min in a 1000 cc soil suspension was 25.0. The effective depth, H_e for R_H of 25 is 12.13 cm. Taking specific gravity of solids as 2.75 and viscosity of water as 0.01 poise, calculate the coordinates of the point on the grain-size curve.

[6 marks]

Sol:

$R_c = 25;\ \mu = \frac{0.01}{981}$ gs/cm^2; $G_s = 2.75;\ H_e = 12.13$ cm, $t = 2$ min,

$\gamma_s = G_s\gamma_w = 2.75$ g/cc, $W_s = 50$ g

Substituting the values,
$$D = \sqrt{\frac{18\mu}{(\gamma_s - \gamma_w)} \times \frac{H_e}{60t}} = \sqrt{\frac{18 \times 0.01}{(2.75-1) \times 980.7} \times \frac{12.13}{60 \times 2}}$$

$$= 0.0033 \text{ cm or } 0.033 \text{ mm}$$

The percentage of particles finer than 0.033 mm can be calculated as

$$N = \left(\frac{G_s}{G_s - 1}\right)\frac{R_c}{W_s} \times 100 = \left(\frac{2.75}{2.75-1}\right)\frac{25}{50} \times 100 = 78.6\%$$

Level-2

2.6 Five different particle sizes are mixed in proportions shown below and water is added to make the volume of soil suspension exactly equal to 1000 cc.

Particle size, mm	Weight, *g*
0.060	5
0.020	15
0.010	20
0.005	4
0.001	6

The particles have a specific gravity of 2.65 and the temperature of the soil suspension was 25°C. γ_w can be taken as 1.0 g/cc and dynamic viscosity of water as 8.95 millipoise at 25°C.
The soil suspension was thoroughly shaken and sedimentation allowed.

(a) What is the largest particle size present at a depth of 10 cm after 8 minutes of the start of sedimentation?

(b) What is the specific gravity of the soil suspension at a depth of 10 cm after 8 minutes from the start of sedimentation?

(c) How long after the start of sedimentation will all particles settle down below a depth of 10 cm?

[20 marks]

Sol:

(a) Given at 25°C, $\mu = 8.95$ millipoise

$$= \frac{8.95 \times 10^{-3}}{10} \text{ Ns/m}^2 = \frac{8.95 \times 10^{-3}}{9.81 \times 10} \times 10^3 \times \frac{1}{10^4}$$

$$= 9.123 \times 10^{-6} \text{ g-s/cm}^2$$

$$\gamma_s = 2.65 \text{ g/cc}$$

$$\gamma_w = 1.0 \text{ g/cc}$$

$$H_e = 10 \text{ cm}$$

$$t = 8 \text{ min.}$$

From Stoke's law,

$$D = \sqrt{\frac{18\mu}{\gamma_s - \gamma_w} \cdot \frac{H_e}{60t}}$$

$$= \sqrt{\frac{18}{(2.65-1)} \times 9.123 \times 10^{-6} \times \frac{10}{60 \times 8}}$$

$$= 1.44 \times 10^{-3} \text{ cm}$$

$$= 0.0144 \text{ mm}$$

This is the largest size of the particle that can be present, theoretically, at a depth of 10 cm after 8 min. of sedimentation. However, as per the sizes of particles actually present in the soil suspension, the largest size of particle will be 0.01 mm.

(b) Since particles of equal size have equal velocities of fall, the concentration of soil suspension at 10 cm depth after 8 minutes will be of all those particles equal to or smaller in size than 0.01 mm.

Weight of soil particles present at 10 cm depth

$$= 20 + 4 + 6 = 30 \text{ g}$$

$$\text{Volume of soil particles} = \frac{30}{2.65 \times 1} = 11.3 \text{ cc}$$

$$\text{Volume of water} = 1000 - 11.3 = 988.7 \text{ cc}$$

$$\text{Weight of water} = 988.7 \times 1 = 988.7 \text{ g}$$

$$\text{Weight of soil suspension} = 988.7 + 30 = 1018.7 \text{ g}$$

$$\text{Unit wt. of soil suspension} = \frac{1018.7}{1000} = 1.0187 \simeq 1.019 \text{ g/cc}$$

Hence, specific gravity of the soil suspension = 1.019

(c)

$$v = \frac{\gamma_s - \gamma_w}{18\mu} D^2$$

$$\Rightarrow \quad \frac{H_e}{t} = \frac{\gamma_s - \gamma_w}{18\mu} D^2$$

For all particles to settle below 10 cm depth, D would have to be the smallest particle size present; that is, 0.001 mm or 0.0001 cm

$$\therefore \quad \frac{10}{t} = \frac{2.65 - 1.0}{18 \times 9.123 \times 10^{-6}} \times 0.0001^2$$

$$\Rightarrow \quad t = 99523.63 \text{ sec}$$

$$\Rightarrow \quad t = 27 \text{ hr } 38 \text{ min } 43.64 \text{ sec}$$

$$\simeq 27 \text{ hr } 39 \text{ min}$$

3. Soil Compaction

Level-1

3.1 In order to determine the relative density of a sand sample, its natural moisture content and bulk density were determined in the field and were found to be 7% and 1.61 gm/cc respectively. Sample of this soil were then compacted in a proctors mould of $\frac{1}{30}$ Cu.ft. Capacity, at the loosest and the densest states.

The following data were obtained

Weight of empty mould = 2100 gm

Weight of mould + soil in the loosest state = 3365.2 gm

Weight of mould + soil in the densest state = 3860.5 gm

Moisture content of the sample used in tests = 10.80%

Determine the relative density of sand.

[15 Marks]

Sol:

$$\text{Volume of the mould} = \frac{1}{30}\,Cu.ft. = \frac{1}{30}\times(12\times 2.54)^3 = 943.895 \text{ cc}$$

In the loosest state:

$$\text{Bulk density} = \frac{3365.2-2100}{943.895} = 1.34 \text{ gm/cc}$$

$$\text{Minimum dry density, } (\gamma_d)_{min} = \frac{1.34}{1+0.1080} = 1.209 \text{ gm/cc}$$

In the densest state,

$$\text{Bulk density} = \left(\frac{3860.5-2100}{943.895}\right) = 1.865 \text{ gm/cc}$$

$$\text{Max dry density of the soil } (\gamma_d)_{max} = \left(\frac{1.865}{1+\frac{10.80}{100}}\right) = 1.683 \text{ gm/cc}$$

$$\text{In-situ density of the soil } (\gamma) = 1.61 \text{ gm/cc}$$

$$\text{Water content} = 7\%$$

$$\text{In-situ dry density} = \left(\frac{1.61}{1+\frac{7}{100}}\right) = 1.505 \text{ gm/cc}$$

$$\text{Relative density, } R_D = \left(\frac{\frac{1}{\gamma_{min}}-\frac{1}{\gamma}}{\frac{1}{\gamma_{min}}-\frac{1}{\gamma_{max}}}\right)\times 100 = \left(\frac{\frac{1}{1.209}-\frac{1}{1.505}}{\frac{1}{1.209}-\frac{1}{1.683}}\right)\times 100 = 69.83\%$$

3.2 A soil after being compacted was found to have maximum dry density of 1.75 gm/cc. The value of optimum moisture content was 15% and percentage air voids at maximum dry density was found to be 7.5%.

(i) Calculate the theoretical maximum dry density at OMC of 15%. Later, the compactive effort on the soil was increased and the new value of optimum moisture content was found to be 12.5%.

(ii) Calculate the theoretical maximum dry density at OMC of 12.5%, assuming percentage air voids at maximum dry density to be 7.5%.

(iii) Find out the value of maximum dry density observed at OMC of 12.50%.

[10 Marks]

Sol:

$$\gamma_d = 1.75 \text{ gm/cc}$$
$$n_a = 7.5\%$$
$$\gamma_d = \frac{(1-n_a)G\gamma_w}{1+G_w}$$
$$1.75 = \frac{(1-0.075)(G\times 1)}{1+G\times 0.15}$$
$$G = 2.64$$

(i) $$\text{Maximum } (\gamma_d)_{\text{Theoretical}} = \left(\frac{G}{1+Gw}\right)\gamma_w = \left(\frac{2.64}{1+2.64\times 0.15}\right)\times 1$$
$$= 1.891 \text{ gm/cc}$$

(ii) $$w_2 = 12.50\%$$
$$\text{Maximum } (\gamma_d)_{\text{Theoretical}} = \left(\frac{2.64}{1+2.64\times 0.1250}\right)\times 1 = 1.985 \text{ gm/cc}$$

(iii) $$n_a = 7.5\%$$
$$\gamma_d = \frac{(1-0.075)\times 2.64\times 1}{1+2.64\times 0.1250} = 1.836 \text{ gm/cc}$$

3.3 (i) Explain the term "optimum moisture content". How is it affected by compacting effort?

(ii) State the factors affecting field compaction of soil.

[7 + 5 Marks]

Sol:

(i) Moisture plays an important role in compaction. In fact, when moisture (water) is added to a dry soil, the soil grains get surrounded by a film of absorbed water. If more moisture is added, then this film becomes thicker and soil grains surrounded by such films can slide over each other more easily. The grains thus become capable to occupy closer spacing more easily on application of external loading. The moisture thus acts only as a lubricant in the compaction process and a closer packing of soil grains is achieved on rolling by the expulsion of air from the voids.

If more water is continued to be added, then a stage comes when moisture becomes excess and starts occupying the space which otherwise could have been occupied by soil grains under compaction. Evidently, moisture then starts hindering the closer packing of grains. In the limiting stage, there occurs a particular amount of water which causes maximum lubrication without becoming excess to cause hinderance in compaction by occupying space for its own molecules. This limiting moisture which is most useful content, the soil would be compacted maximum possible and would have the maximum density or unit weight i.e.. the weight of soil grains in a unit volume of compacted soil mass would be maximum.

It may however be noted that maximum dry unit weight so obtained is only for a given compactive effort and method of compaction. It is also not necessarily the maximum dry unit weight that can be obtained in the field.

For a given type of compactive effort, the higher the maximum dry unit weight and lower the OMC. However, as the moulding water content increases, the influence of compactive effort on dry unit weight tends to diminish. Again the maximum dry unit weight does not go on increasing as the compactive effort is increased. The margin of increase becomes smaller and smaller even on the dry side of OMC, while on the wet side of OMC, there is hardly any increase at all. If the peaks of compaction curves for different compactive efforts are joined together, a line of optimums is obtained which is nearly parallel to the zero air void curve.

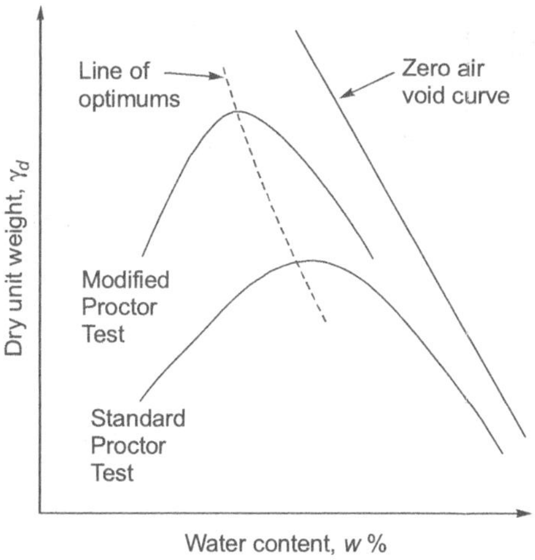

This brings out the fact that even a higher compactive effort does not result in a higher efficiency of compaction. The degree of saturation remains almost the same in all cases.

(ii) **The factors affecting field compaction of soil are:**

(a) Water content
(b) Compactive effort
(c) Type of soil
(d) Method of compaction

3.4 **List out the any 5 equipment used for compaction in the field. Also indicate their suitably for soil type and nature of project in which equipments are used.**

[8 Marks]

Sol:

Type of equipment	Suitability for soil type	Nature of project
1. Rammers or tampers	All soils	In confined area such as fills behind retaining walls, basement walls etc.
2. Smooth wheeled rollers	Crushed rocks, gravels, sands	Road construction etc.
3. Pneumatic tyred rollers	Sands, gravels, silts, clayey soils, not suitable for uniformly graded soil	Base, sub base and embankment compaction for highways, airfields, etc.
4. Sheeps foot rollers	Clayey soil	Core of earth dams
5. Vibratory rollers	Sands	Embankments for oil storage tanks etc.

Level-2

3.5 **In the construction of a levee, the compaction specification (dry density) required was 95% of the Proctor MDD at a field moisture content within 2% of the OMC. The MDD and OMC obtained in the laboratory from the Proctor Test were 1.94 gm/cm^3 and 13.5%, respectively. A field supervisor conducted sand-cone tests at two locations and obtained the results as presented below. The sand in the sand bottle was found to have a density of 1.87 gm/cm^3. Check whether the specification was satisfied.**

Location No.	Mass of Soil Removed (g)		Mass of Sand Used (g)
	Wet	Dry	
1.	42.86	38.46	39.61
2.	37.18	32.12	32.19

[12 marks]

Sol:

Given that MDD (ρ_{dmax}) = 1.94 g/cm^3. To achieve 95% of the Proctor MDD in the field, minimum dry density required to be achieved is

$$\rho_{dmin} = 0.95 \times 1.94 = 1.843 \text{ g/cm}^3$$

Now OMC = 13.5%. To achieve field moisture content within 2% of OMC, minimum permissible field moisture content is

$$w_{nmin} = 13.5 - 2 = 11.5\%$$

Maximum permissible field moisture content

$$w_{nmax} = 13.5 + 2 = 15.5\%$$

Location no. 1: Mass of wet soil (M_1) = 42.86 g; mass of dry soil (M_2) = 38.46 g. Field moisture content is

$$w_{n1} = \frac{M_1 - M_2}{M_2} \times 100 = \frac{42.86 - 38.46}{38.46} \times 100 = 11.44\%$$

This is less than w_{nmin}. Hence the specification was not satisfied. Volume of the wet soil will be

$$V_1 = \frac{\text{Mass of sand used}}{\text{Density of sand}} = \frac{39.61}{1.87} = 21.182 \text{ cm}^3$$

Field dry density of soil will be,

$$\rho_{d1} = \frac{M_2}{V_1} = \frac{38.46}{21.182} = 1.816 \text{ g/cc}$$

This is less than ρ_{dmin}. Hence compaction specification was not satisfied.

Location no.2: Mass of wet soil (M_3) = 37.18 g, mass of dry soil (M_4) = 32.12 g. Field moisture content

$$w_{n2} = \frac{M_3 - M_4}{M_4} \times 100$$

$$= \frac{37.18 - 32.12}{32.12} \times 100 = 15.75$$

This is more than w_{nmax}; Hence the compaction specification was not satisfied. Volume of wet soil will be

$$V_2 = \frac{\text{Mass of sand used}}{\text{Density of sand}} = \frac{32.19}{1.87} = 17.214 \text{ cm}^3$$

Field dry density will be $$\rho_{d2} = \frac{M_4}{V_2} = \frac{32.12}{17.214} = 1.866 \text{ g/cm}^3$$

This is more than ρ_{dmin} = 1.843 g/cm^3.
Hence the compaction specification is satisfied.

3.6 The following results were obtained from a standard compaction test on a sample of soil:

Water content (in fraction)	0.12	0.14	0.16	0.18	0.20	0.22
Mass of wet soil (kg)	1.68	1.85	1.91	1.87	1.87	1.85

The volume of the mould used was 950 ml. Make necessary calculations and plot the compaction curve and obtain the maximum dry density and the optimum moisture content. Also calculate the void ratio, the degree of saturation and the theoretical maximum dry density (*G* = 2.70)

[20 marks]

Sol:

Calculations are shown in tabular form :

Water Content (w)	0.12	0.14	0.16	0.18	0.20	0.22
Mass of wet soil (M) (kg)	1.68	1.85	1.91	1.87	1.87	1.85
Bulk density $\rho = \frac{M}{V} = \frac{M}{0.950}$ (g/ml)	1.77	1.95	2.01	1.97	1.97	1.95
Dry density $\rho_d = \frac{\rho}{(1+w)}$ (g/ml)	1.58	1.71	1.73	1.67	1.64	1.60
Void ratio $e = \frac{G\rho_w}{\rho_d} - 1$	0.71	0.58	0.56	0.62	0.65	0.69
Degree of saturation $S = \frac{wG}{e}$	0.46	0.65	0.77	0.78	0.83	0.86
Theoretical maximum dry density $(\rho_d) = \frac{G\rho_w}{1+Gw}$ (g/ml)	2.04	1.96	1.89	1.82	1.75	1.69

Figure below shows the compaction curve.

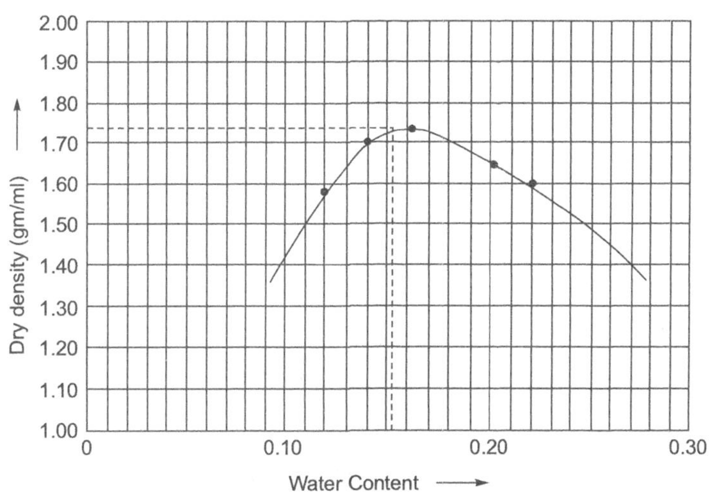

From the plot, $(\rho_d)_{max} = 1.74$ gm/ml at O.M.C. = 15.2%

4. Capillarity, Permeability, Effective Stress and Seepage Analysis

Level-1

4.1 A pumping test was carried out at a site where 10 m thick clay stratum overlies a fine sand stratum 2 m thick. Below the sand stratum lies on impermeable rock stratum. After a steady state was established. The pumping rate from a well-boring was measured at 15000 cc/s. The average water level in the two observation wells made at 5 m and 15 m radius were respectively 5 m and 4.2 m below ground surface. Determine

(i) The coefficient of permeability of sand.

(ii) The effective pressure during pumping at the bottom of the sand stratum at 5 m radius, if the total stress at the bottom of clay was 150 kN/m^2.

Assume γ_{sat} for sand = 20 kN/m^3.

[12 Marks]

Sol:

(i)

For confined aquifer,
$$a = \frac{2\pi kD(h_2 - h_1)}{\ln\left(\frac{r_2}{r_1}\right)}$$

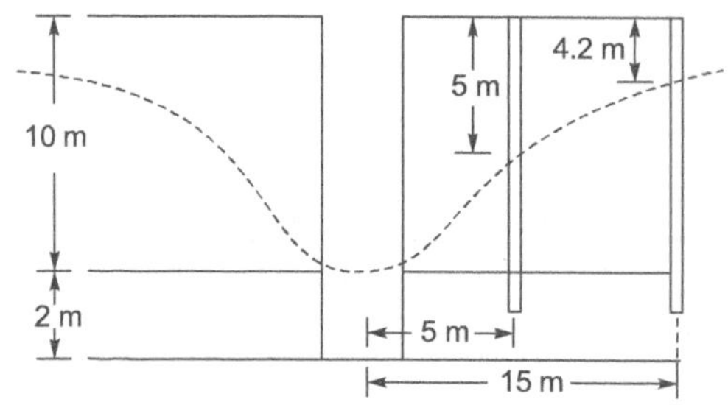

$$h_1 = 10 + 2 - 5 = 7 \text{ m}$$
$$h_2 = 10 + 2 - 4.2 = 7.8 \text{ m}$$
$$r_2 = 15 \text{ m}$$
$$r_1 = 5 \text{ m}$$
$$a = 15000 \times 10^{-6} \text{ m}^3\text{/s}$$
$$15000 \times 10^{-6} = \frac{(2\pi k)(2)(7.8 - 7)}{\ln\left(\frac{15}{5}\right)}$$
$$k = 1.639 \times 10^{-3} \text{ m/s}$$

(ii)

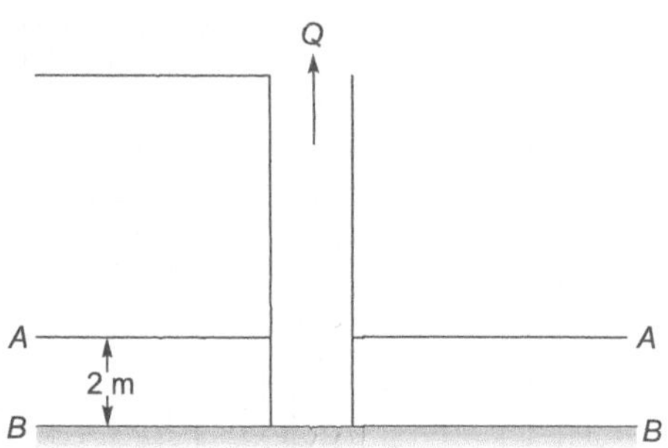

$$\sigma_{A\text{-}A} = 150 \text{ kN/m}^2$$
$$\sigma_{B\text{-}B} = 150 + 2 \times 20 = 190 \text{ kN/m}^2$$
$$\sigma'_{B\text{-}B} = 190 - 5 \times 9.81 = 140.95 \text{ kN/m}^2$$

4.2 In a constant head permeameter test, the following observations were taken:
Distance between piezometric tappings = 15 cm
Difference of water level in piezometers = 40 cm
Diameter of test sample = 5 cm
Quantity of water collected = 500 ml
Duration of test = 900 sec
Find out the coefficient of permeability of the soil. If dry mass of the 15 cm long sample is 495 gm and the specific gravity of the solids is 2.65. Calculate the seepage velocity of water during test.

[12 Marks]

Sol:

We know that,
$$k = \left(\frac{QL}{Ah}\right)$$

where, Q = Discharge flowing, L = Length of sample, A = Area of cross-section of sample, h = Head causing flow

$$Q = \frac{500 \text{ m}l}{900 \text{ sec}} = 0.556 \frac{\text{m}l}{\text{sec}}$$

L = 15 cm, h = 40 cm

$$A = \left(\frac{\pi}{4} \times 5^2\right) = 19.635 \text{ cm}^2$$

$$k = \left(\frac{0.556 \times 15}{40 \times 19.635}\right) = 10.62 \times 10^{-3}\,\text{cm/sec}$$

$$V = ki$$

where, V = Discharge velocity, i = Hydraulic gradient = $\left(\frac{h_L}{L}\right)$, $V = 10.62 \times 10^{-3} \times \frac{40}{15} = 0.02832$ cm/sec

$$\text{Dry mass of sample} = 495 \text{ gm}$$

$$\text{Volume of sample} = A \times L = 19.635 \times 15 = 294.525 \text{ cm}^3$$

$$\text{Dry density } (\gamma_d) = \frac{G\gamma_w}{1+e}$$

$$\frac{495}{294.525} = \frac{2.65 \times 1}{1+e}$$

$$\mathbf{e = 0.577}$$

$$n = \left(\frac{e}{1+e}\right) = \left(\frac{0.577}{1+0.577}\right) = 0.366$$

$$\text{Seepage velocity} = \left(\frac{V}{n}\right) = \left(\frac{0.02832}{0.366}\right) = 0.077 \text{ cm/sec}$$

4.3 **For sub-soil conditions shown in the figure draw total, effective and neutral stress diagram upto a depth of 8 m. Neglect capillary rise. Take γ_w = 10 kN/m^3.**

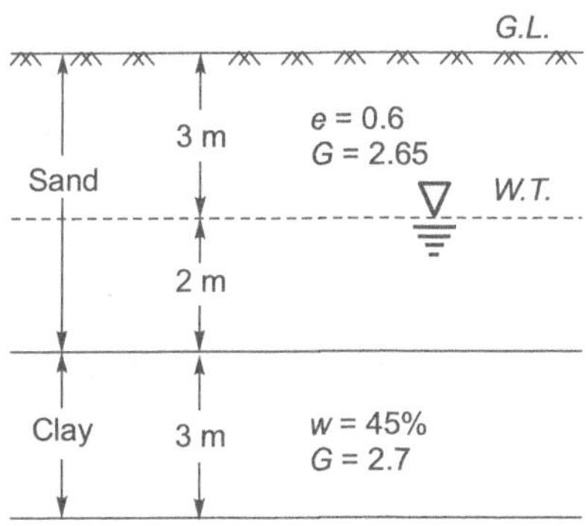

[12 Marks]

Sol:

Assuming sand above W.T. to be dry,

$$\therefore \quad \gamma_d = \left(\frac{G}{1+e}\right)\gamma_w = \frac{2.65}{1.6} \times 10 = 16.5625 \text{ kN/m}^3$$

$$\gamma_{sat} = \left(\frac{G+e}{1+e}\right)\gamma_w = \frac{(2.65+0.6)}{1.6} \times 10 = 20.3125 \text{ kN/m}^3$$

For clay,

$$\gamma_{sat} = \left(\frac{G+e}{1+e}\right)\gamma_w$$

$$S.e = w \cdot G$$

$$\therefore \quad e = \frac{0.45 \times 2.7}{1} = 1.215$$

$$\therefore \quad \gamma_{sat} = \frac{2.7 + 1.215}{2.215} \times 10 = 17.675 \text{ kN/m}^3$$

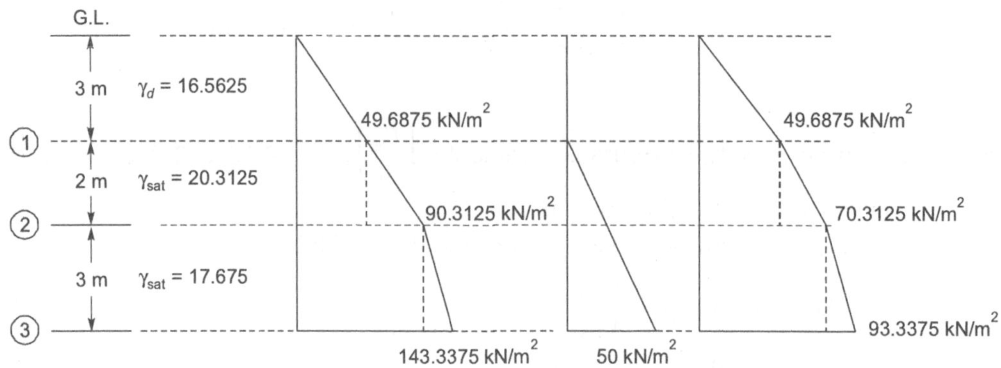

(i) (Total stress) (ii) (Neutral stress) (iii) (Effective stress)

At (1) – (1):

Total stress, $\sigma_{(1)-(1)} = \gamma_d \times 3 = 49.6875 \text{ kN/m}^2$

Neutral stress, $u_{(1)-(1)} = 0$

Effective stress, $\bar{\sigma}_{(1)-(1)} = \sigma_{(1)-(1)} - u_{(1)-(1)} = 49.6875 \text{ kN/m}^2$

At (2) – (2):

Total stress,
$$\sigma_{(2)-(2)} = 49.6875 + \gamma_{sat} \times 2$$
$$= 49.6875 + 20.3125 \times 2 = 90.3125 \text{ kN/m}^2$$

Neutral stress, $u_{(2)-(2)} = \gamma_w \times 2 = 20 \text{ kN/m}^2$

Effective stress, $\bar{\sigma}_{(2)-(2)} = \sigma_{(2)-(2)} - u_{(2)-(2)} = 90.3125 - 20 = 70.3125 \text{ kN/m}^2$

At (3) – (3):

Total stress,
$$\sigma_{(3)-(3)} = 90.3125 + \gamma_{sat} \times 3$$
$$= 90.3125 + 17.675 \times 3 = 143.3375 \text{ kN/m}^2$$

Neutral stress, $u_{(3)-(3)} = \gamma_{sat} \times (2 + 3) = 50 \text{ kN/m}^2$

Effective stress, $\bar{\sigma}_{(3)-(3)} = \sigma_{(3)-(3)} - u_{(3)-(3)} = 143.3375 - 50 = 93.3375 \text{ kN/m}^2$

4.4 A foundation trench is to be excavated in a stratum of stiff clay, 10 m thick, underlain by a bed of coarse sand. In a trial borehole the ground water was observed to rise on elevation of 6.5 m. However, there is a fluctuation in water table of ±0.5 during the various season of the year. Determine the depth upto which an excavation can be safely carried out without the danger of the bottom becoming unstable under the aetesian pressure in the sand stratum. The specific gravity of the clay particles is 2.65 and void ratio is 0.75.

[10 Marks]

Sol:

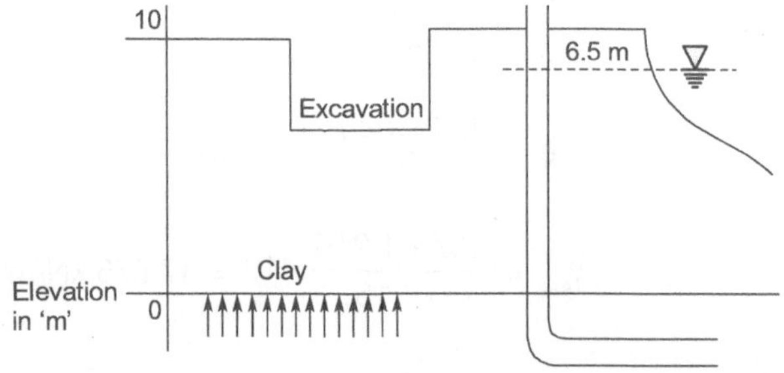

$$\gamma_{sat} = \left(\frac{G+e}{1+e}\right)\gamma_w = \left(\frac{2.65+0.75}{1+0.75}\right)\times 9.81 = 19.06 \text{ kN/m}^3$$

For safe excavation at the site, $\bar{\sigma} > 0$

Hence, the head causing flow or pore water pressure in the soil should be small.

Therefore, the critical condition is the one in which water table rises by 0.5 m above the level.

Let the safe depth of excavation = x m

Thickness of clay overlying sand stratum = $(10 - x)$ m

At elevation 0 in clay stratum

$$\bar{\sigma} = (10-x)\gamma_{sat} - (6.5-0.5)\gamma_w$$

For quick sand condition, $\bar{\sigma} = 0$

$$(10-x)\gamma_{sat} = (7)\gamma_w$$

$$(10-x)\times 19.06 = 7\times 9.81$$

$$x = \mathbf{6.40\ m}$$

Hence, the safe depth of excavation is 6.40 m.

4.5 **Calculate the seepage through an earth dam resting on an impervious foundation. The relevant data are given below:**

Height of the dam = 60.0 m | **Upstream slope = 2.75 : 1 (H : V)**
Downstream slope = 2.50 : 1 (H : V) | **Free board = 2.5 m**
Crest width = 8.0 m | **Length of drainage blanket = 120.0 m**
Coefficient of permeability of the embankment material in
X-direction = 8×10^{-7} m/s | **Y-direction = 2×10^{-7} m/s**

[10 Marks]

Sol:

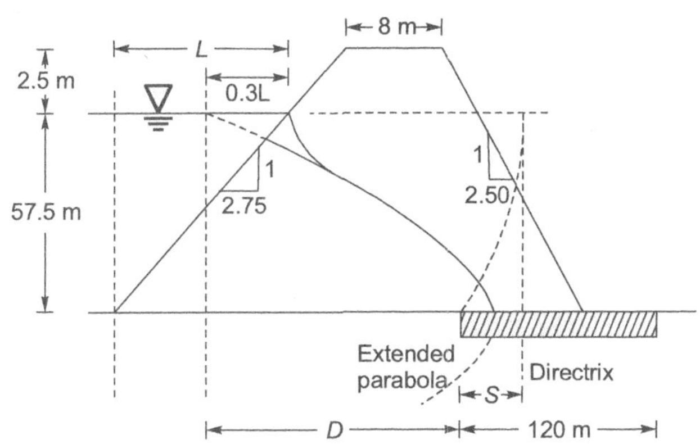

$L = 57.5 \times 2.75 = 158.125$ m and $H = 60 - 2.5 = 57.5$ m

From the above diagram, we have

$$D = (0.3\times 158.125) + (2.75\times 2.5) + 8 + (2.5\times 60 - 120) = 92.3125 \text{ m}$$

$$\because \quad X = X_T\sqrt{\frac{k_x}{k_y}} \text{ therefore } D_T = D\sqrt{\frac{k_y}{k_x}}$$

$$\Rightarrow \quad D_T = D\sqrt{\frac{k_y}{k_x}} = 92.3125\sqrt{\frac{2\times 10^{-7}}{8\times 10^{-7}}} = 46.1562 \text{ m}$$

$$S = \sqrt{D_T^2 + H^2} - D_T = \sqrt{(46.1562)^2 + 57.5^2} - 46.1562 = 27.5774 \text{m}$$

Now $$k_{eq} = \sqrt{k_x k_y} = \sqrt{8\times10^{-7}\times2\times10^{-7}} = 4\times10^{-7}\text{ m/s}$$

∴ Seepage discharge per unit length of dam,

$$q = k_{eq}S = 4\times10^{-7}\times27.5774 = 1.1031\times10^{-5}\text{ m}^3\text{/sec/m}$$

4.6 The subsoil at a site consists of a 15 m thick homogenous layer of dense sand having the following properties. γ_d = 1.62 gm/cc, G = 2.68 and ϕ = 35°.
The natural ground water table lies at 5 m below the ground surface.

(i) Determine the shear strength of soil along a Horizontal plane through the middle of sand layer.

(ii) If during monsoon the water table rises to the ground level, how will be shear strength along the same plane change?

Assume that the soil is dry above water table.

[12 Marks]

Sol:

Given, γ_d = 1.62 gm/cc

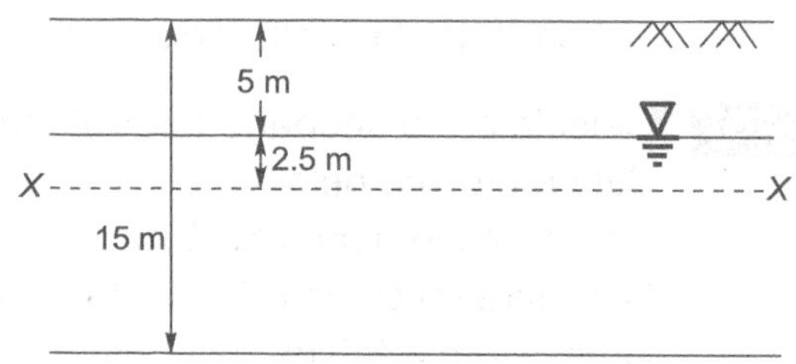

$$\gamma_d = \left(\frac{G\gamma_w}{1+e}\right)$$

$$e = \left(\frac{2.68\times1}{1.62}-1\right)$$

$$e = 0.654$$

$$\gamma_{sat} = \left(\frac{G+e}{1+e}\right)\gamma_w = \left(\frac{2.68+0.654}{1+0.654}\right)\times1$$

$$\gamma_{sat} = 2.02\ gm/cc$$

$$\gamma_{sub} = \gamma_{sat}-\gamma_w = 2.02-1 = 1.02\text{ gm/cc} = 1.02\times9.81 = 10.006\text{ kN/m}^3$$

The Horizontal plane X-X, under consideration is at a depth of 7.5 m below the G.L.

(i) The normal effective stress on given plane ($\sigma_{xx} = \gamma_d\times5+\gamma_{sub}\times2.5$)

$$\sigma_{xx} = (1.62\times9.81\times5+10.006\times2.5) = 104.476\text{ kN/m}^2$$

Shear strength of soil at this plane

$$\tau_{xx} = c+\bar{\sigma}\tan\phi = 0+104.476\times\tan35° = 73.15\text{ kN/m}^2$$

(ii) In the 2nd case, the entire soil is submerged

$$\sigma_{xx} = \gamma_{sub}\,z$$
$$= 10.006\times7.5 = 75.045\text{ kN/m}^2$$
$$\tau_{xx} = 0+75.045\times\tan35° = 52.55\text{ kN/m}^2$$

4.7 An excavation is made in a soil whose porosity is 30% and the specific gravity of soil solids = 2.6. A 1.5 m thick layer of this soil is subjected to an upward seepage head of 2 m. Find out what factor of safety exists against the piping (boiling). If a factor of safety of 2 is needed against the piping, what depth of gravel is required to be placed above the soil layer? Assume the unit weight of gravel to be the same as that of the soil and negligible loss of head in the gravel layer.

[10 marks]

Sol:

$$e = \frac{n}{1-n} = \frac{0.3}{1-0.3} = 0.428$$

$$i_c = \frac{G-1}{1+e} = \frac{2.6-1}{1+0.428} = 1.12$$

$$i = \frac{2m}{1.5m} = 1.33$$

$$F.O.S = \frac{i_c}{i} = \frac{1.12}{1.33} = 0.84$$

Now, for F.O.S = 2

$$i = \frac{i_c}{F.O.S} = \frac{1.12}{2} = 0.56$$

for $i = \frac{h}{L}$; $L = \frac{2}{0.56} = 3.57\text{m}$

Gravel layer required = 3.57 – 1.5 = 2.07 m

4.8 A capillary permeability test was conducted in two stages under a head of 60 cm and 180 cm, respectively at the entry end. In the first stage, the wetted surface moved from 1.5 cm to 7 cm in 7 minutes. In the second stage, it advanced from 7 cm to 18.5 cm in 24 minutes. The degree of saturation at the end of the test was 85% and the porosity was 35%. Determine the capillary head and the coefficient of permeability of soil.

[12 marks]

Sol:

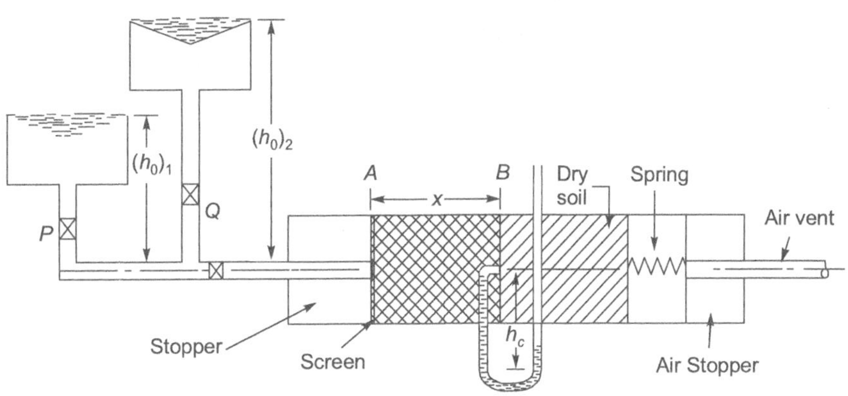

$h_{0(1)}$ = 60 cm, $h_{0(2)}$ = 180 cm, h_c = height of capillary rise

For first stage,

$$\frac{x_2^2 - x_1^2}{t_2 - t_1} = \frac{2k}{S.n}(h_{0_1} + h_c)$$

$$\Rightarrow \quad \frac{(7^2 - 1.5^2)}{7} = \frac{2k}{S \times 0.35}(60 + h_c)$$

$$\Rightarrow \quad 1.169 = \frac{k}{S}(60 + h_c) \quad \text{...(i)}$$

For second stage,

$$\frac{x_2^2 - x_1^2}{t_2 - t_1} = \frac{2k}{S.n}(h_{0_2} + h_c)$$

$$\Rightarrow \quad \frac{18.5^2 - 7^2}{24} = \frac{2k}{0.35S}(180 + h_c)$$

$$\Rightarrow \quad 2.14 = \frac{k}{S}(180 + h_c) \quad \text{...(ii)}$$

Solving (i) and (ii),

$$\frac{1.169}{2.14} = \frac{60 + h_c}{180 + h_c}$$

$$\Rightarrow \quad 0.546\,(180 + h_c) = 60 + h_c$$

$$\Rightarrow \quad 98.28 + 0.546\,h_c = 60 + h_c$$

$$\Rightarrow \quad h_c = 84.32 \text{ cm}$$

Substituting h_c in (ii), we get

$$2.14 = \frac{k}{S}(180 + h_c)$$

$$\Rightarrow \quad 2.14 = \frac{k}{0.85}(180 + 84.32)$$

$$\Rightarrow \quad k \doteq 6.88 \times 10^{-3} \text{ cm/min}$$

Level-2

4.9 **A stratified soil deposit is shown in figure along with coefficient of permeability of the individual strata. Determine the ratio of k_H to k_V. Assuming an average hydraulic gradient of 0.3 in both horizontal and vertical seepage, find (i) discharge value and discharge velocities in each layer for horizontal flow and (ii) hydraulic gradient and loss in head in each layer for vertical flow.**

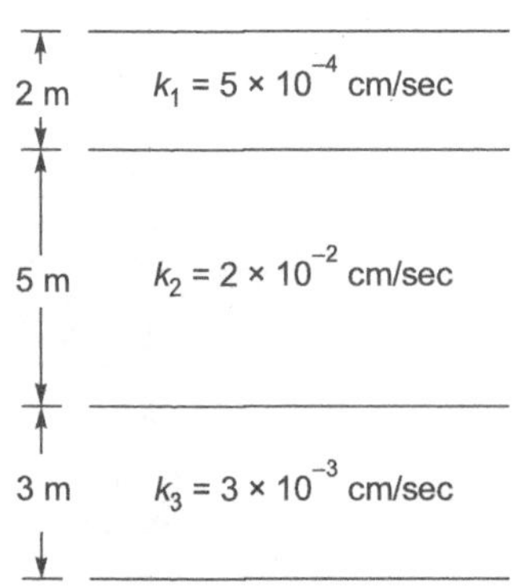

[20 Marks]

Sol:

$$k_H = \left(\frac{\Sigma k_i H_i}{\Sigma H_i}\right)$$

$$= \left(\frac{5\times10^{-4}\times200 + 2\times10^{-2}\times500 + 3\times10^{-3}\times300}{1000}\right)$$

$$= 1.1 \times 10^{-2} \text{ cm/sec}$$

$$k_V = \left(\frac{\Sigma H_i}{\Sigma \dfrac{H_i}{k_i}}\right) = \left(\frac{1000}{\dfrac{200}{5\times10^{-4}} + \dfrac{500}{2\times10^{-2}} + \dfrac{300}{3\times10^{-3}}}\right) = 1.905 \times 10^{-3} \text{ cm/sec}$$

$$\left(\frac{k_H}{k_V}\right) = \left(\frac{1.1\times10^{-2}}{1.905\times10^{-3}}\right) = 5.77$$

(i) Horizontal flow,

$$q = k_H i A = 1.1 \times 10^{-2} \times 0.3 \times 10 \times 10^2 = 3.3 \text{ cc/s/cm width}$$
$$V_1 = k_1 i = 5 \times 10^{-4} \times 0.3 = 1.5 \times 10^{-4} \text{ cm/s}$$
$$q_1 = V_1 A = (1.5 \times 10^{-4} \times 2 \times 100) = 0.03 \text{ cc/s/cm width}$$
$$V_2 = k_2 i = 2 \times 10^{-2} \times 0.3 = 6 \times 10^{-3} \text{ cm/s}$$
$$q_2 = V_2 A = 6 \times 10^{-3} \times 5 \times 100 = 3 \text{ cc/s/cm width}$$
$$V_3 = k_3 i = 3 \times 10^{-3} \times 0.3 = 9 \times 10^{-4} \text{ cm/s}$$
$$q_3 = V_3 A = 9 \times 10^{-4} \times 3 \times 100 = 0.27 \text{ cc/s/cm width}$$

[For horizontal flow, the gradient of flow is the same but the velocity of flow is different for different layer]

(ii) Vertical flow,

$$V = k_V i = V_1 = k_1 i_1 = V_2 = k_2 i_2 = V_3 = k_3 i_3$$

[For vertical flow, velocity of flow is the same because of continuity of flow, but gradients are different for different layers]

$$V = 1.9 \times 10^{-3} \times 0.3 = 5.7 \times 10^{-4} \text{ cm/s}$$
$$i_1 = \left(\frac{5.7 \times 10^{-4}}{5 \times 10^{-4}}\right) = 1.14$$
$$i_2 = \left(\frac{5.7 \times 10^{-4}}{2 \times 10^{-2}}\right) = 0.0285$$
$$i_3 = \left(\frac{5.7 \times 10^{-4}}{3 \times 10^{-3}}\right) = 0.19$$

h_1 (head loss in layer 1) $= i_1 H_1 = 1.14 \times 200 = 228$ cm

Similarly, $h_2 = 0.0285 \times 500 = 14.25$ cm

$h_3 = 0.19 \times 300 = 57$ cm

Check : Total head loss $= 228 + 14.25 + 57 = 229.25 \text{ cm} \simeq 0.3\,(i) \times 1000\,(H)$

4.10 In the experimental setup shown in figure, flow takes place under a constant head through the soils A and B.

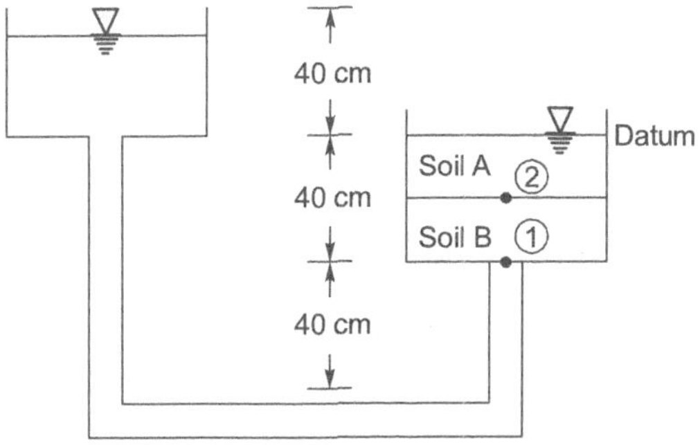

(i) Determine the piezometric head at point 1.

(ii) If 40% of the excess hydrostatic pressure is lost in flowing through soil *B*, what are the hydraulic head and piezometric head at point 2.

(iii) If the coefficient of permeability of soil *B* is 0.06 cm/sec, determine the same for soil *A*.

(iv) What is the discharge per unit area?

[20 Marks]

Sol:

(i) Total head at point 1 = Piezometric head at 1 = 120 – 80 = 40 cm

(ii) Total head at point 1 = 120 + (– 80) = 40 cm

Head loss through soil B = 0.4 × 40 = 16 cm

Total head at point 2 = 40 – 16 = 24 cm

Pressure head at point 2 = 24 – (–40) = 64 cm

(iii) k_B = 0.06 cm/s

Head loss through soil A = 40 – 16 = 24 cm

Discharge through both the soils will be same

$$q_A = q_B$$

$$k_A i_A = k_B i_B$$

$$k_A = \frac{0.06 \times \left(\frac{16}{40}\right)}{\left(\frac{24}{40}\right)} = 0.04 \text{ cm/sec}$$

(iv) Discharge through soil B per unit area = Discharge through soil A per unit area

$$= (0.06) \times \frac{16}{40} \times 1 = 0.024 \text{ cm}^3\text{/s} = 0.024 \text{ m}l\text{/s}$$

4.11 **The following data were recorded in a constant head permeability test:**

Internal diameter of permeameter = 7.5 cm

Head lost over a sample length of 18 cm = 24.7 cm

Quantity of water collected in 60s = 626 ml

Porosity of the soil sample was 44% at 25°C

Calculate the coefficient of permeability of the soil. Also determine the discharge velocity and the seepage velocity during the test. The test was carried out at a temperature of 25°C. Estimate the permeability of the soil for a porosity of 39% at 20°C. Neglect change in γ_w due to temperature variation.

[Take, viscosity of water at 25°C (μ_1) = 8.95 millipoise and at 20°C, μ_2 = 10.09 millipoise]

[12 marks]

Sol:

$$\because \quad k = \frac{Ql}{Ah} = \frac{V \cdot l}{Ath}$$

$$A = \frac{\pi}{4} \times 7.5^2 = 44.18 \text{ cm}^2$$

$$\therefore \quad k = \frac{626 \times 18}{44.18 \times 60 \times 24.7} = 0.172 \text{ cm/s} \qquad [\because \ 1 \text{ m}l = 1 \text{ cm}^3]$$

Discharge velocity, $v = ki = 0.172 \times \dfrac{24.7}{18} = 0.236$ cm/s

Seepage velocity, $v_s = \dfrac{v}{n} = \dfrac{0.236}{0.44} = 0.536$ cm/s

$\because \quad n_1 = 44\%$

$$\therefore \quad e_1 = \frac{n_1}{1-n_1} = 0.79, \quad \therefore \quad \frac{e_1^3}{1+e_1} = 0.275$$

$$n_2 = 39 = 0.39\%$$

$$e_2 = \frac{n_2}{1-n_2} = 0.64, \therefore \frac{e_2^3}{1+e_2} = 0.16$$

At 25°C, viscosity of water $\mu_1 = 8.95$ millipoise

At 20°C, $\mu_2 = 10.09$ millipoise

$$k = C\left(\frac{\gamma_w}{\mu}\right)\left(\frac{e^3}{1+e}\right)D^2$$

$$\therefore \frac{k_{25°C}}{k_{20°C}} = \frac{\frac{1}{\mu_{25°C}}\left(\frac{e^3}{1+e}\right)_{25°C}}{\frac{1}{\mu_{20°C}}\left(\frac{e^3}{1+e}\right)_{20°C}} = \frac{\frac{1}{8.95}(0.275)}{\frac{1}{10.09}(0.16)} = 1.9377$$

$$\therefore k_{20°C} = \frac{k_{25°C}}{1.9377} = \frac{0.172}{1.9377} = 0.0888 \text{ cm/s} = 8.88\times10^{-2} \text{cm/s}$$

4.12 Consider the upward flow of water through a layer of sand in a tank as shown in the figure. For the sand, the following are given: void ratio (e) = 0.52 and specific gravity of solids = 2.67.

(i) Calculate the total stress, pore water pressure and effective stress at points A and B.

(ii) What is the upward seepage force per unit volume of soil?

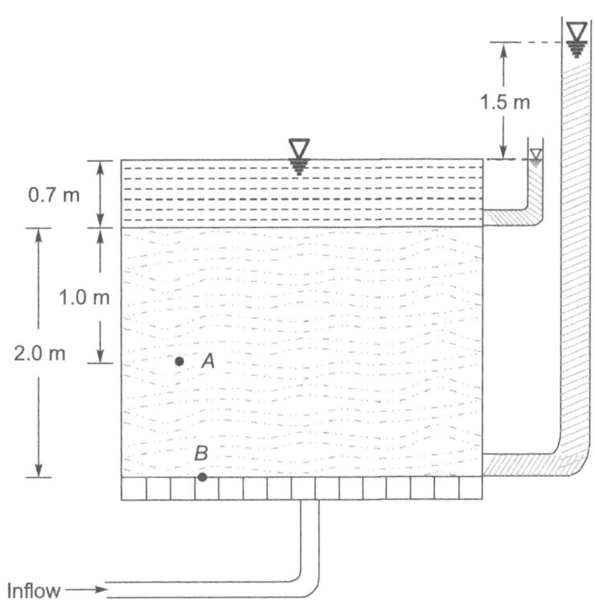

[12 marks]

Sol:

(i) Given, Void ratio, $e = 0.52$

Specific gravity of soil solid, $G_s = 2.67$

$$\gamma_{sat} = \frac{(G_s+e)\gamma_w}{1+e} \quad (\because S = 1)$$

$$= \frac{(2.67+0.52)9.81}{1+0.52} = 20.59 \text{ kN/m}^3$$

At point A,

Total stress, $\sigma = 0.7\gamma_w + 1\,\gamma_{sat} = 0.7 \times 9.81 + 1 \times 20.59 = 27.46\text{ kN/m}^2$

Pore water pressure, $u = \left[1 + 0.7 + \left(\frac{1.5}{2}\right) \times 1\right]\gamma_w = 2.45 \times 9.81 = 24.03\text{ kN/m}^2$

∴ Effective stress, $\sigma' = \sigma - u = 27.46 - 24.03 = 3.43\text{ kN/m}^2$

At point B,

Total stress, $\sigma = 0.7\,\gamma_w + 2\gamma_{sat} = 0.7 \times 9.81 + 2 \times 20.59 = 48.05\text{ kN/m}^2$

Pore water pressure, $u = (2 + 0.7 + 1.5)\gamma_w = 4.2 \times 9.81 = 41.2\text{ kN/m}^2$

Effective stress $\sigma - u = 48.05 - 41.2 = 6.85\text{ kN/m}^2$

(ii) Hydraulic gradient (i) $= \frac{\Delta h}{L} = \frac{1.5}{2} = 0.75$

Seepage force per unit volume $= i\gamma_w = 0.75 \times 9.81 = 7.36\text{ kN/m}^3$

5. Compressibility and Consolidation

Level-1

5.1 **It is estimated that additional stress on the clay layer due to weight of building is 160 kN/m² at the top of the clay and 80 kN/m² at the bottom of the clay. Compute the settlement of the clay layer due to construction of building.**

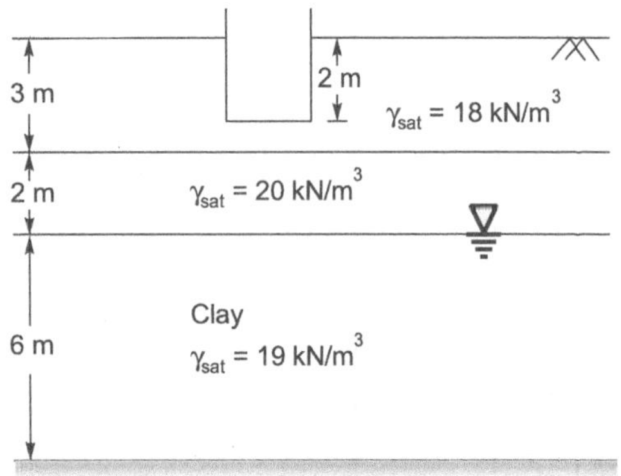

Data of Consolidation Test	
Applied Stress (kN/m²)	**Void ratio**
25	0.89
50	0.882
100	0.865
200	0.830
400	0.798
800	0.765

[12 Marks]

Sol:

Effective stress at the top of the clay layer

$$\bar{\sigma}_1 = 18 \times 3 + (20 - 9.81) \times 2$$

$$\bar{\sigma}_1 = 74.38\text{ kN/m}^2$$

Effective stress at bottom at the clay layer

$$\bar{\sigma}_2 = 18 \times 3 + (20 - 9.81) \times 2 + (19 - 9.81) \times 6$$

$$= 129.52\text{ kN/m}^2$$

Average initial effective stress at the centre of clay layer

$$\bar{\sigma}_0 = \left(\frac{74.38 + 129.52}{2}\right) = 101.95\text{ kN/m}^2$$

Average increment of effective stress in clay layer

$$\Delta\bar{\sigma}_0 = \left(\frac{160+80}{2}\right) = 120 \text{ kN/m}^2$$

$$(\bar{\sigma}_0 + \Delta\bar{\sigma}_0) = 101.95 + 120 = 221.95 \text{ kN/m}^2$$

Void ratio corresponding to 221.95 kN/m^2

$$e_1 = 0.798 + \left(\frac{0.830-0.798}{400-200}\right)(400-221.95)$$

$$e_1 = 0.826$$

Void ratio corresponding to 101.95 kN/m^2

$$e_0 = 0.830 + \left(\frac{0.865-0.830}{200-100}\right)(200-101.95)$$

$$e_0 = 0.864$$

Settlement of clay layer, $$\Delta H = \left(\frac{\Delta e}{1+e_0} \times H\right) = \frac{0.864-0.826}{1+0.864} \times 6000$$

$$\Delta H = 122.32 \text{ mm}$$

5.2 **A flexible footing of 2 m x 2 m size carries a total load of 510 kN, inclusive of its self weight. The footing rests on a sand layer having a modulus of elasticity of 40000 kN/m^2 and a Poisson's ratio of 0.36. Estimate the probable settlement below the centre and below any corner of the footing. Given I_f (corner) = 0.56, I_f (centre) = 1.12.**

[10 marks]

Sol:

We have,

Elastic settlement, $$S_i = \frac{qB(1-\mu^2)}{E_s} \times I_f$$

where q = Intensity of loading

$$= \frac{510}{(2\times 2)} = 127.5 \text{ kN/m}^2$$

$$B = 2 \text{ m}$$

$$\mu = 0.36$$

$$E = 40000 \text{ kN/m}^2$$

Immediate settlement below the corner of footing

$$I_f(\text{corner}) = 0.56$$

$$S_{i,\text{ corner}} = \frac{qB(1-\mu^2)I_f}{E_s} = \frac{127.5\times 2\times(1-0.36^2)\times 0.56}{40000} \text{ m} = 3.107 \text{ mm}$$

Immediate settlement below the centre of footing

$$I_{f(\text{centre})} = 1.12$$

$$S_i = \frac{127.5\times 2\times(1-0.36^2)\times 1.12}{40000} \text{ m}$$

$$= 6.215 \text{ mm}$$

Level-2

5.3 A stratum of saturated clay (G = 2.70) has an initial water content of 44% throughout its depth. After a long dry spell, the water content at the ground surface is 20% and at a depth of 3 m, the water content is 44%. The water content may be taken to vary linearly with depth. The shrinkage limit of the clay is 28%. If the change in height per unit height of the clay ($\Delta H/H$) is 80% of its change in volume per unit volume ($\Delta V/V_0$), what is the amount of settlement of the ground surface due to change in water content which occurred?

[20 Marks]

Sol:

Given, $G_s = 2.70$

Initial water content throughout 3 m of depth = 44%

Now, after a long dry spell,

Water content very linearly with depth

Shrinkage unit of clay = 28%

$$\frac{\Delta H}{H} = 0.8\left(\frac{\Delta V}{V}\right)$$

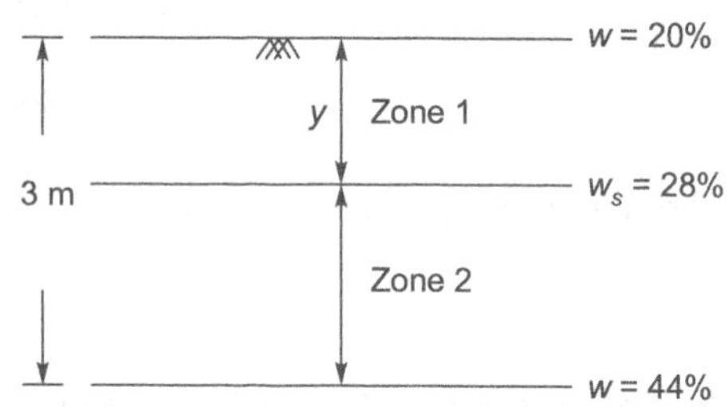

As we know that below shrinkage unit there is no volume change.

As water content carries linearly with depth in 3 m layer soil.

Volume change will takes place after depth 'y' as shown in figure.

$$28\% = 20\% + \frac{44\% - 20\%}{3} \times y$$

$$y = 1 \text{ m}$$

At $w = 44\%$, $e_0 = w.G_s = 0.44 \times 2.7 = 1.188$

$w = 28\%$, $e_1 = w.G_s = 0.28 \times 2.7 = 0.756$

$w - 20\%$; $e = e_1 = 0.756$ {before shrinkage unit there is no volume change}

For zonel

$$\Delta e = 1.188 - 0.756 = 0.432$$

$$\frac{\Delta V}{V_0} = \frac{\Delta e}{1+e_0} = \frac{0.432}{1+1.188} = 0.1974$$

$$\frac{\Delta H_1}{H_0} = 0.8\frac{\Delta V}{V_0} = 0.8 \times (0.1974) = 0.1579$$

$$\Delta H_1 = 0.1579 \times 1 = 0.1579 \text{ m}$$

For zones:

At mid depth:

$$w = \left(\frac{28+44}{2}\right) = 36\%$$

At w = 36%, $e_2 = 0.36 \times 2.7 = 0.972$

$$\Delta e = 1.188 - 0.972 = 0.216$$

$$\frac{\Delta H}{H_0} = 0.8\left(\frac{\Delta V}{V_0}\right)$$

$$\frac{\Delta H_2}{2} = 0.8\,(0.0987)$$

$$\Delta H_2 = 0.0789 \times 2 = 0.1579 \text{ m}$$

Hence, total settlement of the year

$$\Delta H = \Delta H_1 + \Delta H_2$$
$$= 0.1579 + 0.1579 = 0.3158 \text{ m}$$
$$\Delta H = 31.6 \text{ cm}$$

5.4 **A 6.0 m thick layer of clay is located between two layers of free-draining sand. Also, there is a thin drainage layer within the clay at a depth of 1.50 m from its top surface. The average value of C_v is found as 4.92×10^{-2} mm²/sec. If a structure is constructed above the clay layer, how many days would be required for it to attain half its ultimate settlement? Assume that the expression $T = \frac{\pi}{4}U^2$ is applicable for the entire range of consolidation.**

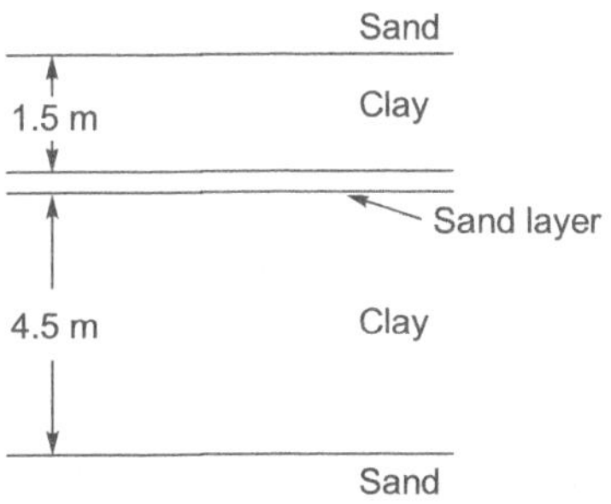

[20 Marks]

Sol:

Let 't' be the time in which half of ultimate settlement takes place.

Given: $H = 6$ m; $C_v = 4.92 \times 10^{-2}$ mm²/sec.

Layer - 1: $H_1 = 1.5$ m

Layer - 2: $H_2 = 4.5$ m

Let U_1 and U_2 be the degree of consolidation of layer 1 and 2 respectively.

Now,
$$\Delta h_1 + \Delta h_2 = \Delta h$$

$$\Rightarrow \quad \frac{\Delta h_1}{H} + \frac{\Delta h_2}{H} = \frac{\Delta h}{H}$$

$$\frac{U_1 \times H_1}{H} + \frac{U_2 \times H_2}{H} = U$$

At time 't', $U = 0.5$

$$\therefore \quad \frac{U_1 \times 1.5}{6} + \frac{U_2 \times 4.5}{6} = 0.5$$

$$U_1 + 3\,U_2 = 2 \qquad \ldots \text{(i)}$$

Drainage path length of layer - 1, $d_1 = \dfrac{H_1}{2}$

Drainage path length of layer-2, $d_2 = \dfrac{H_2}{2}$

$$T_{v1} = \frac{C_v \cdot t}{d_1^2}$$

$$\frac{\pi}{4} \times U_1^2 = \frac{C_v \cdot t}{d_1^2}$$

$$\therefore \quad t = \frac{\pi}{4} \times U_1^2 \times \left(\frac{H_1}{2}\right)^2 \times \frac{1}{C_v} = \frac{\pi}{16} \times \frac{U_1^2 \times H_1^2}{C_v}$$

Similarly, $$t = \frac{\pi}{16} \times \frac{U_2^2 \times H_2^2}{C_v}$$

Equating above two equations,

$$\frac{\pi}{16} \times \frac{U_2^2 \times H_2^2}{C_v} = \frac{\pi}{16} \times \frac{U_1^2 \times H_1^2}{C_v}$$

$$\Rightarrow \quad \frac{U_1}{U_2} = \frac{H_2}{H_1} = \frac{4.5}{1.5} = 3$$

$$\therefore \quad U_1 = 3\,U_2 \qquad \text{... (ii)}$$

From equation (i) and (ii),

$$3\,U_2 + 3\,U_2 = 2$$

$$U_2 = \frac{1}{3}$$

$$\Rightarrow \quad t = \frac{\pi}{16} \times \frac{(1/3)^2 \times (4.5 \times 10^3)^2}{4.92 \times 10^{-2}} \text{ sec.}$$

$$t = 103.928 \text{ days} \approx 104 \text{ days}$$

∴ it would take approximately 104 days to attain half its ultimate settlement.

5.5 **Water content at the end of sample was 28% and G_s = 2.72. Find the void ratio at each applied stress**

Applied Stress (kN/m²)	0	50	100	200	400	800	0
Sample thickness (mm)	19.7	19.3	19.15	18.85	18.56	18.36	18.90

[15 Marks]

Sol:

Let the final void ratio of the sample be e_f.

$$e_f \cdot s = wG$$

$$e_f \cdot 1 = 0.28 \times 2.72$$

$$e_f = 0.762$$

We know, $$\frac{\Delta H}{H_0} = \left(\frac{\Delta e}{1+e_0}\right)$$

$$\frac{18.36 - 18.90}{18.90} = \left(\frac{\Delta e}{1+0.762}\right)$$

$$\Delta e = -0.050$$

Therefore void ratio at 800 kN/m² $= e_f + \Delta e = 0.762 - 0.050 = 0.712$

For 400 - 800 kN/m² stress range:

$$\left(\frac{\Delta e}{1+0.712}\right) = \left(\frac{18.56 - 18.36}{18.36}\right)$$

$$\Delta e = 0.018$$

Void ratio at 400 kN/m², $(e_2) = 0.712 + 0.018 = 0.730$

For 400 - 200 kN/m² stress range:

$$\left(\frac{\Delta e}{1+0.730}\right) = \left(\frac{18.85 - 18.56}{18.56}\right)$$

$$\Delta e = 0.0270$$

Void ratio at 200 kN/m², $(e_3) = 0.730 + 0.027 = 0.757$

For 200 - 100 kN/m² stress range:

$$\left(\frac{\Delta e}{1+0.757}\right) = \left(\frac{19.15-18.85}{18.85}\right)$$

$$\Delta e = 0.02796$$

Void ratio at 100 kN/m², $(e_4) = 0.757 + 0.02796 = 0.785$

For 100 - 50 kN/m² stress range:

$$\left(\frac{\Delta e}{1+0.785}\right) = \left(\frac{19.30-19.15}{19.15}\right)$$

$$\Delta e = 0.014$$

Void ratio at 50 kN/m², $(e_5) = 0.785 + 0.014 = 0.799$

For 50 - 0 kN/m² stress range:

$$\left(\frac{\Delta e}{1+0.799}\right) = \left(\frac{19.70-19.30}{19.30}\right)$$

$$\Delta e = 0.037$$

Void ratio at 0 kN/m², $(e_6) = 0.799 + 0.037 = 0.836$

Applied Stress (kN/m²)	Sample thickness (mm)	Void ratio
0	19.7	0.836
50	19.3	0.799
100	19.15	0.785
200	18.85	0.757
400	18.56	0.730
800	18.36	0.712
0	18.90	0.762

5.6 An oedometer test is performed on a normally consolidated clay stratum that is 350 cm thick an it is found that clays initial void ratio was $e_0 = 0.8$ and its primary compression index $c_c = 0.28$. The in-situ stress at mid clay layer is $\bar{\sigma}_0 = 120$ kN/m² and the secondary compression index $c_s = 0.02$. The building exerts a pressure through the mat foundation of 80 kN/m². The time of completion of the primary settlement is approximately 18 months. What is the total consolidation of 350 cm clay stratum 5 year after the primary consolidation.

[20 Marks]

Sol:

Given, Initial void ratio = 0.8

$$c_c = 0.28$$

$$c_s = 0.02$$

$$\bar{\sigma}_0 = 120 \text{ kN/m}^2$$

$$\Delta\bar{\sigma}_0 = 80 \text{ kN/m}^2$$

Primary settlement of clay layer,

$$\Delta H = \left(\frac{H_0 C_c}{1+e_0}\right)\log\left(\frac{\sigma_0 + \Delta\bar{\sigma}_0}{\bar{\sigma}_0}\right)$$

$$\Delta H = \left(\frac{0.28 \times 350}{1+0.8}\right)\log\left(\frac{200}{120}\right) = 12.078 \text{ cm}$$

Secondary settlement of clay layer,

$$\Delta H = \left(\frac{C_s H_{100}}{1+e_{100}}\right)\log_{10}\left(\frac{t}{t_{100}}\right)$$

t_{100} = time required for completion of primary consolidation

e_{100} = void ratio at end of primary consolidation

H_{100} = thickness of clay layer after primary consolidation

$$e_{100} = e_0 - \Delta e$$

$$\frac{\Delta H}{H} = \left(\frac{\Delta e}{1+e_0}\right)$$

$$\Delta e = \frac{12.078(1.08)}{350} = 0.062$$

$$e_{100} = 0.8 - 0.062 = 0.738$$

$$H_{100} = H_0 - \Delta H = 350 - 12.078 = 337.922 \text{ cm}$$

$$\Delta H_s = \frac{(0.02)(337.922)}{1+0.738}\log_{10}\left(\frac{6.5}{1.5}\right)$$

$$\Delta H_s = 2.476 \text{ cm}$$

Hence, Total settlement $= \Delta H + \Delta H_s$

$= 12.078 + 2.476 = 14.554$ cm

5.7 A 8.5 cm deep compacted fill is to be placed over the soil profile shown in figure below. Consolidation tests on samples from points *A* and *B* produced the following results.

	Sample *A*	Sample *B*
C_c	0.25	0.20
C_r	0.08	0.06
e_0	0.66	0.45
σ_c'	101 kPa	510 kPa

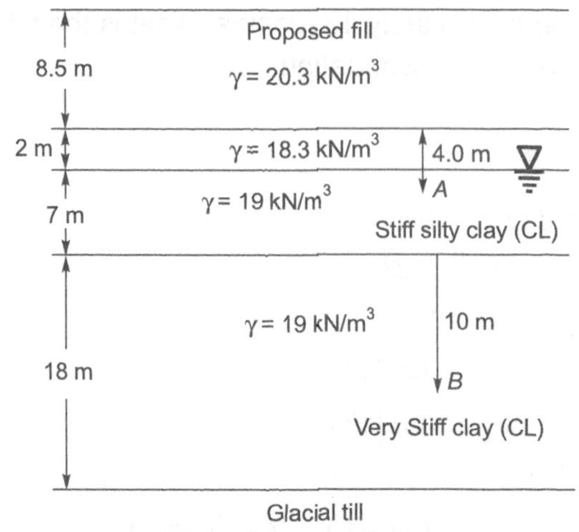

Preconsolidation stress at other points in each layers can be calculated using formula.

$$\sigma'_m = \sigma'_c - \sigma'_{Z_0}$$

$$\sigma'_m = \text{Over consolidation margin}$$

Which is assumed constant (effective)

$$\sigma'_c = \text{Preconsolidated stress}$$

$$\sigma'_{Z_0} = \text{Effective stress at the level under consideration.}$$

Compute the ultimate consolidation due to the weight of thin fill.

[20 Marks]

Sol:

$$\sigma'_m = \sigma'_c - \sigma'_{Z_0}$$

$$\sigma'_c = \sigma'_m + \sigma'_{Z_0}$$

where, σ'_{Z_0} = Effective stress at the level under consideration

For sample *A*

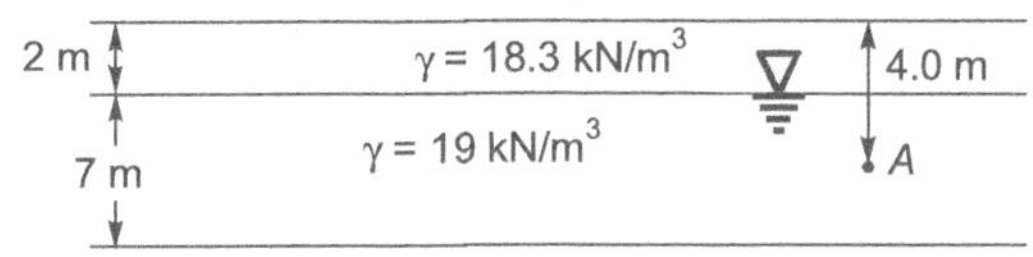

$$\sigma'_{Z_0} = 2 \times 18.3 + 2 \times (19 - 10) = 54.6 \text{ kN/m}^2$$

$$\sigma'_m = (101 - 54.6) = 46.4 \text{ kN/m}^2$$

At the centre of first layer effective stress before application of proposed fill,

$$\sigma'_0 = 2 \times 18.3 + 3.5 \times (19 - 10) = 68.1 \text{ kN/m}^2$$

$$\Delta\sigma'_0 = \text{After placing the proposed fill} = 8.5 \times 20.3 = 172.55 \text{ kN/m}^2$$

$$\text{Final stress at centre} = \sigma'_0 + \Delta\sigma'_0 = 68.1 + 172.55 = 240.65 \text{ kN/m}^2$$

Preconsolidation stress at the point = $\sigma'_m + \sigma'_0 = 68.1 + 46.4 = 114.5 \text{ kN/m}^2$

Settlement of I[st] layer,

$$\Delta H_1 = \frac{C_r H_0}{1+e_0}\log\left(\frac{\sigma_{C_0}}{\sigma'_0}\right) + \left(\frac{C_c H_0}{1+e_0}\right)\log\left(\frac{\sigma'_0 + \Delta\sigma_0}{\sigma_{C_0}}\right)$$

$$\Delta H_1 = \frac{0.08 \times 7}{1.66}\log\left(\frac{114.5}{68.1}\right) + \frac{0.25 \times 7}{1 + 0.66}\log\left(\frac{240.65}{114.5}\right)$$

$$\Delta H_1 = 0.4160 \text{ m}$$

For sample *B*,

$$\sigma'_m = \sigma'_c - \sigma'_{Z_0}$$

$$= 510 - (2 \times 18.3 + (19 - 10) \times 7 + 10 \times (19 - 10)) = 320.4 \text{ kN/m}^2$$

Preconsolidation stress,

$$\sigma'_c = \sigma'_m - \sigma'_0$$

$$= 320.4 + 2 \times 18.3 + 9 \times 7 + 9 \times 9 = 501 \text{ kN/m}^2$$

Final stress at the centre of clay of sample *B*,

$$= \sigma_0' + \Delta\sigma_0' = 180.6 + 8.5 \times 20.3 = 353.15 \text{ kN/m}^2$$

It is evident that, preconsolidation stress is more than final stress at this level. Therefore only preconsolidation settlement will occur.

$$\Delta H_2 = \left(\frac{C_r H_0}{1+e_0}\right)\log\left(\frac{\sigma_0' + \Delta\sigma_0'}{\sigma_0'}\right) = \frac{0.06\times 18}{1.45}\log\left(\frac{353.15}{180.6}\right)$$

$$= 0.2169 \text{ m} \approx 0.217 \text{ m}$$

$$\text{Total settlement, } \Delta H = \Delta H_1 + \Delta H_2$$

$$= 0.116 + 0.217 = 0.632 \text{ m} = 632 \text{ mm}$$

5.8 **The construction of a multistoryed building started in January 1989 and was completed in June 1990. The total consolidation settlement of the building was estimated to be 8 cm. The average settlement of the building was measured in December 1991 and was found to be 2.2 cm. Compute the probable settlement of the building in January 2001.**

[20 Marks]

Sol:

Let C_v be the coefficient of consolidation of the soil in the appropriate pressure range.
Time elapsed from September 1989 to December 1991 = 2.25 years.

$$\text{Degree of consolidation occurred in 15 years} = \left(\frac{2.2\times 100}{8}\right) = 27.5\%$$

$$T_v = \left(\frac{\pi}{4}\right)(U^2) = \frac{\pi}{4}\times(0.275)^2 = (0.059)$$

$$T_v = \left(\frac{C_v x t}{H^2}\right)$$

$$\left(\frac{C_v}{H^2}\right) = \left(\frac{T_v}{t}\right) = \left(\frac{0.059}{2.25}\right)$$

$$\left(\frac{C_v}{H^2}\right) = 0.0262$$

Again time elapsed from September 1989 to January 2001 = 11.25 years.
Let *U* be the corresponding degree of consolidation.

Assume $U > 60\%$

$$T_v = 1.781 - 0.933\log(100 - U)$$

$$T_v = \left(\frac{C_v x t}{H^2}\right) = 0.0262\times 11.25 = 0.2948$$

$\Rightarrow$ $1.781 - 0.933\log(100 - U) = 0.2948$

$\Rightarrow$ $U = 60.83\% > 60\%$ (Hence, our assumption is correct OK)

Amount of consolidation settlement in January 2001.

$$U = \frac{\Delta H}{H}\times 100$$

$$60.83 = \frac{\Delta H}{8}\times 100$$

$$\Delta H = \left(\frac{60.83\times 8}{100}\right) = 4.87 \text{ cm}$$

5.9 A footing 1.5 m square is located 1.5 m below the surface of a uniform soil deposit of density 20 kN/m^3. The void ratio of the soil is 0.8 and its compression index is 0.07. If the total thickness of the deposit, which is underlain by rock strata is 3.5 m, compute the primary consolidation settlement of the footing when it carries a load of 225 kN. Use trapezoidal stress distribution (2 : 1 horizontal is to vertical) and consider four layers.

[20 marks]

Sol:

In this case, the total settlement of clay deposit will be the sum of settlement in individual layer.

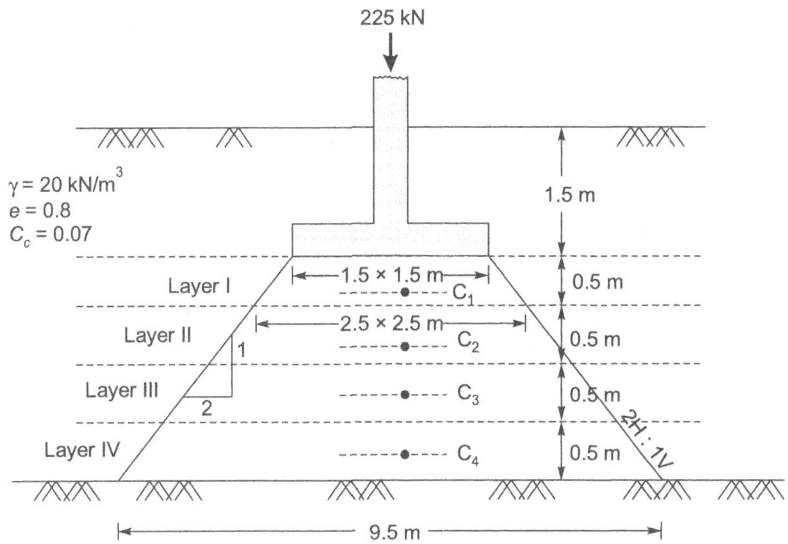

Given, $\gamma = 20$ kN/m^3, $e_0 = 0.8$, $C_c = 0.07$,

Layer I:

Initial overburden pressure at C_1,

$$\bar{\sigma}_1 = 20 \times (1.5 + 0.25) = 20 \times 1.75 = 35 \text{ kN/m}^2$$

Increase in pressure,
$$\Delta\bar{\sigma}_1 = \frac{P}{\text{Base area at } C_1} = \frac{225}{(1.5 + 2 \times 0.5)^2} = 36 \text{ kN/m}^2$$

∴ Primary consolidation settlement in Layer I,

$$\Delta H_1 = \frac{C_c H_0}{1 + e_0} \log_{10}\left[\frac{\bar{\sigma}_1 + \Delta\bar{\sigma}_1}{\bar{\sigma}_1}\right] = \frac{0.07 \times 0.5}{1 + 0.8} \log_{10}\left[\frac{35 + 36}{35}\right]$$

$$= 5.97 \times 10^{-3} \text{ m or } 5.97 \text{ mm}$$

Layer II:

Initial overburden pressure at $C_2 = \bar{\sigma}_2 = 20 \times 2.25 = 45$ kN/m^2

Increase in pressure,
$$\Delta\bar{\sigma}_1 = \frac{225}{\text{Base pressure at } C_2} = \frac{225}{4.5 \times 4.5} = 11.11 \text{ kN/m}^2$$

Primary consolidation settlement in Layer II,

$$\Delta H_2 = \frac{0.07 \times 0.5}{1 + 0.8} \log_{10}\left[\frac{45 + 11.11}{45}\right] = 1.86 \times 10^{-3} \text{ m or } 1.86 \text{ mm}$$

Layer III:

Initial overburden pressure at C_2, $\bar{\sigma}_3 = 20 \times 2.75 = 55$ kN/m^2

Increase in pressure,
$$\Delta\bar{\sigma}_3 = \frac{225}{6.5 \times 6.5} = 5.33 \text{ kN/m}^2$$

Primary consolidation settlement in Layer III,

$$\Delta H_3 = \frac{0.07 \times 0.5}{1+0.8} \log_{10}\left[\frac{55+5.33}{55}\right] = 0.78 \times 10^{-3} \text{ m or } 0.78 \text{ mm}$$

Layer IV:

Initial overburden pressure at C_4, $\bar{\sigma}_4 = 20 \times 3.25 = 65 \text{ kN/m}^2$

$$\text{Increase in pressure, } \Delta\bar{\sigma}_4 = \frac{225}{8.5 \times 8.5} = 3.11 \text{ kN/m}^2$$

Primary consolidation settlement in Layer IV,

$$\Delta H_4 = \frac{0.07 \times 0.5}{1+0.8} \log_{10}\left[\frac{65+3.11}{65}\right] = 0.39 \times 10^{-3} \text{ m or } 0.39 \text{ mm}$$

Total consolidation settlement, $\Delta H = \Delta H_1 + \Delta H_2 + \Delta H_3 + \Delta H_4$

$$= 5.97 + 1.86 + 0.78 + 0.39 = 9 \text{ mm}$$

5.10 **The foundations supporting two columns of a building are shown in figure. An extensive soil investigation was not carried out, and it was assumed in the design of the foundations that the clay layer had a uniform thickness of 1.2 m. Two years after construction, the building settled, with a differential settlement of 10 mm. Walls of the building began to crack. The doors have not jammed, but measuring the out-of-vertical distance of the doors, it is estimated that they would become jammed if the differential settlement exceeded 24 mm. A subsequent soil investigation showed that the thickness of the clay layer was not uniform but varies, as shown in figure. The owners would like to get an estimate of the expected total differential settlement and how long it will take before the doors become jammed.**

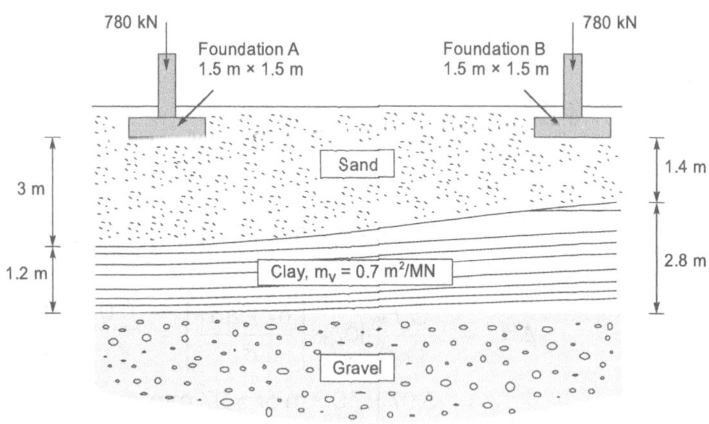

[20 marks]

Sol:

Step-1: Calculate the vertical stress increase at the center of the clay layer under each foundation. Let's use the approximate method.

$$\Delta\sigma_z = \frac{P}{(B+z)(L+z)}$$

$$(\Delta\sigma_z)_A = \frac{780}{(1.5+3.6)(1.5+3.6)} \simeq 30 \text{ kPa}$$

$$(\Delta\sigma_z)_B = \frac{780}{(1.5+2.8)(1.5+2.8)} = 42.2 \text{ kPa}$$

Step-2: Calculate the primary consolidation settlement.

Use $\rho_{pc} = H_o m_v \Delta\sigma$ to calculate the primary consolidation settlement.

$$(\rho_{pc})_A = 1.2 \times 0.7 \times 10^{-3} \times 30 = 25.2 \times 10^{-3} \text{ m} = 25.2 \text{ mm}$$

$$(\rho_{pc})_B = 2.8 \times 0.7 \times 10^{-3} \times 42.2 = 82.7 \times 10^{-3} \text{ m} = 82.7 \text{ mm}$$

Step-3: Calculate the differential settlement.

Differential settlement:

$$\delta = (\rho_{pc})_B - (\rho_{pc})_A = 82.7 - 25.2 = 57.5 \text{ mm}$$

Step-4: Calculate the time for 24 mm differential settlement to occur

Current differential settlement:

$$\delta_c = 10 \text{ mm}$$

Degree of consolidation:
$$U = \frac{\delta_c}{\delta} = \frac{10}{57.5} = 0.174 < 0.6$$

$\therefore$
$$T_v = \frac{\pi}{4}U^2 = \frac{\pi}{4} \times (0.174)^2 = 0.02378$$

$\therefore$
$$C_v = \frac{T_v H_{dr}^2}{t} = \frac{0.02378 \times \left(\frac{2.8}{2}\right)^2}{2} = 0.0233 \text{ m}^2/\text{year}$$

For 24-mm differential settlement:

$$U = \frac{24}{57.5} = 0.417 < 0.6$$

$\therefore$
$$T_v = \frac{\pi}{4} \times (0.417)^2 = 0.1365$$

$\therefore$
$$t = \frac{T_v H_{dr}^2}{C_v} = \frac{0.1365 \times \left(\frac{2.8}{2}\right)^2}{0.0233} = 11.48 \text{ years}$$

Therefore, in the next 9.48 years, the total differential settlement would be 24 mm.

6. Shear Strength of Soil

Level-1

6.1 A consolidated undrained triaxial compression test is performed on a sample of saturated NC clay. The sample has been isotropically consolidated so that at the start of the test, $\bar{\sigma}_1 = \bar{\sigma}_3 = 25$ kPa. The slope of *p* - *q* plot line is 30°. Use the stress path method to determine the value of axial stress ($\bar{\sigma}_1$ at the start plus $\bar{\sigma}_v$) when the effective stress path intersects the *p* - *q* plot line to cause shearing (failure) of sample. Assume skemptons pore pressure coefficient of failure, *A* as 1.

[12 Marks]

Sol:

Given, $\bar{\sigma}_1 = \bar{\sigma}_3 = 25$ kPa (at the start of test)

Slope of *p* - *q* plot line = 30° = α

Also for *p* - *q* line, $\tan\alpha = \sin\phi$

$$\sin\phi = \tan 30°$$

$$\sin\phi = 0.5773$$

When deviator stress $\Delta\sigma_v$ is applied, pore pressure starts building up

$$\text{pore pressure} = \Delta v = A(\Delta\sigma_v) = \Delta\sigma_v$$

At the failure,

$$\bar{\sigma}_{1F} = (\bar{\sigma}_1)_{\text{at start}} + \Delta\sigma_v - \Delta U = 25 + \Delta\sigma_v - \Delta\sigma_v$$

$$= 25 \text{ kPa}$$

$$\bar{\sigma}_{3F} = (\bar{\sigma}_3)_{\text{at start}} - \Delta U$$

$$= 25 - \Delta\sigma_v$$

Using Mohr circle,

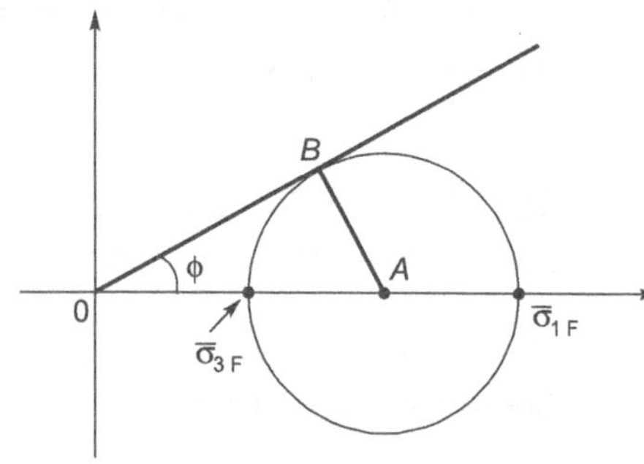

$$AB = \left(\frac{\bar{\sigma}_{1F} - \bar{\sigma}_{3F}}{2}\right)$$

$$OA = \left(\frac{\bar{\sigma}_{1F} + \bar{\sigma}_{3F}}{2}\right)$$

$$\sin\phi = \left(\frac{AB}{OA}\right) = \frac{(\bar{\sigma}_{1F} - \bar{\sigma}_{3F})/2}{(\bar{\sigma}_{1F} + \bar{\sigma}_{3F})/2} = \frac{(\bar{\sigma}_{1F} - \bar{\sigma}_{3F})}{(\bar{\sigma}_{1F} + \bar{\sigma}_{3F})}$$

$$\Rightarrow \quad 0.5773 = \frac{25 - (25 - \Delta\sigma_v)}{25 + (25 - \Delta\sigma_v)}$$

$$\Rightarrow \quad \Delta\sigma_v = 18.3 \text{ kPa}$$

$$\text{Value of axial stress} = \bar{\sigma}_1 + \Delta\sigma_v = 25 + 18.3 = 43.3 \text{ kPa}$$

6.2 **A cylindrical soil specimen of sand was subjected to axial principal stresses (σ_1) and radial principal stresses (σ_3'). The soil could not support additional stresses when $\sigma_1' = 300$ kPa and $\sigma_3' = 100$ kPa.**

(a) Determine the friction angle and inclination of slip plane to horizontal.

(b) Determine the stresses on failure plane.

(c) Determine the maximum shear stress.

(d) If maximum shear stress is equal to failure shear stress? Comment on the reason behind the failure. (Assume no significant dilation effects)

[12 Marks]

Sol:

Given, $\sigma_1' = 300$ kPa and $\sigma_3' = 100$ kPa

Assuming $c = 0$ (for sand)

(a) We know that,

$$\sigma_1' = \sigma_3' \tan^2\alpha + 2c\tan\alpha$$

$$300 = 100\tan^2\alpha + 0$$

$$\Rightarrow \quad \alpha = 60°$$

$$\Rightarrow \quad \alpha = 45 + \frac{\phi}{2} = 60°$$

$$\Rightarrow \quad \phi = 2(60 - 45) = 30°$$

frictional angle = 30°

inclination of slip plane to horizontal = α = 60°

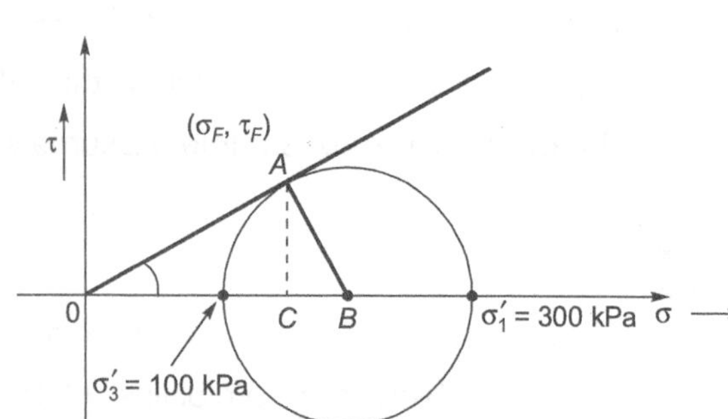

(b) Le σ_F, τ_F be the stresses on failure plane

$$OC = OB - CB$$

$$CB = AB\sin\phi$$

$$CB = \left(\frac{300-100}{2}\right)\sin 30^\circ = 50 \text{ kPa}$$

$$OC = \left(\frac{300+100}{2}\right) - 50 = 150 \text{ kPa}$$

Normal stress on failure plane = 150 kPa

$$AC = AB\cos\phi = \left(\frac{300-100}{2}\right)\cos 30^\circ = 86.6 \text{ kPa}$$

Shear stress on failure plane = 86.6 kPa

(c) Maximum shear stress $= \dfrac{\sigma_1' - \sigma_3'}{2} = \dfrac{300-100}{2} = 100$ kPa

(d) Maximum shear stress is greater than failure shear stress but failure occurs on plane at which shear stress acting is equal to shear strength of soil. But on plane of maximum shear stress, shear strength of soil is greater than shear stress acting and hence there is some reserve of shear strength.

6.3 What is the significance of pore water pressure in the triaxial shear test? Explain the pore water pressure parameters as given by A.W. Skempton.

[12 Marks]

Sol:

When the applied stresses on a soil mass are increased, the full load is not immediately transferred to the soil grains, but a part of it is transferred to the water, standing in the soil pores. This increased hydrostatic pressure in the soil pores is nothing but the increased pore pressure. This pore pressure reduces the effective shear strength of the soil hence it becomes necessary to assess the value of pore pressure. This pore water pressure can be measured by a pore pressure measuring apparatus connecting to the pressure line of the triaxial cell after closing the valve in the drainage line. The pore water pressure is measured under conditions of no flow either into the specimen or out of the specimen. In a saturated soil , when the test is undrained , there will be no volume change in the soil sample.

If it is not possible to determine pore water pressure practically, it can be calculated by using theoretical approach given by A.W. Skempton. The pore pressure parameter represent the response of pore pressure due to change in the total stress under undrained conditions. Such parameters are *A* and *B*.

The pore water pressure can be classified in two stages:

(i) Consolidation stage or cell pressure stage

(ii) Shear stage or deviator stress stage

The value of B can be given as

$$B = \frac{\Delta u_c}{\Delta \sigma_c} = \frac{\Delta u_c}{\Delta \sigma_3}$$

Here B is the ratio of pore pressure developed to the change in confining pressure. Δu_c represents increase in pore pressure due to increase in cell pressure, $\Delta\sigma_c$. For a fully saturated soil $B = 1$ and for a fully dried soil $B = 0$. The value of *B* can also be determined by using soil properties. Thus, the value of *B* can be given as

$$B = \frac{1}{1 + n\dfrac{C_v}{C_c}}$$

Where n is porosity, C_v is coefficient of consolidation, C_c is coefficient of compression

A is defined as $\bar{A} = AB$

where $\bar{A}$ is related to change in pore water pressure due to change in deviator stress i.e.

$$\bar{A} = \frac{\Delta u_d}{\Delta \sigma_d} = \frac{\Delta u_d}{\Delta \sigma_1 - \Delta \sigma_3}$$

where Δu_d is change in pore water pressure due to change in deviator stress.

Now $\Delta u = \Delta u_c + \Delta u_d = B \cdot \Delta \sigma_3 + \bar{A} \cdot \Delta \sigma_d = B \cdot \Delta \sigma_3 + AB \cdot \Delta \sigma_d = B \cdot \Delta \sigma_3 + AB(\Delta \sigma_1 - \Delta \sigma_3)$

$\therefore$ $\Delta u = B[\Delta \sigma_3 + A(\Delta \sigma_1 - \sigma_3)]$

The value of A may be as high as 2 to 3 for saturated fine sand and may be as low as –0.5 for heavily over consolidated clays. For a given soil A depends upon the strain, anisotropy, sample disturbance and the over consolidation ratio.

6.4 **In a laboratory vane shear test a vane 100 mm long and 60 mm diameter was pressed into the soft cohesive soil (G = 2.72). A torque of 40 kN-mm was required to achieve the failure. Same soil when remoulded required 15 kN-mm to achieve failure. Calculate the cohesion in both cases and value of sensitivity. Take the void ratio of soil as 30%.** **[12 Marks]**

Sol:

Given that: d = 60 mm, h = 100 mm, G = 2.72, e = 30%, T_1 = 40 kN-mm, T_2 = 15 kN-mm

$\because$
$$s = c_u + \bar{\sigma}_n \tan\phi = \frac{T}{\pi d^2 \left(\frac{h}{2} + \frac{d}{6}\right)}$$

Since $\phi = 0$ in a cohesive soil, thus

$$s = c_u = \frac{T}{\pi d^2 \left(\frac{h}{2} + \frac{d}{6}\right)}$$

$$\Rightarrow \quad T = c_u \pi d^2 \left[\frac{h}{2} + \frac{d}{6}\right]$$

Assuming that both upper and lower end of the vane shear the soil.

In undisturbed condition, $T_1 = c_u \pi d^2 \left[\frac{h}{2} + \frac{d}{6}\right]$

$$\Rightarrow \quad 40 \times 10^{-3} = c_u \times \pi \times (0.06)^2 \times \left[\frac{0.1}{2} + \frac{0.06}{6}\right]$$

$$\Rightarrow \quad c_u = \frac{40 \times 10^{-3}}{6.786 \times 10^{-4}} = 58.95 \text{ kN/m}^2$$

Since the soil is soft clay, the unit undrained shear resistance will be the shear strength of the soil. Therefore shear strength of soil in undisturbed state is 58.95 kN/m^2.

In remoulded condition,

$$T_2 = c_{ur} \pi d^2 \left[\frac{h}{2} + \frac{d}{6}\right]$$

$$\Rightarrow \quad 15 \times 10^{-3} = c_{ur} \times \pi \times (0.06)^2 \times \left[\frac{0.1}{2} + \frac{0.06}{6}\right]$$

$$\Rightarrow \quad c_{ur} = \frac{15 \times 10^{-3}}{6.786 \times 10^{-4}} = 22.1 \text{ kN/m}^2$$

Thus shear strength of soft clay in remoulded condition is 22.1 kN/m^2

$$\textbf{Sensitivity of soil} = \frac{\text{Undisturbed soil strength}}{\text{Remoulded soil strength}} = \frac{58.95}{22.1} = 2.67$$

Thus the soil is normal as the value of sensitivity lies between 1 and 4

6.5 An embankment is being constructed of soil whose effective shear strength peremeters c' and ϕ' are rspectively, $c' = 51$ kN/m^2 and $\phi' = 21°$. The unit weight of soil is 5.7 kN/m^3. The pore pressure parameters as determined from tri-axial test are $A = 0.5$ and $B = 0.9$. Find the shear strength of the soil at the base of the embankment just after the height of fill has been raised from 3 m to 6 m. Assume that the dissipation of pore pressure during this stage of construction is negligible, and that the lateral pressure at any point is one-half of the vertical pressure.

[12 marks]

Sol:

The equation for pore pressure is

$$\Delta u = B[\Delta\sigma_3 + A(\Delta\sigma_1 - \Delta\sigma_3)]$$

$$\Delta\sigma_1 = \text{Vertical pressure due to 3 m of fill}$$

$$= 3 \times \gamma = 3 \times 5.7 = 17.1 \text{ kN/m}^2$$

$$\Delta\sigma_3 = \frac{17.1}{2} = 8.55 \text{ kN/m}^2$$

Therefore, $$\Delta u = 0.9[8.55 + 0.5 \times (17.1 - 8.55)] = 11.54 \text{ kN/m}^2$$

Original pressure, $$\sigma_1 = 3 \times 5.7 = 17.1 \text{ kN/m}^2$$

Therefore, $$\sigma' = \sigma_1 + \Delta\sigma_1 - \Delta u = 17.1 + 17.1 - 11.54 = 22.66 \text{ kN/m}^2$$

Shear strength, $$S = c' + \sigma' \tan\phi = 51 + 22.66 \times \tan 21°$$

$$= 59.69 \text{ kN/m}^2 \simeq 59.7 \text{ kN/m}^2$$

Level-2

6.6 A saturated clay sample was obtained in field without allowing the water content and void ratio to change. The sample was subjected to the following stresses in field.

$\sigma = 240$ kN/m^2, $u = 140$ kN/m^2, $\bar{\sigma} = 100$ kN/m^2

(i) On sampling what are the stresses acting on it.

(ii) Sample was placed in a triaxial cell and cell pressure applied = 100 kN/m^2 under undrained condition. What are the stress acting on it.

(iii) Cell pressure was raised to 320 kN/m^2 with drainage valve closed. What are the $\sigma, u, \bar{\sigma}$.

(iv) The drainage valve are opened and soil allowed to consolidate under the cell pressure. What are $\sigma, u, \bar{\sigma}$.

(v) Comment on nature of change in water content and degree of saturation after consolidation.

(vi) Drainage value was closed and additional axial stress applied with cell pressure held constant. When additional stress was 100 kN/m^2, the pore water pressure was found to be 80 kN/m^2. What is the value of A at this stage.

(vii) When an additional axial stress of 200 kN/m^2 had been applied, the sample failed and at that stage, the pore pressure coefficient $A = 1$.

(a) What is $\bar{\sigma}_{1F}$?

(b) What was ϕ' value of the soil if it had $c' = 0$?

[20 Marks]

Sol:

(i) On sampling total stress = 0

But as void ratio does not change, $\bar{\sigma}$ is constant.

$$\sigma = 0$$

$$\bar{\sigma} = 100 \text{ kN/m}^2$$

$$u = -100 \text{ kN/m}^2$$

(ii) Total stress (σ) = 100 kN/m²

Pore water pressure = $-100 + 100 = 0$

Effective stress = 100 kN/m²

(iii) Total stress = 320 kN/m²

Pore water pressure = $0 + 220 = 220$ kN/m²

Effective stress = 100 kN/m²

(iv) Pore water pressure dissipates to zero. Hence total stress becomes effective.

$$\sigma = 320 \text{ kN/m}^2$$

$$u = 0$$

$$\bar{\sigma} = 320 \text{ kN/m}^2$$

(v) Water content will change and degree of saturation remains constant

(vi) We know that, $\Delta U = B\Delta\sigma_3 + A(\Delta\sigma_1 - \Delta\sigma_3)$

$$\Delta\sigma_3 = 0$$

For complete saturation, $B = 1$

$$A = \frac{\Delta U}{\Delta\sigma_1 - \Delta\sigma_3} = \left(\frac{80}{100-0}\right) = 0.8$$

(vii)

$$\frac{\Delta U}{\Delta\sigma_1 - \Delta\sigma_3} = A$$

As confining pressure does not change, $\Delta\sigma_3 = 0$

$$\Delta U = A\Delta\sigma_1 = 1 \times 200 = 200 \text{ kN/m}^2$$

Pore water pressure becomes $0 + 200 = 200$ kN/m²

$$\text{Total stress } (\sigma_1) = 320 + 200 = 520 \text{ kN/m}^2$$

$$\bar{\sigma}_{1F} = 520 - 200 = 320 \text{ kN/m}^2$$

Note at failure condition, $\bar{\sigma}_{1F}$ is same as before application of additional axial stress because drainage not allowed.

$$C = 0$$

$$\frac{\bar{\sigma}_{1F}}{\bar{\sigma}_{3F}} = \frac{1+\sin\phi'}{1-\sin\phi'}$$

$$\frac{320}{320-200} = \frac{1+\sin\phi'}{1-\sin\phi'}$$

$$\Rightarrow \quad \phi' = 27.03^\circ$$

6.7 **The results of a CU triaxial test conducted on a soil sample are as follows:**

Cell Pressure (kPa)	1000	1800
Additional Stress at Failure (kPa)	1600	2200
Pore Pressure at Failure (kPa)	400	800

Find the shear strength parameters with respect to

(i) Total Stress

(ii) Effective Stress

[20 marks]

Sol:

(i)

Here, cell pressure = σ_3. So the additional stress at failure will be

$$\sigma_d = \sigma_1 - \sigma_3$$

Pore pressure = u, therefore, major principal stress will be

$$\sigma_1 = \sigma_3 + \sigma_d$$

Major principal effective stress will be

$$\sigma_1' = \sigma_1 - u$$

Effective cell pressure will be $\sigma_3' = \sigma_3 - u$

The calculation of total and effective major and minor principal stresses is shown in the following table:

σ_3	σ_d	σ_1	u	σ_1'	σ_3'
1000	1600	2600	400	2200	600
1800	2200	4000	800	3200	1000

Case I : Calculation of total shear strength parameter:

Consider total stresses σ_1 and σ_3, from the table, so

$$\sigma_1 = \sigma_3 \tan^2\alpha + 2c\tan\alpha$$

$$4000 = 1800\tan^2\alpha + 2c\tan\alpha \quad \text{...(i)}$$

$$2600 = 1000\tan^2\alpha + 2c\tan\alpha \quad \text{...(ii)}$$

After subtracting (ii) from (i), we get

$$1400 = 800\tan^2\alpha$$

$$\Rightarrow \quad \tan^2\alpha = \frac{1400}{800} = 1.75$$

$$\Rightarrow \quad \tan\alpha = \sqrt{1.75} = 1.323$$

$$\Rightarrow \quad \alpha = \tan^{-1} 1.323 = 52.913^\circ$$

Substituting the value of $\tan^2\alpha$ and $\tan\alpha$ in eq. (i), we get

$$4000 = 1800 \times 1.75 + 2c \times 1.323$$

$$\Rightarrow \quad 2c \times 1.323 = 4000 - 1800 \times 1.75$$

$$\Rightarrow \quad 2.646\,c = 850$$

$$\Rightarrow \quad c = \frac{850}{2.646} = 321.24 \text{ kPa}$$

Also, $$\alpha = 45^\circ + \frac{\phi}{2} = 52.913$$

$$\Rightarrow \quad \frac{\phi}{2} = 52.913 - 45 = 7.913$$

$$\Rightarrow \quad \phi = 2 \times 7.913 = 15.83^\circ$$

Case II : Calculation of effective shear strength parameter:

Considering effective stresses, we get

$$\sigma_1' = \sigma_3' \tan^2\alpha' + 2c'\tan\alpha'$$

$$3200 = 1000\tan^2\alpha' + 2c'\tan\alpha' \quad \text{...(iii)}$$

$$2200 = 600\tan^2\alpha' + 2c'\tan\alpha' \quad \text{...(iv)}$$

After subtracting (iv) from (iii),

$$1000 = 400 \tan^2 \alpha'$$

$$\Rightarrow \quad \tan^2 \alpha' = \frac{1000}{400} = 2.5$$

$$\Rightarrow \quad \tan \alpha' = 1.581$$

$$\Rightarrow \quad \alpha' = \tan^{-1}(1.581) = 57.69°$$

Substituting the value of $\tan^2 \alpha'$ and $\tan \alpha'$ in eq. (iii), we get

$$3200 = 1000 \times 2.5 + 2c' \times 1.581$$

Therefore,

$$2c' \times 1.581 = 3200 - 1000 \times 2.5$$

$$\Rightarrow \quad 3.162\, c' = 700$$

$$\Rightarrow \quad c' = \frac{700}{3.162} = 221.38 \text{ kPa}$$

and

$$\alpha' = 45° + \frac{\phi'}{2}$$

$$\Rightarrow \quad 57.69 = 45° + \frac{\phi'}{2}$$

$$\Rightarrow \quad \frac{\phi'}{2} = 57.69 - 45 = 12.69$$

$$\Rightarrow \quad \phi = 2 \times 12.69 = 25.38°$$

Thus, the total shear strength parameters are

$$c = 321.24 \text{ ka}; \qquad \phi = 15.83°$$

Effective shear strength parameters are

$$c' = 221.38 \text{ kPa}; \qquad \phi' = 25.38°$$

6.8 **The following results were obtained from an undrained triaxial test on a compacted soil sample using a cell pressure of 300 kPa. Before the application of the cell pressure, the pore water pressure within the sample was zero.**

Strain (%)	σ_1 (kPa)	u (kPa)
0.0	300	120
2.5	500	150
5.0	720	150
7.5	920	120
10.0	1050	80
15.0	1200	10
20.0	1250	–60

(i) Determine the value of the pore pressure coefficient B and state whether or not the soil was saturated.

(ii) Plot the variation of deviator stress with strain.

(iii) Plot the variation of the pore pressure coefficient A with strain.

[20 marks]

Sol:

$$B = \frac{\Delta u_a}{\Delta \sigma_3} = \frac{120}{300} = 0.4$$

The soil was partially saturated as B is less than 1.0.

Strain (%)	Δu_d	$(\Delta\sigma_1 - \Delta\sigma_3)$	$\bar{A}$	$A\left[\frac{\bar{A}}{\bar{B}}\right]$
—	—	—	$\frac{30}{200} = 0.15$	0.375
2.5	30	200		
5.0	30	420	0.071	0.178
7.5	0	620	0	0
10.0	–40	750	–0.053	–0.135
15.0	–110	900	–0.122	–0.305
20.0	–180	950	–0.189	–0.4725

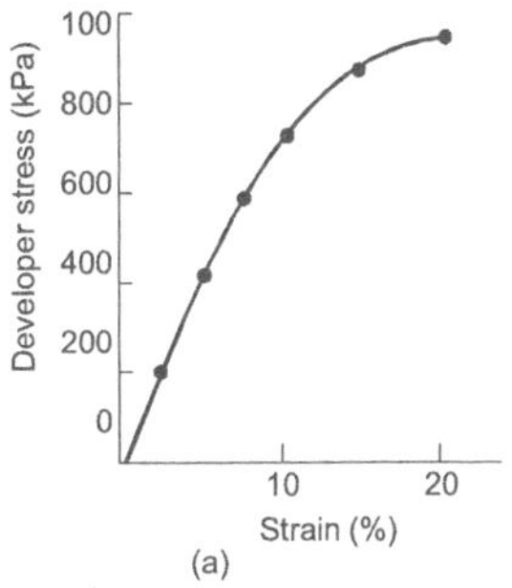

(a)

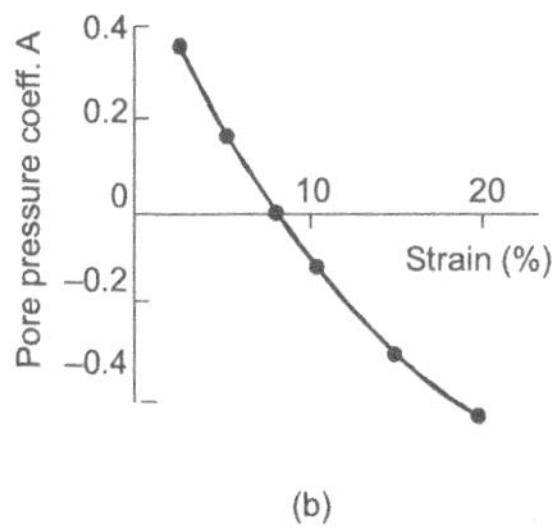

(b)

6.9 A rectangular foundation 4 m × 5 m transmits a total load of 5 MN to a deep, uniform deposit of stiff, overconsolidated clay with an OCR = 4 and γ_{sat} = 18 kN/m^3 (Figure below). Groundwater level is at 1 m below the ground surface. A CU test was conducted on a soil sample taken at a depth 5 m below the center of the foundation. The results obtained are $(s_u)_p$ = 40 kPa, ϕ'_p = 28° (peak state), and ϕ'_{cs} = 24° (critical state). Determine if the soil will reach the failure state for short term and long-term loading conditions. If the soil does not reach the failure state, what are the factors of safety for short-term and long-term loading conditions? Assume the soil above the groundwater level is saturated. The vertical stress increase at z = 5 m under the centre of rectangular foundation is given as, 71.1 kPa in z and 5.1 kPa in x direction respectively.

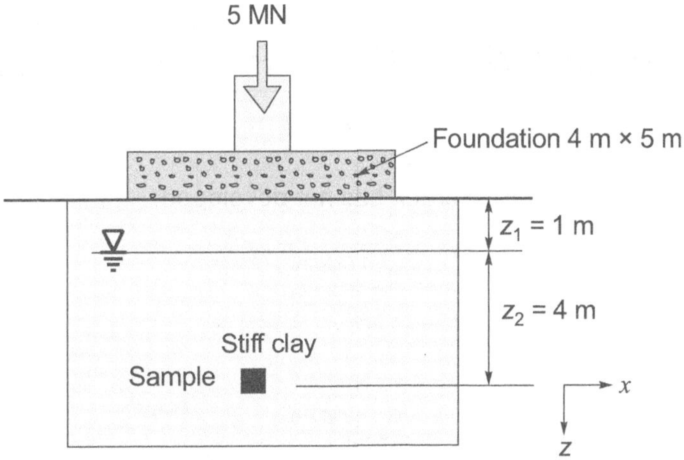

[20 Marks]

Sol:

Step 1: Determine the initial stresses.

$$\sigma'_z = \sigma'_1 = \gamma_{sat} z_1 + (\gamma_{sat} - \gamma_w) z_2$$

$$= (18 \times 1) + (18 - 9.81)4 = 50.76 \text{ kPa}$$

$$K_o^{oc} = K_o^{nc}(OCR)^{0.5} = (1-\sin\phi'_{cf})(OCR)^{0.5} = (1-\sin 24°)(4)^{0.5} = 1.2$$

$$\sigma'_3 = K_o^{oc}\sigma'_z = 1.2 \times 50.76 \simeq 61 \text{kPa}$$

$$\sigma_1 = \sigma'_z + z_2\gamma_w = 50.76 + 4 \times 9.81 = 90 \text{ kPa}$$

$$\sigma_3 = \sigma'_3 + z_2\gamma_w = 61 + 4 \times 9.81 = 100.24 \text{ kPa}$$

Step 2: The vertical stress increases at $z = 5$ m under the center of the rectangular foundation.

$$\Delta\sigma_z = 71.1 \text{ kPa}, \Delta\sigma_x = 5.1 \text{ kPa}$$

Step 3: Determine imposed shear stress for short-term loading.

Current vertical total stress: $(\sigma_1)_T = \sigma_1 + \Delta\sigma_z = 90 + 71.1 = 161.1$ kPa

Current horizontal total stress: $(\sigma_1)_T = \sigma_3 + \Delta\sigma_z = 100.24 + 5.1 = 105.34$ kPa

$$\text{Current shear stress, } \tau_u = \frac{(\sigma_1)_T - (\sigma_3)_T}{2} = \frac{161.1 - 105.34}{2} \simeq 28 \text{ kPa} < 40 \text{ kPa}$$

The soil will not reach the peak stress state.

Step 4: Determine failure shear stress for long-term loading.

For long-term loading, we will assume that all the excess porewater pressure has dissipated.

$$\text{Final effective stresses: } (\sigma'_1)_F = \sigma'_1 + \Delta\sigma_z = 50.76 + 71.1 = 121.86 \text{ kPa}$$

$$(\sigma'_3)_F = \sigma'_3 + \Delta\sigma_x = 61 + 5.1 = 66.1 \text{kPa}$$

$$\text{Angle of friction mobilized: } \phi' = \sin^{-1}\left(\frac{\sigma'_1 - \sigma'_3}{\sigma'_1 + \sigma'_3}\right) = \sin^{-1}\left(\frac{121.86 - 66.1}{121.86 + 66.1}\right) = 17.27°$$

The critical state angle of friction is 24°, which is greater than ϕ'. Therefore, failure would not occur.

Factor of safety based on critical state:

$$FS = \frac{\tan 24°}{\tan 17.27°} = 1.43$$

Factor of safety based on peak state:

$$FS = \frac{\tan 28°}{\tan 17.24°} = 1.71$$

7. Retaining Wall / Earth Pressure Theories

Level-1

7.1 Find active thrust P_A when tension crack has not developed.

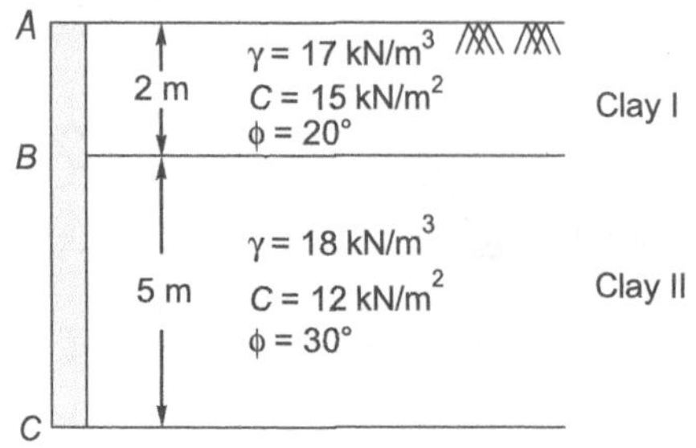

[12 Marks]

Sol:

$$k_{a_1} = \left(\frac{1 - \sin 20°}{1 + \sin 20°}\right) = 0.490$$

$$k_{a_2} = \left(\frac{1-\sin 30^\circ}{1+\sin 30^\circ}\right) = \left(\frac{1}{3}\right)$$

$$p_A = k_{a_1}\gamma_1 z_1 - 2c_1\sqrt{k_a}$$

$$= 0.490 \times 17 \times 0 - 2 \times 15 \times \sqrt{0.490} = -21 \text{ kN/m}^2$$

$$p_{B_1} = -21 + 0.490 \times 17 \times 2 = -4.34 \text{ kN/m}^2$$

$$p_{B_2} = (17 \times 2) \times \frac{1}{3} - 2 \times 12 \times \sqrt{\frac{1}{3}} = -2.523 \text{ kN/m}^2$$

$$p_C = -2.523 + 18 \times 5 \times \frac{1}{3} = 27.477 \text{ kN/m}^2$$

Pressure distribution diagram:

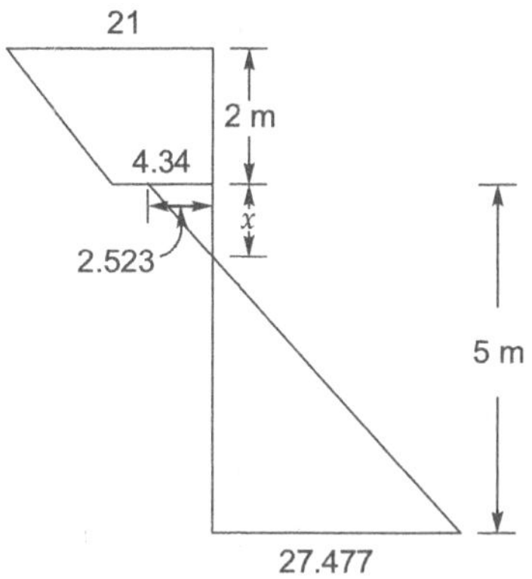

Calculation of x:

$$-2.523 + x \times 18 \times \frac{1}{3} = 0$$

$$x = 0.42 \text{ m}$$

When tension crack has not yet appeared, total force will be

$$p_A = \left(\frac{1}{2}\right)(27.477) \times (S - 0.42) - \left(\frac{1}{2}\right)(2.523)(0.42) - \left(\frac{1}{2}\right)(21 + 4.34) \times 2$$

$$= 37.053 \text{ kN/m}$$

7.2 In the same question, find the active thrust p_A when tension crack has developed.

[5 Marks]

Sol:

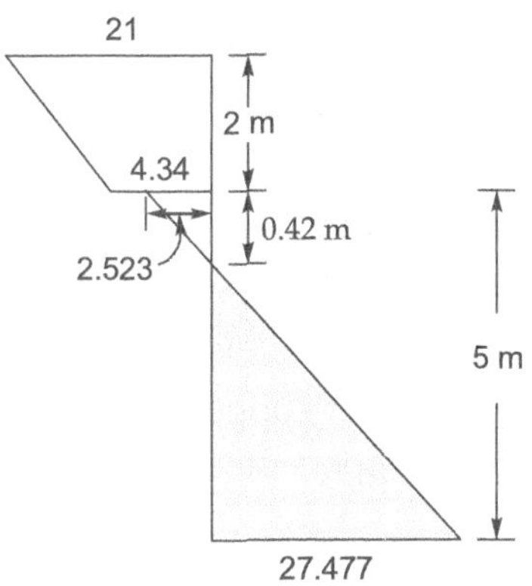

When the tension active thrust will be only due to shaded portion shown in the figure.

$$p_A = \left(\frac{1}{2}\right)(27.477)(5 - 0.42) = 62.92 \text{ kN/m}$$

7.3 A retaining wall 9 m high retains granular fill weighing 18 kN/m^3 with level surface. The active thrust on the wall is 180 kN/m length of the wall. The height of wall is to be increased and keep the force on the wall within allowable limit. The backfill in top half of the depth is removed and replaced by cinders. If cinders are used as backfill even in the additional height, what height may be allowed if the thrust on the wall is to be limited to its initial value? The unit weight of cinders is 9 kN/m^3. Assume the friction angle for cinder the same as that for soil.

[12 Marks]

Sol:

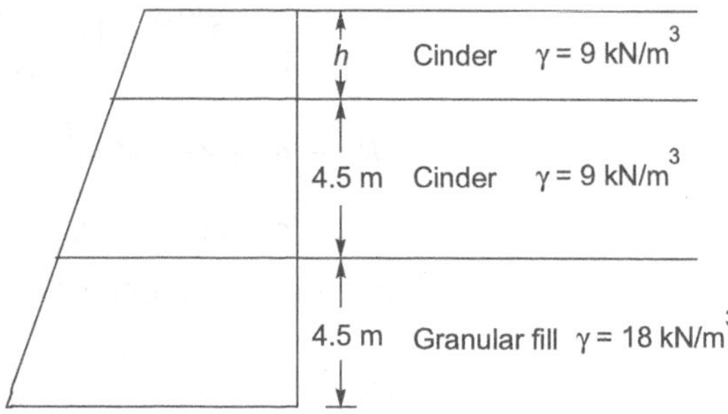

$$P_a = \frac{1}{2} k_a \gamma H^2$$

$$180 = \left(\frac{1}{2}\right)(k_a)(18)(9)^2$$

$$k_a = 0.247$$

(k_a) for cinder is 0.247.

[ϕ is same for both cinder and granular fill]

Let the increase in height of wall will be h.

Intensity of pressure at height $(h + 4.5)$ m $= (0.247)(h + 4.5) \times 9$

$= 2.223 (h + 4.5)$ kN/m^2

Intensity of pressure at base $= 2.223 (h + 4.5) + 0.247 \times 18 \times 4.5$

$= 20.007 + 2.223 (h + 4.5)$

$$\text{Total thrust} = 180 = \left(\frac{1}{2}\right)(2.223)(h+4.5)^2 + \frac{1}{2}[20.007 + 2.223(h+4.5) + 2.223(h+4.5)] \times 4.5$$

$$\text{Total thrust} = \left(\frac{1}{2}\right)(2.223)(h+4.5)^2 + \left(\frac{1}{2}\right)[20.007 + 2.223(h+4.5) + 2.223(h+4.5)] \times 4.5$$

$$1.112h^2 + 20h - 67.5 = 0$$

$$\Rightarrow \quad h = 2.9 \text{ m}$$

Thus, the height of wall may be increased by 2.9 m without increasing the thrust.

7.4 **List the factors that influence the choice of the side-support system to be used for vertical excavation in soil.**

[12 marks]

Sol:

The factors influencing the choice of side-support system are as follows:

(a) **Shape of cross-section of excavation:** If the cross-section is circular then ribs with lagging or segmental lining is often used. This is not so for square or rectangular cross-section of excavation.

(b) **Presence of groundwater :** If ground water is present and water tight side support is needed, then only sheet piles, diaphragm walls, secant piles or segmental lining with water tight gaskets are used. Wooden lagging and soil nails are not that much effective.

(c) **Temporary or permanent support :** If support is to be permanent, then concrete is usually the preferred material. For temporary support, steel and wooden members are used since they can be reused.

(d) **Size of the excavation :** For small sized excavations, wooden members are used and for large excavations, steel sheet piles and concrete members are used. In narrow width excavations, struts are used for bracing whereas in large width excavations, rackers or tiebracks are used.

(e) **Depth of excavation :** For deep excavations, strong steel and concrete members are used.

(f) **Installation technique :** Driven supports are avoided close to buildings due to noise and vibrations. Diaphragm walls are avoided if trench excavation causes settlement of adjacent footing.

7.5 **A soil was consolidated in an oedometer to a vertical stress of 100 kPa and then unloaded incrementally to 50 kPa. The excess pore water pressure is 0. If the frictional soil constant ϕ' is 25° then determine the lateral stress**

[12 marks]

Sol:

The soil in the case given becomes over consolidated. The past maximum vertical effective stress is 100 kPa and the current effective stress is 50 kPa.

The lateral pressure coefficient for normally consolidated soil is

$$K_o^{nc} = 1-\sin\phi' = 1 - \sin 25° = 0.58$$

Over consolidation ratio, $$OCR = \frac{100}{50} = 2$$

The lateral pressure coefficient for over consolidated soil is

$$K_0^{oc} = k_o^{nc} \times (OCR)^{1/2} = 0.58 \times \sqrt{2} = 0.82$$

∴ Lateral effective stress is

$$\sigma' = K_o^{oc} \times \bar{\sigma}_r = 0.82 \times 50 = 41 \text{ kPa}$$

Lateral total stress is

$$\sigma = \sigma' + \Delta u = 41 + 0 = 41 \text{ kPa}$$

Level-2

7.6 **For the retaining wall shown in the figure, plot the distribution of active earth pressure and determine the magnitude and point of application of the resultant active thrust.**

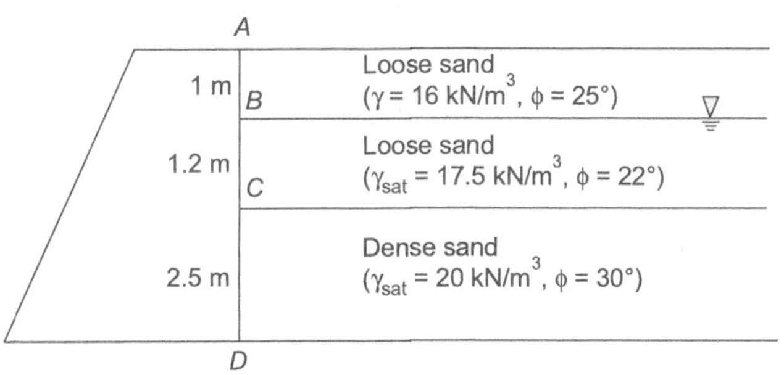

[16 Marks]

Sol:

Active earth pressure coefficient,

$$k_a = \frac{1-\sin\phi}{1+\sin\phi}$$

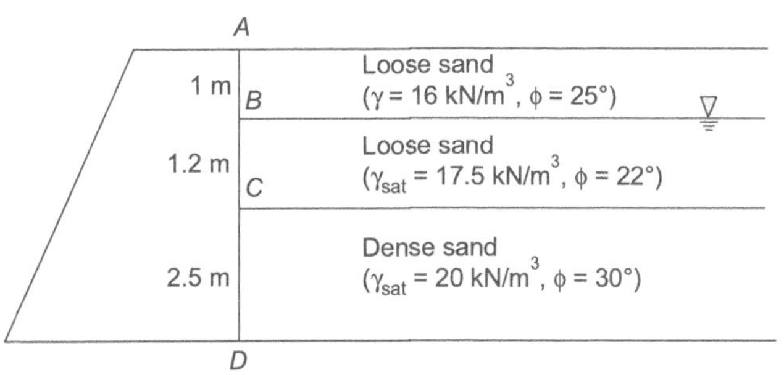

to find the earth pressure thrust end its point of application

For layer 1 : $$k_a = \frac{1-\sin 25°}{1+\sin 25°} = 0.406$$

Active earth pressure, $p_a = k_a\bar{\sigma}_z - 2c\sqrt{k_a}$ Here, $c = 0$

$\therefore$ $p_a = k_a\bar{\sigma}_z$

$A + A$, $\bar{\sigma}_z = 0$

$\therefore$ $p_a = 0$

A + B (just above)

$$\bar{\sigma}_z = \gamma \times 1 = 16 \times 1 = 16 \text{ kN/m}^2$$

$$p_a = 0.406 \times 16 = 6.496 \text{ kN/m}^2$$

For layer 2 : $k_a = \dfrac{1-\sin\phi}{1+\sin\phi} = 0.455$

A + B (just below), $\bar{\sigma}_z = 16 \text{ kN/m}^2$

$$p_a = k_a\bar{\sigma}_z = 0.455 \times 16 = 7.28 \text{ kN/m}^2$$

At C (Just above)

$$\bar{\sigma}_z = r \times 1 + \gamma_{sub}^2 \times 1.2$$

$$= 16 \times 1 + (17.5 - 9.81) \times 1.2 = 25.228 \text{ kN/m}^2$$

$\therefore$ $p_a = k_a\bar{\sigma}_z + \gamma_w z$

$$= 0.455 \times 25.228 + 9.81 \times 1.2 = 23.25 \text{ kN/m}^2$$

For layer 3 :

$$k_a = \frac{1-\sin 30^\circ}{1+\sin 30^\circ} = \frac{1}{3}$$

At C (Just below)

$$\bar{\sigma}_z = 25.228 \text{ kN/m}^2$$

$$p_a = k_a\bar{\sigma}_z + \gamma_w z = \frac{1}{3} \times (25.228) + 9.81 \times 1.2 = 20.181 \text{ kN/m}^2$$

At D,

$$\bar{\sigma}_z = \gamma \times 1 + \gamma_{sub}^2 \times 1.2 + \gamma_{sub}^3 \times 2.5$$

$$= 16 \times 1 + (17.5 - 9.81) \times 1.2 + (20 - 9.81) \times 2.5 = 50.703 \text{ kN/m}^2$$

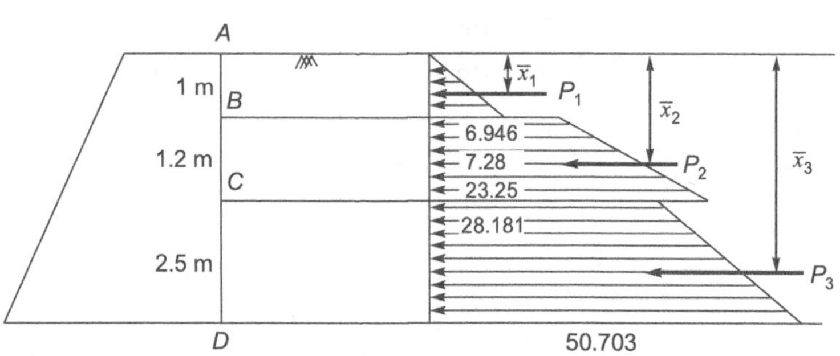

Pa (in kN/m²)
Active Earth Pressure Diagram

Active earth pressure thrust = Area under earth pressure diagram

$$P_1 = \frac{1}{2} \times 6.496 \times 1 = 3.248 \text{ kN/m} @ \bar{x}_1 = 2 \times \frac{1}{3} = \frac{2}{3} \text{ m from top}$$

$$P_2 = \frac{1}{2} \times (7.28 + 23.25) \times 1.2$$

$$= 18.318 \text{ kN/m at } \bar{x}_2 = 1 + \left(\frac{2 \times 23.25 + 7.28}{23.25 + 7.28}\right) \times \frac{1.2}{3} = 1.701 \text{ m from top}$$

$$P_3 = \frac{1}{2} \times (20.181 + 50.703) \times 2.5$$

$$= 88.605 \text{ kN/m at } \bar{x}_3 = 2.2 + \left(\frac{2 \times 50.703 + 20.181}{50.703 + 20.181}\right) \times \frac{2.5}{3} = 3.63 \text{ m from top}$$

∴ Total active earth pressure thrust

$$P = P_1 + P_2 + P_3 = 18.318 + 88.605 = 110.171 \text{ kN/m}$$

Resultant line of action of active earth pressure thrust

$$\bar{x} = \frac{P_1\bar{x}_1 + P_2\bar{x}_2 + P_3\bar{x}_3}{P_1 + P_2 + P_3}$$

$$= \frac{3.248 \times \frac{2}{3} \times 18.318 \times 1.704 + 88.605 \times 3.63}{110.171}$$

$$= 3.22 \text{ m from top}$$

7.7 **An anchored sheet pile founded in a cohesive soil is to retain a sandy backfill as shown in the below figure. Calculate the depth of embedment of the sheet pile and the pull in the anchor.**

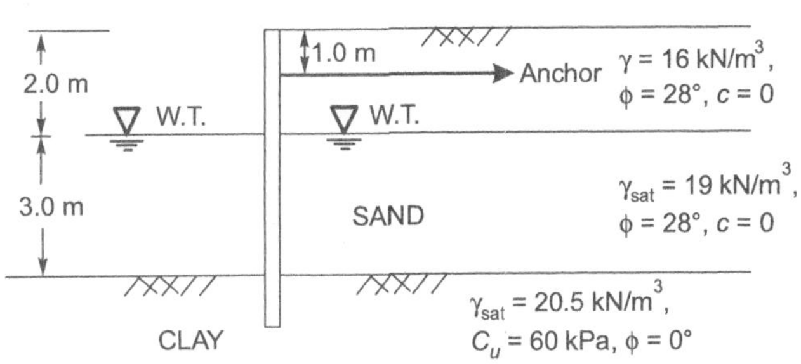

[20 Marks]

Sol:

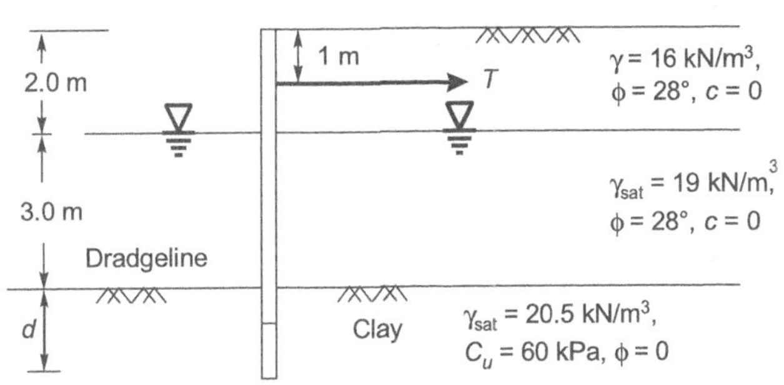

Let the depth of embedment of sheet pile wall = d

$$k_a = \tan^2\left(45 - \frac{28}{2}\right) = 0.36$$

For Clay,

$$k_P = \frac{1}{k_a}$$

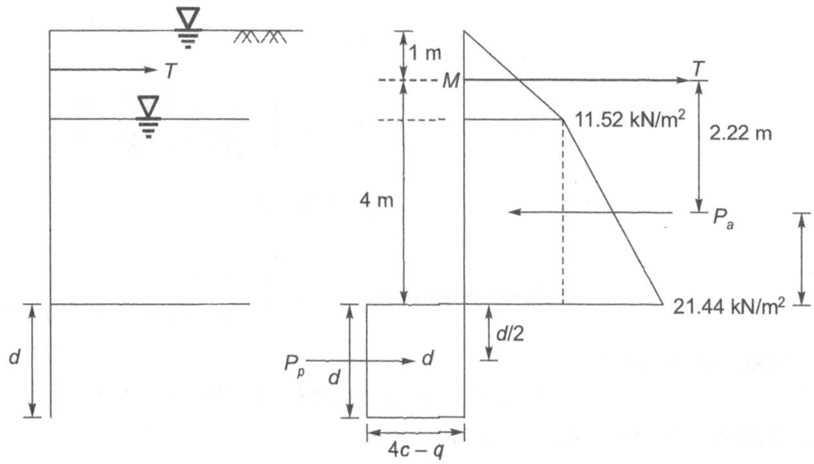

$$P_a = \frac{1}{2}\times 11.52\times 2 + 11.52\times 3 + \frac{1}{2}\times 3\times(21.44-11.52) = 60.96 \text{ kNm}$$

$$\bar{y} = \frac{11.52\times 3.67 + 34.56\times 1.5 + 14.88\times 1}{60.96} = 1.78 \text{ m}$$

$$P_p = (4c - q)\times d$$

Taking moment about M

$$\Rightarrow 60.96\times 2.22 - (4\times 60 - 59.57)d\left[4+\frac{d}{2}\right] = 0$$

$$\Rightarrow \quad 0.5d^2 + 4d - 0.75 = 0$$

$$d = 0.18 \text{ m}$$

Increase, the above d, 20 to 30% [let 25%]

$$\Rightarrow \quad P_{provided} = 1.25\times 0.18 = 0.225 \text{ m}$$

$$\Rightarrow \quad P_p = (4\times 60 - 59.57)\times 0.225 = 40.59 \text{ kN/m}$$

Pull in Anchor (let T)

equilibrium of horizontal forces

$$P_p + T - P_a = 0$$

$$\Rightarrow \quad T = P_a - P_p = 60.96 - 40.59 = 20.36 \text{ kN}$$

7.8 **A retaining wall with a smooth vertical backface has to retain a backfill of c-ϕ soil upto 5 above G.L. The surface of the backfill is horizontal and it has the following properties. $\gamma = 1.8$ t/m³, $C = 1.5$ t/m², $\phi = 20°$.**

(i) Plot the distribution of active earth pressure on the wall.

(ii) Determine the magnitude and point of application of active thrust.

(iii) Determine the depth of zone of tension cracks.

(iv) Determine the intensity of uniform surcharge which if placed over the backfill, can prevent the formation of tension cracks.

(v) Compute the resultant active thrust after placing the surcharge.

[20 Marks]

Sol:

Given : $c = 1.5$ t/m²; $\gamma = 1.8$ t/m³; $\phi = 20°$

(i)

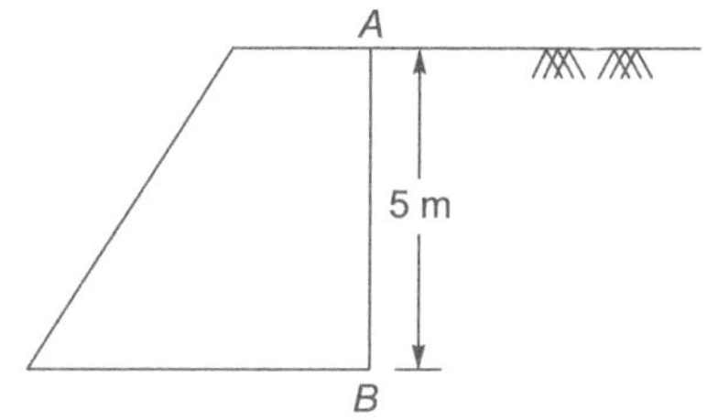

$$k_a = \left(\frac{1-\sin 20°}{1+\sin 20°}\right) = 0.490$$

$$P_A = -2c\sqrt{k_a}$$

$$= -2 \times 1.5 \times \sqrt{0.490} = -2.1 \text{ t/m}^2$$

$$P_B = k_a \gamma z - 2c\sqrt{k_a}$$

$$= 0.490 \times 1.8 \times 5 - 2 \times 1.5 \times \sqrt{0.490} = 2.31 \text{ t/m}^2$$

Active earth pressure distribution,

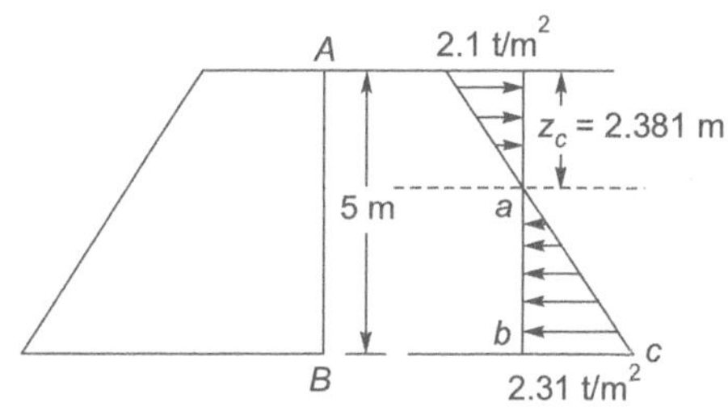

The depth of tension crack,

$$P = 0 = k_a \gamma z_c - 2c\sqrt{k_a}$$

$$Z_c = \left(\frac{2c}{\gamma\sqrt{k_a}}\right) = \left(\frac{2 \times 1.5}{1.8 \times \sqrt{0.490}}\right) = 2.381 \text{ m}$$

(ii)

The resultant active thrust is given by the part *abc* of the pressure distribution diagram

$$P_A = \left(\frac{1}{2}\right)(2.31) \times (5 - 2.381) = 3.025 \text{ kN/m}$$

Point of application is located at $\frac{(5-2.381)}{3} = 0.873$ m above the base of the wall.

(iii) Depth of tension crack = 2.381 m

(iv) The formation of tension cracks can be prevented by placing a surcharge q on the backfill which can neutralize the negative pressure.

$$(qk_a) - 2.1 = 0$$

$$q = \left(\frac{2.1}{k_a}\right) = \left(\frac{2.1}{0.490}\right) = 4.286 \text{ t/m}^2$$

(v) After placing the surcharge,

$$P_A = 0$$

$$P_B = 2.31 + 4.286 \times 0.490 = 4.410 \text{ t/m}^2$$

$$\text{Resultant thrust} = \left(\frac{1}{2}\right)(4.41) \times 5 = 11.025 \text{ t/m}$$

$$\text{Point of action} = \left(\frac{5}{3}\right) = 1.667 \text{ m above the base of wall}$$

7.9 **A 6 m deep excavation in sand is supported by a smooth vertical wall. The backfill is horizontal and supports a surcharge of 10 kN/m² on its surface.**

(i) Determine the active thrust on the wall if the water table is far below the bottom of the excavation.

(ii) What should be the embedment depth if the wall in the above case is cantilever sheet pile wall?

Properties of the sand are: $c' = 0$, $\phi' = 30$ degree and $\gamma_{bulk} = 16$ kN/m³

[20 Marks]

Sol:

(i)

$$\phi' = 30°, \gamma_b = 16 \text{ kN/m}^3$$

$$k_a = \frac{1-\sin\phi'}{1+\sin\phi'} \Rightarrow k_a = \left(\frac{1}{3}\right)$$

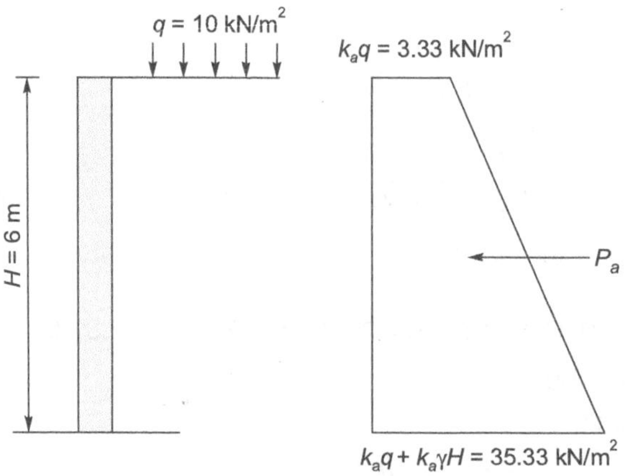

At top, $p_a = k_a q = 3.33\ \text{kN/m}^2$

At bottom $p_b = k_a q + k_a \gamma H = 35.33\ \text{kN/m}^2$

Active thrust, $P_a = \frac{1}{2} \times (3.33 + 35.33) \times 6$ ..(i.e., area of pressure diagram)

$\Rightarrow$ $P_a = 116\ \text{kN/m}$

(ii)

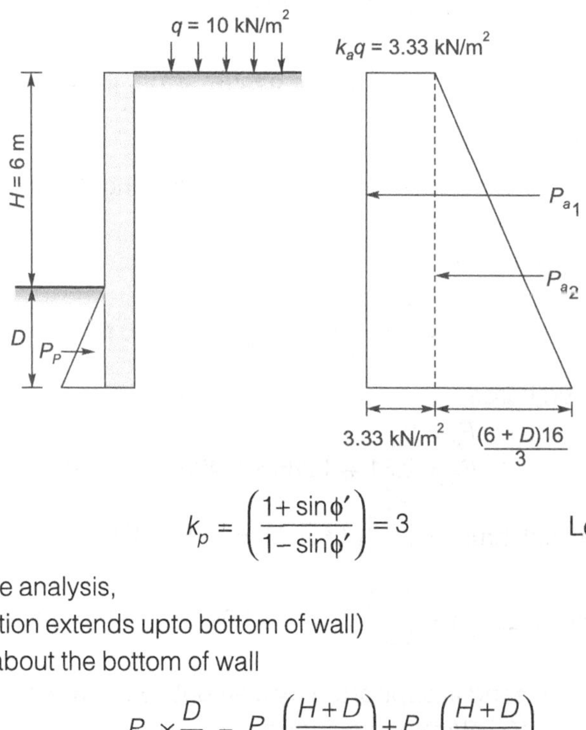

$$k_p = \left(\frac{1 + \sin\phi'}{1 - \sin\phi'}\right) = 3$$

Let D = depth of embankment

Using approximate analysis,

(Pressure distribution extends upto bottom of wall)

Taking moments about the bottom of wall

$$P_p \times \frac{D}{3} = P_{a_1}\left(\frac{H+D}{2}\right) + P_{a_2}\left(\frac{H+D}{3}\right)$$

$$\Rightarrow \quad k_p \gamma \frac{D^2}{2}\frac{D}{3} = k_a q (H+D)\left(\frac{H+D}{2}\right) + \frac{k_a \gamma}{2}(H+D)^2\left(\frac{H+D}{3}\right)$$

$$\Rightarrow \quad 3 \times 16 \times \frac{D^3}{6} = \frac{10(6+D)^2}{6} + \frac{16}{3}\frac{(6+D)^3}{6}$$

$$\Rightarrow \quad 48D^3 = 10(6+D)^2 + \frac{16}{3}(6+D)^3$$

On solving $D = 6.106\ \text{m}$

8. Stability of Slopes

Level-1

8.1 A long natural slope of 5 m depth in an over consolidated clay ($c' = 20$ kN/m^2, $\phi' = 30°$, $\gamma_{sat} = 20$ kN/m^3) is inclined at 10° to the horizontal. Determine factor of safety for

(i) Water table at surface and seepage is parallel to slope.

(ii) Fully submerged case (no seepage).

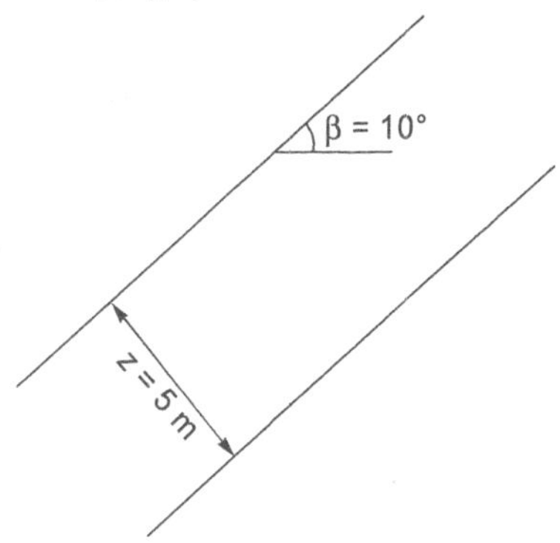

[10 Marks]

Sol:

$$\gamma_{sub.} = \gamma_{sat.} - \gamma_w = 20 - 9.81 = 10.19 \text{ kN/m}^2$$

Case 1: When seepage is parallel to slope.

$$\text{FOS for cohesive soil} = \frac{c + \gamma_{sub} \times z\cos^2\beta\tan\phi}{\gamma_{sat} z \times \sin\beta \times \cos\beta}$$

$$\text{FOS} = \frac{20 + 10.19 \times 5 \times \cos^2 10° \times \tan 30°}{20 \times 5 \times \sin 10° \times \cos 10°}$$

$$\text{FOS} = 2.84$$

Case 2: For completely submerged conditions.

$$\text{FOS} = \frac{c + \gamma_{sub} \times z\cos^2\beta\tan\phi}{\gamma_{sat} z \times \sin\beta \times \cos\beta}$$

$$\text{FOS} = \frac{20 + 10.19 \times 5 \times \cos^2 10° \times \tan 30°}{10.19 \times 5 \times \sin 10° \times \cos 10°}$$

$$\text{FOS} = 5.570$$

8.2 (a) Determine the factor of safety (in regard to cohesion only) for a submerged embankment 30.5 m high whose upstream face has an inclination of 30°. The soil properties are $\gamma_{sat} = 19$ kN/m^3, c = 38 kN/m^2, $\phi = 10°$. For $\phi = 10°$ and $i = 30°$, $s_n = 0.075$ (Assume seepage occurs)

(b) Also find the factor of safety if embankment experiences the effects of sudden drawdown, soil properties are same as that of the above conditions. [For $\phi = 4.8°$ and $i = 30°$, $s_n = 0.11$]

[8 Marks]

Sol:

(a) For $\phi = 10°$ and $i = 30°$, $s_n = 0.075$

$$\text{FOS} = \left(\frac{C}{\gamma_{sub.} \times H s_n}\right)$$

$$FOS = \frac{38}{(19-9.81)\times 30.5\times 0.075} = 1.808$$

(b) When drawdown occurs

$$\phi' = \frac{\gamma_{sub}}{\gamma_{sat}}\phi = \left(\frac{19-9.81}{19}\right)\times 10 = 4.8°$$

$\Rightarrow$ $s_n = 0.11$

$$FOS = \left(\frac{C}{\gamma_{sat.}Hs_n}\right) = \left(\frac{38}{19\times 30.5\times 0.11}\right) = 0.596$$

8.3 **Find the FOS (i) with respect to shear strength, and (ii) with respect to height along the indicated sliding surface *AB* in the figure below.**

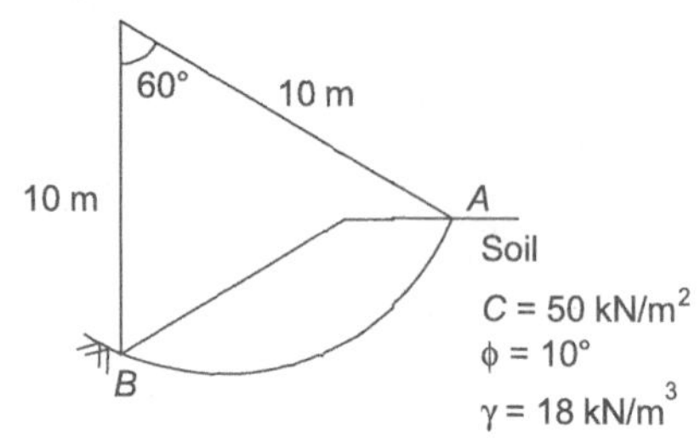

Normal effective pressure on sliding surface *AB* is 255 kN/m².
Downward tangential disturbing force along *AB* is 840 kN.

[8 Marks]

Sol:

(i) With respect to shear strength,

$$C_m = 50 \text{ kN/m}^2;\quad \Sigma N = 255 \text{ kN/m}^2;\quad \Sigma T = 840 \text{ kN/m}^2;\quad r = 10 \text{ m};\quad \theta = 60° = \frac{\pi}{3}$$

$$FOS = \frac{C_m\cdot r\theta + \Sigma N\tan\phi}{\Sigma T} = \frac{(50\times 10\times \pi/3)+(255\times \tan 10°)}{840}$$

$$= 0.68$$

(ii) With respect to height,

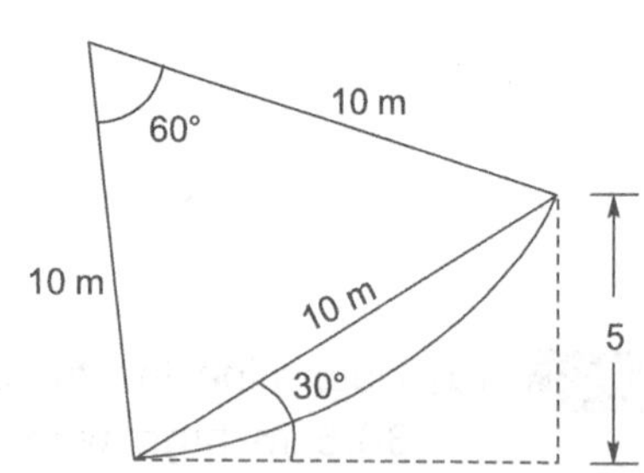

$$i = 30°$$

$$H_c = \frac{C\cdot \sec^2 i}{\gamma(\tan i - \tan\phi)} = \frac{50\times \sec^2(30°)}{18(\tan 30° - \tan 10°)} = 9.23 \text{ m}$$

$$H = 10\sin(i) = 10\sin 30° = 5 \text{ m}$$

$$FOS = \frac{H_c}{H} = \frac{9.23}{5} = 1.846$$

8.4 **What is construction pore pressure? How upstream and downstream slopes fail due to it.**

[5 Marks]

Sol:

Construction pore pressure : Many slope fails at the end of construction, if construction is rapid and there is no sufficient drainage system, pore water pressure is not allowed to be dissipated, this excess pore water pressure is called construction pore pressure.

Upstream and downstream slope failure occurs due to construction pore pressure:

When a dam is built of relatively impervious compressible soil, the drainage is extremely slow and excess pore pressure develop during and immediately after construction. When the permeability is low, there may be no substantial drop in pore pressure in central zone of the dam by the end of construction if this lies within the usual range of 2 - 4 years. An initial pore pressure upto 100% weight of the soil above the pint considered has been measured in the field. Under these circumstances, the construction stage in many cases may be more critical from the point of stability and the slopes may fail. Usually there are two kinds of construction slides. The first type of slide occurs slowly and continues at a uniform rate for a period usually of one or two weeks. The second type of slide occurs suddenly and rapidly.

Level-2

8.5 **Figure below gives details of an embankment to be made of cohesive soil with c_u = 20 kPa. The unit weight of the soil is 19 kN/m^3.**

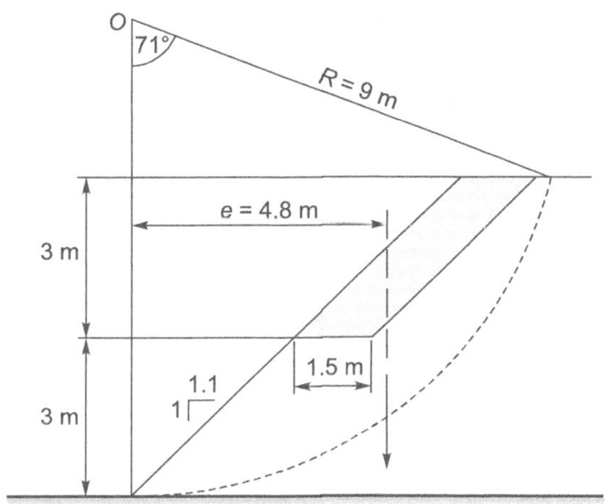

For the trial circle shown, determine the factor of safety against sliding soon after construction. The weight of the sliding sector is 329 kN acting at an eccentricity of 4.8 m from the centre of rotation. What would the factor of safety be if the shaded portion of the embankment were removed? In both cases, assume that no tension crack develops.

[12 marks]

Sol:

Case-1: Factor of safety against sliding soon after construction

Disturbing moment = 329 × 4.8 = 1579.2 kNm

Restoring moment = $c_u R^2 \theta$

$$= 20 \times 9^2 \times \frac{71}{180} \times \pi$$

$$= 2007.5 \text{ kNm}$$

∴ FOS, $F = \dfrac{2007.5}{1579.2} = 1.27$

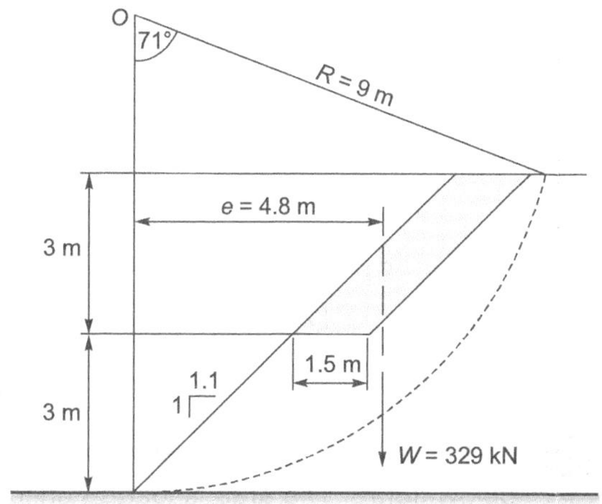

Cas-2: FOS is the shaded portion of the embankment were removed:

(considering unit thickness of soil mass)

Area of portion removed = 1.5 × 3 = 4.5 m^2

Weight of portion removed=4.5 × 19 = 85.5 kN

$$\text{Eccentricity from } O = 3.3 + \frac{3.3 + 1.5}{2} = 5.7 \text{ m}$$

∴ Reduction in of disturbing moment

$$= 5.7 \times 85.5 = 487.35 \text{ kNm}$$

∴ FOS,
$$F = \frac{2007.5}{1579.2 - 487.35} = 1.84$$

9. Vertical stresses

Level-1

9.1 **A rectangular foundation 3.0 m x 1.50 m carries a uniform load of 40 kN/m². Determine the vertical stresses at *p* which is 3 m below the ground surface. Use equivalent point load method.**

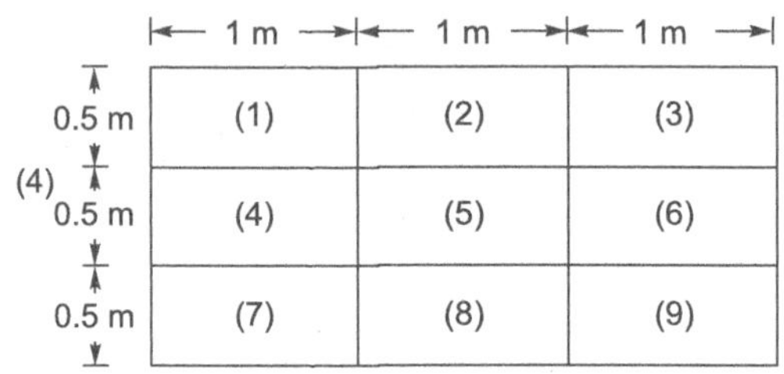

[8 Marks]

Sol:

Divide the loaded area into 9 small area of size 0.5 m × 1 m.

Load on each area = 40 × (0.5 × 1) = 20 kN

The stresses at point *P* can be determined using Boussinesq's solution for 9 point loads.

$$\text{For loads (1) and (4), } r = \sqrt{1.5^2 + 0.25^2} = 1.521$$

$$\text{For loads (2), (3), (5) and (6), } r = \sqrt{0.5^2 + 0.25^2} = 0.559$$

$$\text{For loads (8) and (9), } r = \sqrt{0.75^2 + 0.5^2} = 0.901$$

$$\text{For load (7), } r = \sqrt{1.5^2 + 0.75^2} = 1.677$$

$$\sigma_z = \Sigma \frac{3Q}{2\pi (z)^2} \times \frac{1}{\left[1 + \left(\frac{r}{z}\right)^2\right]^{5/2}}$$

$$\sigma_z = \left[\frac{3 \times 20}{2\pi \times (3)^2}\right]\left[\frac{2}{\left[1+\left(\frac{1.521}{3}\right)^2\right]^{5/2}} + \frac{4}{\left[1+\left(\frac{0.559}{3}\right)^2\right]^{5/2}} + \frac{2}{\left[1+\left(\frac{0.901}{3}\right)^2\right]^{5/2}} + \frac{1}{\left[1+\left(\frac{1.677}{3}\right)^2\right]^{5/2}}\right]$$

$$\sigma_z = 1.061\,[1.129 + 3.6741 + 1.612 + 0.507] = 7.34 \text{ kN/m}^2$$

9.2 **Compute the vertical stress on a horizontal plane situated at a depth of 2 m below point O in the below figure. The area is loaded uniformly to an intensity of 300 kN/m².**

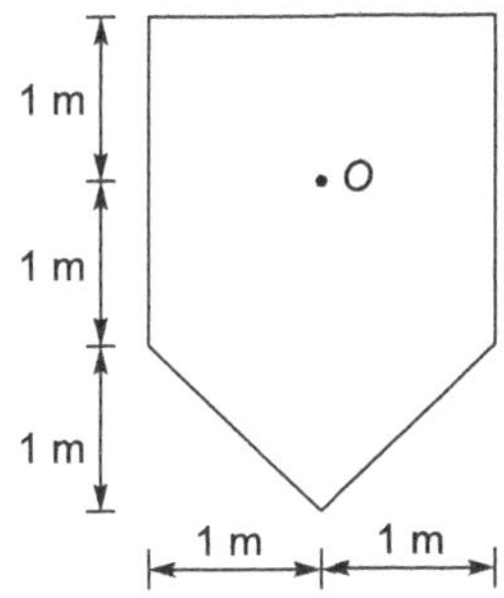

[10 Marks]

Sol:

Assuming that given loading can be taken as equivalent point load at P

where equivalent load, Q = Load intensity × Area

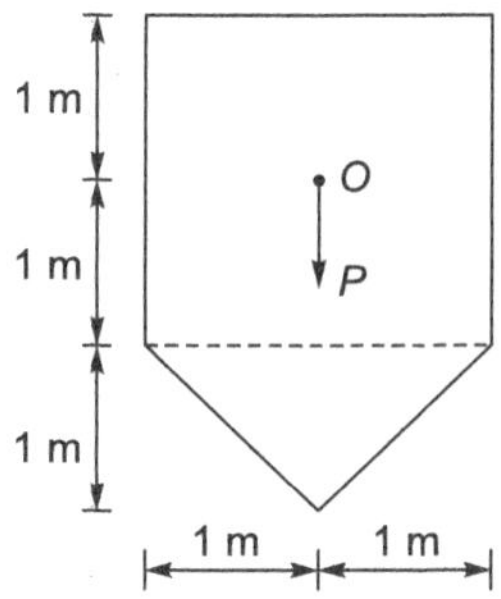

$$= 300\times\left(2\times2+\frac{1}{2}\times2\times1\right) = 1500 \text{ kN}$$

$$OP = r = \frac{300\times4\times0+300\times1\times\left(1+\frac{1}{3}\right)}{1500} = 0.267 \text{ m}$$

at depth, $z = 2$ m

Using Boussinesq's formula

$$\sigma_z = \frac{3Q}{2\pi z^2}\left(\frac{1}{1+\left(\frac{r}{z}\right)^2}\right)^{5/2} = \frac{3\times1500}{2\times\pi\times2^2}\left(\frac{1}{1+\left(\frac{0.267}{2}\right)^2}\right)^{5/2}$$

$$= 171.31 \text{ kN/m}^2$$

Level-2

9.3 **A 200 kN concentrated point load acts on the surface of soil mass. Determine the vertical stress 3 m below the ground surface at locations directly beneath the point of load application, 3 meters horizontally from the load and 6 meters horizontally from the load.**

(a) Assume Boussinesq condition apply.

(b) Assume westergaard condition apply.

[12 Marks]

Sol:

(a) By Boussinesq theory →

at point below 3 m:

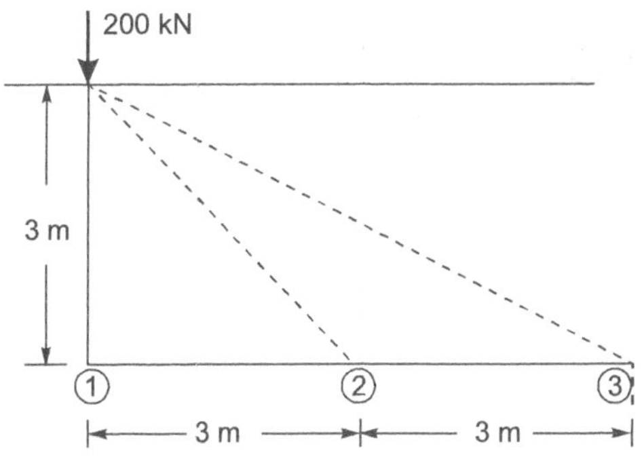

$$r = 0,$$

$$\Rightarrow \quad \frac{r}{z} = 0$$

$$\sigma_z = \frac{3Q}{2\pi z^2}\left(\frac{1}{1+\left(\frac{r}{z}\right)^2}\right)^{5/2}$$

$$\sigma_z = \frac{3\times200}{2\pi\times3^2}\left(\frac{1}{1+0}\right)^{5/2} = 10.61 \text{ kN/m}^2$$

- At point 3 m below and 3 m away from load

$$\sigma_z = \left(\frac{3\times200}{2\pi\times3^2}\right)\left(\frac{1}{1+\left(\frac{3}{3}\right)^2}\right)^{5/2} = 1.88 \text{ kN/m}^2$$

At point 3 m below and 6 m away from load,

$$\sigma_z = \frac{3\times200}{2\pi\times3^2}\left[\frac{1}{1+\left(\frac{6}{3}\right)^2}\right]^{5/2} = 0.190 \text{ kN/m}^2$$

(b) By westergaard theory,

- At 3 m depth; $r = 0$

$$\sigma_z = \frac{200}{(3)^2\times\pi}\left[\frac{1}{1+2(0)^2}\right]^{3/2} = 7.073 \text{ kN/m}^2$$

- At 3 m depth and 3 m away from load,

$$\sigma_z = \frac{200}{(3)^2\times\pi}\left[\frac{1}{1+2\left(\frac{3}{3}\right)^2}\right]^{3/2} = 1.361 \text{ kN/m}^2$$

- At 3 m depth and 6 m away from load,

$$\sigma_z = \frac{200}{(3)^2\times\pi}\left[\frac{1}{1+2\left(\frac{6}{3}\right)^2}\right]^{3/2} = 0.262 \text{ kN/m}^2$$

9.4 **The 1.0 m × 1.5 m footing shown in figure below supports a vertical load of 475 kN. Compute the magnitude of σ_z at a point *P*, considering both geostatic and induced stress.**

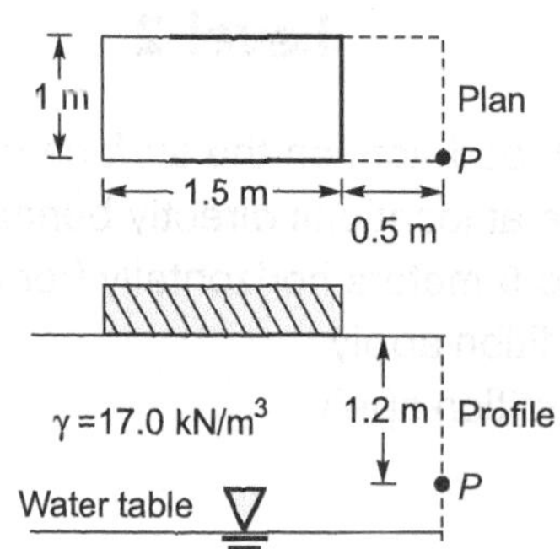

[12 Marks]

Sol:

The total stress at point *P* is the sum of geostatic stress and induced stress due to the load.

Total stress = Geostatic stress + Induced stress

Geostatic stress = $\gamma \times z = 17 \times 1.2 = 20.4$ kN/m²

Load intensity, $q = \frac{\text{Load}}{\text{Area}} = \frac{475}{1.0\times1.5} = 316.67$ kN/m²

Vertical stress below the corner of a uniformly loaded rectangular area is given by

$$\sigma_z = \frac{q}{4\pi}\left[\left(\frac{2mn\,(m^2+n^2+1)^{1/2}}{m^2+n^2+1+m^2n^2}\right)\times\left(\frac{m^2+n^2+2}{m^2+n^2+1}\right)+\tan^{-1}\left(\frac{2mn(m^2+n^2+1)^{1/2}}{m^2+n^2+1-m^2n^2}\right)\right]$$

where $m = \frac{L}{z}$ and $n = \frac{B}{z}$

(i) When $L = 2.0$ m and B = 1 m and z = 1.2 m

$$m = \frac{L}{z} = \frac{2}{1.2} = 1.67$$

$$n = \frac{B}{z} = \frac{1}{1.2} = 0.83$$

$$\therefore \quad \sigma_{z_1} = \frac{316.67}{4\pi}\left[\left(\frac{2\times1.67\times0.83\times(1.67^2+0.83^2+1)^{1/2}}{1.67^2+0.83^2+1+1.67^2\times0.83^2}\right)\times\left(\frac{1.67^2+0.83^2+2}{1.67^2+0.83^2+1}\right) + \tan^{-1}\left(\frac{2\times1.67\times0.83\times(1.67^2+0.83^2+1)^{1/2}}{1.67^2+0.83^2+1-1.67^2\times0.83^2}\right)\right]$$

$$\Rightarrow \quad \sigma_{z_1} = \frac{316.67}{4\pi}[1.1214+1.1598] = 57.48 \text{ kN/m}^2$$

(ii) When $L = 0.5$ m and $B = 1$ m and z = 1.2 m

$$m = \frac{L}{z} = \frac{0.5}{1.2} = 0.42$$

$$n = \frac{B}{z} = \frac{1}{1.2} = 0.83$$

$$\therefore \quad \sigma_{z_2} = \frac{316.67}{4\pi}\left[\left(\frac{2\times0.42\times0.83\times(0.42^2+0.83^2+1)^{1/2}}{0.42^2+0.83^2+1+0.42^2\times0.83^2}\right)\times\left(\frac{0.42^2+0.83^2+2}{0.42^2+0.83^2+1}\right) + \tan^{-1}\left(\frac{2\times0.42\times0.83\times(0.42^2+0.83^2+1)^{1/2}}{0.42^2+0.83^2+1-0.42^2\times0.83^2}\right)\right]$$

$$\Rightarrow \quad \sigma_{z_2} = \frac{316.67}{4\pi}[0.7362+0.4998]$$

$$\Rightarrow \quad \sigma_{z_2} = 31.15 \text{ kN/m}^2$$

$\therefore$ Induced stress,

$$\sigma_z = \sigma_{z_1} - \sigma_{z_2} = 57.48 - 31.15 = 26.33 \text{ kN/m}^2$$

$\therefore$ **Total stress = 20.4 + 26.33 = 46.76 kN/m²**

10. Shallow Foundation

Level-1

10.1 Using peck Hensons equation, determine the net allowable bearing pressure for a permissible settlement of 40 mm. The SPT is conducted at 1.5 m interval as shown in figure. The observed SPT values are 10, 15, 20 and 25 at depths 1.5 m, 3 m, 4.5 m and 6 m. Assume water table is at 3 m and $\gamma_w = 10$ kN/m³.

[12 Marks]

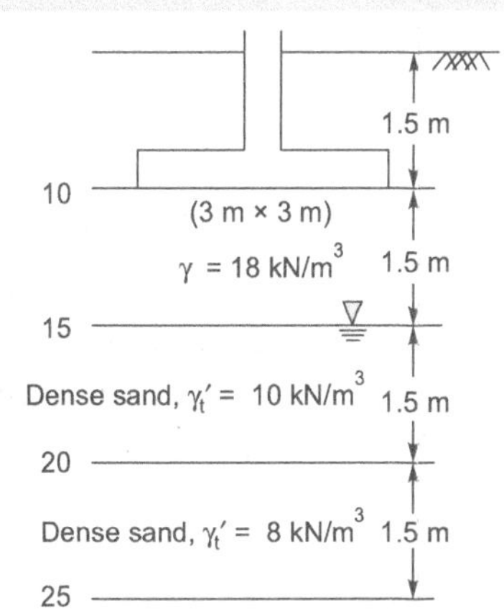

Sol:

By pack Hansen, equation

$$(q_a)_{net} \text{ kN/m}^2 = 0.44\, c_w N S_a$$

S_a = Permissible allowable settlement (in mm) = 40 mm

$$C_w = 0.5\left(1+\frac{D_w}{D_f+B}\right);\ C_W \ngtr 1$$

$$= 0.5\left(1+\frac{0.3}{1.5+3}\right) = 0.833$$

N = Average correct SPT 'Number'

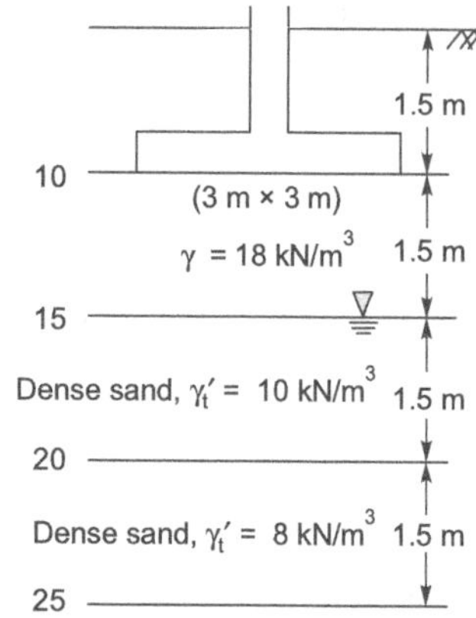

1. By overburden correction

$$N_1 = N_0 \cdot \frac{350}{\bar{\sigma}+70} \qquad \bar{\sigma} \ngtr 280\ \text{kN/m}^2$$

2. By dilatancy correction, $N_2 = 15+\frac{1}{2}(N_1-15)$

S.No.	SPT No.	$\bar{\sigma}(\text{kN/m}^2)$	N_1	N_2
1	10	$18\times1.5=27$	$\frac{350}{27+70}\times10=36.08$	36.08
2	15	$27+18\times1.5=54$	42.34	28.67
3	20	$54+\gamma'_{sub}\times1.5=54+10\times1.5=69$	50.36	32.68
4	25	$69+\gamma'_{sub}\times1.5=69+8\times1.5=81$	57.95	36.48

$$N_{avg} = \frac{36.08+28.67+32.68+36.48}{4} = 33.48$$

$$N_i \ngtr 1.5\,N_{avg} = 50.22 \qquad \text{(OK)}$$

$$\therefore \quad (q_0)_{net} = 0.44\times0.833\times33.48\times40$$

$$= 490.84\ \text{kN/m}^2$$

10.2 A multistorey, building with a basement is to be constructed. Top 4 m consists of loose silt, below which dense sand layer is present up to a great depth. Ground water table is at the surface. The foundation consists of the basement slab of 6 m width which will rest on the top of dense sand as shown in the figure. For dense sand, saturated unit weight = 20 kN/m³ and bearing capacity factor N_q = 40 and N_γ = 45. For loose silt, saturated unit weight = 18 kN/m³, N_q = 15 and N_γ = 20. Effective cohesion C is zero for both soils. Unit weight of water is 10 kN/m³. Neglect shape factor and depth factor. Average elastic modulus E and Poisson's ratio μ of dense sand is 60 × 10³ kN/m² and 0.3 respectively.

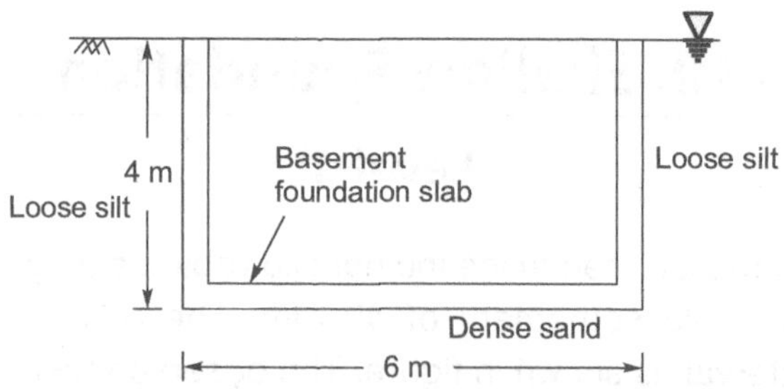

(i) Using factor of safety = 3, the safe bearing capacity (in kN/m²) of the foundation is

(ii) The foundation slab is subjected to vertical downward stresses equal to net safe bearing capacity derived in the above question (i). Using influence factor I_F = 2.0, and neglecting embedment depth and rigidity corrections, calculate the immediate settlement of the dense sand layer.

[10 Marks]

Sol:

(i) The safe bearing capacity, $q_s = \left[\frac{q_{nu}}{FOS} + \gamma_{sub.} D_F\right]$

$$q_{nu} = CN_C + q(N_q - 1) + 0.5B\gamma_{sub.}N$$
$$= 0 + (18 - 10) \times 4 \times (15 - 1) + 0.5 \times 6 \times (20 - 10) \times 45 = 1798 \text{ kN/m}^2$$

Net safe bearing capacity, $q_{ns} = \frac{1798}{3} = 599.33 \text{ kN/m}^2$

Safe bearing capacity, $q_s = \left(\frac{1798}{3} + 8 \times 4\right) = 631.33 \text{ kN/m}^2$

(ii) $$S_i = \frac{q_{ns}B(1-\mu^2)I_F}{E_s} = \frac{599.33 \times 6 \times (1-0.3^2) \times 2}{60 \times 10^3} = 109.07 \text{ mm}$$

10.3 A plate load test was conducted on a uniform deposit of sand and the size of the plate was 750 mm × 750 mm and that of the pit was 3.75 × 3.75 × 1.5 m. The failure stress is 500 kN/m². A square footing 2 m × 2 m is to be founded at 1.5 m depth in this soil. Assuming the factor of safety as 3 and maximum permissible settlement as 40 mm. Determine the allowable bearing pressure. Use Terzaghi formula ($N_q = 50$, $\gamma = 20$ kN/m³)

Pressure (kN / m²)	100	200	300	400	500	600
Settlement (mm)	2	4	7.5	12.5	20	40

[12 Marks]

Sol:

q_u of plate →

$$q_u = 0.4B\gamma N_\gamma$$
$$500 = 0.4 \times 0.75 \times 20\, N_\gamma$$

⇒ $$N_\gamma = 83.33$$
$$N_q = 50$$

For footing → $$q_{nu} = 0.4B\gamma N_\gamma + \gamma D_F(N_q - 1)$$
$$= 0.4 \times 20 \times 2 \times 83.33 + 20 \times 1.5 \times (50 - 1) = 2803.28 \text{ kN/m}^2$$

$$q_{safe} = \left(\frac{q_{nu}}{FOS}\right) = \left(\frac{2803.33}{3}\right) = 934.44 \text{ kN/m}^2$$

For settlement criteria,

$$\left(\frac{S_f}{S_p}\right) = \left[\left(\frac{B_F}{B_p}\right)\left(\frac{B_P + 0.3}{B_F + 0.3}\right)\right]^2$$

$$S_P = 40 \times \left[\frac{0.75(2+0.3)}{2(0.75+0.3)}\right]^2 = 27 \text{ mm}$$

Pressure for settlement of 27 mm = P

$$\left(\frac{40-27}{20}\right) = \left(\frac{600-P}{100}\right)$$

$$P = 535 \text{ kN/m}^2$$

Allowable bearing pressure is the smaller of the values from two criteria.

Hence, $$q_a = 535 \text{ kN/m}^2$$

10.4 A square footing for a column is to carry a loading of 1500 kN. Determine the footing dimension when bearing on the surface of a sand having properties indicated on the sketch. Use a factor of safety 3 for terzaghi bearing capacity equation.

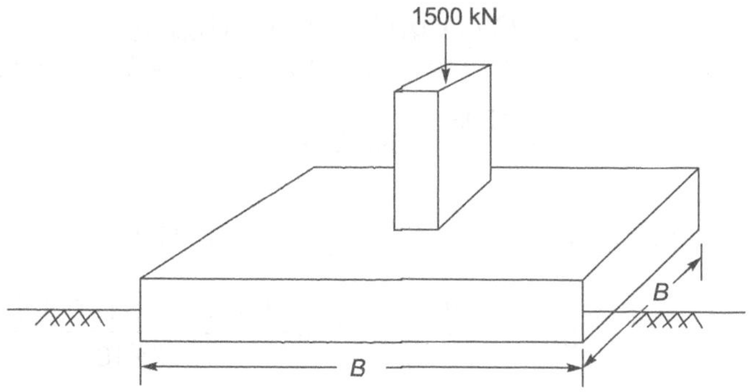

Sand: γ = 20 kN/m^3 ϕ = 36°

c = 0 N_y = 50, N_q = 43, N_c = 60

[8 Marks]

Sol:

As per Terzaghi capacity equation for sand

$$q_{nu} = \bar{\sigma}(N_q - 1) + 0.5B\gamma N_y\left(1 - 0.2 \times \frac{B}{L}\right)$$

q_{nu} = Net ultimate bearing capacity

i.e. here, $\bar{\sigma} = 0 \quad \because$ Surface loading

$$q_{nu} = 0.5B\gamma N_y\left(1 - 0.2 \times \frac{B}{B}\right) = (0.5) \times B \times 20 \times 50 \times (1 - 0.2) = 400\,B \text{ kN/m}^2$$

$$q_{nu} = \left(\frac{q_{nu}}{FOS}\right) = \left(\frac{400B}{3}\right) \text{ kN/m}^2$$

$$\frac{400B}{3} = \frac{1500}{B^2}$$

$$B = 2.24 \text{ m}$$

$\therefore$ Width of footing = 2.24 m

10.5 The footing of a wall is 1.5 m wide at the base and is located in a homogeneous cohesive soil at a depth of 1 m below G.L. The soil has γ = 17.6 kN/m^3, C = 36 kN/m^2, ϕ = 20°. Assuming the soil as soft which is likely to fail under local shear, determine the safe bearing capacity of soil per metre length of the footing. [Take FOS = 3]

ϕ = 20°	N_C = 17	N_q = 7	N_γ = 4.5
ϕ = 13.6°	N_C = 11	N_q = 4	N_γ = 1.6

[10 marks]

Sol:

Soil fails under local shear,

$\therefore$ Mobilised parameters c_m and ϕ_m should be used

$$c_m = \frac{2}{3} \times c = \frac{2}{3} \times 36 = 24 \text{ kN/m}^2$$

$$\tan \phi_m = \frac{2}{3}\tan\phi$$

$\Rightarrow$ $$\phi_m = \tan^{-1}\left(\frac{2}{3}\tan 20°\right) = 13.6°$$

$\therefore$ $$N_c = 11,\ N_q = 4,\ N_\gamma = 1.6$$

$$D_f = 1 \text{ m},\quad B = 1.5 \text{ m}$$

Ultimate bearing capacity, $$q_u = c_m N_c + qN_q + 0.5B\gamma N_\gamma$$

$\Rightarrow$ $$q_u = (24 \times 11) + (17.6 \times 1 \times 4) + (0.5 \times 1.5 \times 17.6 \times 1.6)$$

$\Rightarrow$ $$q_u = 355.52 \text{ kN/m}^2$$

$\therefore$ $$q_{nu} = q_u - \gamma D_f = 355.52 - (17.6 \times 1) = 337.92 \text{ kN/m}^2$$

$\therefore$ $$q_{ns} = \frac{q_{nu}}{F} = \frac{337.92}{3} = 112.64 \text{ kN/m}^2$$

$\therefore$ Safe bearing capacity, $$q_s = q_{ns} + \gamma D_f$$

$$q_s = 112.64 + (17.6 \times 1) = 130.24 \text{ kN/m}^2$$

$\therefore$ Safe bearing capacity per metre length of footing,

$$Q_s = q_s \times B = 130.24 \times 1.5 = 195.36 \text{ kN/m}$$

10.6 The base of a long retaining wall is 3 m wide and is 1 m below the ground surface in front of the wall: the water table is well below base level. The vertical and horizontal components of the base reaction are 282 kN/m and 102 kN/m, respectively. The eccentricity of the base reaction is 0.36 m. Appropriate shear strength parameters for the foundation soil are $c' = 0$ and $\phi' = 35°$, and the unit weight of the soil is 18 kN/m^3. Determine the factor of safely against shear failure.

[Take, for $\phi = 35°$, $N_q = 33$, $N_\gamma = 41$, take $i_q = 0.46$ and $i_\gamma = 0.21$]

[10 Marks]

Sol:

(ii) The effective width of the base is given by

$$B' = B - 2e = 3 - 2 \times 0.36 = 2.28 \text{ m}$$

For $\phi' = 35°$, the bearing capacity factors are $N_q = 33$ and $N_\gamma = 41$. The inclination factors are:

$$i_q = 1 - \left(\frac{1.5 \times 102}{282}\right) = 0.46$$

$$i_\gamma = 0.46^2 = 0.21$$

The ultimate bearing capacity is given by

$$q_f = \gamma D N_q i_q + \frac{1}{2}\gamma B' N_\gamma i_\gamma$$

$$= (18 \times 1 \times 33 \times 0.46) + \left(\frac{1}{2} \times 18 \times 2.28 \times 41 \times 0.21\right)$$

$$= 273.24 + 176.68 \simeq 450 \text{ kN/m}^2$$

$\therefore$ $$q_{nf} = q_f - \gamma D = 450 - 18 \times 1 = 432 \text{ kN/m}^2$$

The net base pressure is $$q_n = \frac{282}{2.28} - 18 \simeq 106 \text{ kN/m}^2$$

Then the factor of safety is $$F = \frac{q_{nf}}{q_n} = \frac{432}{106} = 4.08 \simeq 4.1$$

Level-2

10.7 **A concrete strip footing rectangular in cross-section is located at ground level and extends 1.2 m below the ground level. It carries UDL of 15000 kg/m. The soil profile consists of homogeneous clay 6 m thick overlying rock. The clay properties are as under:**
Saturated unit bulk weight = 1750 kg/m^3
Shear strength (undrained) = 8500 kg/m^2
Compressibility = 1 × 10^{-4} m^2/100 kg
Determine,
(i) width of footing for factor of safety, F = 2
(ii) ultimate consolidation settlement
Assume bulk unit weight of concrete = 2500 kg/m^3
Neglect the spread of load beneath the footing and any side cohesion on the foundation.

[15 Marks]

Sol:

(i) As per **Terzaghi's bearing capacity equation** for strip footing on clayey soil,

$$q_u = 5.7\,c_u + \gamma D_f$$

Now, the net ultimate bearing capacity of the soil is

$$q_{nu} = 5.7\,c_u$$

Net safe bearing capacity,

$$q_{ns} = \frac{q_{nu}}{F},$$

and thus safe bearing capacity,

$$q_s = q_{ns} + \gamma D_f = \frac{q_{nu}}{F} + \gamma D_f$$

Given γ = 1750 kg/m^3; c_u = 8500 kg/m^2; m_v = 1 × 10^{-4} m^2/100 kg

Factor of safety, $F = 2$

Unit weight of concrete = 2500 kg/m^3

Depth of footing, D_f = 1.2 m

Now UDL = 15000 kg/m

Self weight of concrete = 1.2 × B × 2500 = 3000 B kg/m

Total load = 15000 + 3000 B

Thus, total load carried by the footing should be equal to the safe bearing capacity of soil

$$\therefore \quad 15000 + 3000B = \left(\frac{q_{nu}}{F} + \gamma_{sat} D_f\right) \times B$$

$$\Rightarrow \quad 15000 + 3000B = \left(\frac{5.7c_u}{F} + \gamma_{sat} D_f\right) B$$

$$\Rightarrow \quad 15000 + 3000B = \left[\frac{5.7 \times 8500}{2} + (1750 \times 1.2)\right] \times B$$

$$\Rightarrow \quad B = 0.643\text{ m}$$

(ii) **Initial effective overburden pressure at the level of C-C,**

$$\bar{\sigma}_1 = 3.6 \times 1750 = 6300\text{ kg/m}^2$$

Final effective overburden pressure at the level C-C,

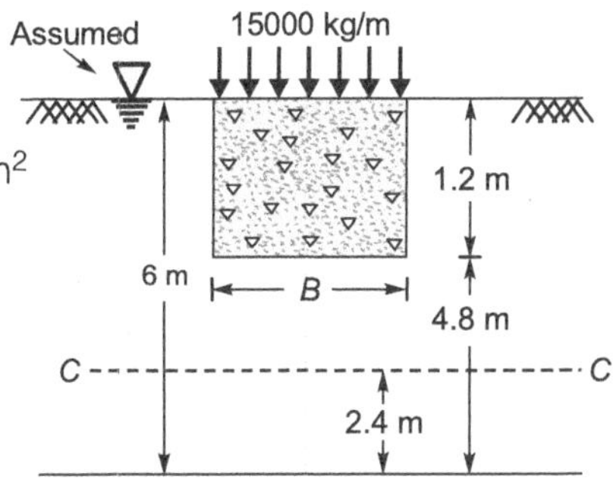

$$\bar{\sigma}_2 = 1.2 \times 2500 + 2.4 \times 1750 + \frac{15000}{B} - 3.6 \times 1000$$

$$= 3000 + 4200 + \frac{15000}{0.643} - 3600 = 26928.14 \text{ kg/m}^2$$

∴ Change in effective overburden pressure,

$$\Delta\bar{\sigma} = \bar{\sigma}_2 - \bar{\sigma}_1 = 26928.14 - 6300 = 20628.15 \text{ kg/m}^2$$

Ultimate consolidation settlement,

$$\Delta H = m_v H_0 \Delta\bar{\sigma}$$

$$= \frac{1 \times 10^{-4}}{100} \times 4.8 \times 20628.15 = 0.099 \text{ m} = 99.01 \text{ mm}$$

10.8 **Determine the allowable bearing capacity of a 2 m x 2 m square footing founded at a depth of 1.5 m below the G.L in a deep stratum of silty clay having the following avg. properties $\gamma = 1.8$ t/m^3, $c = 3$t/m^2$\phi = 0°$; $C_c = 0.259$, $e_0 = 0.85$**
The maximum permissible settlement of the footing is 7.5 cm the highest position of the WT at the site is at a depth of 1.0 m below G.L.

[15 Marks]

Sol:

Computation of bearing capacity

We know from skempton's equation

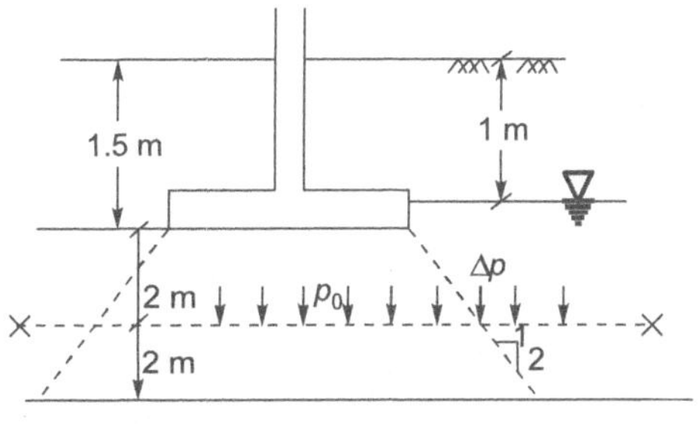

$$q_{nu} = cN_c$$

$$\therefore \quad \frac{D}{B} = \frac{1.5}{2} = 0.75 < 2.5$$

$$\therefore \quad N_c = 5\left(1 + 0.2\frac{D}{B}f\right)\left(1 + 0.2\frac{B}{L}\right)$$

$$= 5(1 + 0.2 \times 0.75)\left(1 + 0.2 \times \frac{2}{2}\right) = 5 \times 1.15 \times 1.2 = 6.9$$

$$\therefore \quad q_{nu} = cN_c$$

$$= 6.9c = 6.9 \times 3 = 20.7 \text{ t/m}^2 \quad \text{...(i)}$$

∴ For a FOS of 2.5, the net safe bearing capacity is given by

$$q_{nu} = \left(\frac{q_{ns}}{FOS}\right) = \left(\frac{20.7}{2.5}\right) = 8.28 \text{ t/m}^2$$

Computation of settlement

As the underlying soil is saturated silty clay, only consolidation settlement will take place.

- The zone of influence below the base of footing is extended to max depth of twice the width of footing, i.e. 4 m below the base.
- X-X is a horizontal plane through the middle of thin consolidation layer.

$$(\sigma_0)_{XX} = (\gamma_t z_1 + \gamma_{sub} z_2)$$

$$= (1.8 \times 1.0) + (1.8 - 1) \times (0.5 + 2.0) = 3.8 \text{ t/m}^2 = 0.38 \text{ kg/cm}^2$$

Using 2:1 dispersion method, stress increment at X-X

$$(\Delta\sigma)_{XX} = \frac{(8.28 \times 2.0 \times 2.0)}{(2.0 + 2.0)^2}$$

(Assuming the footing to be loaded with 8.28 t/m^2)

$$= 2.07 \text{ t/m}^2 \approx 0.207 \text{ kg/cm}^2$$

$$\Delta H = \frac{H.C_c}{(1+e_0)} \log_{10}\left(\frac{\sigma_0 + \Delta\sigma}{\sigma_0}\right)$$

$$= \frac{400 \times 0.259}{(1+0.85)} \log 10\left(\frac{0.38+0.207}{0.38}\right) = 10.58 \text{ cm} \quad \text{...(ii)}$$

As the Estimated Settlement is greater than the max. permissible limit of 7.5 cm. The allowable bearing capacity of the footing should be less than 8.28 t/m^2.

$$\frac{HC_c}{(1+e_0)} \log_{10}\left(\frac{(\sigma_0)_{XX} + \Delta\sigma}{\sigma_0}\right) = 7.5$$

$$\Rightarrow \frac{400 \times 0.259}{(1+0.85)} \log_{10}\left(\frac{0.38 + \Delta\sigma}{0.38}\right) = 7.5$$

$$\Rightarrow \frac{0.38 + \Delta\sigma}{0.38} = 1.3612$$

$$\Rightarrow \Delta\sigma = 0.1372 \text{ kg/cm}^2 = 1.372 \text{ t/m}^2$$

$$\Rightarrow \Delta\sigma = \frac{qBL}{(B+Z)\times(L+Z)} = \frac{qBL}{(B+Z)^2} = \frac{q \times (2)^2}{(2+2)^2}$$

$$\Rightarrow 1.372 = \frac{q \times 4}{16}$$

$$q = (1.372 \times 4)$$

$$q = 5.49 \text{ t/m}^2 = 5.5 \text{ t/m}^2$$

Hence a loading intensity of 5.5 t/m^2 will result in a consolidation settlement of 7.5 cm. Therefore, the required allowable bearing capacity of the footing = 5.5 t/m^2.

10.9 **A 3.0 m square footing is located in a dense soil and at a depth of 2.0 m. Determine the ultimate bearing capacity for the following water table positions:**

(i) at ground surface,

(ii) at footing level and

(iii) at 1m below the footing.

The moist unit weight of sand above the water table is 18 kN/m^3 and the saturated unit weight is 20 kN/m^3, ϕ = 35°, c = 0, N_q = 33 and N_γ = 34.0.

[15 Marks]

Sol:

Given data: Width of square footing (B) = 3.0 m, Depth of footing (D_f) = 2.0 m, Bulk unit weight of sand (γ_t) = 18 kN/m^3, Saturated unit weight of sand (γ_{sat}) = 20 kN/m^3

$$\phi = 35°;\ c = 0;\ N_q = 33 \text{ and } N_\gamma = 34.0$$

We know that for a square footing, ultimate bearing capacity is

$$q_u = 1.3\, cN_c + \gamma D_f N_q + 0.4\, \gamma B N_\gamma$$

For the given soil which is sand $c = 0$

$$\therefore \quad q_u = \gamma D_f N_q + 0.4\, \gamma B N_\gamma$$

The value of γ in the above equation is susceptible to the location of water table. Thus for each water table position, the ultimate bearing capacity will be different. So, the ultimate bearing capacity will be given as $q_u = \gamma_1 D_f N_q + 0.4 \gamma_2 B N_\gamma$

(i) **When water table is at the ground surface**

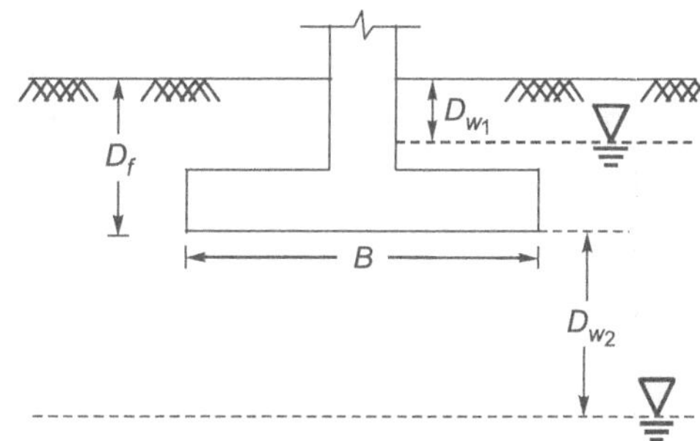

Now for the given condition,

$D_{w_1} = 0$ and D_{w_2} is negative, hence neglected.

$$\gamma_1 = \gamma' + \frac{D_{w_1}}{D_f}(\gamma - \gamma') \text{ and } \gamma_2 = \gamma' + \frac{D_{w_2}}{B}(\gamma - \gamma')$$

$$\therefore \quad \gamma_1 = (\gamma_{sat} - \gamma_w) + \frac{0}{2}(\gamma - \gamma') \text{ and } \gamma_2 = \gamma_{sat} - \gamma_w$$

$$\gamma_1 = 20 - 9.81 \text{ and } \gamma_2 = 20 - 9.81$$

$$\gamma_1 = 10.19 \text{ kN/m}^3 \text{ and } \gamma_2 = 10.19 \text{ kN/m}^3$$

$$\therefore \quad q_u = \gamma_1 D_f N_q + 0.4\gamma_2 B N_\gamma$$

$$= 10.19 \times 2 \times 33 + 0.4 \times 10.19 \times 3 \times 34 = 1088.292 \text{ kN/m}^2$$

(ii) **When the water table is at the footing level**

$$D_{w_1} = D_f = 2.0 \text{ m and } D_{w_2} = 0$$

$$\therefore \quad \gamma_1 = \gamma' + \frac{2}{2}(\gamma - \gamma') \text{ and } \gamma_2 = \gamma' + \frac{0}{3}(\gamma - \gamma')$$

$$\Rightarrow \quad \gamma_1 = \gamma' + (\gamma - \gamma') \text{ and } \gamma_2 = \gamma'$$

$$\Rightarrow \quad \gamma_1 = \gamma \text{ and } \gamma_2 = \gamma_{sat} - \gamma_w$$

$$\gamma_1 = 18 \text{ kN/m}^3 \text{ and } \gamma_2 = 20 - 9.81 = 10.19 \text{ kN/m}^3$$

$$\therefore \quad q_u = \gamma_1 D_f N_q + 0.4\gamma_2 B N_\gamma = 18 \times 2 \times 33 + 0.4 \times 10.19 \times 3 \times 34$$

$$= 1603.75 \text{ kN/m}^2$$

(iii) **When the water table is at 1 m below the footing**

$$\gamma_1 = \gamma \text{ and } \gamma_2 = \gamma' + \frac{1}{3}(\gamma - \gamma')$$

$$\Rightarrow \quad \gamma_1 = 18 \text{ kN/m}^3 \text{ and } \gamma_2 = \gamma' + \frac{\gamma}{3} - \frac{\gamma'}{3}$$

$$= \frac{2\gamma'}{3} + \frac{\gamma}{3} = \frac{1}{3}[2(\gamma_{sat} - \gamma_w) + 18] = \frac{1}{3}[2(20 - 9.81) + 18] = 12.79 \text{ kN/m}^3$$

$$q_u = 18 \times 2 \times 33 + 0.4 \times 12.79 \times 3 \times 34 = 1709.83 \text{ kN/m}^2$$

11. Deep Foundation and Machine Foundation

Level-1

11.1 A 30 cm circular concrete pile, 15 m long is driven in a deposit of medium dense sand ($\phi = 36°$, $N_\gamma = 40$ and $N_q = 42$). The unit weight of sand is 15 kN/m^3. What is the allowable load with a factor of safety 30. Assume lateral earth pressure coefficient = 1.6.

[8 Marks]

Sol:

Given, length of pile (L) = 15 m

Diameter of pile (D) = 300 mm = 30 cm = 0.3 m

Concrete pile in dense sand, $\phi = 36°$

$$\therefore \quad \delta = \frac{3}{4}\phi = \frac{3}{4} \times 36 = 27°$$

Lateral earth pressure coefficient,

$$k = 1.6$$

$$N_q = 42;\ \text{unit weight of sand } (\gamma) = 15\ \text{kN/m}^3$$

$$\text{Critical depth for dense sand} = 20D = 20 \times 0.3 = 6\ \text{m}$$

$$L > 20D$$

$\bar{\sigma}$ variation for dense sand,

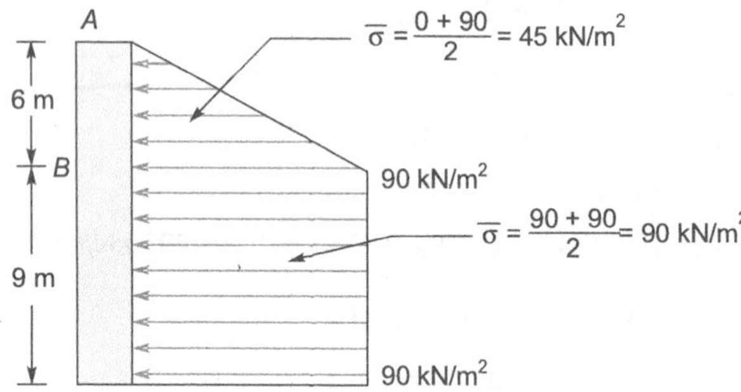

At 6 m,

$$\therefore \quad q_0 = \gamma_t D_f = 6 \times 15 = 90\ \text{kN/m}^2$$

∴ Ultimate load on the pile,

$$Q_u = q_0 N_q A_b + (k\bar{\sigma}_1 \tan\delta A_{s_1}) + (k\bar{\sigma}_2 \tan\delta A_{s_2})_{BC}$$

$$= 90 \times 42 \times \frac{\pi}{4} \times (0.3)^2 + (1.6 \times 45 \times \tan 27° \times \pi \times 0.3 \times 6) + (1.6 \times 90 \times \tan 27° \times \pi \times 0.3 \times 9)$$

$$Q_u = 1097\ \text{kN}$$

∴Allowable load on the pile,

$$Q_e = \frac{Q_u}{\text{FOS}} = \frac{1097}{3} = 365.67\ \text{kN}$$

11.2 **A bored concrete pile of 400 mm diameter and having an overall length of 12.5 m is embedded in a saturated stratum of $c - \phi$ soil having the following properties**
$c = 15\ \text{kN/m}^2$, $\phi = 20$; $\gamma_{sat} = 18\ \text{kN/m}^3$. Determine the safe bearing capacity of the pile. Given for $\phi = 20°$
The bearing capacity factors are
$N_c = 26$, $N_q = 10$, $N_\gamma = 4$
Assume reasonable values for all other factors.

[8 Marks]

Sol:

For Pile Embedded in $c - \phi$ soil

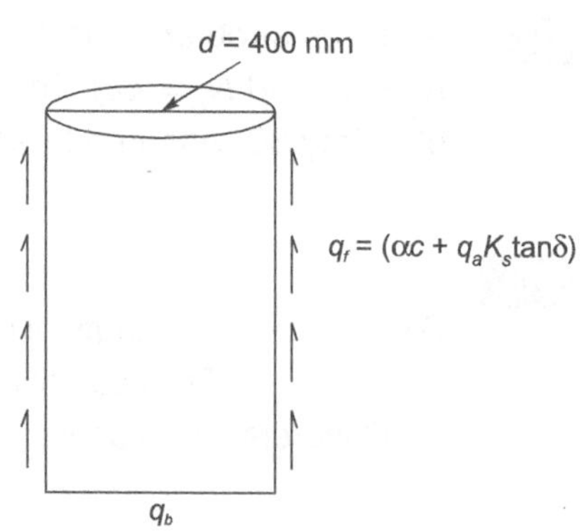

$$q_b = cN_c + \gamma' D(N_q - 1) + \frac{1}{2}(B\gamma' N_\gamma)$$

$$= 15 \times 26 + (18 - 10) \times 12.5(10 - 1) + \frac{1}{2} \times 0.40 \times 4(18 - 10)$$

$$= 1296.4\ \text{kN/m}^2$$

$$q_f = (\alpha_c + \bar{q}_a K_s \tan\delta)$$

Assume $\alpha = 0.5,\ K = 0.5,\ \dfrac{\delta}{\phi} = 0.80$

$$\delta = (0.80 \times 20) = 16°$$

$\Rightarrow$ $$q_f = (0.5 \times 15) + \left(\frac{\gamma' H}{2}\right) \times K_s \tan\delta = 7.5 + \frac{(18-10) \times 12.5}{2} \times 0.5 \times \tan 16 = 14.67 \text{ kN/}$$

m2

Again $$A_f = (\pi d l) = (\pi \times 0.40 \times 12.5) = 15.71 \text{ m}^2$$

$$A_b = \frac{\pi}{4} d^2 = \frac{\pi}{4} \times (0.4)^2 = 0.126 \text{ m}^2$$

$$Q_u = (14.67 \times 15.71) + (1296.4) \times 0.126 = 393.79 \text{ kN}$$

$$Q_s = \frac{Q_u}{(FOS)} = \frac{393.79}{3} = 131.26 \text{ kN}$$

11.3 A single under-reamed pile is installed in a soft clay deposit. The centre of the underream is located at a depth of 15 m from the ground surface. The diameters of the pile shaft and bulb are respectively 1.0 and 2.5 m. Determine the allowable load with a factor of safety of 2.5.
The underained shear strength of the soil obtained from the vane shear test is given by the relation

$$c_u = 65 + 7D$$

where c_u is in kN/m^2 and D is the depth in metres. Assume $\alpha = 1.0$ [8 Marks]

Sol:

The ultimate load Q_u is given by

$$Q_u = \frac{\pi}{4} B^2 \times 9c + \frac{\pi}{4}\left(B_1^2 - B^2\right) \times 9c' + \alpha c_a A_s$$

In this case taking, $$c' = c = 65 + 7 \times 15 = 170 \text{ kN/m}^2$$

and $$c_a = \frac{65 + 170}{2} = 117.5 \text{ kN/m}^2$$

$$Q_u = \frac{\pi}{4}(1)^2 \times (9 \times 170) + \frac{\pi}{4}\left(2.5^2 - 1\right) \times (9 \times 170) + 1.0 \times 117.5 \times (\pi \times 1 \times 15)$$

$$= 1201.66 + 6308.71 + 5537.05 = 13047.43 \text{ kN}$$

$$Q_a = \frac{Q_u}{2.5} = \frac{13047.43}{2.50} = 5218.97 \text{ kN}$$

11.4 A machine having a total weight of 20,000 kN has a unbalance such that it is subjected to a force of amplitude 5,000 kN at a frequency of 600 rpm. What should be the spring constant for the supporting springs if the maximum for transmitted into the foundation due to the unbalance is to be 500 kN? Assume that the damping can be neglected. [12 Marks]

Sol:

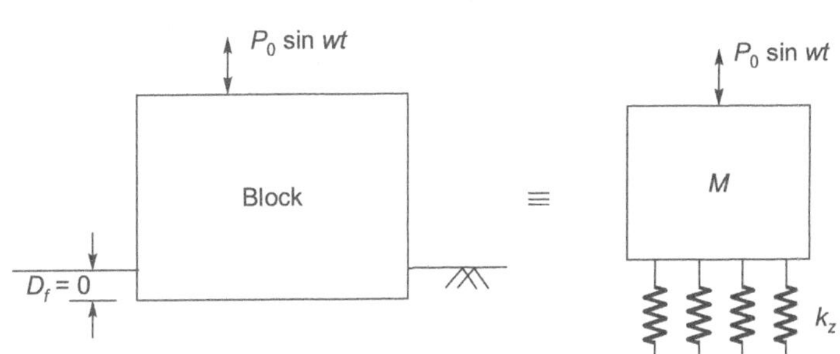

$$w_{nz} = \sqrt{\frac{k_z}{m}}; \quad A_z = \frac{P_0}{m\left(w_{nz}^2 - w^2\right)}$$

Now,
$$m = \frac{20{,}000 \times 10^3}{9.81} \text{ kg} = 2038735.984 \text{ kg}$$
$$P_0 = 5000 \text{ kN}$$
$$A_z = 500 \text{ kN}$$
$$w = \frac{2\pi N}{60} = \frac{2\pi \times 600}{60} = 20\pi$$
$\Rightarrow$
$$500 = \frac{5000}{2038735.984\left(w_{nz}^2 - w^2\right)}$$
$\therefore$
$$w_{nz} = 62.832$$
$$\sqrt{\frac{k_z}{m}} = 62.832$$
or,
$$k_z = 8048644698 \text{ kg/m} = 78957204.49 \text{ kN/m}$$
$\therefore$
$$k_z = 7.89 \times 10^7 \text{ kN/m}$$

11.5 An SPT test was conducted in a sand deposit at 12 m depth. The unit weight of sand is 20 kN/m³ and water table is at 2 m below natural ground surface. If the observed SPT value is 25 then what will be the corrected N value? Use the figure below to apply overburden correction.

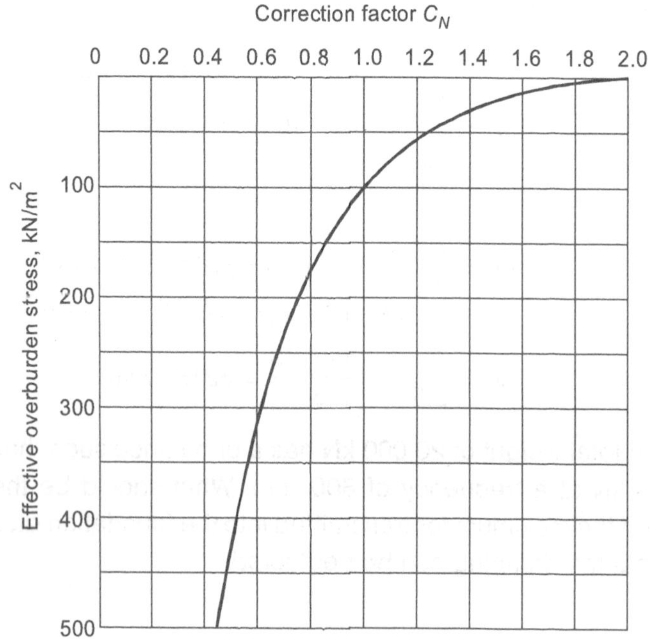

Correction factor for N-values for overburden correction

[12 marks]

Sol:

Given:
$$N_{obs} = 25$$
$$\gamma = 20 \text{ kN/m}^3$$
$$\gamma_w = 9.81 \text{ kN/m}^3$$

(i) Overburden correction

Effective stress at 12 m depth $= 12 \times 20 - 10 \times 9.81 = 141.9 \text{ kN/m}^2$
$$N' = C_N \cdot N_{obs}$$

C_N from figure corresponding to $\bar{\sigma}$ = 141.9 kN/m^2 is

$$C_N = 0.85$$

$$\therefore \quad N' = 0.85 \times 25 = 21.25 > 15$$

(ii) Dilatency correction $\quad N'' = 15 + \frac{(N'-15)}{2} = 15 + \frac{(21.25-15)}{2} = 18.125$

So corrected N value is 18.

Level-2

11.6 **Determine the safe load that can be carried by a pile having a gross weight of 1.5 *t*, using he modified Hiley's Formula. Given,**

Weight of hammer = 2.0 tonn **Height of free fall = 91 cm**

Hammer efficiency = 75% **Avg. penetration under the last 5 blows = 10 mm**

Length of pile = 22 mm **Diameter of pile = 300 mm**

Coefficient of Restitution = 0.55 %

[15 Marks]

Sol:

From the Modified Hiley's Formula

We know that, the Ultimate load on pile

$$Q_u = \frac{(\eta_h W H \eta_b)}{(S + C/2)} \quad \text{...(i)}$$

Where η_h = Efficiency of hammer = 75% = 0.75

W = 2.0 tonn

H = 91 cm

S = Avg. penetration under the last 5-blows = 10 mm = 1 cm

eP = 0.55 × 1.5 = 0.825 tonn

$\therefore$ W > eP

$$\Rightarrow \quad \eta_b = \frac{W + e^2 P}{(W + P)} = \frac{2 + (0.55)^2 \times 1.5}{(2 + 1.5)} = 0.701$$

In order to find out the value of Q_u, assume as a first approximate,

C = 2.5 cm

$$Q_u = \frac{\eta_h W H \eta_b}{(S + C/2)} = \left(\frac{0.75 \times 2 \times 91 \times 0.701}{1 + 2.5/2}\right) = 42.52 \text{ tonn}$$

Now using,

$$C_1 = 1.77\left(\frac{Q_u}{A_p}\right) = 1.77 \times \frac{42.52}{\frac{\pi}{4}(30)^2} = 0.106 \text{ cm}$$

$$C_2 = 0.657\left(\frac{Q_u L}{A_p}\right) = 0.657 \times \frac{(42.52 \times 22)}{(\pi/4)(30)^2} = 0.869$$

$$C_3 = 3.55 \times \left(\frac{Q_u}{A_p}\right) = 3.55 \times \frac{42.52}{(\pi/4)(30)^2} = 0.213 \text{ cm}$$

$$C = (C_1 + C_2 + C_3) = 1.1885 < 2.5 \text{ cm}$$

Let Q_u = 50 tonn

$$\therefore \quad C = \frac{1.188 \times 50}{(42.52)} = 1.397$$

$$\therefore \quad Q_u = \frac{(0.75\times2\times91\times0.701)}{(1+1.397/2)} = 56.33 \text{ tonn}$$

Let $\quad Q_u = 55$ tonn

$$\therefore \quad C = \frac{1.188\times55}{(42.52)} = 1.537$$

$$\therefore \quad Q_u = \frac{0.75\times2\times91\times0.7}{(1.0+1.537/2)} = 54.03 \text{ tonn}$$

In the second iteration, the assumed and computed values of Q_u are quite close. Hence, the ultimate load bearing capacity of the pile is 54 tonn.

∴ Therefore, the safe bearing capacity

$$Q_s = \left(\frac{Q_u}{F}\right) = \left(\frac{54}{2.5}\right) = 21.6 \text{ tonn}$$

11.7 **A 3.3 m x 2.2 m group consists of 9 m long piles. It carries a load of 1500 kN. The piles penetrate a sand layer (bulk unit weight = 16 kN/m³) 1 m thick and rest in a normally consolidated clay having a saturated unit weight of 19 kN/m³, specific gravity of 2.7 and liquid limit of 50. Clay extends to a depth of 12 m below the top of sand layer, ground water is at depth of 1 m below the ground surface. Bulk unit weight of sand is 16 kN/m³. Compute the settlement of clay if pile cap rests on the top of sand. Use 2 : 1 stress distribution to estimate the average increase of stress in the clay layer.**

[15 Marks]

Sol:

For given data:

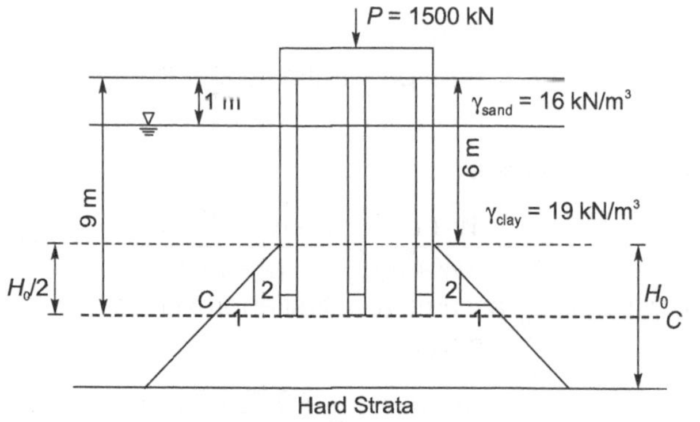

Unit weight of sand,	γ_{sand}	$= 16$ kN/m³
Effective unit weight of clay,	γ'_{clay}	$= 19 - 9.81 = 9.19$ kN/m³
Compression index of clay,	C_c	$= 0.009\,(w_L - 10) = 0.009\,(50 - 10) = 0.36$

For clay using,
$$\gamma_{sat} = \left(\frac{G+e}{1+e}\right)\gamma_w$$

$$19 = \frac{2.7+e_o}{1+e_o}\times 9.81$$

Initial void ratio, $\quad e_0 = 0.815$

For footing size $\quad (B\times D) = 3.3 \times 2.2$ m

Depth of top of compressible layer $= \frac{2}{3}\times 9 = 6$ m

Height of compressible layer, $H_0 = 12 - 6 = 6$ m

Overburden pressure at centre of compressible layer,

$$\sigma_0 = 16 \times 1 + 9.19 \times 8 = 89.52 \text{ kN/m}^2$$

Increase in stress due to load, assuming 2V : 1 H stress distribution,

$$\Delta\sigma = \frac{P}{\left(B+\frac{H_o}{2}\right)\left(D+\frac{H_o}{2}\right)} = \frac{1500}{\left(2.2+\frac{6}{2}\right)\times\left(3.3+\frac{6}{2}\right)} = 45.79 \text{ kN/m}^2$$

Settlement,

$$\Delta H = \frac{C_c H_o}{1+e_o}\log_{10}\left(\frac{\sigma_o+\Delta\sigma}{\sigma_o}\right) = \frac{0.36\times 6}{1+0.815}\log_{10}\left(\frac{89.52+45.79}{89.52}\right)$$

$$= 0.2135 \text{ m} = 213.5 \text{ mm}$$

11.8 **The results of a soil investigation at a particular site are given below. The water table is at 1 m below the surface:**

Depth (m)	Soil	Properties
0.00 – 4.0	Very soft clay	$N_{av} = 1$, $C_u = 6$ kN/m^2, $\gamma = 15$ kN/m^3
4.0 – 13.0	Loose sand	$N_{av} = 8$, $\gamma = 17$ kN/m^3
13.0 – 32.0	Med. stiff silty clay	$N_{av} = 7$, $C_u = 13.5$ kN/m^2, $\gamma = 16$ kN/m^3
32.0 – 40.0	Dense sand	$N_{av} = 45$, $\gamma = 19$ kN/m^3

Compute the safe capacity of a 500 mm diameter, 35 m long bored cast in situ pile.

[For $N = 8, \phi = 28°$

$N = 45, \phi = 40°, N_\gamma = 109$ and $N_q = 130$]

Any other data if required, can be assumed suitably. **[20 Marks]**

Sol:

Using static formula:

Let us assume adhesion factor $\alpha = 0.7$, 0.5, 0.4 and 0.3 for soft clay, loose sand, stiff clay and dense sand respectively.

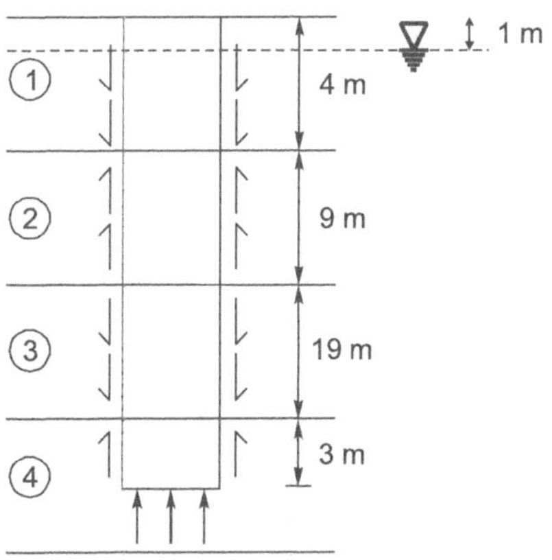

$$Q_{up_1} = \alpha \cdot C \cdot \pi d \times 4$$

$$= 0.7 \times 6 \times \pi \times 0.5 \times 4 = 26.34 \text{ kN}$$

$$Q_{up_2} = \left(\tan\phi \cdot K \cdot \frac{\gamma L}{2}\right)\cdot \pi d \times 9; \quad L \le z_c$$

$$z_c = 10\, d = 5 \text{ m}$$

As, $L = 9 \text{ m} > z_c \Rightarrow$ we will use $L = z_c = 5$ m

$$Q_{up_2} = (\tan 28°)(1-\sin 28°)\times\frac{17\times 5}{2} = 11.99 \text{ kN}$$

Q_{up_2} is negative skin friction

$$Q_{up_3} = \alpha \cdot c \cdot \pi d \times 19$$

$$= 0.4 \times 13.5 \times \pi \times 0.5 \times 19 = 161.16 \text{ kN}$$

$$Q_{up_4} = (q \cdot N_q)A_s + \left(\tan\phi \cdot K \cdot \frac{\gamma L}{2}\right)\pi d \times 3$$

Now,

$$q = (4 \times 15) + (9 \times 17) + (19 \times 16) + (3 \times 19) - 34\,\gamma_w$$

$$= 240.46 \text{ kN/m}^2$$

$$z_c = 10\,d = 5 \text{ m} > 3 \text{ m}$$

$$\therefore \quad Q_{up_4} = 240.46 \times 130 \times \frac{\pi \times 0.5^2}{4} + \tan 40°(1 - \sin 40°) \times \frac{19 \times 3}{2} \times \pi \times 0.5 \times 3$$

$$= 6178.10 \text{ kN}$$

$$\therefore \quad Q_{up} = Q_{up_1} - Q_{up_2} + Q_{up_3} + Q_{up_4}$$

$$= 26.34 - 11.99 + 161.16 + 6178.10 = 6353.61 \text{ kN}$$

Using N **value** for bored piles:

$$Q_{up_1} = 0.67 \times \bar{N} \times A_s = 0.67 \times 1 \times \pi \times 0.5 \times 4 = 4.21 \text{ kN}$$

$$Q_{up_2} = 0.67 \times \bar{N} \times A_s = 0.67 \times 8 \times \pi \times 0.5 \times 9 = 75.77 \text{ kN}$$

$$Q_{up_3} = 0.67 \times \bar{N} \times A_s = 0.67 \times 7 \times \pi \times 0.5 \times 19 = 139.97 \text{ kN}$$

$$Q_{up_4} = 133 \cdot N \cdot A_b + 0.67 \times \bar{N} \times A_s$$

$$Q_{up_4} = 133 \times 45 \times \frac{\pi \times 0.5^2}{4} + 067 \times 45 \times \pi \times 0.5 \times 3 = 1317.23 \text{ kN}$$

$$\therefore \quad Q_{up} = Q_{up_1} - Q_{up_2} + Q_{up_3} + Q_{up_4}$$

$$= 1385.64 \text{ kN} < 6353.61 \text{ kN}$$

$\Rightarrow$ Safe capacity of piles

$$= \frac{1385.64}{F} = \frac{1385.64}{2.5} = 554.256 \text{ kN}$$

11.9 **Design a square pile group to carry a load of 3000 kN including the weight of the pile cap at site where the soil is uniform clay to a depth of 20 m, underlain by rock. Average unconfined compressive strength of the clay is 70 kN/m². The clay may be assumed to be of normal sensitivity and normally loaded with liquid limit 60%. A factor of safety 3 is required against shear failure.**
[Take adhesion factor = 0.8, initial length of pile = 10 m and diameter of pile = 0.5 m]

[20 marks]

Sol:

$$C_u = \frac{q_u}{2} = \frac{70}{2} = 35 \text{ kN/m}^2$$

$$\text{Permissible, } C_u = \frac{35}{3} \text{ kN/m}^2$$

Length of pile = 10 m

Let the diameter of pile = 0.5 m = 50 cm

Let Spacing of piles = $3d = 3 \times 50 = 150$ cm

Let the number of piles = n

Considering the piles to act individually, the load at failure is given by

$$Q_{safe} = \frac{nQ_{up}}{F} = \frac{n\left[CN_c\frac{\pi}{4}d^2 + \alpha C\pi dL\right]}{3}$$

$$\Rightarrow \quad 3000 = n\times\left[\frac{35}{3}\times 9\times\frac{\pi}{4}\times 0.5^2 + 0.8\times\frac{35}{3}\times\pi\times 0.5\times 10\right]$$

$$\Rightarrow \quad n = \frac{3000}{167.23} = 17.94$$

For square arrangement keep $n = 16$

The modified length L will then have to be increased by the ratio $\frac{17.94}{16}$

$$\therefore \quad L' = 10\times\frac{17.94}{16} = 11.21\text{ m} \qquad \text{Adopt, } L = 12\text{ m}$$

Check for group action

$$B = 3s + d = 3\times 150 + 50 = 500\text{ cm} = 5\text{ m}$$

$$\therefore \quad \text{Load taken by group action} = 4BL\cdot C + A_b\cdot CN_c$$

$$A_b = B^2 = (5)^2 = 25\text{ m}^2,\quad N_c = 9\text{ (for clay)}$$

$$\therefore \quad Q_{vg} = \left(4\times 5\times 12\times\frac{35}{3}\right) + 25\times\frac{35}{3}\times 9$$

$$= 5425\text{ kN} > 3000\text{ kN} \qquad \text{Hence safe}$$

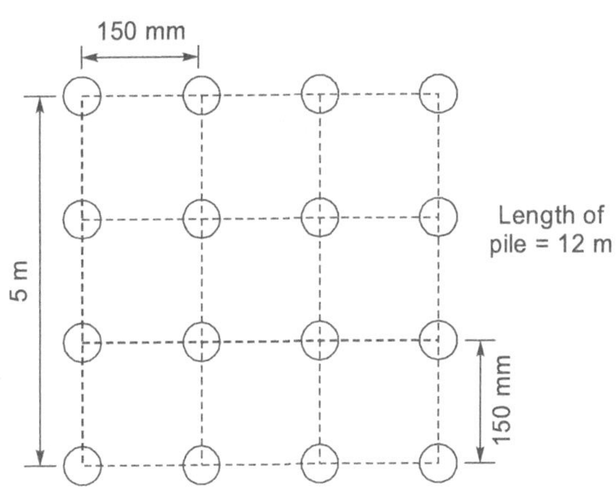

11.10 A 500 mm diameter bored concrete pile is to be formed in the soil profile as shown in figure. The ground conditions are as follows:

Granular soil: $\gamma = 20\text{ kN/m}^3$ $\quad \phi' = 22.5°$ $\quad K_s = 1.0$

Dense gravel: $\gamma = 21\text{ kN/m}^3$ $\quad \phi' = 26.25°$ $\quad K_s = 2.0$

Glacial clay: $\gamma = 20\text{ kN/m}^3$

c_u at 7.0 m = 120 kPa

c_u at 8.0 m = 145 kPa

c_u at 11.0 m = 220 kPa

Adhesion factor, $\alpha = 0.6$

$N_c = 9.0$

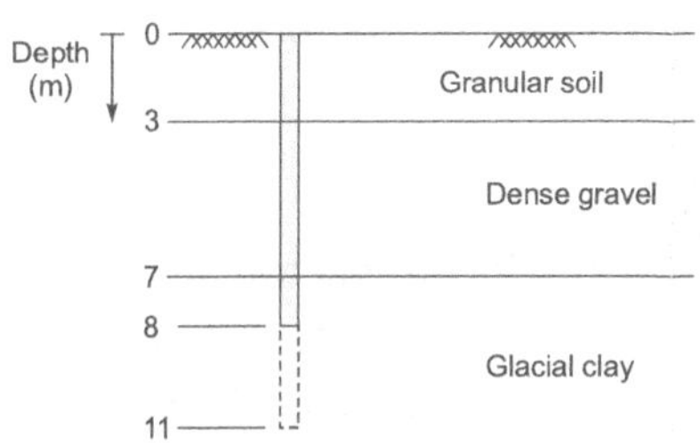

Determine the ultimate bearing capacity of the pile for:
(a) Embedded length of 8 m
(b) Embedded length of 11 m [20 marks]

Sol:

Granular soil:

$$\gamma = 20 \text{ kN/m}^3$$

$$\phi' = 22.5° \; ; K_s = 1.0$$

$$Q_{sf} = q_s A_s = K(\bar{\sigma}_v)_{Avg} \tan\delta A_s = K\left[\frac{(\bar{\sigma}_v)_A + (\bar{\sigma}_v)_B}{2}\right]\tan\delta A_s$$

$$= 1\left[\frac{0 + 20\times 3}{2}\right]\tan(22.5)\times\pi(0.5)\times 3 = 58.6 \text{ kN}$$

Dense gravel:

$$\gamma = 21 \text{ kN/m}^3; \; \phi' = 26.25° \; ; \; K_s = 2.0$$

$$Q_{sf} = q_s A_s = K(\bar{\sigma}_v)_{Avg} \tan\delta A_s = K\left[\frac{(\bar{\sigma}_v)_B + (\bar{\sigma}_v)_C}{2}\right]\tan\delta A_s$$

$$= 2\left[\frac{(60) + (60 + 4\times 21)}{2}\right]\tan(26.25)(\pi)0.5\times 4 = 632.1 \text{ kN}$$

Clay (when length = 8 m)

$$Q_s = \alpha\times\bar{c}_u\times A_s = 0.6\times\left(\frac{120+145}{2}\right)\times(\pi\times 0.5\times 1.0) = 124.9 \text{ kN}$$

$$Q_b = c_b\times N_c\times A_b = 145\times 9.0\times\left(\pi\times\frac{0.5^2}{4}\right) = 256.2 \text{ kN}$$

Clay (when length = 11 m)

$$Q_s = \alpha\times\bar{c}_u\times A_s = 0.6\times\left(\frac{120+220}{2}\right)\times(\pi\times 0.5\times 4.0) = 640.9 \text{ kN}$$

$$Q_b = c_b\times N_c\times A_b = 220\times 9.0\times\left(\pi\times\frac{0.5^2}{4}\right) = 388.8 \text{ kN}$$

Ultimate bearing capacity:

$$Q_u = Q_b + Q_s$$

(a) At 8 m; $Q_u = 256.2 + (58.6 + 632.1 + 124.9) = 1071.8$ kN
(b) At 11 m; $Q_u = 388.8 + (58.6 + 632.1 + 640.9) = 1720.4$ kN

12. Soil Exploration and Soil Stabilization

Level-1

12.1 **Explain differential free swell test. Also state the relationship between degree of expansiveness and DFS, percent.**

[8 Marks]

Sol:

- In this test two sample of dry soil weighing 10 gm each, passing through 425 μ sieve are taken.
- One of those sample is poured into a 50 cm^3 graduated cylinder containing kerosene oil (non polar liquid) while other is poured in a similar cylinder containing distilled water.

- Their volume are noted after 24 hours.
- The differential free swell is defined as

$$DFS = \frac{\text{Soil volume in water} - \text{Soil volume in Kerose}}{\text{Soil volume in Kerosene}} \times 100\%$$

The relationship between degree of expansiveness and DFS, percent is stated below in the table.

Degree of expansiveness	DFS, percent
Low	Less than 20
Moderate	20 - 35
High	35 - 50
Very High	Greater than 50

12.2 List various measure that can be used to eliminate possibility of swelling of soil.

[5 Marks]

Sol:

The elimination of possible soiling can theoretically be brought about by following measures:

1. Pre-wetting the soil mass to a moisture content equal to the equilibrium moisture content.
2. Providing large enough external loads, which exceeds the swelling pressure.
3. Chemical stabilization with lime, as lime stabilization is effective in reducing the liquid limit and plastic limit of the soil, hence swelling potential is reduced.

12.3 What is the purpose of Soil Stabilization?

[6 Marks]

Sol:

Soil-stabilization is Practiced in road construction with one or more of the following objectives:

(i) To improve the strength of sub-bases, and, in the case of low-cost roads, surface courses.
(ii) To reduce compressibility and thereby settlements.
(iii) To eliminate or improve certain undesirable properties of soils, such as excessive swelling or shrinkage, high plasticity, difficulty in compacting etc.
(iv) To facilitate compaction and increase load-bearing property.
(v) To improve permeability characteristics.
(vi) To reduce frost susceptibility.
(vii) To make use of locally available soils and other materials which are otherwise inferior.
(viii) To bring about economy in the cost of a road.
(ix) To control dust.

12.4 Describe stationary piston sampler giving special features of this sampler.

[4 Marks]

Sol:

A stationary piston sampler is a special type of a thin walled sampler, not provided with air relief hole or valve, but provided with a piston. The piston is connected to a rod, called the piston rod which passes through the sampler head and runs inside the hollow boring rod.

Keeping this piston lowered to the bottom of the tube and keeping them locked, the sampler is lowered into the bore hole. This piston will not allow any air or water to enter the tube while pushing the sampler into the soil below the bottom of the bore hole. The piston is made to touch the ground, and only tube is pushed down into the soil, until the sampler head meets the top of the piston. The sampler is then taken out, keeping the piston locked at the top of the tube over the sample. The vacuum between the piston and the sample would help to retain the soil in the tube. This sampler is an excellent tool for obtaining very fine undisturbed samples especially in soft sensitive clays. It can also be used for silts and silty sands, having some cohesion.

12.5 **What is the principle of stone column method of soil improvement? What are the merits and demerits of this method?**

[12 marks]

Sol:

Stone column method is used to increase the load-bearing capacity of shallow foundations on soft clay layers.

Principle : The stabilization of soils by displacing the soil radially, with the help of a deep vibrator, refilling the resulting space with granular material and compacting the same with the vibrator.

Merits:

(i) Cost effective.

(ii) It improves the bearing capacity and reduces the settlement of weak soil strata.

(iii) It accelerate the dissipation of excess pore water pressure.

(iv) It improve stiffness of subsoil.

(v) There is no waiting period after installation unlike PVD.

Demerits:

(i) Stone column method cannot be used in clays having sensitivity greater than 4 because these soils cannot adequately regain their shear strength.

(ii) Stone column when installed at a distance less than 3.5 m can cause high lateral pressures and displacement of adjacent structures.

(iii) Stone column installation in extremely cohesive clays and silts is suitable only if pre-loading facility is available.

12.6 **A soil sampler has the following dimensions:**
Inside diameter of cutting edge = 69 mm
Outside diameter of cutting edge = 73 mm
Inside diameter of sampling tube = 70 mm
Outside diameter of sampling tube = 72 mm
Compute inside clearance, outside clearance and area ratio.
What is permissible value of "area ratio" for soft sensitive clays as per I.S. Code?

[10 marks]

Sol:

Given: $D_1 = 69$ mm; $D_2 = 73$ mm; $D_3 = 70$ mm; $D_4 = 72$ mm

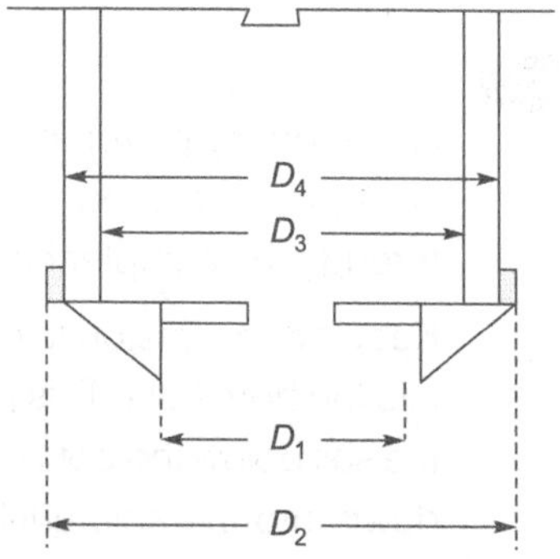

(i) Inside clearance, $C_i = \dfrac{D_3 - D_1}{D_1} \times 100 = \dfrac{70 - 69}{69} \times 100 = 1.4493\%$

(ii) outside clearance, $C_o = \dfrac{D_2 - D_4}{D_4} \times 100 = \dfrac{73 - 72}{72} \times 100 = 1.3889\%$

(iii) Area ratio, $A_R = \dfrac{A_2 - A_1}{A_1} \times 100 = \dfrac{\left(\frac{\pi}{4} D_2^2\right) - \left(\frac{\pi}{4} D_1^2\right)}{\frac{\pi}{4} D_1^2} \times 100$

$$= \frac{D_2^2 - D_1^2}{D_1^2} \times 100 = \frac{73^2 - 69^2}{69^2} \times 100 = 11.9303\ \%$$

As per IS Code, area ratio should be less than 10% but in no case it should exceed 20%.

Level-2

12.7 **(i) What are Geosynthetics? What are different types of Geosynthetics? Explain them briefly.**

(ii) Briefly describe various geological or geo-technical consideration for the site selection of bridges? **[8 + 7 Marks]**

Sol:

(i)

Geosynthetics : Geosynthetics are human-made materials from various types of polymers used to enhance, augment and make possible cost effective environmental, transportation and geotechnical engineering construction projects. They are used to provide one or more of the following function; separation, reinforcement, filtration, drainage or liquid barrier.

Categories of Geosynthetics:

1. **Geotextiles :** flexible, textile-like fabrics of controlled permeability used to provide filtration, separation or reinforcement in soil, rock and waste materials.
2. **Geomembranes :** essentially impermeable polymetric sheets used as barriers for liquid or solid waste containment.
3. **Geogrids :** stiff or flexible polymer grid-like sheets with large apertures used primarily as reinforcement of unstable soil and waste masses.
4. **Geonets :** stiff polymer net-like sheets with in-plane openings used primarily as a drainage material within landfills or in soil and rock masses.
5. **Geosynthetic clay liners :** prefabricated bentonite clay layers incorporated between geotextiles and/or geomembranes and used as a barrier for liquid or solid waste containment.
6. **Geopipes :** perforated or solid wall polymeric pipes used for the drainage of various liquids.
7. **Geocomposites :** hybrid systems of any, or all, of the above geosynthetic types which can function as specifically designed for use in soil, rock, waste and liquid related problems.

(ii)

The role of the bridge abutments is to transfer the load of the structure vertically into the ground. Similar to the dam, the only thing that needs extra consideration is that the loads are transferred within a relatively smaller area to a larger depth. As a result, the bridge foundation is affected more significantly because of heterogeneity in soil and rock. So, a very through subsurface geotechnical investigation is a must at the location of all the abutments and foundations before a bridge construction can be taken up. If the investigation is inadequate or does not go to sufficient depth, the bridge construction would have to encounter inordinate delay because of constructability problems.

Following geological consideration shall be made before site selection.

(i) A study should be performed to get the stability of the river, location of faults, their activity and their likely repercussions on a major structure to be put up and particulars of past earthquake in the site vicinity.

(ii) A study should be performed to get soil classification, grain size and depth at which hard strata is likely to be met with.

(iii) Mainly augur boring test is used to collect soil samples and further to study its engineering properties.

12.8 **How would you design the foundation of a structure in swelling soils?**

[12 marks]

Sol:

The rational methods of foundation design which are being used now to reduce or prevent the effects of swelling can be grouped into three categories, namely,

(a) isolating the structure from the swelling soil, (b) designing a structure to withstand the effects of swelling and (c) preventing the swelling.

Belled piers which are popular in USA for use in expansive soil conditions are isolated structures on swelling soils. The underreamed pile-beam construction is similar to that of belled piers and are designed for use under Indian conditions. The principal involved is to transfer the load of a building through the piles to a depth beyond the zone of seasonal variation in moisture content. The underreamed piles are bored cast in situ piles laid to a depth beyond the zone of seasonal variation in moisture content. The underreamed piles are bored cast in situ piles with their lower portions enlarged or reamed in the form of a bulb. The piles are connected at their top by plinth beams of reinforced concrete which support the super structure.

Designing a structure which is strong enough and rigid enough to withstand the effects of swelling may prove to be highly uneconomical except in the case of very small structures where even if the loads are supported by a central area or peripheral area much smaller than the plan area, the bearing pressures are within limits. Obviously, it is futile to attempt designing a structure that will remain undamaged in spite of swelling, because the cost of providing the structure with extra strength and rigidity would be much more as compared to the cost of underreamed pile or piers foundation.

Swelling can often be controlled, if not eliminated, by providing an impervious apron around the structure. By providing an apron, the moisture gradient between the centre of the structure and its edges is minimized and hence the differential swelling is controlled.

Elimination of possible swelling can theoretically be brought about by (a) pre-wetting the ground to a moisture content equal to the equilibrium moisture content, (b) making downward loads large enough to exceed swelling pressures and (c) chemical stabilization.

12.9 Briefly describe any two geophysical methods of soil exploration. [20 marks]

Sol:

Geo-physical methods:

These method consists of identification of underground material by measuring change in the character of surface material due to presence of different types of sub-surface materials.

These method are very useful to explore large areas much more rapidly and economically than boring methods.

The various geophysical methods of sub surface exploration are:

(i) Seismic refraction method

(ii) Electrical resistivity method

(i) Seismic refraction method

- This method is based on the fact that seismic waves have different velocities in different type of soils and rock and refracted when they cross the boundary between different types of soil.
- Velocity of seismic wave is faster in denser or consolidated medium.
- In this method, the shock waves are created into the soil at their ground level or certain depth below by exploding small charge of high velocity dynamite or by striking a plate on the soil with a hammer.
- The radiating shock waves are received by a device called geophone, where the time of travel gets recorded.
- To record shock waves, a number of geophones are installed at known intervals on the ground along a line of produced shock waves.

- From the sources, three kinds of waves are produced; primary waves, secondary waves and surface wave. In general only primary waves are observed. These primary waves are of three types; direct waves, reflected waves, refracted waves.
- Direct waves travel directly from the shock source along the ground surface in the direction of geophones and other waves travel in a downward direction at various angles to the horizontal. Later waves get reflected and refracted when they meet a boundary separating media of different seismic velocities.
- If underlying layer is denser, the refracted waves travel much faster. As the distance between the shock source and the geophones increases, the refracted wave reach the geophones earlier than the direct wave.
- The direct waves travelling with velocity, V_T in the top layer arrive first at the geophones nearer to the shot point. The wave with the higher velocity V_B, reach the farther geophones first by travelling in the lower velocity material (top layer), horizontally in the higher velocity material and return to the surface.
- The arrival time is plotted against the distance between the source and the geophone. The point B in distance-time graph represents the point of simultaneous arrival of direct and refracted waves.
- The depth of the boundary between the two strata can be estimated from the equation

$$D = \frac{d}{2}\sqrt{\left(\frac{V_B - V_T}{V_B + V_T}\right)}$$

The above equation is under assumption that thickness of upper stratum is unvarying.

(ii) Electrical resistivity method

- This method is based on the fact that different types of soil offer different resistance to the flow of an electrical current.
- The resistance to flow of current is determined by measurement of resistivity (ρ).
- The test is conducted by driving four equally spaced electrodes into the ground at a straight line. Current (I) from a battery, flow through the soil between the two outer electrodes, producing an electrical field within the soil. The potential difference E between the two inner electrodes is then measured. The apparent resistivity, ρ is given by the equation

$$\rho = \frac{2\pi DE}{I}$$

Where D = Electrode spacing

The two methods are adopted for exploration.

(i) Resistivity profiling

(ii) Resistivity sounding

(i) Resistivity profiling

- This method is used for investigation of lateral variation of soil types:
- In this method, four electrodes are kept at a constant spacing and moves across the area and profile lines are plotted with station point on horizontal axis and resistivity along the vertical axis.

(ii) Resistivity sounding

- This method is used to get information on the variation of sub-surface materials with depth.
- In this method, the electrode spacing D is progressively increased to pick up changes in resistivity with depth.

■■■■

06 Surveying

1. Fundamental Concepts of Surveying and Linear Measurement

Level-1

1.1 *AB* is a chain line crossing a lake. *A* and *B* are on the opposite sides of the lake. *A* line *AC*, 800 m long is ranged to the right of *AB* clear of the lake. Similarly another line *AD*, 1000 m long is ranged to the left of *AB* clear of the lake such that the points *C*, *B*, and *D* are collinear. The length *BC* and *BD* are 400 m and 600 m respectively. If the chainage at *A* is 1262.44 m, calculate the chainage of *B*.

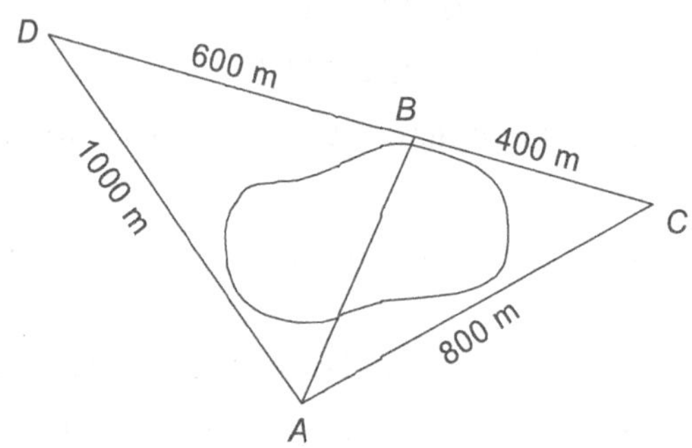

[10 Marks]

Sol:

From triangle *ACD*, $\cos D = \dfrac{CD^2 + AD^2 - AC^2}{2AD \cdot CD} = \dfrac{1000^2 + 1000^2 - 800^2}{(2)(1000)(1000)} = 0.68$

From ΔADB, $\cos D = \dfrac{BD^2 + AD^2 - AB^2}{2DB \cdot AD}$

$$\Rightarrow \quad AB^2 = BD^2 + AD^2 - 2DB \cdot AD \times \cos D$$

$$= 600^2 + 1000^2 - (2)(600)(1000)(0.68) = 544000$$

$$\therefore \quad AB = 737.5636 \text{ m} \simeq 737.56 \text{ m}$$

$$\text{Chainage at } B = 1262.44 + 737.56 = 2000.00 \text{ m}$$

1.2 A line AB between the stations A and B was measured as 348.28 m using a 20m tape, too short by 0.05 m. Determine :

(a) Correct length of AB

(b) Reduced horizontal length of AB if AB lay on a slope of 1 in 25.

[12 Marks]

Sol:

(a) Since tape is short by 0.05 m, actual length would be less than the measured length.

$$\text{Actual length} = \text{Measured length} \times \frac{\text{Actual length of tape}}{\text{Designated length of tape}}$$

$$= \frac{20 - 0.05}{20} \times 348.28 = 347.41 \text{ m}$$

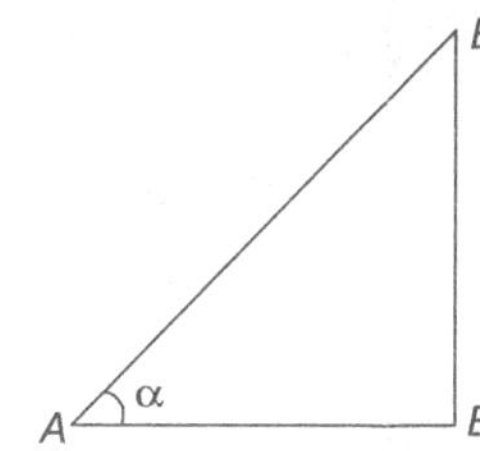

(b) Given $\tan\alpha = \frac{1}{25}$,

$$\alpha = \tan\alpha = \left(\frac{1}{25}\right) = 2° \ 17' \ 26.2$$

$$\text{Reduced horizontal length} = AB^1$$

$$AB' = AB \cos\alpha$$

$$AB' = 347.41 \cos (2° \ 17' \ 26.2'') = 347.13 \text{ m}$$

Enlarged view of *AB*.

AB' is the horizontal projection of *AB*.

1.3 **While reading on old plan it is found that there is no graphical scale shown, but it is written that $RF = \frac{1}{2000}$. Distance between two points A and B on the plan is found to be 6cm but actual measurement on the ground is 125 m.**

(a) Calculate Shrinkage factor and shrunk scale.

(b) If area of an estate as measured a plan is found to be 24cm², find actual area.

[8 Marks]

Sol:

Calculated length of AB on the ground

$$= 6 \times 2000 = 12000 \text{ cm} = 120 \text{ m.}$$

$$\text{Actual length of AB} = 125 \text{ m.}$$

$$\text{Shrinkage factor} = \frac{\text{Calculated length}}{\text{Actual length}}$$

$$\frac{120}{125} = 0.96$$

$$\text{Shrunk Scale} = 0.96 \times \frac{1}{2000} = \frac{1}{2083.33}$$

(b)

$$\text{Area Measured} = 240 \text{ cm}^2$$

$$\text{Actual Area} = 240 \times (2083.33)^2 \text{ cm}^2 = 240 \times (2083.33)^2 \times 10^{-4} \text{ m}^2 = 104166.7 \text{ m}^2$$

1.4 **An offset is laid out at an angle of 4° out from its true direction in the field. Find the resulting displacement of the plotted point on the paper for an offset length of 8 m. The scale of plotting was 6 m to 1 cm:**

(a) In the direction parallel to chain line

(b) In the direction perpendicular to chain line

[8 Marks]

Sol:

Displacement of the point on ground parallel to chain line

$$DL = l \sin\alpha = 8 \sin 4° = 0.558 \text{ m}$$

$$\text{Displacement of point on plan} = \frac{0.558}{6} = 0.093 \text{cm}$$

(b) Displacement of point on ground perpendicular to chain line

$$BD = l\,(1 - \cos\alpha) = 8\,(1 - \cos 4°) = 0.01948 \text{ m.}$$

$$\text{Displacement point on plan} = \frac{0.01948}{6} = 0.003 \text{cm}$$

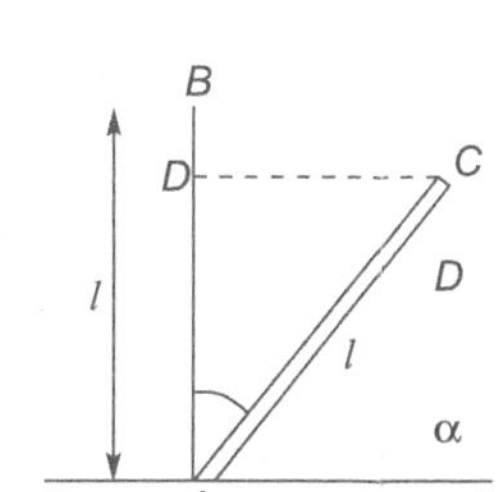

Chain line

1.5 An old map was plotted to a scale of 40 m to 1 cm. Over the years, this map has been shrinking and a line originally 20 cm long is only 19.5 cm long at present. Also a 20 m chain was 5 cm too long. If the present area of the map measured by planimeter is 125.50 cm², then determine the true area of land surveyed.

[12 marks]

Sol:

19.5 cm on the map was originally 20 cm,

$\therefore$ 1 cm² on the map was originally $\frac{20^2}{19.5^2}$ cm² = 1.05194 cm²

$$\therefore \quad 125.5 \text{ cm}^2 \text{ area on map was originally} = 125.5 \times 1.05194 = 132.01847 \text{ cm}^2$$

$$\text{Since scale of map was 1 cm} = 40 \text{ m}$$

$$\Rightarrow \quad 1 \text{ cm}^2 = 1600 \text{ m}^2$$

$$\therefore \quad \text{Actual area on the ground} = 1600 \times 132.01847 = 211229.552 \text{ m}^2$$

Since the chain was 0.05 m too long.

$$\therefore \quad \text{True area} = \frac{20.05^2}{20^2} \times \frac{211229.552}{10^4} = 21.23 \text{ hectares}$$

Level-2

1.6 Two cross sections *AB* and *CD* each perpendicular to the base line *AC* of 250 m length are established for measuring the velocity of flowing water in a river. When the float was on the section *AB*, the angle *AEB* measured from a point *E* on the base line, 100 m from *A*, was 50°30′40″ and the angle *CED* was 45°35′20″. If the time taken by the float to travel the distance *BD* was 90 seconds, calculate the velocity of water.

[12 Marks]

Sol:

The statement of the given problem can be represented through a diagram as shown below:

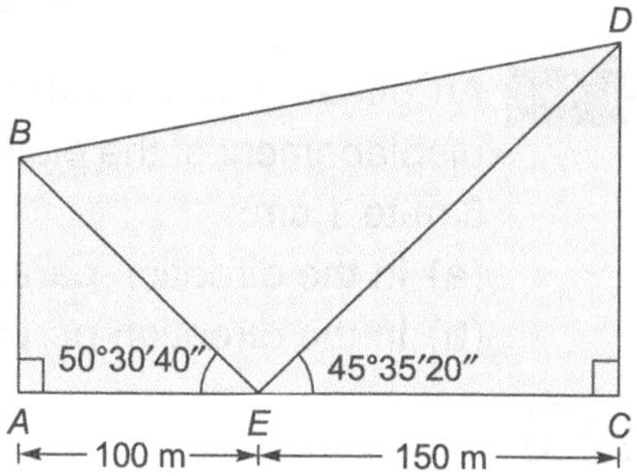

$$\angle BED = 180° - (\angle AEB + \angle CED)$$

$$= 180° - (50°\,30'\,40'' + 45°\,35'\,20'')$$

$$= 83°\,54'$$

$$\text{In } \Delta\, AEB, \cos 50°\,30'\,40'' = \frac{AE}{BE}$$

$$\text{But} \quad AE = 100 \text{ m}$$

$$\therefore \quad BE = \frac{100}{\cos 50°\,30'\,40''} = 157.25 \text{ m}$$

$$\text{In } \Delta\, DCE, \cos 45°\,35'\,20'' = \frac{EC}{DE}$$

$$\text{But} \quad EC = 150 \text{ m}$$

$$\therefore \quad DE = \frac{150}{\cos 45°\,35'\,20''} = 214.35 \text{ m}$$

Now applying cosine formula in Δ *BED*, we get

$$\cos(\angle BED) = \frac{BE^2 + DE^2 - BD^2}{2 \times BE \times DE}$$

$$\Rightarrow \quad \cos 83°54' = \frac{(157.25)^2 + (214.35)^2 - BD^2}{2 \times 157.25 \times 214.35}$$

$$\Rightarrow \quad BD^2 = (157.25)^2 + (214.35)^2 - 7163.59$$

$$\Rightarrow \quad BD = 252.01 \text{ m}$$

$$\therefore \quad \text{Velocity of water} = \frac{BD}{\text{Time taken by float to travel distance } BD} = \frac{252.01}{90} = 2.8 \text{ m/s}$$

1.7 **A base line was measured by tape suspended in category under a pull of 145 N, mean temperature being 14°C. the lengths of various segments are given below.**

Span	Length (in m)	Difference in level (n)
1	29.988	+0.346
2	29.895	–0.214
3	29.838	+0.309
4	29.910	–0.106

If tape was standardized on the flat surface under 95 N at 18°c, determine correct length of line. Take
Cross sectional area of tape = 3.35 nm^2
Mass of tape = 0.025 kg/m
Coefficient of linear expansion = 0.9 × 10^{-6} per °c
Young's modules = 14.8 × 10^4 MN/m^2
Mean height of line above MSL = 51.76 m
Radius of earth = 6370 km **[20 Marks]**

Sol:

Given: P_o = 95 N, P = 145 N, t_o=18°c, t_m = 14°c, A = 3.35 mm^2, α = 0.9 × 156/°c , w = mg = 0.025 × 9.81 = 0.245 N/m. E =14.8 × 10^4 N/mm^2, H = 51.76 m, R = 6370 km.

$$\text{Total length of tape} = 29.988 + 29.895 + 29.838 + 29.910$$

$$L = 119.631 \text{ m}$$

$$\text{Temp. Correction } C_E = \alpha\,(t_m - t_o)\,L = 0.09 \times 10^{-6}\,(14 - 18) \times 119.631 = -0.0004 \text{ m}$$

$$\text{Pull Correction } C_P = \frac{(P - P_O)\,L}{AE} = \frac{(145 - 95)}{3.35 \times 14.8 \times 10^4} = 0.0121 \text{ m.}$$

$$\text{Sag Correction } C_g = \frac{-1}{24}\sum_{i=1}^{4}\left(\frac{-w_i^2 l_i^3}{24P^2}\right)$$

$$= \frac{-0.245^2}{24 \times 145^2}\left[29.988^3 + 29.895^3 + 29.838^3 + 29.910^3\right] = -0.0128 \text{ m}$$

$$\text{Slope Correction} = C_s = \frac{-n^2}{2L} = \frac{-1}{2}\left[\frac{0.346^2}{29.988} + \frac{0.214^2}{29.895} + \frac{0.309^2}{29.838} + \frac{0.106^2}{29.910}\right] = -0.0045 \text{ m}$$

$$\text{Mean sea level correction} = C_{msl} = \frac{-HL}{R}$$

$$= \frac{-51.76 \times 119.631}{6370 \times 1000} = -0.001 \text{m.}$$

$$\text{Total Correction} = C_t + C_p + C_g + C_s + C_{msl}$$

$$= -0.0004 + 0.0121 - 0.0128 - 0.0045 - 0.001 = -0.0066 \text{m}$$

$$\text{Correct length} = 119.631 - 0.0066 = \mathbf{119.624 \text{ m}}$$

2. Compass Surveying, Theodolites and Traverse Surveying

Level-1

2.1 List the fundamental lines of a theodolite and explain briefly the desired relationships between these lines and effects if such relationships are not maintained. Also describe the checks need to be applied in an open traverse. [10 Marks]

Sol:

Fundamental lines of a theodolite: The fundamental lines of a theodolite are the vertical axis, axis of plate levels, the line of collimation, the horizontal axis and the bubble line of altitude. The following relationships are desired:

1. The axis of the plate levels must be at perpendicular to the vertical axis.
2. The line of collimation must be at right angles to the horizontal axis.
3. The horizontal axis must be perpendicular to the vertical axis.
4. The axis of the telescope level must be parallel to the line of collimation.

Errors: The following errors arises if above mentioned relationships are not maintained:

1. **Vertical axis error (α) :** Axis not vertical in an observations either from imperfect plate level adjustment or settlement of instrument.
2. **Lateral collimation error (β) :** Line of collimation not perpendicular to the horizontal axis.
3. **Horizontal axis error (γ) :** Horizontal axis not perpendicular to the vertical axis.
4. **Vertical collimation error (δ) :** Line of altitude bubble not parallel to the line of collimation when the verniers of vertical circle read zero.

Checks for an open traverse: In an open traverse, method to check the measurements as a whole are not available. Though linear measurements cannot be determined, the angular errors may be found by indirect method as follows:

1. **Well defined point:**
   (i) A well-defined point such as *O* is chosen and bearing of the object *O* is measured from different stations say *a*, *f*, *h* etc.
   (ii) Knowing the coordinates of *O* and *h*, the bearing of line *hO* can be calculated. The traverse from *a* to *h* can be checked by comparing observed and computed bearing of *hO*.

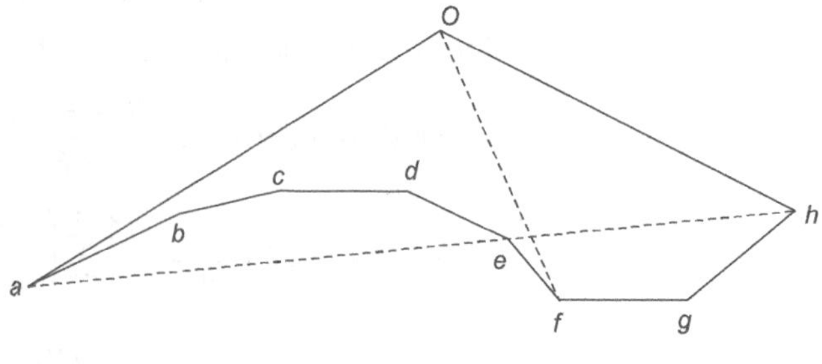

   (iii) The limitation of the method is that if any discrepancy is found, it is not possible to say exactly where the error lies.
2. **Cutoff lines:** By the cutoff lines, the different parts of the traverse can be checked as the work progresses.

2.2 Compute the bearings of a regular hexagon given the starting bearing of side *AB* is S 50° 10′ E (Station *C* is towards east from *B*). [12 marks]

Sol:

The sum of internal angles of a regular hexagon are

$$= (n-2)\,180° = (6-2)\,180° = 720°$$

Each angle therefore, is $\frac{720°}{6} = 120°$. The rough sketch of the hexagon is shown in figure.

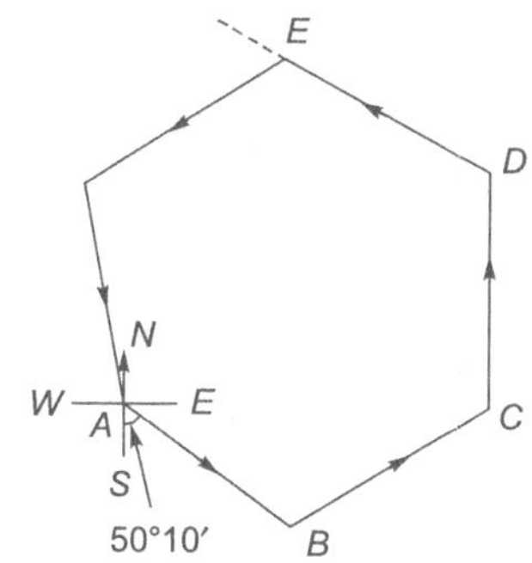

Whole circle bearing of AB

$$= 180° - 50°10' = 129°50'$$

Bearing of BC $= 129°50' - 60° = 69°50'$

Bearing of CD $= 69°50' - 60° = 9°50'$

Bearing of DE $= 9°50' - 60° = -50°10'$

Negative bearing means it has crossed the N-S line in the anticlockwise direction by 50°10′ (see figure).

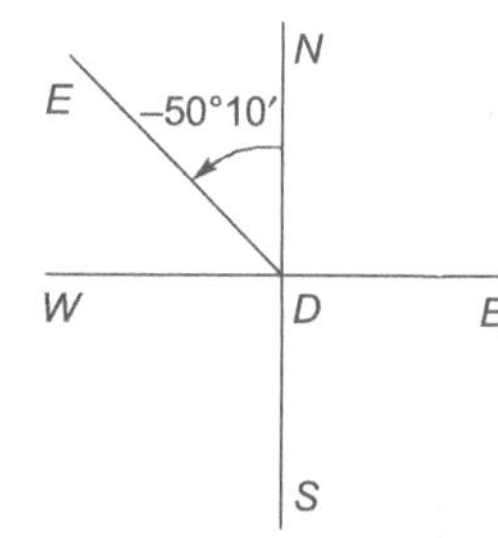

$\therefore$ *WCB* of *DE*360° – 50°10′

$$= 309°50'$$

Bearing of EF $= 309°50' - 60°$

$$= 249°50'$$

Bearing of FA $= (249°50' - 180°) + 120°$

$$= 189°50'$$

Bearing of AB $= 9°50' + 120° = 129°50'$ as before.

Alternatively

Bearing of AB $= 129°50'$

We know that:

Fore bearing of next line = *BB* of previous line + Included angle

$\Rightarrow$ Fore bearing of BC $= (129°50' + 180°) + 120° - 360° = 69°50'$

Fore bearing of CD $= (69°50' + 180°) + 120° - 360° = 9°50'$

Fore bearing of DE $= (9°50' + 180°) + 120° = 309°50'$

Fore bearing of EF $= (309°50' - 180°) + 120° = 249°50'$

Fore bearing of FA $= (249°50' - 180°) + 120° = 189°50'$

Fore bearing of AB $= (189°50' - 180°) + 120° = 129°50'$

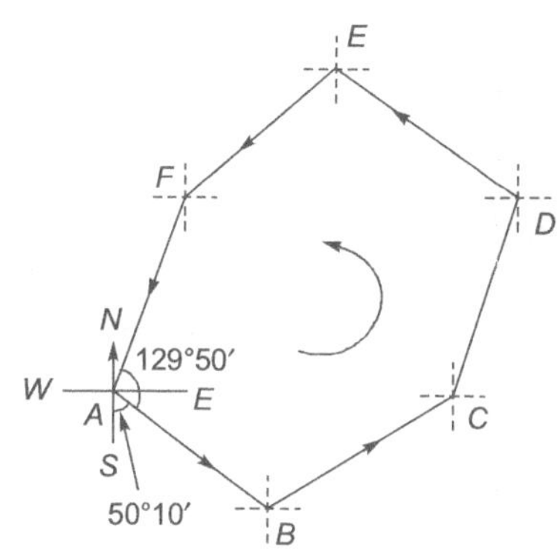

2.3 The length and bearings of four sided traverse are as follows:

AB = 320 m, BC = 440 m, CD = 390 m, and DA = 513.80 m

AB = N 45°15′E, BC = S71°30′E, CD = S30°15′W, DA = N66°44′W

Calculate area of the traverse.

[10 Marks]

Sol:

Convert the bearing of lines into coordinates of point *A*, *B*, *C*, *D*.

We know,

Easting, Departure = $l \sin \theta$ (θ is angle with vertical)

Northing, Latitude = $l \cos \theta$

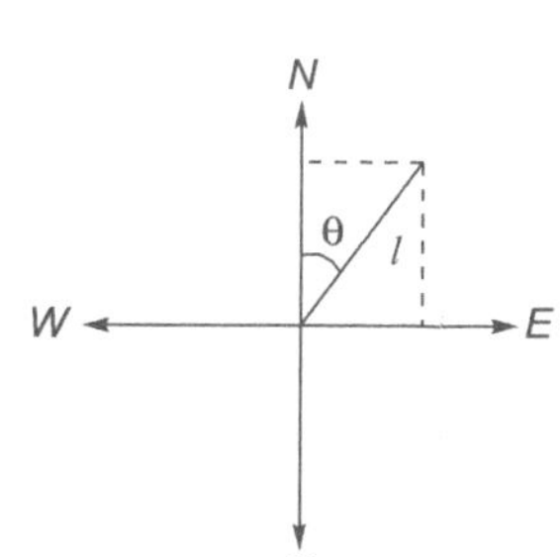

Length	Easting (Departure)	Northing (Latitude)
AB	227.26	225.28
BC	417.262	–139.614
CD	–196.47	–336.895
DA	–472.016	+202.96

Arrange the coordinates in following manner:

	A	B	C	D	A	B
Easting →	227.26	417.262	−196.47	−472.016	227.26	
Northing →	225.28	−139.614	−336.895	+202.96	225.28	

$$\text{Area} = \frac{1}{2}|\Sigma P - \Sigma Q|$$

$$\Sigma P = 227.26 \times (-139.614) + 417.262 \times (-336.895) + (-196.47) \times 202.96 + (-472.016) \times 225.28$$

$$= -318513.47$$

$$\Sigma Q = 225.28 \times 417.262 + (-139.614) \times (-196.47) + (-336.895) \times (-472.016) + 202.96 \times 227.26$$

$$= 326575.26$$

$$A = \frac{1}{2}\left| -318513.47 - 326575.26 \right| = 322544.37 \text{ m}^2 = 32.25 \text{ Ha}$$

2.4 Following are the observed bearing of the line of a traverse ABCDEA with the compass in a place where local attraction was suspected. Find the correct bearings of the lines.

Line	FB	BB
AB	191°45′	13°0′
BC	39°30′	222°30′
CD	22°15′	200°30′
DE	242°45′	62°45′
EA	330°15′	147°45′

[8 Marks]

Sol:

From the table it can be observed that FB and BB of line DE differ by 180°. So both stations D and E are free from local attractions and readings taken from these stations are correct i.e. FB of EA and BB of AB are correct.

FB of EA = 330° 15′

BB of EA = 330° 15′ – 180° = 150° 15′

Correction in BB of EA = 150° 15′ – 147° 45′

= 2° 30′ correction at A.

Correct FB of AB = 191° 45′ + 2° 30′ = 194° 15′

Correct BB of AB = 194° 15′ – 180° = 14° 15′

Correction in BB of AB = 14° 15′ – 13° = 1° 15′

Correct FB of BC = 39° 30′ + 1° 15′ = 40° 45′

Correction in BB of BC = 180° + 45° 45′ = 220° 45′

Correction in BB of BC = 220° 45′ – 222° 30′ = – 1° 45′

Correct FB of CD = 22° 15′ – 1° 45′ = 20° 30′

Correct BB of CD = 180° + 20° 30′ = 200° 30′ (check)

Line	Corrected Bearings	
	FB	BB
AB	194°15′	14°15′
BC	40°45′	220°45′
CD	20°30′	200°30′
DE	242°45′	62°45′
EA	330°15′	150°15′

2.5 The following are the bearings observed in traversing with a compass, an area where local attraction was suspected. Calculate interior angles of the traverse and correct then if necessary.

[12 Marks]

Sol:

Line	FB	BB
AB	150°0′	330°0′
BC	230°30′	48°0′
CD	306°15′	127°45′
DE	298°00′	120°0′
EA	49°30′	229°30′

$$\text{Interior } \angle A = \text{BB of EA} - \text{FB of AB} = 229^\circ 30' - 150^\circ = 79^\circ 30'$$
$$\text{Interior } \angle B = \text{BB of AB} - \text{FB of BC} = 330^\circ - 230^\circ 30' = 99^\circ 30'$$
$$\text{Interior } \angle C = \text{BB of BC} - \text{FB of CD} = 48^\circ - 306^\circ 15' + 360^\circ = 101^\circ 45'$$
$$\text{Interior } \angle D = \text{BB of CD} - \text{FB of DE} = 127^\circ 45' - 298^\circ + 360 = 189^\circ 45'$$
$$\text{Interior } \angle A = \text{BB of DE} - \text{FB of EA} = 120^\circ - 49^\circ 30' = 70^\circ 30'$$
$$\text{Sum of interior angles} = 79^\circ 30' + 99^\circ 30' + 101^\circ 45' + 189^\circ 45' + 70^\circ 30' = 541^\circ$$

But sum of interior angles should be

$$= (N-2) \times 180^\circ = (5-2) \times 180 = 540^\circ$$
$$\text{Correction} = 540 - 541 = -1^\circ = 60^\circ$$
$$\text{Correction in each interior angles} = \frac{-60'}{5} = -12'$$

Angle	Calculated value	Corrected value
∠A	79°30′	79°18′
∠B	99°30′	99°18′
∠C	101°45′	101°33′
∠D	189°45′	189°33′
∠E	71°30′	70°18′

$$\angle A + \angle B + \angle C + \angle D + \angle E = 540^\circ \text{ check}$$

2.6 The following data were collected while running a closed traverse ABCDA. Calculate missing data.

Line	Length (in m)	Bearing
AB	300	181°25′
BC	—	89°50′
CD	411	355°00′
DA	827	—

[12 Marks]

Sol:

Let length of BC be 'l' and bearing of DA be θ

$$D_{AB} = L_{AB} \sin\theta_{AB} = 330 \sin 181^\circ 25' = -8.159 \text{ m}$$
$$L_{AB} = L_{AB} \cos\theta_{AB} = 330 \cos 181^\circ 25' = -329.899 \text{ m.}$$

Similarly

$$D_{BC} = 411 \times \sin 355^\circ = -35.821 \text{ m}$$

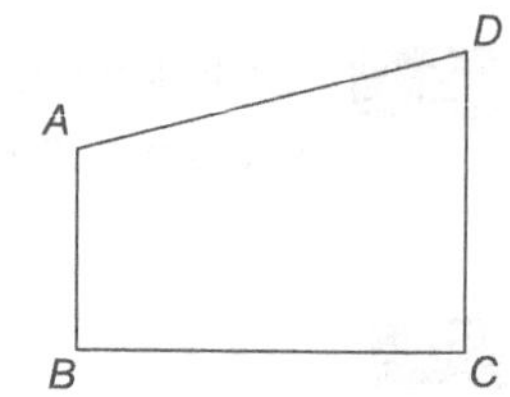

$$L_{CD} = 411 \cos 355° = 409.436 \text{ m}$$
$$D_{DA} = 827 \sin\theta$$
$$L_{DA} = 827 \cos\theta$$

For a closed traverse, $\Sigma D = 0 \ \& \ \Sigma L = 0$

$$\Sigma D = 0 = -8.159 + l \sin(89° 50') - 35.821 = 827 \sin\theta$$

$$\Rightarrow \quad 827 \sin\theta = 0.999l - 43.980 \quad ....(1)$$

$$\Sigma L = 0 = -329.899 + l \cos(89° 50') + 409.436 = 827\cos\theta$$

$$\Rightarrow \quad 827 \cos\theta = 79.537 + 000.3l \quad ....(2)$$

Squaring and adding equation 1 and 2, we get,

$$= 0.998\, l^2 - 87.395\, l - 675668.625 = 0$$

On solving above equation, we get

$$l = 867.76 \text{ m}, -780.19 \text{ m (neglect)}$$

Substituting $l = 867.76$ in equation-1, we get

$$\theta = 84° 18' 3.22''$$

2.7 **P, Q, R and S are four stations whose coordinates are given below:**

Station	Eastings (m)	Northing (m)
P	1000	1000
Q	1180.94	1075.18
R	1021.98	1215.62
S	939.70	1102.36

Another station X is to be fixed at the intersection of lines PR and QS. What are the coordinate of X?

[8 Marks]

Sol:

Coordinates of all Points are shown in fig.

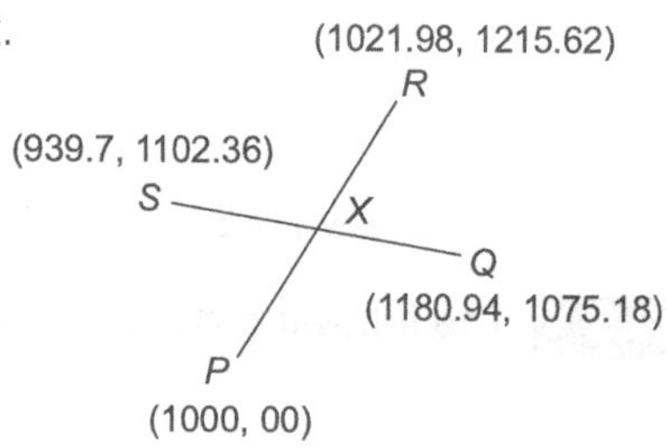

Equation of line PR, $y = \left(\dfrac{y_2 - y_1}{x_2 - x_1}\right) + c_1$ where c_1 is const.

$$\Rightarrow \quad y = \frac{1215.62 - 1000}{1021.98 - 1000} x + c_1$$

$$\Rightarrow \quad y = 9.81x + c_1$$

Lines passes through (1000,1000), substituting in above

Equation, $1000 = 9.81 \times 1000 + c_1$

$$C_1 = -8810$$

So Equation of PR $= y = 9381\,x - 8810 \quad ....(1)$

Similarly for SQ, $y = \dfrac{1102.36 - 1075.18}{939.7 - 1180.940} x + c_2$

$$\Rightarrow \quad y = -0.113\,x + c_2$$

Passing through (1180.94, 1075.8),

$$\Rightarrow \quad c_2 = 1208.23$$

Equation of Sθ $\quad y = -0.113\,x + 1208.23 \quad ....(2)$

Solving Equation 1 & 2, we get coordinate of station X

as $x = 1009.597$ m

$y = 1094.146$ m

2.8 An open traverse was run from A to E in order to obtain length and bearing of the line AE which could not be measured directly, with the following results:

Line	AB	BC	CD	DE
Length (in m)	1025	1087	925	1250
WCB	261°41′	9°6′	282°22′	71°31′

Find length and bearing of line AE. [8 Marks]

Sol:

$$\Sigma\text{Latitude} = \Sigma l \cos\theta$$
$$= 1025 \cos(261°\,41') + 1087 \cos(9°\,6') + 925 \cos(282°\,22') + 1250 \cos(71°\,31') = 1519.45 \text{ m}$$
$$\Sigma\text{Departure} = \Sigma l \sin\theta$$
$$= 1025 \sin(261°\,41') + 1087 \sin(9°\,6') + 925 \sin(282°\,22') + 1250 \sin(71°\,31') = -560.32 \text{ m}$$
$$\text{Length of AE} = \sqrt{(\Sigma\text{Latitude})^2 + (\Sigma\text{Depature})^2} = 1619.47 \text{ m.}$$
$$\text{Bearing of AE} = 360° - \tan^{-1}\left(\frac{560.32}{1519.45}\right) = 339°45'28''$$

2.9 A traverse ACDB was surveyed by theodolite and tape. The lengths and bearings of the lines AC, CD and DB are given below:

Line	AC	CD	DB
Length (in m)	480.6	292.0	448.1
WCB	25°19′	37°53′	301°00′

If the coordinates of A are $x = 0$, y = 0 and those of B are $x = 0$, y = 897.05, adjust the traverse and determine the coordinates of C and D. coordinates of A and B must not be altered. [12 Marks]

Sol:

$$\Sigma\text{Latitude} = \Sigma l\cos\theta = 480.6 \cos(25°\,19') + 292 \cos(37°\,53') + 448.1 \cos(301°) = 0.8 \text{ m}$$
$$\text{Error in departure} = 0.8 - 0 = 0.8 \text{ m}$$
$$\text{Calculated latitude of AC} = 480.6 \cos(25°\,19') = 434.44 \text{ m}$$
$$\text{Calculated departure of AC} = 480.6 \sin(25°\,19') = 205.51 \text{ m}$$
$$\text{Correction in latitude of AC} = \frac{1.35 \times 480.6}{480.6 + 292 + 448.1 + 897.05} = 0.31 \text{ m}$$
$$\text{Corrected y} = \text{coordinate of C} = 434.44 + 0.31 = 434.75 \text{ m}$$

Similarly corrected x – coordinate of C

$$= 205.51 - \frac{0.8 \times 480.6}{2117.75} = 205.83 \text{ m}$$

Similarly for D, corrected, latitude of CD

$$= 292.0 \cos(37°\,53') + \frac{1.35}{2117.75} \times 292 = 230.62 \text{ m}$$
$$\text{corrected y–coordinate of D} = 434.75 + 230.65 = 665.4$$
$$\text{corrected departure of CB} = 292 \sin(37°\,53') - \frac{0.08 \times 292}{2117.75} = 179.3 \text{ m}$$
$$\text{corrected } x \text{ coordinate of D} = 205.33 + 179.3 = 384.62 \text{ m.}$$

2.10 The following are the observed bearings of the lines of a traverse ABCDEA with a compass at the place where local attraction was suspected.

Line	FB	BB
AB	191°45′	13°00′
BC	39°30′	222°30′
CD	22°15′	200°30′
DE	242°45′	62°45′
EA	330°15′	147°45′

Find the correct bearings of the lines. [10 marks]

Sol:

(a) On verifying the observed bearing, it is found that the FB and BB of line DE differ by exactly 180°. So, the stations D and E are free from local attraction and the observed FB of EA and observed BB of CD are correct.

(b) The actual BB of EA should be

$$330°15' - 180°0' = 150°15'$$

But the observed BB of EA 147°45′.

So, a correction of (150°15′ – 147°45′) = +2°30′ should be applied at A.

(c) Correct FB of AB = 191°45′ + 2°30′ = 194°15′

Therefore, the actual correct BB of AB should be

$$194°15' - 180°00' = 14°15'$$

But Observed BB of AB = 13°0′

So, a correction of (14°15′ – 13°0′) = +1°15′ should be applied at B.

(d) Correct FB of BC = 39°30′ + 1°15′ = 40°45′

∴ Correct BB of BC should be = 40°45′ + 180°0′ = 220°45′

But observed BB of BC = 222°30′

So, a correction of (220°45′ – 222°30′) = –1°45′ should be applied at C.

(e) Correct FB of CD = 22°15′ – 1°45′ = 20°30′

Therefore, the correct BB of CD should be

$$20°30' + 180°0' = 200°30'$$

which is equal to the observed BB of CD.

So, D is free from local attraction, which also tallies with the remark made at the beginning.

The result is tabulated as follows:

Line	Observed		Correction	Correct		Remarks
	FB	BB		FB	BB	
AB	191°45′	13°00′	+2°30′ at A	194°15′	14°15′	
BC	39°30′	222°30′	+1°15′ at B	40°45′	220°45′	
CD	22°15′	200°30′	–1°45′ at C	20°30′	200°30′	
DE	242°45′	62°45′	0° at D	242°45′	62°45′	Station D is free from local attraction
EA	330°15′	147°45′	0° at E	330°15′	150°15′	Station E is also free from local attraction

Level-2

2.11 The following bearings were observed in running a closed traverse.

Line	F.B.	B.B.
AB	71°05′	250°20′
BC	110°20′	292°35′
CD	161°35′	341°45′
DE	220°50′	40°05′
EA	300°50′	121°10′

Determine the correct magnetic bearings of the traverse lines. [20 marks]

Sol:

As there is no traverse line whose *FB* and *BB* differ exactly by 180°. However the *FB* and *BB* of line *CD* differ by 180°10′. The difference being only +10′which is lowest among all other traverse lines. Hence the correct FB of *CD* is obtained by adding half the difference.

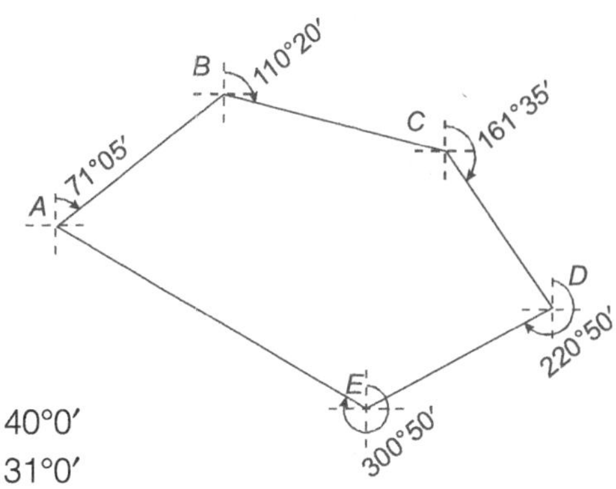

Hence, corrected FB of $CD = 161°35′ + 5′ = 161°40′$

and corrected *BB* of $CD = 341°45′ - 5′ = 341°40′$

$$\angle ABC = 250°20′ - 110°20′ = 140°0′$$
$$\angle BCD = 292°35′ - 161°35′ = 131°0′$$
$$\angle CDE = 341°45′ - 220°50′ = 120°55′$$
$$\angle DEA = 300°50′ - 40°05′ = 260°45′ \text{ (Exterior)} = 99°15′ \text{ (Interior)}$$
$$\angle EAB = 121°10′ - 71°05′ = 50°05′$$
$$\text{Sum of all interior angles} = 541°15′$$
$$\text{Theoretical sum} = (2N - 4) \times 90° = 540°$$
$$\text{Error} = 541°15′ - 540° = 1°15′$$
$$\therefore \quad \text{Correction for each angle} = -15′$$

Hence the corrected interior angles are

$$\angle ABC = 140°0′ - 15′ = 139°45′$$
$$\angle BCD = 131°0′ - 15′ = 130°45′$$
$$\angle CDE = 120°55′ - 15′ = 120°40′$$
$$\angle DEA = 99°15′ - 15′ = 99°$$
$$\angle EAB = 50°05′ - 15′ = 49°50′$$

The corrected bearings of all the lines are obtained from the included angles and the corrected bearing of traverse line *CD*.

$$\text{Corrected } FB \text{ of } DE = 341°40′ - 120°40′ = 221°$$
$$BB \text{ of } DE = 221°00′ - 180° = 41°$$
$$FB \text{ of } EA = 41° + (360° - 99°) = 302°00′$$
$$BB \text{ of } EA = 302°00′ - 180° = 122°$$
$$FB \text{ of } AB = 122° - 49°50′ = 72°10′$$
$$BB \text{ of } AB = 72°10′ + 180° = 252°10′$$
$$FB \text{ of } BC = 252°10′ - 139°45′ = 112°25′$$
$$BB \text{ of } BC = 112°25′ + 180° = 292°25′$$
$$FB \text{ of } CD = 292°25′ - 130°45′ = 161°40′$$

2.12 For the following traverse, compute the length of line *CD* so that *A*, *D* and *E* lie in one straight line:

Line	Length(m)	Bearing
AB	110	83°12′
BC	165	30°42′
CD	?	346°06′
DE	212	16°18′

[12 Marks]

Sol:

In given traverse *ABCDE*, the points *A*, *D* and *E* are on the same line. Let *CA* be the imaginary line so as to form a close traverse *ABC*.

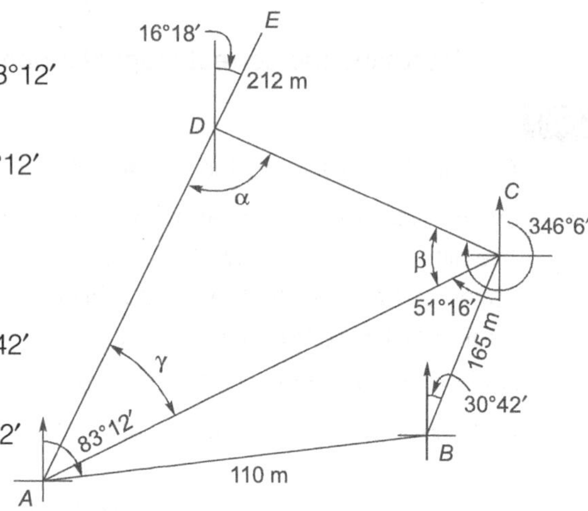

Line *AB*:

$$\text{Length of line } AB = 110\text{ m}$$
$$\text{Latitude of the line } AB = +110\cos 83°12'$$
$$= +13.03$$
$$\text{Departure of line } AB = +110\sin 83°12'$$
$$= +109.23$$

Line *BC*:

$$\text{Length of line } BC = 165\text{ m}$$
$$\text{Latitude of the line } BC = 165\cos 30°42'$$
$$= +141.88$$
$$\text{Departure of the line } BC = 165\sin 30°42'$$
$$= +84.24$$

Now, for the traverse *ABC*

$$\Sigma\text{Latitude} = 0$$

So, the latitude of *CA* $= 0 - 13.03 - 141.88 = -154.91$

and $\Sigma\text{departure} = 0$

Thus, departure of *CA* $= 0 - 109.23 - 84.24 = -193.47$

Since, the latitude and departure are both negative, the line *CA* lies in the third quadrant. Its bearing is given by:

$$\tan\theta = \left|\frac{\text{departure}}{\text{Latitude}}\right| = \frac{193.47}{154.91}$$

or, $\theta = S51°\,18'\,W$ (CA lies in 3rd quadrant)

$$\text{Length of } CA = \sqrt{Lat_{CA}^2 + dep_{CA}^2} = \sqrt{(154.91)^2 + (193.47)^2} = 247.78\text{ m}$$

Since *A*, *D* and E are on the same line,

∴ bearing of *AD* = bearing of *DE* = 16°18′

From triangle *ACD*,

$$\alpha = (360° - 346°06') + 16°18' = 13°54' + 16°18' = 30°12'$$
$$\beta = 346°06' - (51°18' + 180°) = 114°48'$$
$$\gamma = 180° - (144°48' + 30°12') = 35°$$

From the sine rule in ΔALD,

$$\frac{CD}{\sin\gamma} = \frac{CA}{\sin\alpha}$$

$$CD = CA\frac{\sin\gamma}{\sin\alpha} = 247.78\frac{\sin 35°}{\sin 30°12'} = 282.53\text{ m}$$

2.13 Adjust the following traverse by Bowditch method and determine the coordinates of the stations relative to $A(O_{10})$. Also determine the length and bearing of line BE.

Line	Length (in m)	WCB	Departure (D) (m)	Latitude (L) (m)
AB	560.5		0	–560.5
BC	901.5		795.4	–424.3
CD	557.0		–243.0	–501.2
DE	639.8		488.7	412.9
EF	679.5	293°59′		
FA	467.2	244°42′		

[15 Marks]

Sol:

For line EF, departure = $\angle EF\sin\theta = 679.5 \times \sin(293°59') = -620.8$ m.

Latitude = $\angle EF\cos\theta = 679.5 \times \cos(293°\ 59') = 276.2$ m.

Similarly for line FA, departure = $467.2 \sin(244°\ 42') = -422.4$ m.

Latitude = $467.2 \cos(244°\ 42') = -199.7$ m.

ΣLatitude = $-560.5 - 424.3 + 501.2 + 412.9 + 276.2 - 199.7 = 5.8$ m.

ΣDeparture = $0 + 795.4 - 243 + 488.7 - 620.8 - 422.4 = -2.1$ m.

Total correction in latitude = -5.8 m

Total correction in departure = 2.1 m

In Bowditch method, correction is proportional to length of side

$$\text{Correction in latitude} = \frac{-5.8}{\text{Total length}} \times \text{length of side}$$

$$= \frac{-5.8}{560.5+901.5+557+639.8+679.5+467.2} \times \text{side length}$$

$$= \frac{-5.8}{3805.5} \times \text{side length}$$

$$\text{Similarly, Correct in departure} = \frac{2.1}{3805.5} \times \text{side length}$$

Line	Length (in m)	Departure (D) (m)	Latitude (L) (m)	ΔD correction	ΔL correction	Corrected values D	Corrected values L
AB	560.5	0	–560.5	0.3	–0.8	0.3	–561.3
BC	901.5	795.4	–424.3	0.5	–1.4	795.9	–425.7
CD	557.0	–243.0	–501.2	0.3	–0.9	–242.7	–500.3
DE	639.8	488.7	412.9	0.3	–1	489	411.9
EF	679.5	–620.8	276.2	0.4	–1	–620.4	275.2
FA	467.2	–422.4	–199.7	0.3	–0.7	–422.1	–200.4

Station	*x* Coordinate of station Σ Departure of previous stations	*y* Coordinate of station Σ Latitude of previous stations
B	0.3	–561.3
C	796.2	–987.0
D	553.5	–486.7
E	1042.5	–74.8
F	422.1	200.4
A	0	0

$$\text{Length of BE} = \sqrt{(x_B - x_E)^2 (y_B - y_E)^2} = \sqrt{(+0.3 - 1042.5)^2 + (-561.3 - (-74.8))^2} = 1150.1 \text{ m.}$$

$$\text{Bearing of BE} = \tan^{-1}\left(\frac{x_E - x_B}{y_E - y_B}\right) = \tan^{-1}\left(\frac{1042.5 - 0.3}{-74.8 - (-561.3)}\right) = 64°59'$$

2.14 The angles at the stations of a closed traverse ABCDEFA were observed as given below :

Traverse Station	Included Angle
A	120°35′00″
B	89°23′40″
C	131°01′00″
D	128°02′20″
E	94°54′40″
F	155°59′20″

Adjust the angular error in the observations and calculate the bearings of the traverse lines in whole circle and quad rental bearing system. WCB of AB is 42°. [20 Marks]

Sol:

Sum of internal angles of sir sided polygon = (6 – 2) × 180° = 720°

Sum of observed internal angles = 120° 35′ + 89°23′40″ + 131°01′ + 128°02′20″ + 94°54′40″ + 155°59′20″ = 719°56′

Total error = 719° 56′ – 720° = –4′

Total Correction = +4′

$$\text{Correction in each angle} = \frac{4 \times 60}{6} = 40''$$

Traverse Station	Included Angle	Corrected Value
A	120°35′00″	120°35′40″
B	89°23′40″	89°24′20″
C	131°01′00″	131°1′40″
D	128°02′20″	128°3′
E	94°54′40″	156°
F	155°59′20″	94°55′20″

WCB of AB = 42° (given)

WCB of BA = 180 + 42 = 222°

WCB of BC = WCB of BA – ∠B

= 222° – 89° 24′ 20″ = 132° 35′ 40″

WCB of CB = 180° + 132° 35′ 40″ =312° 35′ 40″

WCB of CD = WCB of CB – ∠C

= 312° 35′ 40″ – 131° 1° 40″

= 181° 34′

WCB of DC = 180° + 181° 34′= 361° 34′ – 360° = 1°34′

WCB of DE = WCB of DC – ∠D = 1°34′ –128° 3′ + 360° = 233°31′

WCB of ED = 180° + 233° 31′ – 360° = 53° 31′

WCB of EF = 53° 31′ – 94° 55′ 22″ + 360° = 318° 35′ 40″

WCB of FE = 318° 35′ 40″ – 180 = 138° 35′ 40″

WCB of FA = 138° 35′ 40″ – 156° + 360° = 342° 35′ 40″

WCB of AF = 180° + 342° 35′ 40″ – 360 = 162° 35′ 40″

WCB of AB = 162° 35′ 40″ – 120° 35′ 40″ = 42°

Line	WCB	Quadrantal Bearing
AB	42°	N42°E
BC	132°35′40″	S47°24′20″E
CD	181°34′00″	S1°34′W
DE	233°31′00″	S53°31′W
EF	318°35′40″	N41°24′20″W
FA	342°35′40″	N17°24′20″W

3. Levelling, Contouring and Plane Table Surveying

Level-1

3.1 The following observations were made in running fly levels from a bench mark of *RL* 60.65 m. Five pegs at 20 m intervals are to be set on a falling gradient of 1 in 100 m from the last position of the instrument. The first peg is to be at *RL* 60 m. Work out the staff readings required for setting the pegs and prepare the page of the level book.

[10 Marks]

Back sight	Fore sight
0.964	0.948
1.632	1.153
1.105	1.984
0.850	

Sol:

The page of the level book can be prepared as follows:

Station	*BS*	*IS*	*FS*	*HI*	*RL*	Remarks
1.	0.964	-	-	61.614	60.65	*BM*
2.	1.632	-	0.948	62.298	60.666	-
3.	1.105	-	1.153	62.250	61.145	-
4.	0.850	-	1.984	61.116	60.266	-

Now pegs are to be set on a falling gradient of 1 in 100 at 20 m interval.

$\therefore$ Difference in elevation between two consecutive pegs $= \frac{1}{100} \times 20 = 0.2$ m

Staff readings:

Station	*BS*	*IS*	*FS*	*HI*	*RL*	Remarks
4	0.850	-	-	61.116	60.266	*BM*
P_1	-	-	1.116	-	60.00	-
P_2	-	-	1.316	-	59.80	-
P_3	-	-	1.516	-	59.60	-
P_4	-	-	1.716	-	59.40	-
P_5	-	-	1.916	-	59.20	-

3.2 The staff reading taken from an instrument when bubble was at centre on a staff kept at 250 m distance was 1.62 m. The bubble was the moved by 6 division and staff reading changed to 1.95 m. If length of one division of bubble tube is 2.5mm. calculate

[8 Marks]

(i) Sensitivity of bubble tube.

(ii) Radius of curvature of bubble tube.

Sol:

D = 250 m, n = 6, S_1 = 1.62 m, L = 2.5 mm, S_2 = 1.95m, $S = S_2 - S_1 = 1.95 - 1.62 = 0.33$ m.

(i) Sensitiveness
$$\alpha = \frac{S}{nD} = \frac{0.33}{6 \times 250} \times 206265 \text{ seconds} = 45.4''$$

(ii) Also
$$\alpha = \frac{l}{R} \times 206265$$

$$R = \frac{l}{\alpha} \times 206265 = \frac{2.5 \times 10^{-3}}{45.4} \times 206265 = 11.36 \text{ m.}$$

3.3 The height of an inaccessible tower is to be determined by ranging rods. The two ranging rods of 2.50 m and 1.50 m are placed in such a way that their tops are in line with the top of the tower. The distances between the ranging rods is 10 m. The two ranging rods are then placed at different setting on the same ranged out line such that longer rod is placed 110 m away from its first position and distance between the ranging rods is 30 m. If their tops are again in line with the top of tower, determine the height of the tower.

[12 Marks]

Sol:

The problem can be represented diagrammatically as below:

Let h be the height of tower.

Given, $$x + y = 110 \text{ m}$$

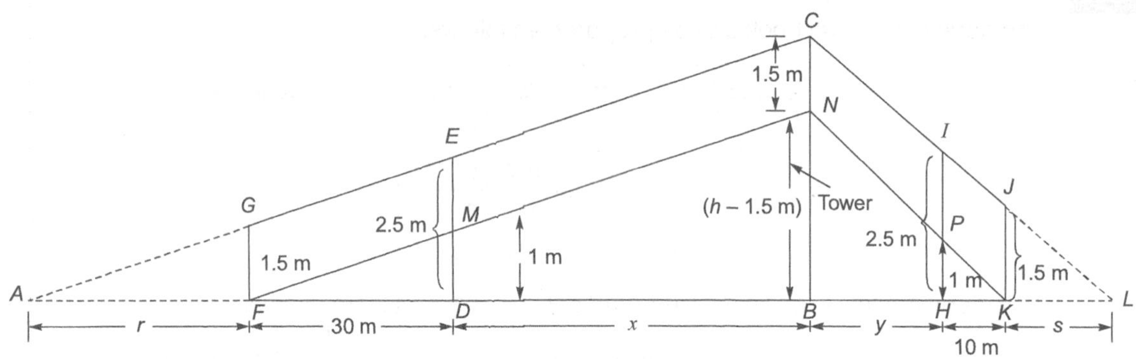

Using property of similar triangles;

In ΔFMD and ΔFNB;

$$\Rightarrow \quad \frac{1}{h-1.5} = \frac{30}{30+x} \quad \text{...(i)}$$

In ΔKPH and ΔKNB,

$$\Rightarrow \quad \frac{1}{h-1.5} = \frac{10}{10+y} \quad \text{...(ii)}$$

Equating eq. (i) and (ii); we get,

$$\frac{30}{30+x} = \frac{10}{10+y}$$

$$\Rightarrow \quad 300 + 30y = 300 + 10x$$

$$\Rightarrow \quad 10x - 30y = 0 \quad \text{...(iii)}$$

Also $$x + y = 110 \quad \text{(Given)} \quad \text{...(iv)}$$

Solving Eq. (iii) and (iv), we get,

$$\Rightarrow \quad x = 82.5 \text{ m} \quad \text{and} \quad y = 27.5 \text{ m}$$

Put value of $x = 82.5$ m in equation (i);

$$\Rightarrow \quad \frac{1}{h-1.5} = \frac{30}{30+82.5}$$

$$\Rightarrow \quad h - 1.5 = 3.75 \text{ m}$$

or, $$h = 3.75 + 1.5 = 5.25 \text{ m}$$

The height of tower is 5.25 m.

3.4 From an instrument station B.S is 2.0m taken to the B.M of R.L 250 m. and F.S is taken to the inverted staff at station as 1.8m. After that instrument is shifted to another location and B.S of 1.2m is taken on same station Q and F.S of 1.5m taken at a ground station R. Calculate RL of station R.

[8 Marks]

Sol:

For first instrument station

$$\text{H.I} = \text{RL of BM} + \text{B.S}$$
$$= 250 + 2 = 252.0 \text{ m}$$
$$\text{RL of station Q} = \text{HI} + \text{FS at Q}$$
$$= 252 - (-1.8) \quad \text{(negative FS reading since staff is inverted)}$$
$$= 253.8 \text{ m}$$

For 2nd instrument station

$$\text{HI}_2 = \text{RL of Q} + \text{BS at Q}$$
$$= 253.8 + 1.2$$
$$= 255 \text{ m}$$
$$\text{RL of station R} = \text{HI} - \text{FS at R}$$
$$= 255 \text{ m} - 1.5 \text{ m} = 235 \text{ m}$$

3.5 Reciprocal leveling was conducted across a wide river to determine the difference in level of points A and B, A situated on one bank of the river and B on the other. the following results on the staff held vertically at A and B from level stations 1 and 2, respectively were obtained. The level station 1 was near to A and station 2 was near to B.

Instrumental at	Staff reading on	
	A	B
1	1.485	1.725
2	1.190	1.415

(a) If RL of B is 55.18m, find RL of A.

(b) Assuming atmospheric conditions remain unchanged calculate

(i) Combined curvature and refraction correction if distance between AB is 315m.

(ii) Collimation error

[12 Marks]

Sol:

$a_1 = 1.485$ m, $a_2 = 1.725$ m, $b_1 = 1.190$ m, $b_2 = 1.415$ m

(a) Difference in hence Δh

$$= \frac{(a_2 - a_1) + (b_2 - b_1)}{2} = \frac{(1.725 - 1.485) + (1.415 - 1.190)}{2}$$
$$= 0.233 \text{ m}$$

Station B is below Station A

So $$\text{RL of A} = \text{RL of B} + \Delta h = 55.18 + 0.233 = 55.413 \text{ m}$$

(b)

(i) Combined curvature and refraction error

$$e_{cr} = 0.0673d^2 = 0.0673\,0 \times 0.315^2 = 0.007 \text{ m}$$

(ii) $$\text{Total Error} = e_{cr} + \text{collimation error}$$

$$\text{Total Error} = \frac{(1.725 - 1.485) - (1.415 - 1.190)}{2} = 0.008 \text{ m}$$

$$\text{Collimation error} = 0.008 - 0.007 = 0.001 \text{ m}$$

3.6 **Four stations C, A, B and D were set out in a straight line such that CA = AB = BD = 30 m. A level was set up at C and readings of 2.135 and 1.823 were observed on vertically held staff at A and B respectively, when bubble was at centre of its run. The level was the set up at D and readings of 2.026 and 1.768 were again observed at A and B respectively. Determine collimation error of the level and correct difference in level of A and B.** **[12 Marks]**

Sol:

Difference in level of A and B when instrument at C

$$\Rightarrow \quad \Delta h_1 = 2.135 - 1.823 = 0.312 \text{ m.}$$

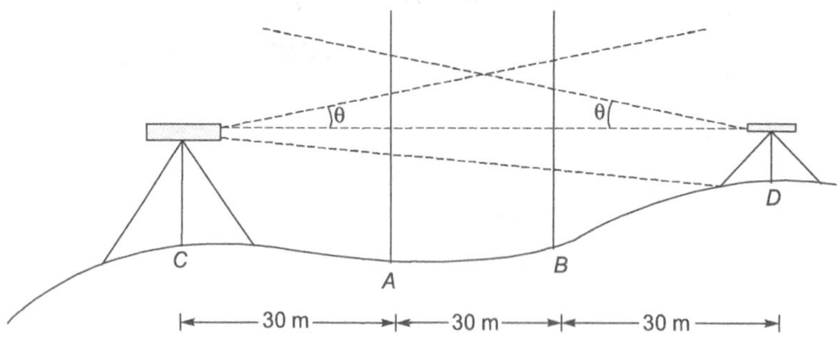

difference in level of A & B when instrument at D

$$\Rightarrow \quad \Delta h_2 = 2.026 - 1.768 = 0.258\text{m.} \neq \Delta h_1$$

Let inclination of line of collimation is θ with the horizontal upwards as shown in figure.

Correct difference in level are

$$\Delta h_1 = (2.135 - 30\theta) - (1.823 - 60\theta) = 0.312 + 30\theta \quad ....(1)$$

$$\Delta h_2 = (2.026 - 60\theta) - (1.768 - 30\theta) = 0.258 - 30\theta \quad ....(2)$$

Equating equation (1) and (2)

$$0.312 + 30\theta = 0.258 - 30\theta$$

$$\Rightarrow \quad \theta = -0.0009 \text{ rad}$$

$$= -3'5.64'' \quad \text{(Collimation error is in down ward direction)}$$

Substituting in equation 1, we get

$$\Delta h_1 = 0.312 + 30 \times (-0.0009) = 0.285 \text{ m.}$$

3.7 **Fig. shows rectangle ABCD, in which A, B and C are stations where staff readings were obtained with a level set up at E and D. The observed readings and given in table below.**

Level at	Staff reading on		
	A	B	C
E	1.856	0.809	—
D	2.428	1.369	1.667

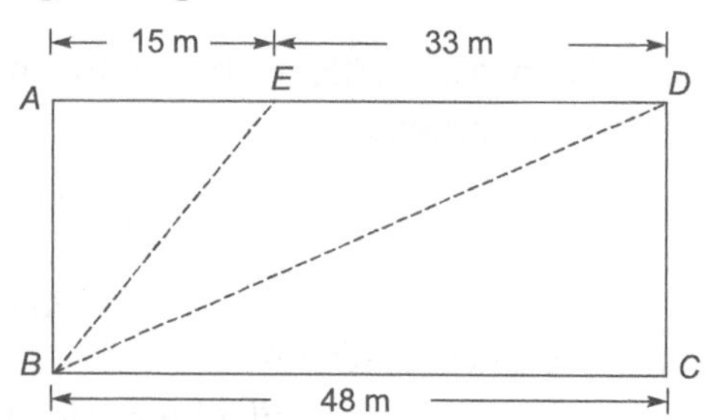

If A is B.M having RL 150 m. Calculate correct elevations of B and C. **[12 Marks]**

Sol:

$$\text{EB} = \sqrt{15^2 + 20^2} = 25 \text{ m}$$

$$\text{DB} = \sqrt{20^2 + 48^2} = 52 \text{ m}$$

Assuming line of sight is inclined upwards by angleQ. Correct staff reading on A when line at E.

$$= 1.856 - 15\,\theta$$

Correct staff reading on B when level at E = 0.890 – 25θ
Correct staff reading on A when level at D = 2.428 – 48θ
Correct staff reading on B when level at D = 1.369 – 52θ
Correct staff reading on C when level at D = 1.667 – 20θ
Δh_{AB} from both stations must be equal.
Therefore,

$$(1.856 - 15\theta) - (0.809 - 25\theta) = (2.428 - 48\theta) - (1.369 - 52\theta)$$

$$\theta = 0.002 \text{ rad}$$

$$\text{Correct difference } \Delta h_{AB} = (1.856 - 15 \times 0.002) - (0.809 - 25 \times 0.002) = 1.067 \text{ m}$$

$$\text{RL of B} = 150 + 1.067 = 151.067 \text{ m}$$

$$\text{Correct staff reading at C} = 1.667 - 20\theta = 1.1667 - 20 \times 0.0002 = 1.627$$

$$\Delta h_{AC} \text{ Correct} = (2.428 - 48 \times 0.002) - 1.627 = 0.705 \text{ m}$$

$$\text{RL of C} = 150 + 0.705 = 150.705 \text{ m}$$

Level-2

3.8 **The following readings were taken with a level and 4 m staff. Draw up a level book and reduce the levels by height of instrument method.**
0.578 (BM = 58.250 m), 0.933, 1768, 2.450, (2.005 and 0.567) C.P, 1.888, 1.181, (3.697 and 0.612) C.P, 0.705, 1.810. **[20 Marks]**

Sol:

Fifth and eight station is change point.
Let h_i denotes Reduced level at station *i*.
Height of instrument at station 1 HI= h_1 + Back sight

$$HI = 58.250 + 0.578 = 58.828 \text{ m}$$

$$h_2 = HI - \text{Intermediate sight at station 2}$$

Similarly,

$$h_3 = 58.828 - 9.768 = 57.060 \text{ m}$$

$$h_4 = 58.828 - 2.450 = 56.378 \text{ m}$$

$$h_5 = 58.858 - \text{fore sight} = 58.828 - 2.005 = 56.823 \text{ m}$$

Now instrument is shifted

$$HI_5 = h_5 + \text{Back sight at station 5} = 56.823 + 0.657 = 57.390 \text{ m}$$

$$h_6 = 57.390 - 1.888 = 55.502 \text{ m}$$

$$h_7 = 57.390 - 1.181 = 56.209 \text{ m}$$

$$h_8 = 57.390 - 3.679 = 53.711 \text{ m}$$

Now instrument is shifted

$$HI_8 = 53.711 + 0.612 = 54.323 \text{ m}$$

$$h_9 = 54.323 - 0.705 = 53.618 \text{ m}$$

$$h_{10} = 54.323 - 1.810 = 52.513 \text{ m}$$

Station	BS	IS	FS	HI	RL	Remarks
1	0.578			58.828	58.250	BM = 58.250 m
2		0.933			57.895	
3		1.768			57.060	
4		2.450			56.378	
5	0.567		2.005	57.390	56.823	Change point
6		1.888			55.502	
7		1.181	3.679	53.323	56.209	
8	0.612				53.711	Change point
9		0.705	1.810		53.618	
10					52.513	
Sum	**1.757**	**8.925**	**7.494**			

Check ⇒ $\Sigma BS - \Sigma FS$ = RL of last station – RL of first.

LHS = 1.757 – 7.494 = –5.737

and RHS = 52.513 – 58.250 = – 5.737 (Checked)

3.9 **The following consecutive reading were taken with a level on continuously sloping ground at a common interval of 20m. Last station has an elevation of 155.272 m. Calculate**
(a) R.L. of points by rise and fall method (b) Gradient of line joining first and last point.
Reading: 0.420, 1.115, 2.265, 2.900, 3.615, 0.535, 1.470, 2.815, 3.505, 4.445, 0.605, 1.925, 2.885
[20 Marks]

Sol:

Since the ground is continuous sloping, any sudden decrease in reading indicate change point as observed in 6th and 11th reading. Therefore 6th and 11th readings are back sights and 5th and 10th readings are fore sights.

Calculation of rise and fall

Instrument at station – 1

Fall $f_2 = IS_2 - BS_1 = 1.115 - 0.420 = 0.695$

$f_3 = 2.265 - 1.115 = 1.150$

$f_4 = 2.900 - 2.265 = 0.635$

$f_5 = 3.615 - 2.900 = 0.715$

Instrument at section – 2

Fall $f_6 = IS_6 - BS_5 = 1.470 - 0.535 = 0.935$

$f_7 = 2.818 - 1.470 = 1.345$

$f_8 = 3.505 - 2.818 = 0.690$

$f_9 = 4.445 - 3.505 = 0.940$

Instrument at section – 3, $f_{10} = 1.925 - 0.0605 = 1.320$

$f_{11} = 2.885 - 1.925 = 0.960$

let h_i indicate RL at i station.

Given $h_{11} = 155.272$ m.

$h_{10} = h_{11} + f_{11} = 155.72 + 0.960 = 156.232$ m.

$h_9 = h_{10} + f_{10} = 156.232 + 1.320 = 157.552$ m.

$h_8 = 157.552 + 0.940 = 158.492$ m.

$h_7 = 158.492 + 0.690 = 159.182$ m.

$h_6 = 159.182 + 1.345 = 160.527$m.

$h_5 = 160.527 + 0.935 = 161.462$ m.

$h_4 = 161.462 + 0.715 = 162.177$ m.

$h_3 = 162.177 + 0.635 = 162.812$ m.

$h_2 = 162.812 + 1.15 = 163.962$ m.

$h_1 = 163.962 + 0.695 = 164.657$ m.

Station	Chainage (m)	BS	IS	FS	Rise	Fall	RL	Remarks
1	0	0.420					164.657	BM = 58.250 m
2	20		1.115			0.695	163.962	
3	40		2.265			1.150	162.812	
4	60		2.900			0.635	162.177	
5	80	0.535		3.615		0.715	161.462	Change point
6	100		1.470			0.935	160.527	
7	120		2.815			1.345	159.182	
8	140		3.505			0.690	158.492	
9	160	0.605		4.445		0.940	157.552	Change point
10	180		1.925			1.320	156.232	
11	200			2.885		0.960	155.272	Given
Sum		**1.560**		**10.945**	**0**	**9.385**		

Check

$\Rightarrow$ $\Sigma BS - \Sigma FS = 1.560 - 10.945 = -9.385$

$\Sigma_{Rise} - \Sigma_{Fall} = 0 - 9.385 = -9.385$

Last RL – First RL = 155.272 – 164.657 = – 9.385 (checked)

(b) Calculation of gradient

$$\text{Gradient of line 1 – 11 is} = \frac{\text{difference in elevation}}{\text{distance between points}} = \frac{155.272 - 164.657}{200} = \frac{-1}{21.3}$$

So gradient is 1 in 21.3 falling.

4. Calculation of Area, Volume and Theory of Errors

Level-1

4.1 **The probable error of a single observation which when repeated on the same angle gave values of 43°20′ plus: 10″, 30″, 10″, 20″, 00″, 40″, 00″, 10″, 30″, 00″, 10″.** **(8 Marks)**

Solution:

Observed value	Deviation from mean	Squares of deviation
43°20′10″	–04.5″	20.25
30″	+15.5″	240.25
10″	–04.5″	20.25
20″	+04.5″	20.25
00″	–14.5″	210.25
40″	+25.5″	650.25
00″	–14.5″	210.25
10″	–04.5″	20.25
30″	+15.5″	240.25
00″	–14.5″	210.25
10″	–4.5″	20.25
Mean value = 43°20′14.5″	Σ = 0	Σ 1862.75

Portable error of a single observation

$$= \pm 0.6745\sqrt{\frac{1862.75}{10}} = \pm 9.20''$$

4.2 **The best value of a quantity which measured four times by a method having a probable error of 3 units, gave an average value of 1800 units and measured nine times by a method having a probable error of 6 units gave an average value of 1808 units of (L.U.)**

(5 Marks)

Solution:

Probable error of mean by 1st method

$$= \pm\frac{3}{\sqrt{4}} = \pm\frac{3}{2} \text{ units}$$

Probable error of mean by 2nd method

$$= \pm\frac{6}{\sqrt{9}} = \pm 2 \text{ units}$$

Weights are inversely proportional to squares of probable error

$$w_1 : w_2 = \frac{4}{9} : \frac{1}{4}$$

$$\text{Most probable value} = \text{Weighted mean}$$

$$= \frac{\frac{4}{9}\times 1800 + \frac{1}{4}\times 1808}{\frac{4}{9}+\frac{1}{4}} = 1802.88 \text{ units}$$

4.3 **The following are the observed values of an angle and their weightage :**
Find :

(i) Probable error of single observation of unit weight.
(ii) Probable error of weighted arithmetic mean.
(iii) Probable error of single observation of weight 3.

Angle	Weightage
30° 24′ 20″	2
30° 24′ 18″	2
30° 24′ 19″	3

[12 Marks]

Sol:

Since the error is in seconds only, the degrees and minutes of the quantities have not been included in the tabulation. The computations are arranged in the tabular form below:

S.No.	Value	Weight(w)	V	V^2	wV^2
1.	20″	2	1″	1	2
2.	18″	2	–1″	1	2
3.	19″	3	0	0	0
		$\Sigma w = 7$			$\Sigma wv^2 = 4$

Weighted arithmetic mean of the seconds readings of the observed angles

$$= \frac{20''\times 2 + 18''\times 2 + 19''\times 3}{2+2+3} = 19''$$

∴ Weighted arithmetic mean of the angle = 30° 24′ 19″

$$\therefore \quad v_1 = 20'' - 19'' = 1''$$

$$v_2 = 18'' - 19'' = -1''$$

$$v_3 = 19'' - 19'' = 0$$

(i) Probable error of single observation unit weight is given by

$$E_s = \pm 0.6745\sqrt{\frac{\Sigma wv^2}{n-1}} = \pm 0.6745\sqrt{\frac{4}{3-1}} = \pm 0.95'' \quad \text{...(i)}$$

(ii) Probable error of the arithmetic mean is given by

$$E_m = \pm 0.6745\sqrt{\frac{\Sigma wv^2}{\Sigma w(n-1)}} = \pm 0.6745\sqrt{\frac{4}{7(3-1)}} = \pm 0.36''$$

(iii) Probable error of single observation of weight 3 is given by

$$E_w = \frac{E_s}{\sqrt{w}} = \pm 0.6745\sqrt{\frac{\Sigma w v^2}{w(n-1)}}$$

$$= \pm\frac{0.95}{\sqrt{3}} \text{ (substuting } E_s \text{ from (i))} = \pm 0.55''$$

4.4 An excavation has been made as shown in fig. calculate the quantity of earth executed using
(i) Trapezoidal rule
(ii) Simpson's rule

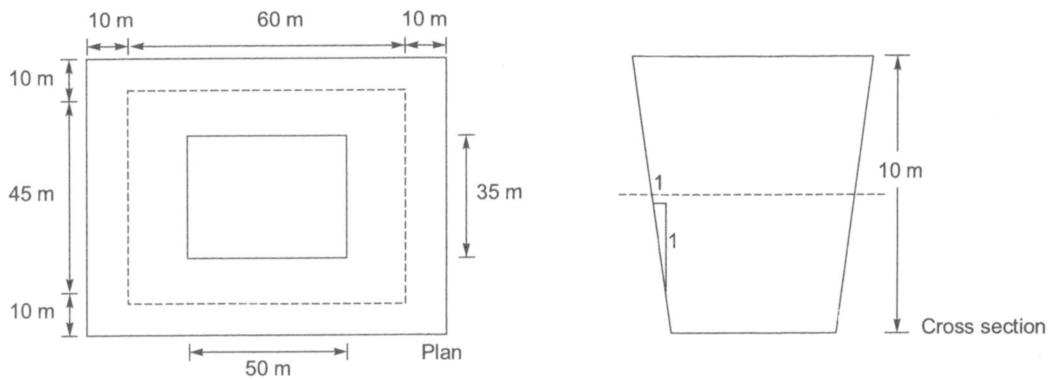

[12 Marks]

Sol:

$$A_1 = 50 \times 35 = 1750 \text{ m}^2 \qquad d = 5 \text{ m}$$
$$A_2 = 60 \times 45 = 2700 \text{ m}^2$$
$$A_3 = 80 \times 65 = 5200 \text{ m}^2$$

(i) Using Trapezoidal rule

$$V = d\left[\frac{A_1 + A_3}{2} + A_2\right] = 5\left[\frac{1750 + 5200}{2} + 2700\right] = 30875 \text{ m}^3$$

(ii) Using Simpsion's rule

$$V = \frac{d}{3}[A_1 + A_3 + 4A_2]$$

$$= \frac{5}{3}[1750 + 5200 + 4 \times 2700] = 29583.33 \text{ m}^3$$

4.5 A traverse of land has three straight boundaries AB, BC and CD. Fourth boundary DA is irregular. Measured length are as under : -
AB = 135 m, BC = 191 m, CD = 126 m, BD = 255 m

The offsets measured outside the boundary DA to the irregular boundary at regular interval of 30 m from D are as below

Distance from D (m)	0	30	60	90	120	150	180
Offsets (m)	0	3.7	4.9	4.2	2.8	3.6	0

Determine are of traverse.

12 Marks]

Sol:

$$\text{Area of traverse} = \text{Area } \Delta ADB + \text{Area } \Delta DBC + \text{Area of irregular fig}$$

$$\text{Area of triangle} = (s(s-a)(s-b)(s-c))^{1/2}$$

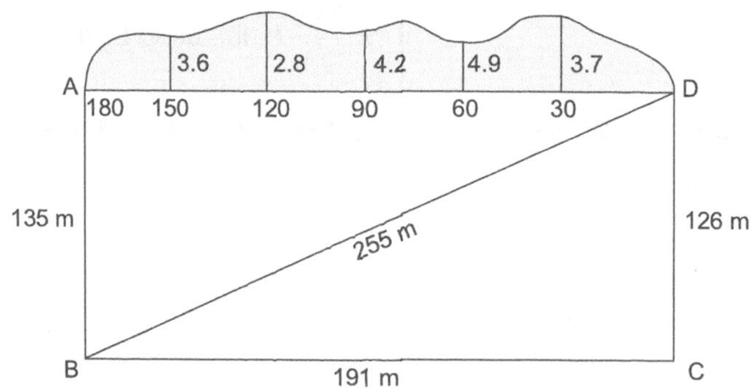

$$\text{Where } S = \frac{a+b+c}{2}$$

For ΔADB $\quad S = \dfrac{135+180+255}{2} = 285\,m.$

Area $\quad \Delta ADB = \sqrt{285(285-255)(285-180)(285-135)} = 11604.42 \text{ m}^2$

For ΔDBC, $\quad S = \dfrac{255+191+126}{2} = 286$

Area ΔDBC, $\quad = \sqrt{286(286-255)(286-191)(286-126)} = 11608.76 \text{ m}^2$

For area of irregular figure, using Simpson's rule.

$$\text{Area} = \frac{d}{3}[A_1 + A_n + 4(A_2 + A_4.....) + 2(A_3 + A_5.....)]$$

$$= \frac{30}{3}[0+0+4(3.7+4.2+3.6)+2(4.9+2.8)]$$

$$= 614 \text{ m}^2$$

Hence, total area of traverse = $11604.42 \text{ m}^2 + 11608.76 \text{ m}^2 + 614 \text{ m}^2 = 23827.18 \text{ m}^2$

4.6 **Figure shows the distribution of 12 spot heights with a regular 20m spacing covering a rectangular area which is to be graded to form a horizontal plane of elevation h.**

(a) Calculate volume of earth for h = 10 m.

(b) Calculate h such that cut and fill are balanced.

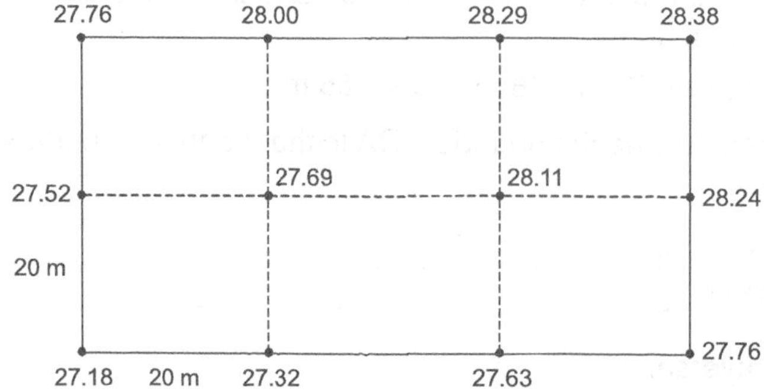

[12 Marks]

Sol:

Horizontal plane has elevation 10 m.

Ordinates of each point has to be reduced by (d –10) when d = elevation of each point.

Volume of each section = Area of base × average height of elevation to be cut

$$\text{Area of base} = 20 \times 20 = 400 \text{ m}^2$$

$$\Sigma d_1 = (27.76 + 28.38 + 27.76 + 27.18 - 10 \times 4) = 71.08 \text{ m}$$

$$\Sigma d_2 = (27.52 + 28 + 28.29 + 28.24 + 27.63 + 27.32 - 6 \times 10) = 107 \text{ m}$$

$$\Sigma d_3 = 0$$

$$\Sigma d_4 = 27.69 + 28.11 - 2 \times 10 = 35.8 \text{ m.}$$

$$\text{Total Volume} = A\left[\frac{\Sigma d_1 + 2\Sigma d_2 + 3\Sigma d_3 + 4\Sigma d_4}{4}\right]$$

$$= 400\left[\frac{71.08 + 107 \times 2 + 0 + 35.8 \times 4}{4}\right] = 42828 \text{ m}^3$$

(b) Let design level be h.

$$\Sigma d_1 = 27.18 + 27.76 + 28.38 + 27.76 - 4h = 111.08.4 \text{ h}$$

$$\Sigma d_2 = 27.52 + 28 + 28.89 + 28.24 + 27.63 + 27.32 - 6 \text{ h} = 167 - 6 \text{ h}$$

$$\Sigma d_3 = 0$$

$$\Sigma d_4 = 27.69 + 28.11 - 2h = 55.8 - 2 \text{ h}$$

Total Volume = 0 since cut and fill are balanced.

$$0 = \frac{400}{4}\left[(111.08 - 4h) + 2(167 - 6h) + 4(55.8 - 2h)\right]$$

$$66.8.28 - 24 \text{ h} = 0$$

$$\text{h} = 27.845 \text{ m.}$$

Level-2

4.7 **Following coordinates were taken for station A and B with their standard deviations. Calculate length and standard deviation of AB.**

Station	Eastings	Northing
A	456.961 m ± 20 mm	573.237 m ± 30 mm
B	724.616 m ± 40 mm	702.443 m ± 50 mm

Length of AB was independently measured as 297.426 m. ± 70 mm and its separate determination by EDM is as 297.155 m ± 15 mm. Calculate most probable length of the line and its standard deviation. **[20 Marks]**

Sol:

$$\text{Length of AB} = \sqrt{(\Delta E)^2 + (\Delta N)^2}$$

$$L = \sqrt{(724.616 - 456.961)^2 + (702.443 - 573.237)^2} = 297.209 \text{ m}$$

Standard deviation σ_E and σ_N for ΔE and ΔN respectively.

$$\sigma_E^2 = \sigma_{EA}^2 + \sigma_{EB}^2 = 20^2 + 40^2 = \sigma_E = \pm 44.7 \text{ mm}$$

$$\sigma_N^2 = 30^2 + 50^2 = \sigma_N = \pm 58.3 \text{ mm}$$

Standard deviation of computed length AB would be σ_{AB}

$$\sigma_{AB}^2 = \left(\frac{\partial L}{\partial(\Delta E)}\sigma_E\right)^2 + \left(\frac{\partial L}{\partial(\Delta N)}\sigma_N\right)^2$$

$$L = \left(\Delta E^2 + \Delta N^2\right)^{\frac{1}{2}}$$

$$\frac{\partial L}{\partial(\Delta E)} = \Delta E\left(\Delta E^2 + \Delta N^2\right)^{\frac{-1}{2}} = \frac{\Delta E}{L};$$

$$\frac{\partial L}{\partial(\Delta N)} = \Delta N\left(\Delta E^2 + \Delta N^2\right)^{\frac{-1}{2}} = \frac{\Delta N}{L}$$

$$\sigma_{AB}^2 = \left[\frac{724.616 - 456.961}{297.209} \times 44.7\right]^2 + \left[\frac{702.443 - 573.237}{297.209} \times 58.3\right]^2$$

$$\sigma_{AB} = \pm 47.6 \text{ mm}$$

Weight is inversely proportional to squared of standard deviation so,

Length measured by	Length	Standard deviation (σ)	Weight, $w = \frac{1}{\sigma^2}$
Tape	297.426	70	$\frac{1}{4900}$
EDM	297.155	15	$\frac{1}{225}$
Calculation	297.209	47.6	$\frac{1}{2266}$

Most probable length = weighted average

$$= \frac{\frac{1}{4900} \times 297.426 + \frac{1}{225} \times 297.155 + \frac{1}{2266} \times 297.209}{\frac{1}{4900} + \frac{1}{225} + \frac{1}{2266}}$$

$$= 297.171 \text{ m.}$$

Weight of most probable length $W = \sum w = \frac{1}{4900} + \frac{1}{225} + \frac{1}{2266} = 0.00509$

$$W = \frac{1}{\sigma^2} = 0.00509$$

$$\sigma = \frac{1}{\sqrt{W}} = \frac{1}{\sqrt{0.00509}} \pm 14\text{mm}$$

4.8 **A base line AB was measured accurately using a substance bar 1 m long. From a point C near the centre of the base, the lengths AC and CB were measured as 9.375 m and 9.493 m respectively. If the standard error is angular measured was ±1″, determine the error in length of line.**

[8 Marks]

Sol:

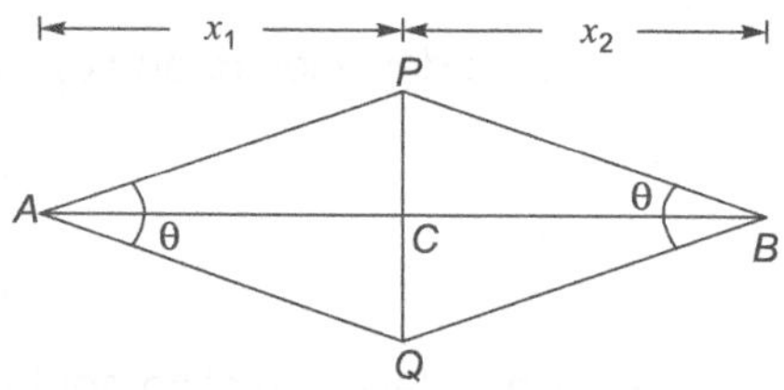

Substance base PQ is at C,

Respectively,

$$PQ = b = 1 \text{ m}$$

$$x_1 = AC = 9.375 \text{ m.}$$

$$x_2 = BC = 9.493 \text{ m.}$$

For substance bar

$$x = \frac{b}{2\tan\left(\frac{\theta}{2}\right)}$$

x = Computed distance, $b = p\theta = 1$ m

For small angle θ, $\tan\frac{\theta}{2} = \frac{\theta}{2}$

$$x = \frac{b}{\theta} \text{ or } \theta = \frac{b}{x}$$

so
$$dx = \frac{-b}{\theta^2}d\theta = \frac{-bd\theta}{\left(\frac{b}{x}\right)^2} = \frac{-x^2}{b}d\theta$$

So standard deviation σ_{AC} and σ_{BC} can be computed.

$$\sigma_{AC} = \frac{x_1^2}{b}\sigma_a = \frac{9.375^2}{1}\times 1'' = \frac{9.375^2}{1}\times\frac{1}{206265} = \pm 0.000426 \text{ m}$$

$$\sigma_{CB} = \frac{x_1^2}{b}\sigma_a = \frac{9.493^2}{1}\times\frac{1}{206265} = \pm 0.000437 \text{ m}$$

$$\sigma_{AB}^2 = \sigma_{AC}^2 + \sigma_{CB}^2$$

$$\sigma_{AB} = \sqrt{(0.000426)^2 + (0.000437)^2} = \pm 0.61 \text{ mm}$$

Measured length = 9.375 + 9.493 = 18.868 m.

So length of AB = 18.868 m ± 0.61 mm.

5. Techeometric, Curve, Hydrographic Survey, Tides and Triangulation

Level-1

5.1 **Explain the objectives of triangulation surveys and explain the criteria for selection of layout of triangles. Also, explain the terms well conditioned triangles, and strength of figure.**

[10 marks]

Sol:

Objectives of triangulation surveys: The triangulation surveys are carried out :

(i) To establish accurate control for plane and geodetic surveys of large areas, by terrestrial methods.

(ii) To establish accurate control for photogrammetric surveys of the large areas.

(iii) To assist in the determination of the size and shape of the earth by making observations for latitude, longitude and gravity.

(iv) To determine accurate locations of points in engineering works such as fixing of centreline and abutments of long bridges over large rivers. etc.

Criteria for selection of layouts of triangles: The criteria while deciding and selecting a suitable layout of triangles are as follows:

1. Triangles should preferrably be equilateral.
2. Braced quadrilaterals should preferrably be equilateral.
3. Centered polygons are regular.
4. The arrangements should be such that computations can be done through two or more independent routes.
5. The arrangements should be such that at least one route and preferrably two routes from well conditioned triangles.

6. No angle of the figure opposite to the known side should be small.
7. The sides of figures should be of comparable lengths. Very long and very short lengths should be avoided.
8. Angles of similar triangles should not be less than 45° and in case of quadrilaterals no angle should be less than 30°.

Well conditioned triangles:

- The triangles of such a shape, in which any error in angular measurement has minimum effect upon the computed lengths is known as well-conditioned triangles.
- The best shape of an a well-conditioned triangle is isosceles triangle is that triangle, whose base angles are 56°14′ each. However, from practical considerations, an equilateral triangle may be treated as well condition triangle. In actual practice, the triangles having an angle less than 30° or more than 120° should not be considered.

Strength of figure: The strength of figure is a factor to be considered in establishing a triangulation system to maintain the computations within the desired degree of precision. It also plays an important role in deciding layout of a triangulation system.

5.2 The stadia readings with a horizontal sight and a vertical staff held 50 m away from a tachometer were 1.284 m and 1.780 m. The focal length of objective glass was 25 cm. Horizontal sight on staff held vertically at another station is 1.380 m and 1.630 m held 25.4 m away from tachometer. calculate

(a) Distance between object glass and tumion axis of tachometer.

(b) Stadia interval. **[8 Marks]**

Sol:

Using relation $D = KS + c.$, $f = 25\text{cm} = 0.25$ m

$S_1 = 1.780 - 1.284 = 0.496$ m

$D_1 = 50$ m.

$S_2 = 1.630 - 1.380 = 0.25$ m

$D_2 = 25.4$ m.

So, $0.496\,K + C = 50$

and $0.25\,K + C = 25.4$

on solving we get $K = 100, C = 0.4$

(a) Additions const. $C = f + d$ (where d is distance below object glass and tumion axis)

So $0.4 = 0.25 + d$

$d = 0.15$ m or 15 cm.

(b) Also multiplying const. $K = \frac{f}{i}$

where i = Stadia interval

$$100 = \frac{0.25}{i}$$

$i = 0.25$ cm.

5.3 The chainage of the intersection point of two straights is 1060 m and angle of intersection is 120° m. If radius of circular centre to be set out is 570 m and peg interval is 30 m, determine

(a) Tangent length

(b) Length of curve

(c) Chainage at beginning and end of curve

(d) Length of long chord

(e) Length of sub-chords and total number of chords. **[12 Marks]**

Sol:

$$\text{Given } \phi = 120°$$
$$\Delta = 180 - \phi = 180 - 120 = 60°$$

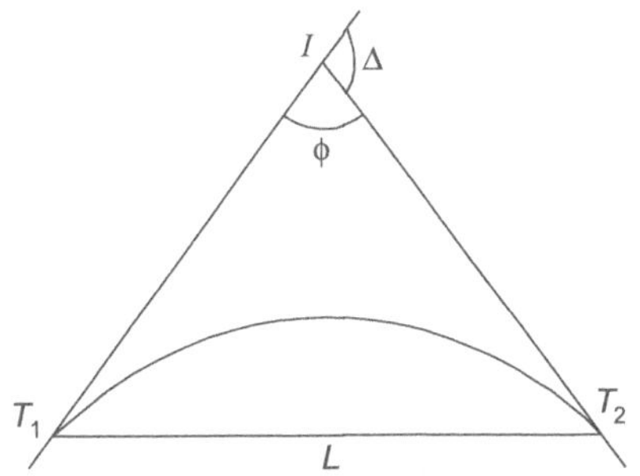

(a) $$\text{Tangent length} = T = R \tan \frac{\Delta}{2}$$

$$T = 570 \tan 30° = 329.09 \text{ m.}$$

(b) $$\text{Length of curve } l = \frac{\pi R \Delta}{180} = \frac{\pi \times 570 \times 60}{180} = 596.9 m.$$

(c) Chainange of Pl = 1060 m.

Chainage of T_1 = 1060 – T = 1060 – 329.09
= 730.91 m.
= 24 × 30 + 10.91 m
= 24 full chain + 10.91 m

Chainage of T_2 = 730.91 + l = 730.91 + 596.9 = 1327.81 m
= 44 full chain + 7.18 m

(d) $$\text{Chord length } L = 2R \sin\frac{\Delta}{2} = 2 \times 570 \sin 30 = 570 \text{ m.}$$

(e) Since chainage of T_1 = 24 full chain + 10.91 m.

Length of first subchord = 7.81 m

Total number of chords = No. of full chords + sub-chords
= (44 – 25) + 2 = 21

5.4 **Angle of elevation measured for a tower erected over a building from a point of instrument located on ground was 12° and 9° for top and bottom of tower. Height of tower is 9 m. If angle of depression of plinth level of building from same instrument is 4°. calculate height of building.**

[12 Marks]

Sol:

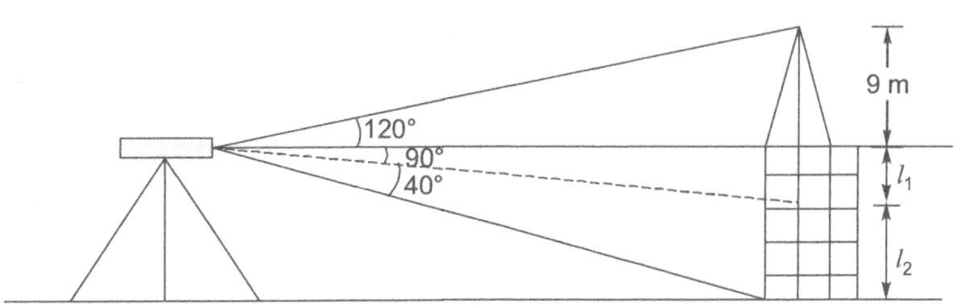

From ΔOAC,

$$\frac{9 + h_1}{D} = \tan 12° \quad ...(1)$$

Also $\frac{h_1}{D} = \tan 9° = D = h_1 \cot 9°$

Substituting in equation (1),

$$\frac{9+h_1}{h_1 \cot\ 9°} = \tan 12°$$

$$h_1 = 26.31 \text{ m}, D = 26.31 \cot 9° = 166.14 \text{ m}.$$

From ΔOCA,

$$\tan 4° = \frac{h_2}{D}$$

$$\Rightarrow \quad h_2 = 166.14 \tan 4° = 11.62 \text{ m}$$

$$\text{Height of building} = h_1 + h_2 = 26.31 + 11.62 = 37.93 \text{ m}$$

Level-2

5.5 **Two straight AB and BC falling to the right at gradients 12% and 6% respectively, are to be connected by a vertical parabolic curve of length 240 m. Chainage and reduced level of point B are 3000 m and 60.00 m respectively. Calculate the chainage and reduced level of the first three and last three points of the curve by tangent correction method. Take peg interval as 20 m.**

[15 Marks]

Sol:

Two straight *AB* and *BC* falling to right.

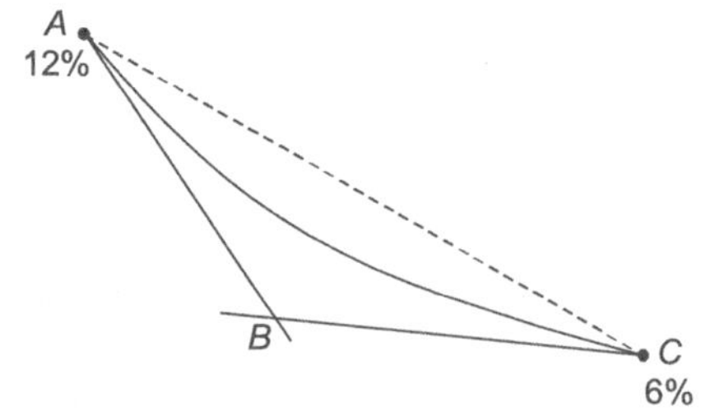

$$N = |n_1 - n_2| = 6\%$$

- As per the figure, it will be a sag curve.
- Length of curve *AC* = 240 m

  No. of chord on one side of apex = $\frac{120}{20} = 6$

  Given chainage of *B* = 3000 m

  RL of *B* = 60 m
- Chainage of point *A* = 3000 – (6 × 20) = 2880 m

  Chainage of point *C* = 3000 + (6 × 20) = 3120 m
- Elevation of tangent AB falls at 12% rate

$$\therefore \quad \text{Fall per peg interval} = -\frac{12}{100}\times 20 \text{ m} = -2.4 \text{ m}$$

Elevation of *A* = Elevation of *B* – (– Rate of fall × No. of chords to *A*)

= 60 + 2.4 × 6 = 74.4 m

- Elevation of tangent *BC* falls at 6% rate

$$\therefore \quad \text{Fall per peg interval} = \frac{-6}{100}\times 20 = -1.2 \text{ m}$$

Elevation of *C* = Elevation of *B* + (–Rate of fall × No. of chords till *C*)

= 60 – 1.2 × 6 = 52.8 m

We know, tangent correction is given by

$$y = kx^2$$

To here,

$$k = \left(\frac{e_1 - e_2}{4N}\right) = \left(-\frac{2.4-(-1.2)}{24}\right) = -0.05 \text{ m}$$

$$y = -0.05x^2$$

y will be additive to tangent elevation to get elevation on the curve.

- Elevation of points on curve = Tangent Elevation + Tangent correction

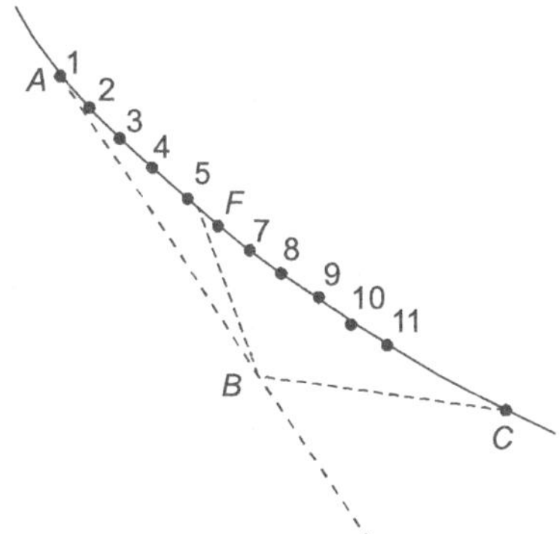

Tangent elevations:

Point-1: $74.4 - 2.4 \times 1 = 72$ m

Point-2: $74.4 - 2.4 \times 2 = 69.6$ m

Point-3: $74.4 - 2.4 \times 3 = 67.2$ m

Point-10: $74.4 - 2.4 \times 10 = 50.4$ m

Point-11: $74.4 - 2.4 \times 11 = 48$ m

Point-12: $74.4 - 2.4 \times 12 = 45.6$ m

Tangent corrections:

$$y = kx^2 = 0.05x^2$$
$$y_1 = 0.05 \times 1 = 0.05 \text{ m}$$
$$y_2 = 0.05 \times 2^2 = 0.2 \text{ m}$$
$$y_3 = 0.05 \times 3^2 = 0.45 \text{ m}$$
$$y_4 = 0.05 \times 4^2 = 0.80 \text{ m}$$
$$y_{10} = 0.05 \times 10^2 = 5.0 \text{ m}$$
$$y_{11} = 0.05 \times 11^2 = 6.05 \text{ m}$$
$$y_{12} = 0.05 \times 12^2 = 7.2 \text{ m}$$

$\therefore$ Elevation of point on curve = Tangent elevation + Tangent correction

Point	Elevation	Chainge
A	74.4m	2880
1.	72.05m	2900
2.	69.8m	2920
10.	55.4m	3080
11.	54.05m	3100
C.	52.80m	3120

5.6 **To determine the distance between two points A and B, a tachometer was set up at P and the following observation were recovered**

	Staff readings	Verticle angle
Staff at A	2.225, 2.605, 2.985	+7°54′
Staff at B	1.640, 1.920, 2.200	−7°46′

Horizontal angle A and B = + 68° 32′ 30″

Elevation of A = 315.600 m, K = 100 and C = 0 m.

Determine distance AB and elevation of B.

[20 Marks]

Sol:

Since C = 0, horizontal distance can be obtained by D = KS $\cos^2\alpha$

Horizontal distance PA = $100\,(2.985 - 2.225)\cos^2(7°\,54')$

= 74.564 m.

Similarly distance PB = $D_B = 100\,(2.2 - 1.64)\cos^2(-\,1°\,46')$

= $D_B = 55.947$ m

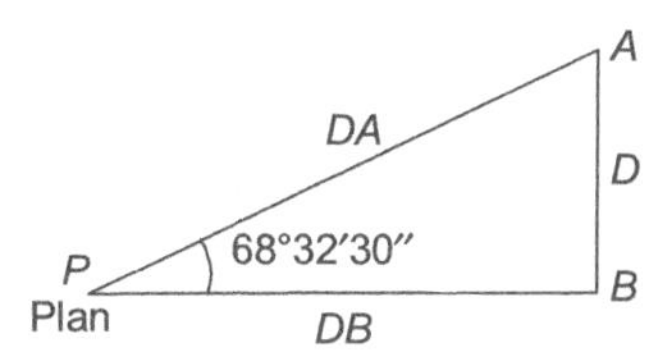

$$\angle APB = \theta = 68° \, 32' \, 30''$$

Let horizontal distance AB be D

From Cosine rule in ΔAPB,

$$D^2 = D_A^2 + D_B^2 - 2D_A D_B \cos\theta$$

$$D^2 = 74.564^2 + 55.947^2 - 2 \times 74.564 \times 55.947 \cos(68° \, 33' \, 30'')$$

$$D = 79.085 \text{ m}$$

Vertical distance V can be obtained by using relation

$$V = KS \sin\alpha \cos\alpha$$

$$= \frac{1}{2} KS \sin 2\alpha$$

Vertical distance of point A

$$V_A = \frac{100}{2} \times (2.985 - 2.225) \sin(2 \times 7°54') = 10.347 \text{ m.}$$

Similarly vertical distance of point B

$$V_B = \frac{100}{2} \times (2.2 - 2.164) \sin(2 \times 1°46') = 1.726 \text{ m}$$

Height of instrument

$$HI = \text{elevation of A + middle hair reading of A} - V_A$$

$$= 315.6 + 2.605 - 10.347 = 307.858 \text{ m.}$$

Elevation of B is $h_B = HI - V_B - \text{middle hair reading of B}$

$$= 307.858 - 1.726 - 1.920$$

$$= 304.212 \text{ m.}$$

5.7 **A circular curve of 800 m radius has been set out connecting two straights with a defection angle of 42°. It is divided, for construction reasons, that mid-point of the curve must be moved 4 m towards the centre i.e. away from intersection point. The alignment of the straights to remain Unaltered. calculate**

(a) Radius of new curve.

(b) Distance of intersection point to new tangent point.

(c) Defection angles required for setting out 30 m. chords of new curve.

(d) Length of final sub chord.

[12 Marks]

Sol:

$t_1 A t_2$ are on original circle.

$T_1 B T_2$ are on new circle. $O_1O_2 = 4$ m

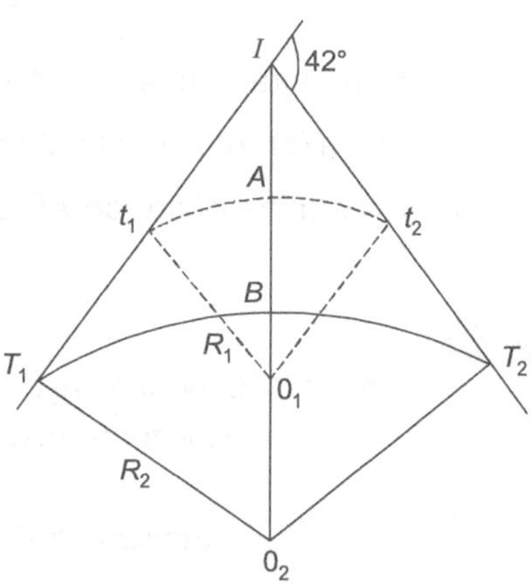

$$IA = R_1\left(\sec\frac{\Delta}{2} - 1\right)$$

$$= 800\left(\sec\left(\frac{42}{2}\right) - 1\right)$$

$$= 56.92 \text{ m.}$$

$$IB = IA + 4 = 60.92 \text{ m.}$$

(a)

$$IB = R_2\left(\sec\frac{\Delta}{2} - 1\right)$$

$$60.92 = R_2(\text{Sec } 21° - 1)$$

$$R_2 = 856\text{ m.}$$

(b) New tangent length $IT_2 = R_2 \tan \Delta/2 = 85° \tan 21°$
$= 328.6$ m.

(c) Defection angle for 30 m chord $= \dfrac{1719}{R} = \dfrac{1719}{856} = 2°0'29''$

(d) Length of curve $= R\Delta = 856 \times \dfrac{42 \times 5}{180} = 627.48\,m$

Since chord length is 30 m.

Nearest full chord is achieved at 600 m.

Length of final sub chord = 627.48 – 600 = 27.48 m

6. Field Astronomy, Photogrammetric Survey, Remote Sensing & Geology

Level-1

6.1 **In a pair of overlapping vertical photographs, the mean distance between two principal points both of which lie on the datum is 6.375 cm. At the time of photography, the aircraft was 600 m above the datum. The camera has a focal length of 150 mm. In the common overlap, a tall chimney 120 m high with its base on the datum surface is observed. Determine difference of parallax for top and bottom of chimney.**

[12 marks]

Sol:

Let s = Scale of the photograph for datum elevation

$$\therefore \quad s = \frac{f}{H} = \frac{\frac{150}{1000}}{600} = \frac{1}{4000}$$

For the datum elevation, we have,

$$\frac{B}{b} = \frac{H}{f}$$

$$\Rightarrow \quad B = \frac{H}{f}b = \frac{b}{s} = 4000 \times \frac{6.375}{100} = 255\text{ m}$$

The parallaxes for the top and the bottom of the chimney are calculated as,

$$p = \frac{Bf}{H-h}$$

For the bottom of the chimney, $h = 0$ (since the bottom of the chimney is on the datum), and hence

$$p_1 = \frac{255 \times 150(\text{mm})}{600} = 63.75\text{ mm}$$

For the top of the chimney, $h = 120$ m

$$\therefore \quad p_2 = \frac{255 \times 150(\text{mm})}{(600-120)} = 79.69\text{ m}$$

Hence difference of parallax is

$$\Delta p = (p_2 - p_1) = 79.69 - 63.75 = 15.94\text{ mm}$$

Check:

$$\Delta h = \frac{H\Delta p}{b+\Delta p} = \frac{600\text{m} \times 15.94(\text{mm})}{63.75(\text{mm})+15.94(\text{mm})} = 120.015 \text{ m} \approx 120 \text{ m}$$

which is the same as the given height of the chimney.

6.2 Explain the suitability of igneous rocks for super-structure and sub-structure.

[12 marks]

Sol:

Among various types of rocks, igneous rocks are inherently very competent and desirable for different civil engineering purposes. No other rock possesses all the desirable properties so as to make it ideally suitable for various types.

The requirement of good building stones are strength, durability (including frost resistance and fire resistance), colour, appearance, workability and its easy availability. These properties of rocks except the last one are, in turn, a function of their minerals, texture, structure, grain size, porosity and permeability.

Minerals: The type of minerals present in the rocks influence colour, durability and to some extent strength. The igneous rocks are composed mainly of silicate minerals. Among various minerals, silicate minerals are the most durable. Further, rocks rich in silica content are pleasingly light coloured as in the case of granite. All minerals of common igneous rocks are relatively very hard (quartz 7, feldspars 6, pyroxenes and amphiboles 5-6) and do not have cleavage to any harmful extent. These characters contribute to the strength of the rock and resistance to abrasion of rocks.

Structure: Since igneous rocks are formed out of solidification of a melt, they are necessarily dense, compact and massive (except vesicular rocks). In other words, these rocks do not have any internal openings or hollow nature. This contributes to the strength and heaviness of these rocks. The same factor is also responsible for non-porous and impermeable character of these rocks which do not contribute to possible decrease in strength and durability. Impervious rocks are less prone to frost-attack. When rocks are porous, they naturally become permeable (with some exceptions). When rocks are saturated with water, it makes the interior of the rock accessible to percolating solutions, thereby making it susceptible to decay and decomposition. Both the factors, i.e., saturation of rocks with moisture and internal decay of rocks, substantially reduce the strength, toughness and durability of rocks. Further, igneous rocks do not have an inherent weakness due to the occurrence of bedding planes (as in many sedimentary rocks) or mineral alignment (as in many metamorphic rocks). From the workability point of view, of course, massive igneous rocks are difficult to use because of their toughness and hardness. However, suitably placed joints are not ruled out to enable these rocks to be worked out with less difficulty. This is so because joints are very common in all types of rocks. Mural joints in granites, columnar joints in basalts and sheet joints in all types of rocks are well known. Rift and grain are the microfractures occurring in association with mural joints in granites. These facilitate the dressing of quarried rocks into smaller blocks.

Igneous rocks by virtue of their texture and minerals present in them have the ability to take very good polish and thus have become increasingly popular for face work too. Above all, constructions made of these rocks have a majestic appearance.

6.3 What do you mean by differential GPS? Also briefly explain the principle of differential GPS.

[10 marks]

Sol:

Differential GPS : Differential GPS is a system in which differences between observed and computed coordinates or ranges called differential corrections, at a particular known point called as the reference station, are transmitted to users to improve the accuracy of the user's receiver position. In fact, DGPS is a way to make GPS even more accurate. It can yield measurements good to a couple of meters in moving applications, and even better in stationary situations. This improved accuracy has a very profound effect on the importance of GPS as a resource and it becomes a universal measurement system, capable of positioning things on a very precise scale. DGPS is mainly a navigation method in which there is a fixed reference station and coordinates of moving platform or rover are determined in real time.

Principle of DGPS :

DGPS is based on the concept that bias error in position of the location is similar to those for all locations in a given local area (say within 100 km). Unfortunately, a user can not just figure out the error once and use it to correct all the measurements made for the rest of the day, because the satellite errors are continuously changing. A user needs to have two receivers working simultaneously to do the job. The reference receiver stays put and continuously monitors the errors and then transmits or records corrections for those errors so that the second receiver (the one that is out roving around doing positioning work) can apply these corrections to its measurements, either as it is making them or some time later. Thus, by having a reference receiver at a fixed location, the user can time up the accuracy of a roving receiver, or for a whole fleet of roving receivers.

6.4 The altitudes of a star at upper and lower transits of a star are 70°20′ and 20°40′ respectively; both the transits being on the north side of zenith of the place. Find the declination of the star and the latitude of the place of observation.

[12 Marks]

Sol:

Let θ be lattitude of the observer

$$\delta = \text{Declination of the star}$$

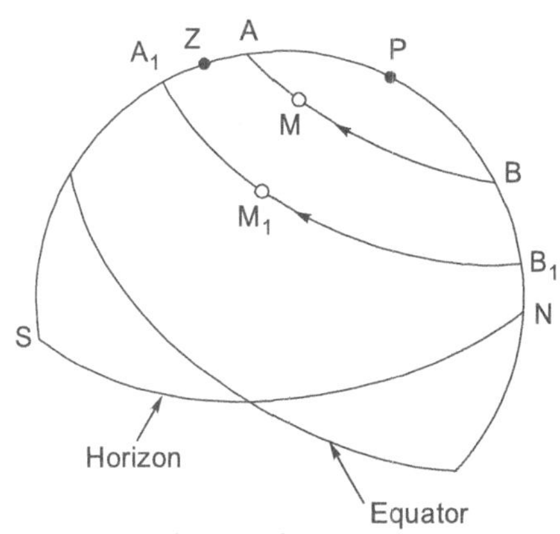

Star at Culmination

Let M be the star having A and B as its upper and lower transits.

At upper transits, zenith distance

$$ZA = ZP - AP$$
$$= (90 - \theta) - (90 - \delta) = \delta - \theta$$

$\therefore$ Altitude of star $= 90° - \text{Zenith distance} = 90 - (\delta - \theta)$

But according to question,

$$70°20' = 90 - (\delta - \theta)$$
$$\delta - \theta = 19°40' \qquad \text{...(i)}$$

At lower transits, the zenith distance of star,

$$ZB = ZP + PB$$
$$= (90 - \theta) + (90 - \delta) = 180 - (\theta + \delta)$$

$\therefore$ Altitude of star $= 90° - \text{Zenith distance}$

$= \theta + \delta - 90°$

But $\theta + \delta - 90° = 20°40'$

$\Rightarrow$ $\theta + \delta = 110°40'$ ...(ii)

Solving (i) and (ii), we get $\delta = 65°10'$

$\theta = 45°30'$

6.5 **What are the sources of errors in GIS? Name only four.** **[6 marks]**

Sol:

Sources of error in GIS

- Error arising from our understanding and modeling of reality.
- Errors due to source data.
- Errors occurring during data input.
- Errors in data analysis and manipulation.

6.6 **The observed altitude of β-ursae Minoris at lower and upper culminations are 29° 58′15″ and 60° 45′3″. Find the latitude of the place of observation assuming the correction for refraction to be equal to 57″ × tangent of apparent zenith distance.**

[10 Marks]

Sol:

Observed altitude of star at lower culmination,

$$\alpha = 29°58'15''$$

Observed altitude of star at upper culmination,

$$\beta = 60°\,45'3''$$

True altitude of star at lower culmination is given by

$$\alpha_1 = \text{Observed altitude} - \text{Refraction correction}$$

$$= 29°58'15'' - 57'' \times \tan z$$

$$= 29°58'15'' - 57'' \times \cot \alpha$$

$$= 29°58'15'' - 57'' \times \frac{1}{\tan\alpha} = 29°58'15'' - 57'' \times \frac{1}{\tan(29°58'15'')}$$

$$= 29°58'15'' - 57'' \times 1.734 = 29°58'15'' - 98.4'' = 29°56'\,36.6''$$

True altitude of star at upper culmination is given by

$$\beta_1 = \beta - 57'' \times \frac{1}{\tan\beta} = 60°45'3'' - 57'' \times \frac{1}{\tan(60°45'3'')}$$

$$= 60°45'3'' - 31.92'' = 60°44'31.08''$$

$\therefore$ **Latitude of the place of observation is given by**

$$\theta = \frac{\alpha_1 + \beta_1}{2} = \frac{29°56'36.6'' + 60°44'31.08''}{2} = 45°20'33.84''$$

6.7 **A aircraft takes photographs at a scale of 1 : 10,000. Photo size is 23 × 23 cm. Overlaps are : longitudinal 6.5% and lateral 30%, The photography consists of 5 strips of 21 photographs each. calculate**

(a) Ground area covered by single photograph.

(b) Ground area covered by first strip.

(c) Ground area covered by whole photography.

[12 Marks]

Sol:

(a) $$\text{Scale} = \frac{1}{10000}$$

$$\text{Size of photo} = 23 \times 23 \text{ cm.}$$

Area covered by single photography

$$= (23 \times 10000)^2 \text{ cm}^2 = (23 \times 10^4)^2 \times 10^{-10} \text{ km}^2 = 5.29 \text{ km}^2$$

(b) For first strip, first photograph covered area 5.29 km². 2nd photo onwards each photo covers reduced area due to overlap = (1–0.65) × 5.29.

$$\text{Total area covered by first strip} = 5.29 + (21-1) \times (0.35 \times 5.29) = 42.32 \text{ km}^2$$

(c) Similarly total ground area covered by where photography

$$= 42.32 + (5-1)(1-0.3) \times 42.32 = 160.816 \text{ km}^2$$

6.8 **Points A and B are at elevations 273 m and 328 m above datum respectively. Photogrammetric coordinates of their images on a vertical photograph a**
x_a = + 68.27 mm, x_b = – 87.44 mm, y_a = – 32.37 mm, y_b = + 26.81 mm
What is the horizontal length of line AB if photo was taken from 3200 m observe datum with a 21 cm focal length camera. Also determine the average scale of photograph. **[12 Marks]**

Sol:

$$\text{Scale for coordinates at Point A} = \frac{21}{100 \times (3200-273)} \quad \left(\frac{f}{H-G_{avg}}\right)$$

$$\text{Scale for Coordinates at Point B} = \frac{21}{100 \times (3200-328)}$$

Coordinates on ground

for point A, $$x_a = \frac{100 \times (3200-273)}{21} \times \frac{(68.27)}{1000} = +951.554\, m.$$

$$y_a = \frac{100}{21} \times (3200-273) \times \frac{(-32.37)}{1000} = -451.176\, m.$$

Similarly for Point B, $$x_b = \frac{100}{21} \times (3200-328) \times \frac{(-87.44)}{1000} = -1195.846\, m.$$

$$y_b = \frac{100}{21} \times (3200-328) \times \frac{26.81}{1000} = 366.659 \text{ m}$$

Length of line AB in horizontal plane

$$L_{AB} = \sqrt{(x_a - x_b)^2 + (y_a - y_b)^2}$$

$$= \sqrt{(951.554 + 195.846)^2 + (-451.176 - 366.659)^2} = 2297.864 \text{ m}$$

$$\text{Average height of elevation } h_{average} = \frac{273 + 328}{2} = 300.5 \text{ m}$$

$$\text{Average Scale} = \frac{F}{H - h_{avg}} = \frac{21}{100(3200 - 300.5)} = \frac{1}{13807.14}$$

6.9 **In a pair of overlapping vertical photographs, the mean distance between two principal points both of which lie on the datum is 6.375 cm. At the time of photography, the aircraft was 600 m above the datum. The camera has a focal length of 150 mm. In the common overlap, a tall chimney 120 m high with its base on the datum surface is observed. Determine difference of parallax for top and bottom of chimney.** **[10 Marks]**

Sol:

Let s = Scale of the photograph for datum elevation

$$\therefore \quad s = \frac{f}{H} = \frac{\frac{150}{1000}}{600} = \frac{1}{4000}$$

For the datum elevation, we have, $\frac{B}{b} = \frac{H}{f}$

$$\Rightarrow \quad B = \frac{H}{f}b = \frac{b}{s} = 4000 \times \frac{6.375}{100} = 255 \text{ m}$$

The parallaxes for the top and the bottom of the chimney are calculated as,

$$p = \frac{Bf}{H-h}$$

For the bottom of the chimney, $h = 0$ (since the bottom of the chimney is on the datum), and hence

$$p_1 = \frac{255 \times 150(\text{mm})}{600} = 63.75 \text{ mm}$$

For the top of the chimney, $h = 120$ m

$$\therefore \quad p_2 = \frac{255 \times 150(\text{mm})}{(600-120)} = 79.69 \text{ m}$$

Hence difference of parallax is $\Delta p = (p_2 - p_1) = 79.69 - 63.75 = 15.94$ mm

Check:

$$\Delta h = \frac{H\Delta p}{b + \Delta p} = \frac{600\text{m} \times 15.94(\text{mm})}{63.75(\text{mm}) + 15.94(\text{mm})} = 120.015 \text{ m} \approx 120 \text{ m}$$

which is the same as the given height of the chimney.

6.10 **Explain the suitability of igneous rocks for super-structure and sub-structure.**

[10 Marks]

Solution:

Among various types of rocks, igneous rocks are inherently very competent and desirable for different civil engineering purposes. No other rock possesses all the desirable properties so as to make it ideally suitable for various types.

The requirement of good building stones are strength, durability (including frost resistance and fire resistance), colour, appearance, workability and its easy availability. These properties of rocks except the last one are, in turn, a function of their minerals, texture, structure, grain size, porosity and permeability.

Minerals: The type of minerals present in the rocks influence colour, durability and to some extent strength. The igneous rocks are composed mainly of silicate minerals. Among various minerals, silicate minerals are the most durable. Further, rocks rich in silica content are pleasingly light coloured as in the case of granite. All minerals of common igneous rocks are relatively very hard (quartz 7, feldspars 6, pyroxenes and amphiboles 5-6) and do not have cleavage to any harmful extent. These characters contribute to the strength of the rock and resistance to abrasion of rocks.

Structure: Since igneous rocks are formed out of solidification of a melt, they are necessarily dense, compact and massive (except vesicular rocks). In other words, these rocks do not have any internal openings or hollow nature. This contributes to the strength and heaviness of these rocks. The same factor is also responsible for non-porous and impermeable character of these rocks which do not contribute to possible decrease in strength and durability. Impervious rocks are less prone to frost-attack. When rocks are porous, they naturally become permeable (with some exceptions). When rocks are saturated with water, it makes the interior of the rock accessible to percolating solutions, thereby making it susceptible to decay and decomposition. Both the factors, i.e., saturation of rocks with moisture and internal decay of rocks, substantially reduce the strength, toughness and durability of rocks. Further, igneous rocks do not have an inherent weakness due to the occurrence of bedding planes (as in many sedimentary rocks) or mineral alignment (as in many metamorphic rocks). From the workability point of view, of course, massive igneous rocks are difficult to use because of their toughness and hardness. However, suitably placed joints are not ruled out to enable these rocks to be worked out with less difficulty. This is so because joints are very common in all types of rocks. Mural joints in granites, columnar joints in basalts and sheet joints in all types of rocks are well known. Rift and grain are the microfractures occurring in association with mural joints in granites. These facilitate the dressing of quarried rocks into smaller blocks.

Igneous rocks by virtue of their texture and minerals present in them have the ability to take very good polish and thus have become increasingly popular for face work too. Above all, constructions made of these rocks have a majestic appearance.

Level-2

6.11 **An area 30 km long in the north-south direction and 24 km in the east-west direction is to be photographed with a lens having 30 cm focal length for the purpose of constructing a mosaic. The photograph size is 20 cm × 20 cm. The average scale is to be 1 : 12,000 effective at an elevation of 400 m above datum. Longitudinal overlap is to be atleast 60% and the side lap is to be at least 30%. An intervalometer will be used to control the interval between exposures. The ground speed of the aircraft will be maintained at 200 km per hour. The flight lines are to be laid out in a north-south direction on an existing map having a scale of 1 : 60,000. The two outer flight lines are to coincided with the east and west boundaries of the area. Determine the data for the flight plan.**

[20 Marks]

Sol:

1. Flying height:

We have,
$$\frac{f}{H} = S$$

$$\frac{0.3}{H} = \frac{1}{12000}$$

$$\Rightarrow \quad H = 12{,}000 \times 0.3 = 3600 \text{ m above ground}$$

$$\therefore \quad \text{Height above datum} = 3600 + 400 = 4000 \text{ m}$$

2. Theoretical ground spacing of flight lines

The ground width covered by each photograph, with 30% side lap is given by

$$W = (1 - P_w)Sw$$

where, w = width of photograph = 20 cm; $S = \text{scale} = \frac{f}{H} = \frac{0.3}{3600} = \frac{1}{12000}$

i.e., 1 cm = 120 m; $P_w = 0.30$

$\therefore \quad W = (1 - 0.3)\, 120 \times 20 = 1680$ m

3. Number of flight lines required:

The number of flight lines is given

$$N_2 = \frac{L_2}{(1-P_w)Sw}+1=\frac{L_2}{W}+1=\frac{24{,}000 \text{ m}}{1680 \text{ m}}+1 = 14.2 + 1 \simeq 16$$

4. Actual spacing of flight lines: Since the number of flight lines is to be an integral number, the actual flight lines = 16 and the number of flight strips or spacings is 15. Hence the actual spacing is given by

   $W=\frac{24{,}000}{15}= 1600$ m, against the theoretically calculated value of 1680 m.

5. Spacing of flight lines on flight map:
   Flight map is on a scale of 1 : 60,000 or 1 cm = 600 m. Hence the distance on the flight map corresponding to a ground distance of 1600 m = $\frac{1600}{600}$ = 2.67 cm.

6. Ground distance between exposures:
   The ground length covered by each photograph in the direction of flight with an overlap of 60% is given by $L = (1 - P_t)$ sl = (1 – 0.6) × 120 × 20 = 960 m.

7. Exposure interval:
   The time interval between exposures is usually the integral number of seconds.

   $V = 200$ km per hour = $\frac{200\times 1000}{60\times 60}$ m/sec = 55.56 m/sec.

   ∴ The required exposure interval is $\frac{960\text{ (m)}}{55.56\text{ (m/sec)}} = 17.3$ sec. $\simeq$ 17 sec.

8. Adjusted ground distance between exposures:
   Keeping the exposure inverval as an integral number of seconds the adjusted ground distance covered by each photograph is given by

   $$L = V\times T = 55.56 \text{ (m/sec)} \times 17.0 \text{ (sec)} \simeq 945 \text{ m.}$$

9. Number of photographs per flight line:
   The number of photographs per flight line is given by

   $$N_1 = \frac{L_1}{(1-P_1)sl}+1=\frac{L_1}{L}+1=\frac{30{,}000}{945}+1=31.7+1\approx 33$$

10. Total number of photographs required is

    $$N = N_1 \times N_2 = 33 \times 16 = 528$$

6.12 **The following data is given for flight planning :**

Photo size 18 × 18 cm.

Focal length = 21 cm.

Scale = $\frac{1}{20{,}000}$

Longitudinal overlap = 60%

Lateral overlap = 20%

East west terrain length = 100 km

North-South terrain width = 50 km

Aircraft velocity = 296 km / hr.

Permissible image movement = 0.02 mm

Wind velocity = 10 m / s from SSE direction

Calculate

(a) Exposure interval **(b) Maximum exposure time**

(c) Ground speed of aircraft **(d) Drift angle.** **[12 Marks]**

Sol:

(a)

$$\text{Length of aircraft} = (1-0.6)\frac{118\times 20000}{100\times 100}\text{km} = 14.4 \text{ km}$$

$$\text{exposure interval} = \frac{\text{airbase length}}{\text{aircraft velocity}} = \frac{14.4\times 60\times 60}{296}\text{S} = 175.1 \text{ seconds}$$

$$\simeq 175 \qquad \text{(since minimum longitude overlap needed.)}$$

(b) Permissible image movement = 0.02 mm

$$\text{Permissible movement on ground} = \frac{0.02}{1000}\times 20000 = 0.4m.$$

Time taken to cover this distance = maximum exposure time

$$\frac{0.4\times 60\times 60}{1000\times 296} = \frac{1}{205.55} \simeq \frac{1}{206}\text{S}$$

(c)

$$\text{Wind Speed} = 10\times\frac{18}{5} = 36\text{km/hr}$$

$$\text{Flying speed} = 296 \text{ km / hr.}$$

$$\text{Net speed of aircraft} = \sqrt{296^2+36^2+2\times 36\times 296\cos(67.5°)} = 311.557 \text{ km / hr}$$

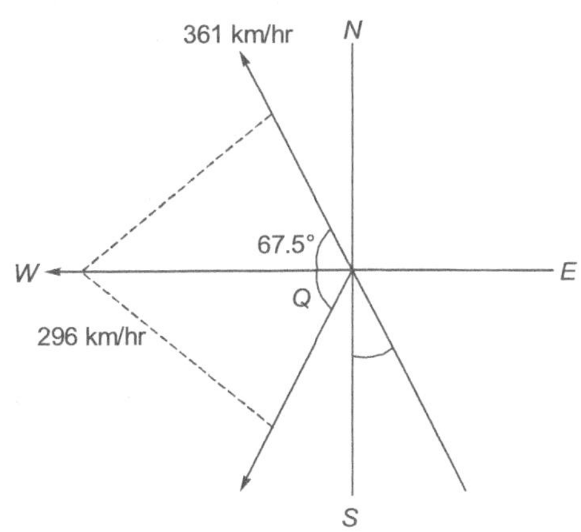

(d) Drift angle (θ)

$$\frac{\text{Wind speed}}{\sin\theta} = \frac{\text{Ground speed}}{\sin 67.5}$$

$$\sin\theta = \frac{36}{311.557}\times\sin 67.5°$$

$$\theta = 6° 7' 41.44''$$

6.13 **Explain with the help of a neat sketch, an idealized remote sensing system.**

[20 marks]

Sol:

An idealised remote sensing system consists of the following stages as shown in figure below.

1. Energy source.
2. Propagation of energy through atmosphere.
3. Energy interaction with earth's surface features.
4. Airborne/space borne sensors receiving the reflected and emitted energy.
5. Transmission of data to earth station and generation of data produce.
6. Multiple-data users.

1. **The energy source:** The uniform energy source provides energy over all wave lengths. The passive RS system relies on sun as the strongest source of EM energy and measures energy that is either reflected and/or emitted from the earth's surface features. However, active RS systems use their own source of EM energy.
2. **Propagation of energy through the atmosphere:** The EM energy, from the source passes through the atmosphere on its way to earth's surface. Also, after reflection from the earth's surface, it again passes through the atmosphere on its way to sensor. The atmosphere modifies the wave length and spectral distribution of energy to some extent and this modification varies particularly with the wave length.
3. **Interaction of energy with surface feature of the earth:** The interaction of EM energy, with earth's surface features generates reflected and/or emitted signals (spectral response patterns or signatures). The spectral response patterns play a central role in detection, identification and analysis of earth's surface material.
4. **Air borne/space borne sensors:** Sensors are electromagnetic instruments designed to receive and record transmitted energy. They are mounted on satellites, aeroplanes or even on balloons. The sensors are highly sensitive to wave length, yielding data on the absolute brightness from the object as a function of wavelength.
5. **Transmission of data to earth station and data product generation:** The data from the sensing system is transmitted to the ground based earth station along with the telemetry data. The real-time (instantaneous) data handling system consists of high density data tapes for recording and visual devices (such as television) for quick look displays. The data products are mainly classified into two categories:
   (i) Pictorial or photographic product (analogue)
   (ii) Digital product.
6. **Multiple data users:** The multiple data users are those who have knowledge of great depth, both of their respective disciplines as well as of remote sensing data and analysis techniques. The same set of data becomes various forms of information for different users with the understanding of their field and interpretation skills.

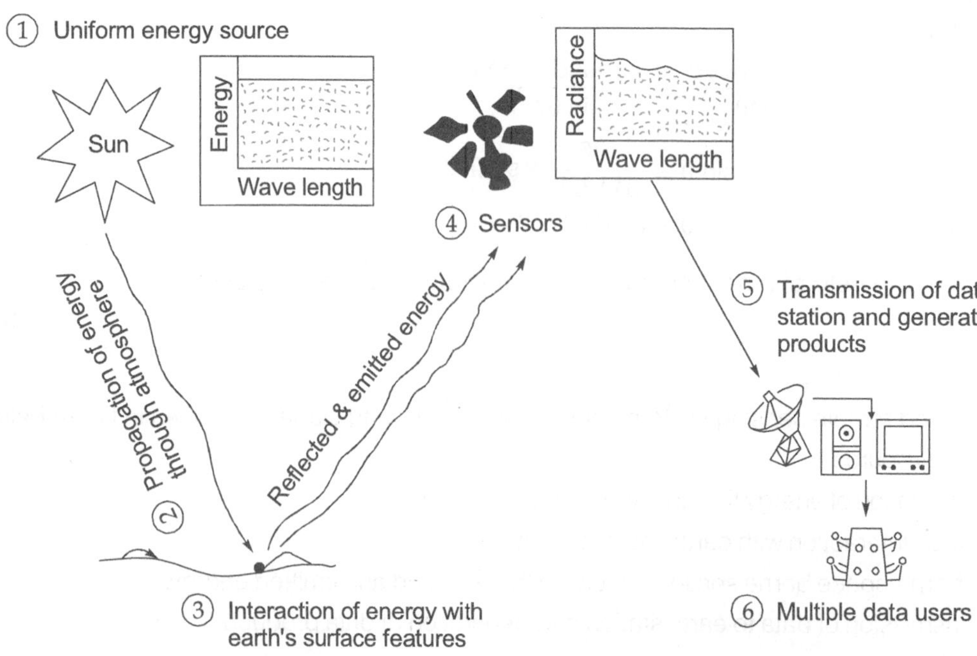

6.14 Write notes on the following:

1. Folds
2. Faults
3. Joints
4. Unconformity

[20 marks]

Sol:

1. **Folds :** Folds are one of the most common geological feature found in rocks. When a set of horizontal layers of rock are subjected to compressive forces, rocks bend either upwards or downwards.

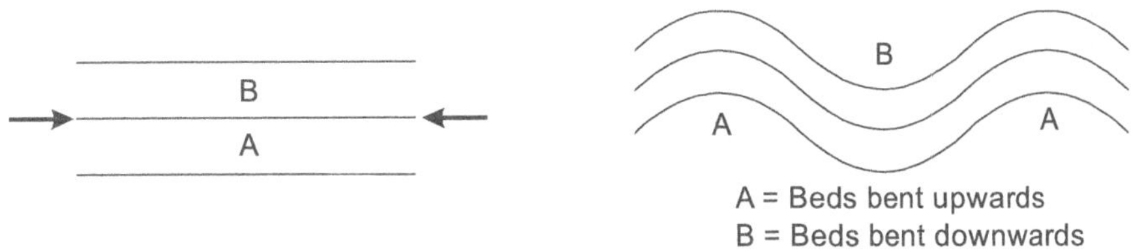

Horizontal Beds   Folded beds

Horizontal and folded beds

These bends in rocks are called as folds. Folds are described variously as wavy or arch-like or curved or warping appearances found in rocks. They are also called flexures or buckling phenomenon of rocks. Generally, they occur in series. In nature, folds found in rocks have a great range of magnitude in terms of their length and breadth. At one extreme, they may be very small, i.e., a fraction of an inch across and a few inches in length. At the other extreme they may be a few miles across and several miles long. The radius of curvature of a fold is small compared to its wavelength and amplitude.

Classification and Types of Folds : Based on different principles, the folds are variously classified. Usually, folds are classified on the basis of : (i) symmetrical character, (ii) upward or downward bend, (iii) occurrence of plunge, (iv) uniformity of bed thickness, (v) behaviour of the fold pattern with depth.

Mechanism of folding : Folding of rocks takes place by different ways of accommodation of stress (caused by tectonic forces). In many cases, slips or shears occur in between the beds. This process is similar to slipping of cards which occurs when the set (deck) is folded. If they are not allowed to slip over one another, folding of the set cannot take place. This is the way in which folding generally occurs in the case of hard and competent rocks like quartzites.

2. **Faults :** Faults are the most unfavourable and undesirable geological features at the site for any given purpose, i.e., for location of reservoir; as foundation site for construction of dams, important bridges or huge buildings, for tunnelling; for laying roads, railway tracks, etc. This is because faults considerably weaken the rocks (in addition to other harmful effects) and render the sites in which they occur as unfavorable places for all constructional purposes. Further, as long as the faults are active, the site is unstable and susceptible to upward, downward or sideward movement along the fault plane, thereby making the place highly hazardous for foundation purposes.

Recurrence of Faulting : Faulting occurs when shearing resistance of the geological formation is overcome by the tectonic forces. The faulting phenomenon may take place suddenly or slowly and may or may not be recurring. Occurrence of faulting is often accompanied by earthquakes (elastic rebound hypothesis) and it is an indication of subsurface (i.e., subterranean) instability of the region concerned. Thus, faulting may be treated as an attempt to reach stability.

Parts of a Fault : The different parts of a fault are shown in figure.

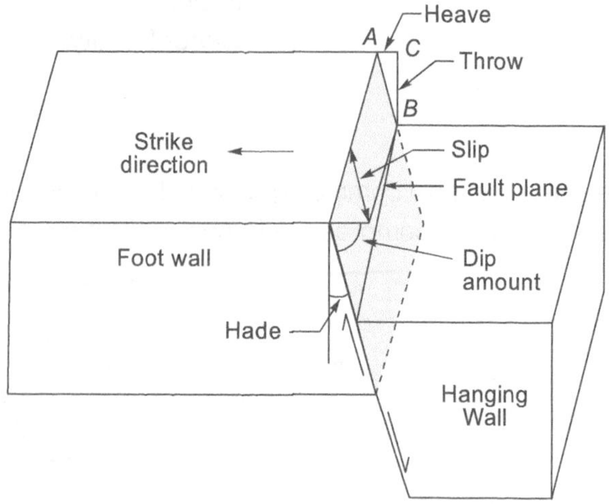

Parts of a fault

3. **Joints :** Joints, are fractures found in all types of rocks. They are cracks or openings formed due to various reasons. Naturally, the presence of joints divides the rock into a number of parts or blocks. These separated blocks may move only perpendicular to the plane of fracture, but do not move past one another as in the case of faults. In simple terms, through the joints may be described as mere cracks in rocks, they differ mutually.

Joints, like cleavage of minerals, occur oriented in a definite direction and as a set, i.e., as a number of parallel planes. The cracks, on the other hand, like the fracture of minerals are random or irregular in their mode of occurrence. Thus, joints occur, generally, as a number of parallel and oriented fractures in rocks. Such a group of fractures is called a joint set.

Effects of Joints : From the civil engineering point of view, joints are important because they split the rocks into a number of pieces which, in turn, reduce the competence of rock mass, increase the porosity and permeability and make them (rocks) susceptible to quick decay and weathering. Joints become avenues for the leakage of water in case of reservoirs. If they are closely spaced on the upstream side, silting problems also arise in reservoirs. Joints may pose ground water problems in tunnelling. The undesirable leakage and other effects introduced by joints in rock may create foundation problems in dam sites.

Parts of Joints : Joints, like faults, refer to the fractures in rocks. Hence, like faults, inclined and vertical joints also can be described by their attitude. However, in joints, the fractured blocks are not named as footwall or hanging wall.

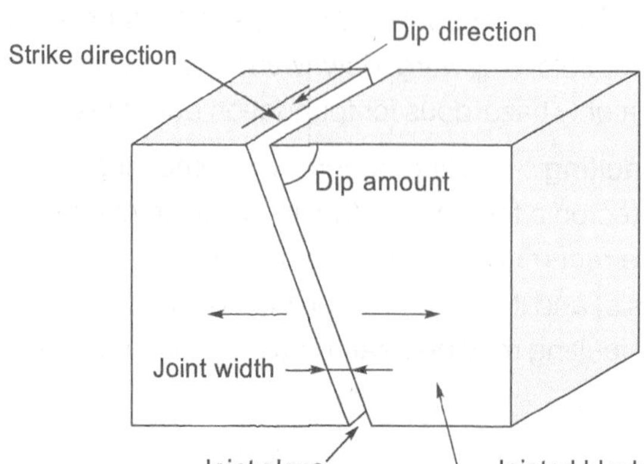

Parts of a joint

4. **Unconformity :** Unconformity is one of the common geological features found in rocks. It is somewhat different from other features like folds, faults and joints in which the rocks are distorted, deformed or dislocated at a particular place. Still, unconformity is a product of diastrophism and involves tectonic activity in the form of upliftment and subsidence of land mass.

   When sedimentary rocks are formed continuously or regularly (i.e., at a stretch) one after another without any major break, they are said to be a set of conformable beds, and this phenomenon is called conformity. All beds belonging to a conformable set shall possess the same strike direction, dip direction and dip amount. On the other hand, if a major break occurs in sedimentation (i.e., if sedimentation does not take place for a long interval) in between two sets of conformable beds, it is called an unconformity. So, an unconformity refers to a period of non-deposition and appears as a plane of contact between two sets of conformable beds.

   Parts of an Unconformity :

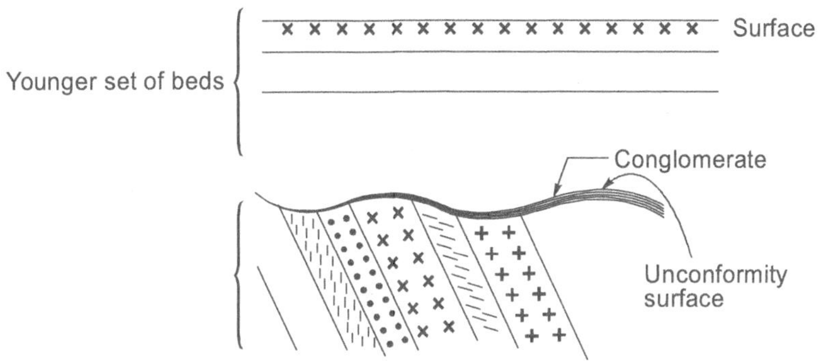

Parts of an unconformity

6.15 **Find the shortest distance between two places *A* and *B* given that the latitude of *A* and *B* are 16°0′N and 13°0′N and their longitude are 50°0′E and 54°0′E.**

[12 Marks]

Sol:

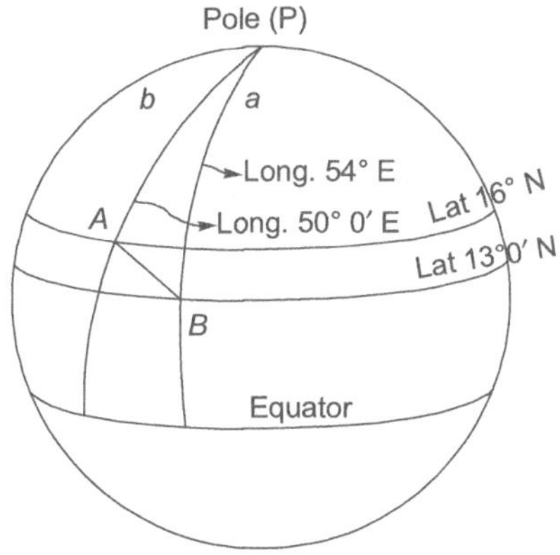

In above figure, the positions of *A* and *B* have been shown.

In spherical triangle *ABP*,

$$b = AP = 90° - \text{Latitude of } A = 90° - 16° = 74°0'$$

$$a = BP = 90° - \text{Latitude of } B = 90° - 13°0' = 77°0'$$

$$P = \angle APB = \text{Difference of longitude} = 54° - 50° = 4°0'$$

The shortest distance between two points is the distance along the great circle passing through the two points.

Knowing the two sides one angle, the third side AB (= P) can be easily computed by the cosine rule.

Thus, $$\cos P = \frac{\cos p - (\cos a)(\cos b)}{\sin a \times \sin b}$$

or,
$$\begin{aligned} \cos p &= \cos P \times \sin a \times \sin b + \cos a \times \cos b \\ &= \cos 4° \times \sin 77° \times \sin 74° + \cos 77° \times \cos 74° \\ &= 0.93434 + 0.0620 = 0.99634 \end{aligned}$$

$\therefore$ $$p = 4°54'' \, 0.92''$$

Now, $$\text{arc} = \text{radius} \times \text{central angle}$$

Assume, $$\text{radius} = 6370 \text{ km}$$

$$= \frac{6370 \times 4.9036 \times \pi}{180°} = 545.164 \text{ km}$$

$\therefore$ Shortest distance, $AB = 545.16$ km

■■■■

07 Transportation Engineering

1. Highway Planning

Level-1

1.1 Compare Macadam and Telford method in brief.

[5 Marks]

Sol:

Macadam Method	Telford Method
• A cross slope of 1 in 36 was provided in subgrade to help its drainage.	• Subgrade was horizontal and hence poor drainage.
• Broken stones of size less than 5 cm make up the bottom layer of pavement or sub-base course.	• About 17 cm heavy foundation stones towards the edges and 22 cm towards the centre were hand packed and prepared to act as sub base course.
• Broken stones of smaller sizes to compacted thickness of 10 cm and 5 cm respectively for base courses and surface courses respectively. The top surface was given a cross slope of in 36.	• Two layers of broken stones were compacted over the foundation stones before laying the wearing course, which is 4 cm thick with cross slope of 1 in 45.
• The total thickness was kept uniform from edge to centre to a minimum value of 25 cm only.	• The total thickness of construction varied from about 35 cm at edge to 41 cm at centre.

1.2 During a five-year plan period, 3 roads, A, B and C are to completed in a district. Find the priority order by maximum utility principle from the data given below. Take utility unit of 1.0 for population between 5000-10000 in a town or for catering for 1000 tonnes of agriculture products/100 tonnes of industrial products.

Road	Length (km)	No. of town with population			Productivity in 1000 tons	
		< 5000	5000 – 10000	> 10000	Agricultural	Industrial
A	45	16	16	16	18	1.5
B	30	18	18	18	15	2.0
C	40	22	22	22	22	0.75

[8 Marks]

Sol:

Road	Length (km)	Total utility units served by the road	Utility per unit length
A	45	(16 × 0.5) + (8 × 1) + (4 × 2) + (18 × 1) + (1.5 × 10) = 57	57/45 = 1.27
B	30	(18 × 0.5) + (8 × 1) + (2 × 2) + (15 × 1) + (2 × 10) = 56	56/30 = 1.87
C	40	(22 × 0.5) + (12 × 1) + (3 × 2) + (22 × 1) + (0.75 × 10) = 58.5	58.5/40 = 1.46

Priority order = B, C, A

Assumed utility units,

Population of town < 5000 = 0.5

Population of town, 5000 – 10000 = 1.0

Population of town > 10000 = 2.0
Agricultural produce per 1000 tons= 1.0
Industrial produce per 100 tons = 10.0

1.3 **The area of a district in Uttar Pradesh is 26000 km² and there are 18 towns as per 2011 census. Determine the lengths of different categories roads to be provided in district as per III road plan's formulas.**

[10 Marks]

Sol:

Given, Area of district = 26000 km², No. of towns = 18

(i) $$\text{Length of NH} = \frac{A(km^2)}{50} = \frac{26300}{50} = 526 \text{ km}$$

(ii) $$\text{Length of SH} = \text{Maximum of} \begin{cases} \dfrac{A(\text{km}^2)}{25} = \dfrac{26300}{25} = 1052 \text{ km} \\ 62.5\text{ N-L}, = 62.5 \times 18 - 526 = 599 \text{ km} \end{cases}$$

(N = No. of towns, L = Length of NH)

= 1052 km

(iii) $$\text{Length of MDR} = \text{Maximum of} \begin{cases} \dfrac{A(\text{km}^2)}{12.5} = \dfrac{26300}{12.5} = 2104 \text{ km} \\ 90 \times N = 1620 \text{ km} \end{cases} = 2104 \text{ km}$$

(iv) Total length of all categories of roads may be assumed to be provide an overall density of road length equal to 82 km per 100 km².

$$\text{NH + SH + MDR + ODR + VR} = 26300 \times \frac{82}{100} = 21566 \text{ km}$$

$\therefore$ $$\text{ODR + VR} = 21566 - (526 + 1052 + 2104) = 17884 \text{ km}$$

$\therefore$ Rural roads are 17884 km in length.

Primary system of NH = 526 km
Secondary system of (SH + MDR) = 3156 km
Tertiary system of rural roads = 17884 km
Total road length = 21566 km

Level-2

1.4 **What are the varies studies and data compiled before execution of new road scheme?**

[15 Marks]

Sol:

For a new road scheme or a new highway planning; field surveys are to be carried out to collect the data required for determining the length of the road system. And finally; preparation of master plan showing phasing of plan in annual or five year plans is to be completed. Field surveys or planning survey consist of the following studies;

(a) **Economic studies:** In these studies; various details useful in estimating the economics involved in highway development programme such as;

(i) Population; its distribution and classification

(ii) Trend of population growth

(iii) Listing of agricultural and industrial products development and future trends.

(iv) Existing facilities with regard to communication, recreation and education etc.

(v) Per capita income, banking, post office etc.

(b) **Financial studies:** To study the various financial aspects such as:

(i) Sources of income and estimated revenue from taxation on road transport.

(ii) Living standards

(iii) Resources at local level, toll taxes vehicle, registration and fines.

(iv) Future trends in financial aspects.

(c) **Traffic or road use studies:** Traffic surveys should be carried out in whole area and on selected routes and locations in order to collect the following particulars:

(i) Traffic volume in vehicles per day, peak and design hourly traffic volume.

(ii) Origin and destination studies, traffic flow pattern.

(iii) Accidents, their cost analysis and causes.

(iv) Future trend and growth in traffic volume and trend in choice of modes.

(d) **Engineering studies:** These studies include:

(i) Topographic survey, soil survey

(ii) Location and classification of existing roads.

(iii) Road life studies.

(iv) Estimation of possible developments in all aspects.

(v) Special problem in drainage, construction and maintenance of roads.

Preparation of Plans:

The above details collected are tabulated and plotted on the maps of area under planning. Usually four drawings are prepared showing the various details of area as under:

Plan I General area plan showing almost all existing details such as topography, existing road network drainage structure, rivers, canals, towns and villages with population, commercial, industrial and agricultural activities.

Plan II Distribution of population groups in accordance with the categories.

Plan III Location of places with their respective quantities of productivity.

Plan IV Existing roads network with traffic flows and desire lines obtained from *O* and *D* studies.

Preparation of Master Plan and its phasing:

- Master plan is the final road development plan. Based on above plans, different possible network of new roads are proposed. In each proposal; the population and productivity of each locality, traffic flow, topography and other details both existing and possible changes in future are kept in view.
- The next step is to compare the various alternate proposals of road system and to select most optimum network under plan period.
- After deciding the optimum road length for a plan period, the final step is the phasing of road development plan by fixing up the priorities for construction of different road links. The priority for each road link may be fixed scientifically based on maximum utility per unit length of road based on population and productivity for each road.

2. Geometric Design of Highways

Level-1

2.1 **The driver of a car, applied brakes and barely avoided hitting an obstacle on the road. The vehicle left skid marks of tyres on the road for a length of 25 m. There was a speed limit restriction of 55 kmph for that road. Was the driver of the car violating the limit if he was travelling on**

(a) level road
(b) a 4% downgrade.
Assume the longitudinal coefficient of friction between the tyres and the road surface as 0.5.

[6 Marks]

Sol:

Given longitudinal coefficient of friction = 0.5

Stopping distance = 25 m

Total reaction time t = 2.5 sec as per IRC

Speed of vehicle = 55 kmph

(a) SSD (only braking distance)

$$= \frac{V^2}{254f} \text{ (on a level road)}$$

$$\therefore \qquad SSD = \frac{55^2}{254 \times 0.5} = 23.82 \text{ m}$$

Which is less than the given breaking distance = 25 m
So the driver of the car was not violating the limit.

(b) SSD (only braking distance)

$$= \frac{V^2}{254(f - n\%)} \text{ (for 4% downgrade)}$$

$$= \frac{55^2}{254 \times (0.5 - 0.04)}$$

$$= 25.89 \text{ m}$$

Which is greater than the given braking distance = 25
So the driver of the car was violating the limit.

2.2 **An ascending gradient of 1 in 45 meets a descending gradient of 1 in 60. A summit curve has to be designed for design speed of 80 km/h so as to provide a safe overtaking sight distance of 230 m. Estimate the length of the summit curve.**

[8 Marks]

Sol:

$$n_1 = \frac{1}{45};\ n_2 = \frac{1}{60}$$

$$N = \frac{1}{45} - \left(-\frac{1}{60}\right) = \frac{4+3}{180} = \frac{7}{180}$$

If $\qquad L > OSD$

Then the length of summit curve

$$L = \frac{NS^2}{9.6}$$

$$= \frac{7}{180} \times \frac{(230)^2}{9.6} = 214.3\text{m}$$

This value is less than

$$OSD = 230 \text{ m}$$

Assuming

$$L < OSD$$

$$L = 2S - \frac{9.6}{N}$$

$$= 2 \times 230 - 9.6 \times \frac{180}{7}$$

$$= 213.14 \text{ m} = 214 \text{ m}$$

It is less than OSD, hence the length of summit curve is 214 m.

2.3 **The speeds of overtaking and overtaken vehicles on a highway are 85 kmph and 70 kmph respectively. Calculate the overtaking sight-distance needed for two way traffic. Assume the acceleration of the overtaking vehicle as 2.5 kmph per second and the speed of the vehicle in the opposite direction as 85 kmph.**

[8 Marks]

Sol:

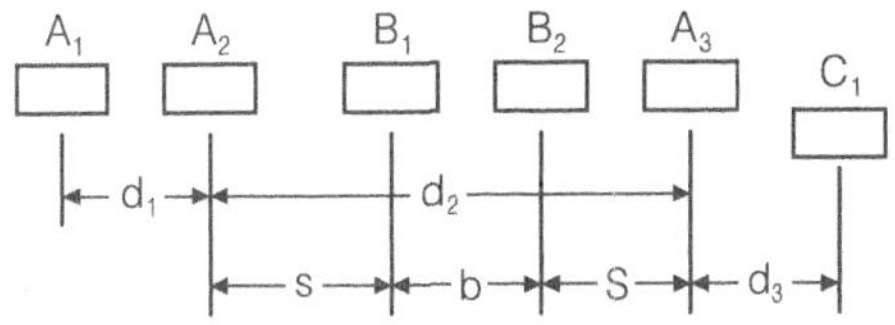

Here

$$t = \text{reaction time} = 2 \text{ sec}$$

$$V_b = \text{Velocity of overtaken vehicle}$$

$$= 70 \text{ kmph} = 19.44 \text{ m/sec}$$

$$S = 0.7\, V_b + 6$$

$$= 0.7 \times 19.44 + 6 = 19.60$$

$$T = \sqrt{\frac{4S}{a}}$$

Where,

$$a = 2.5 \times \frac{5}{18} \text{m/sec}^2$$

$$= 0.694 \text{ m/sec}^2$$

$\therefore$

$$T = \sqrt{\frac{4 \times 19.60}{0.694}} = 10.62 \text{ sec}$$

Now,

$$d_1 = V_b \times t = 19.44 \times 2 = 38.88 \text{ m}$$

$$d_2 = VT = 85 \times \frac{5}{18} \times 10.62$$

$$= 250.75 \text{ m}$$

$\therefore$ Overtaking side distance, OSD

$$= d_1 + d_2 + d_3$$

$$= 38.88 + 245.75 + 250.75$$

$$= 535.38 \text{ m}$$

2.4 A vehicle is maneuvering a horizontal curve of radius R with superelevation θ°. Derive the expression for maximum speed beyond which it would overturn outward if the coefficient of friction between the tyre and the pavement is 'f'. Find the value of this speed if R = 250 m, θ = 5°, f = 0.15, h = 1 m and b = 0.75 m (refer to the following figure)

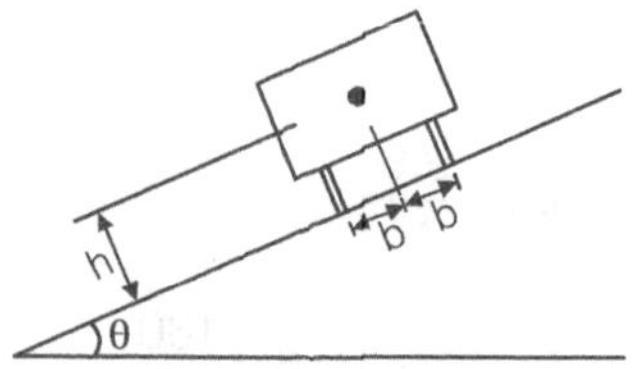

[10 Marks]

Sol:

Superelevation

$$e = \frac{NL}{ML} = \tan\theta$$

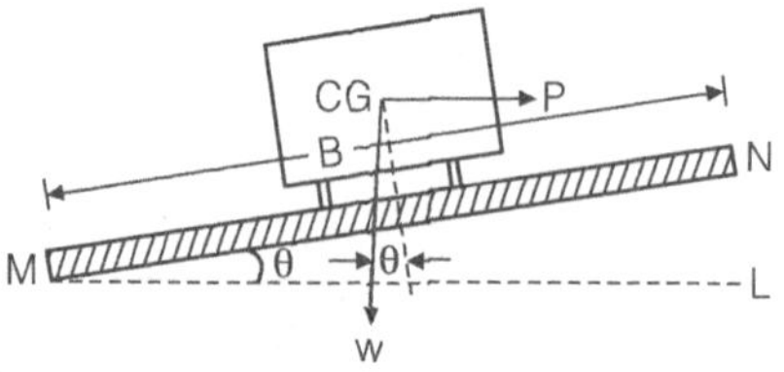

Superelevated Pavement Section

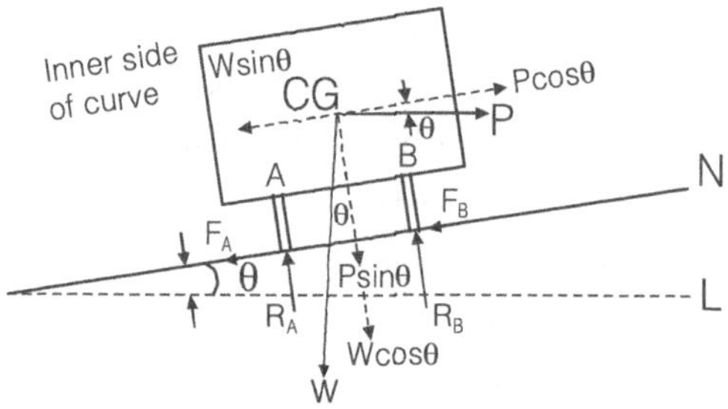

Analysis of Superelevation

For equilibrium condition

$$\begin{aligned} P\cos\theta &= W\sin\theta + F_A + F_B \\ &= W\sin\theta + fR_A + fR_B \\ &= W\sin\theta + f(W\cos\theta + P\sin\theta) \end{aligned}$$

or $P(\cos\theta - f(\sin\theta) = W\sin\theta + fW\cos\theta$

Divided by $\cos\theta$, we get

$$\frac{P}{W}(1 - f\tan\theta) = \tan\theta + f$$

or
$$\frac{P}{W} = \frac{f + \tan\theta}{1 - f\tan\theta}$$

Also
$$\frac{P}{W} = \frac{V^2}{gR} = \frac{f + \tan\theta}{1 - f\tan\theta}$$

But
$$1 - f\tan\theta = 1$$

$\therefore$
$$\frac{V^2}{gR} = f + \tan\bar{\theta}$$

Where, $\tan\theta = e$

$$\therefore \quad \frac{V^2}{gR} = f + e$$

or

$$V = \sqrt{(e+f)gR}$$

$$= \sqrt{(0.15+0.087)\times 9.81\times 250} = 24.14 \text{ m/s}$$

2.5 Calculate the SSD for design speed for SH at rolling terrain for

(i) Two way traffic on two lane road.

(ii) Two way traffic on one lane road.

[8 Marks]

Sol:

Design speed for national and state highway in rolling terrain is 80 kmph.

Note : For assuming longitudinal friction coefficient, use following guidelines:

Speed, kmph	20 to 30	40	50	60	65	80	100
Coefficient	0.40	0.38	0	0	0.36	0.35	0.35

$\therefore$ $f = 0.35$ and reaction time of driver = 2.5 sec

SSD = Lag distance + Braking distance

$$= Vt_R + \frac{V^2}{2gf}$$

$$V = V\times 0.277 = 80\times 0.277 = 22.216 \text{ m/s}$$

$$\text{SSD} = (22.216\times 2.5) + \frac{22.216^2}{2\times 9.81\times 0.35} = 127.41\text{ m}$$

(i) SSD, when there are 2 lanes = SSD × 1 = 127.41 m

(ii) SSD, when there is 1 lane = SSD × 2 = 254.82 m

2.6 A vehicle going towards Lansdowne from Najibabad at an average ascending gradient of 3.2% at the design speed of 80 kmph. Calculate SSD for both sides of the trip. Make assumptions as IRC suggestions.

[8 Marks]

Sol:

V = 80 kmph

$$\therefore \quad v = 80\times 0.277 = 22.16 \text{ m/s}$$

Total reaction time = 2.5 sec

Coefficient, f = 0.35

$$\therefore \quad \text{SSD (going up)} = vt_R + \frac{v^2}{2g(f+n\%)}$$

$$= (22.16\times 2.5) + \frac{22.16^2}{2\times 9.81(0.35+0.032)}$$

$$= 120.92\text{ m}$$

$$\text{SSD (going down)} = vt_R + \frac{v^2}{2g(f-n\%)}$$

$$= (22.16 \times 2.5) + \frac{22.16^2}{2 \times 9.81 \times (0.35 - 0.032)}$$

$$= 134.11 \text{ m}$$

2.7 **Calculate OSD for design speed at NH. Taking assumptions suitably.**

[8 Marks]

Sol:

Given data : V_1 = 100 kmph [Ruling design speed for NH is plain area]

$V_1 = 0.277 \times 100$ m/s = 27.77 m/s, $V_2 = (V_1 - 16)$ = 84 kmph, $V_2 = 0.277 \times 84$ = 23.27 m/s

Taking acceleration, a = 1.92 kmph/sec

= 0.53 m/s^2

t = 2 sec

For one way traffic, OSD = $D_1 + D_2$

For two way traffic, OSD = $D_1 + D_2 + D_3$

$\therefore$ $D_1 = v_2 t = 23.27 \times 2 = 46.54$ m

$D_2 = 2s + v_2 T$

where, $s = 0.2\,V_2 + 6 = 0.2 \times 84 + 6 = 22.8$ m

$$T = \sqrt{\frac{4s}{a}} = \sqrt{\frac{4 \times 22.8}{0.53}} = 13.12 \text{ sec}$$

$\therefore$ $D_2 = 2 \times 22.8 + 23.27 \times 13.12 = 350.90$ m

$\therefore$ OSD for one-way traffic = 46.54 + 350.90 = 397.44 mm $\simeq$ 400 m

OSD for two-way traffic = 46.54 + 350.90 + 364.34 = 761.78 m $\simeq$ 765 m

2.8 **Design the rate of superelevation for a horizontal highway curve of radius 400 m and design speed of 100 kmph.**

[6 Marks]

Sol:

Given data : V = 100 kmph, R = 400 m

$$\therefore \text{ Superelevation, } e \text{ for mixed traffic} = \frac{V^2}{225R} = \frac{100^2}{225 \times 400} = 0.111$$

This value is greater than maximum superelevation of 0.07.

$\therefore$ e is restricted to 0.07 only

Check for lateral friction coefficient developed,

$$f = \frac{V^2}{127R} = e + f$$

$$0.197 = 0.07 + f$$

f = 0.127, which is less than 0.15

Hence, design is safe. Provide e = 0.07 for V = 100 kmph

2.9 **Determine the expression for mechanical extra widening on curves.**

[10 Marks]

Sol:

Due to the rigidity of the wheel base, off-tracking take place. To account for it, a term mechanical extra widening, w_m may be calculated as shown.

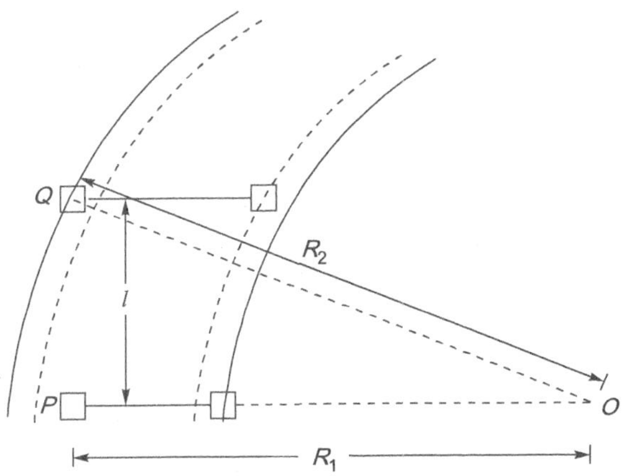

R_1 = Radius required to traverse the path by outer rear wheel in metres.

R_2 = Radius required to traverse the path by outer front wheel in metres.

l = Length of vehicle wheel base in metres.

W_m = Mechanical widening (off-tracking)

In ΔOQP,

$$PQ^2 + OP^2 = OQ^2$$

$$\therefore \quad OP^2 = OQ^2 - PQ^2$$

$$R_1^2 = R_2^2 - l^2$$

$$(R_2 - W_m)^2 = R_2^2 - l^2 \text{; using } R_1 = (R_2 - W_m)$$

$$R_2^2 - W_m^2 - 2R_2W_m = R_2^2 - l^2$$

$$l^2 = W_m(2R_2 - W_m)$$

$$\text{Approximeter } (2R_2 - W_m) = R$$

$$\therefore \quad l^2 = W_m \times R$$

Hence,
$$W_m = \left(\frac{l^2}{2R}\right)$$

If the road has 'n' traffic lanes, total mechanical widening $W_m = \left(\frac{nl^2}{2R}\right)$ because 'n' vehicles may travel simultaneously.

2.10 Design the length of a valley curve where a descending grade of 1 in 30 meets an ascending grade of 1 in 30. Take

Design speed = 90 kmph

Allowable rate of change of centrifugal acceleration, C = 0.55 m/s^3.

[8 Marks]

Sol:

$$N = |n_1 - (n_2)|$$

$$= \left|-\frac{1}{30} - \frac{1}{30}\right| = \left|\left(-\frac{1}{15}\right)\right| = \frac{1}{15}$$

$V = 90$ kmph $= 0.277 \times 90$ m/s $= 24.93$ m/s

Case I : Comfort condition

$$L = 2\left[\frac{NV^3}{C}\right]^{1/2} = 2\sqrt{\frac{1}{15} \times \frac{24.93^3}{0.55}} = 86.67 \text{ m}$$

Case II : Head light sight distance condition

$$L = \frac{NS^2}{2h_1 + 2s\tan\alpha}, \text{ Place } h_1 = 0.75 \text{ m}, \alpha = 1° = \frac{NS^2}{(1.5 + 0.035S)}$$

$$S = \text{SSD} = vt_r + \frac{V^2}{2gf}$$

$$= (24.93 \times 2.5) + \frac{24.93^2}{2 \times 9.81 \times 0.35} = 152.83 \text{ m}$$

If $L >$ SSD

$$L = \frac{NS^2}{1.5 + 0.035S} = \frac{1}{15} \times \frac{152.83^2}{(1.5 + 0.035 \times 152.83)}$$

$$= 227.35 \text{ m} \simeq 230 \text{ m}$$

$\therefore$ $L >$ SSD

Assumption stands valid.

Hence provide length of valley curve as 230 m, maximum from the above 2 conditions.

2.11 **A six-lane divided carriageway has a curve 1000 m long and of radius 500 m. The safe stopping sight distance is 200 m. Calculate the minimum set-back distance from the inner edge of the road to the edge of a building to ensure safe visibility. The pavement width per lane is 3.5 m.**

[8 marks]

Sol:

Given: Six lane carriageway

Radius of curve,	R	$= 500$ m
Length of curve,	L_c	$= 1000$ m
Width of pavement per lane		$= 3.5$ m
Stopping sight distance,	S	$= 200$ m

Here S is less than L i.e., $L > S$

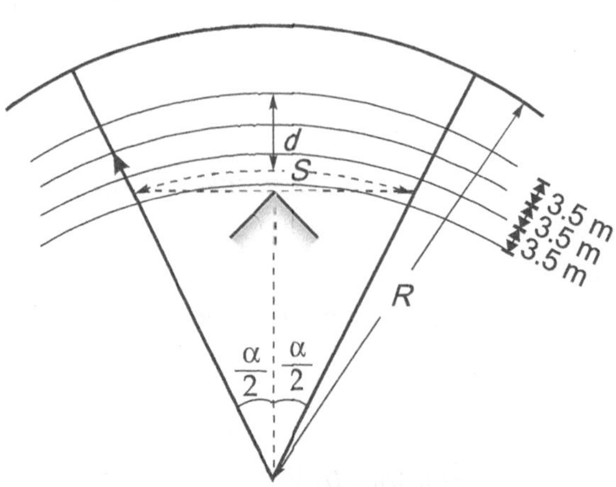

Set-back distance on three lane road

$$\therefore \quad m = R-(R-d)\cos\frac{\alpha}{2}$$

$$d = \text{distance from centre line of road to centre line of inner lane}$$

$$= \left(2\times3.5+\frac{3.5}{2}\right) = 8.75\text{m}$$

$$\frac{\alpha}{2} = \frac{180S}{2\pi(R-d)} = \frac{180\times200}{2\pi[500-8.75]}\text{ degrees}$$

$$= 11.66\text{ degrees}$$

$$\therefore \quad m = 500-(500-8.75)\cos11.66°$$

$$= 500-491.25\cos11.66°$$

$$= 18.89\text{ m (from centre line)}$$

$\therefore$ Distance from inner edge to obstruction as shown in figure

$$M' = 18.89-3.5\times3 = 8.39\text{ m}$$

2.12 **There is a horizontal curve of radius 360 m and length 180 m. Calculate the clearance required from the centre line on the inner side of the curve, so as to provide an overtaking sight distance of 250 m.**

[10 Marks]

Sol:

Given data

Radius of horizontal curve,

$$R = 360\text{ m}$$

Length of the curve

$$L_c = 180\text{ m}$$

Sight distance

$$S = 250\text{ m}$$

Clearance required on centre line of road is set back distance (m)

As $L_c <$ Sight distance (s)

$$\Rightarrow \quad m = \left[R-R\cos\left(\frac{\alpha}{2}\right)\right]+\left(\frac{S-L_c}{2}\right)\sin\frac{\alpha}{2}$$

Here, $\frac{\alpha}{2} = \frac{180L_c}{2\pi R}$

$$\Rightarrow \frac{\alpha}{2} = \frac{180\times180}{2\pi\times360} = \frac{45}{\pi} = 14.32°$$

$$\therefore \quad m = (360-360\cos14.32°) + \left(\frac{250-180}{2}\right)\sin14.32$$

$$= 11.185+8.657$$

$$= 19.842\text{ m}$$

2.13 **An ascending gradient of 1 in 80 meets a descending gradient of 1 in 120. Design a summit curve for national highway in rolling terrain so as to have an OSD = 400 m.**

[12 Marks]

Sol:

Design speed for NH in rolling terrain = 80 kmph

Here, $$n_1 = \frac{1}{80}$$

$$n_2 = -\frac{1}{120}$$

$$N = \frac{1}{80} - \left(-\frac{1}{120}\right) = \frac{1}{48}$$

- If L > OSD

$$L = \frac{NS^2}{\left(\sqrt{2H} + \sqrt{2h}\right)^2}$$

S = Overtaking sight distance, N = Deviation angle, equal to algebric difference in grades, H = Height of eye level of driver above road surface, h = Height of object above the pavement surface.

$$= \frac{NS^2}{9.6} = \frac{1}{48} \times \frac{400^2}{9.6} = 347.22 \text{ m}$$

$L = 347.22 <$ OSD hence assumption is wrong.

- If L < OSD

For OSD; H = h = 1.2 m

$$L = 2S - \frac{\left(\sqrt{2H} + \sqrt{2h}\right)^2}{N} = 2s - \frac{9.6}{N}$$

$$= 2 \times 400 - \frac{9.6}{1/48} = 339.2 \text{ m} \simeq 340 \text{ m} < \text{OSD}$$

Hence, length of curve = 340 m

Level-2

2.14 A horizontal curve with super elevation, e = 0.05 (note, e = tan θ) has a posted speed limit of 60 km/h. For this speed limit, only those vehicles with Y/X ratio (in figure) greater than 2 and travelling at speeds greater than 1.2 times the speed limit, will overturn. However, with recent increase in truck traffic the speed limit needs to be modified such that only those vehicles with Y/X ratio greater than 3 and travelling at more than 1.2 times the speed limit will overturn. What should be the new speed limit? (with respect to the figure, assume that the vehicle is travelling into the plane of the paper).

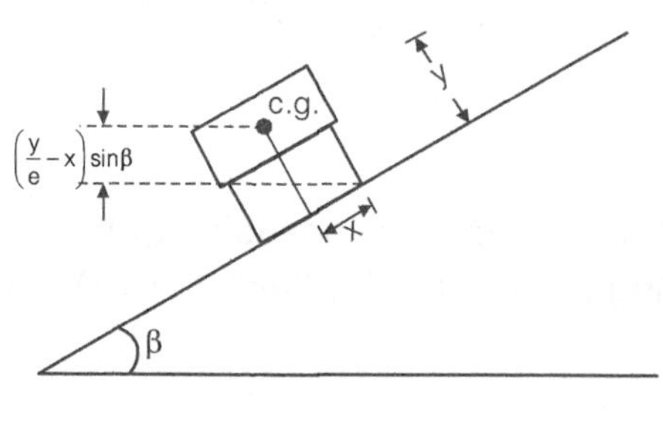

[15 Marks]

Sol:

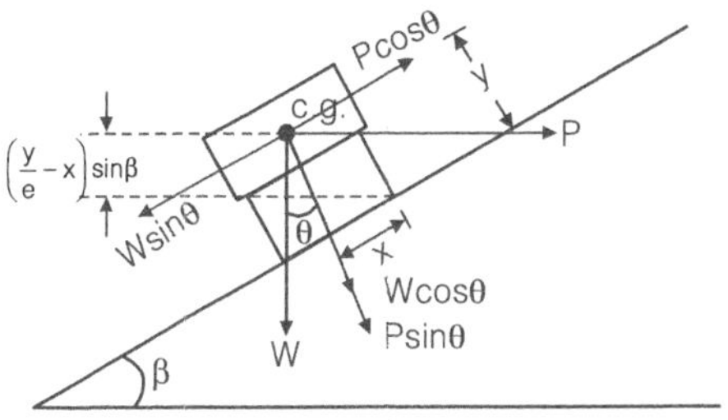

Given,

For $$\frac{y}{x} \geq 2$$

$$\text{speed} = 1.2 \times 60$$
$$= 72 \text{ km/hr} = 20 \text{ m/sec}$$

For $$\frac{y}{x} \geq 3$$

$$\text{speed} = 1.2 \times v$$

For overturning,

$$P\cos\theta \cdot y \geq (W\cos\theta + P\sin\theta)x$$

$$\Rightarrow \quad \frac{y}{x} \geq \frac{W\cos\theta + P\sin\theta}{P\cos\theta}$$

$$\Rightarrow \quad \frac{y}{x} \geq \left(\frac{W}{P} + \tan\theta\right)$$

$$\therefore \quad \tan\theta = e = 0.05$$

$$\Rightarrow \quad \frac{y}{x} \geq \left(\frac{W}{P} + 0.05\right)$$

$\because$ P = centrifugal force = $\frac{mv^2}{R}$

and W = weight of vehicle = mg

$$\Rightarrow \quad \frac{P}{W} = \frac{V^2}{Rg}$$

$$\Rightarrow \quad \frac{W}{P} = \frac{R \cdot g}{V^2}$$

$$\because \quad \frac{y}{x} = 2$$

$$\therefore \quad 2 \geq \frac{gR}{V^2} + 0.05$$

$$\Rightarrow \quad 2 = \frac{gR}{V^2} + 0.05$$

$$\Rightarrow 2 - 0.05 = \frac{9.81R}{(0.278 \times 72)^2} = \frac{9.81R}{(20)^2}$$

$$\Rightarrow \quad R = 79.51 \text{ m}$$

Now, if $\frac{y}{x} = 3$ then

$$3 = \frac{gR}{V^2} + 0.05$$

$$\Rightarrow \quad 3 = \frac{9.81 \times 79.51}{(V)^2} + 0.05$$

$$V = 16.26 \text{ m/sec} = 58.536 \text{ km/hr}$$

$\therefore$ New speed limit

$$= \frac{V}{1.2} = \frac{58.536}{1.2} = 48.78 \text{ km/hr}$$

2.15 **Determine the extra width of pavement and the length of transition curve needed on a horizontal alignment of radius 225 m for a two-lane road, with a design speed of 80 kmph. Assume the wheel base of design vehicle as 6 m.**

[20 Marks]

Sol:

Length of wheel base,

$$l = 6 \text{ m}$$

Radius, $R = 225 \text{ m}$

Number of lanes,

$$n = 2$$

$$V = 80 \text{ kmph}$$

∴ Extra width of pavement

$$W_e = W_m + W_{ps}$$

$$= \frac{nl^2}{2R} + \frac{V}{9.5\sqrt{R}} = \frac{2\times 6^2}{2\times 225} + \frac{80}{9.5\sqrt{225}} = 0.16 + 0.56 = 0.72 \text{ m}$$

(ii)Allowable rate of centrifugal acceleration

$$C = \frac{80}{75+V} = \frac{80}{75+80} = 0.516$$

∴ Length of transition curve based on rate of change of centrifugal acceleration

$$L_s = \frac{0.0215V^3}{CR} = \frac{0.0215\times 80^3}{0.516\times 225} = 94.81 \text{ m}$$

Now, superelevation

$$e = \frac{V^2}{225R} = \frac{80^2}{225\times 225} = 12.64\% > 7\%$$

Hence provide, e = 0.07

Check for coefficient of friction

$$f = \frac{V^2}{127R} - 0.07 = \frac{80^2}{127\times 225} - 0.07 \approx 0.15$$

∴ Provide $e = 0.07$

Total width of pavement

$$= 6 + 0.72 = 6.72 \text{ m}$$

Total raise of outer edge of pavement with respect to centre line

$$= \frac{E}{2} = \frac{eB}{2} = \frac{0.07\times 6.72}{2} = 0.2352 \text{ m}$$

Rate of introduction of superelevation

$$1 \text{ in } N = 1 \text{ in } 150$$

$$\therefore \quad L_s = 0.2352 \times 150 = 35.28 \text{ m}$$

Maximum value of L_s as per IRS

$$L_s = \frac{2.7V^2}{R} = \frac{2.7\times 80^2}{225} = 76.8 \text{ m}$$

∴ Adopt the highest value of the three i.e., 91.81 m or say 95 m as the design length transition curve.

2.16 On a two-lane two-way highway, a car A was following a truck B and both were travelling at a speed of 40 kmph. While looking for a opportunity to overtake the truck, the driver of the car A saw another car C coming from the opposite direction. At that moment, the distance between A and C was 450 m. After an initial hesitation period of two seconds, the driver of car A started the overtaking operation. The distance between A and B at that instance was 30 m. A overtook B by accelerating at an uniform rate of 1.20 m/sec^2. When the overtaking operation was completed, there was a distance of 25 m between B and A. Determine the distance between the two cars (A and C) at the instance of completion of the overtaking action. The distance between different vehicles of given are as measured from the front the bumper of one vehicle to the front bumper of another vehicle. Design speed of the highway is 80 kmph.

[20 Marks]

Sol:

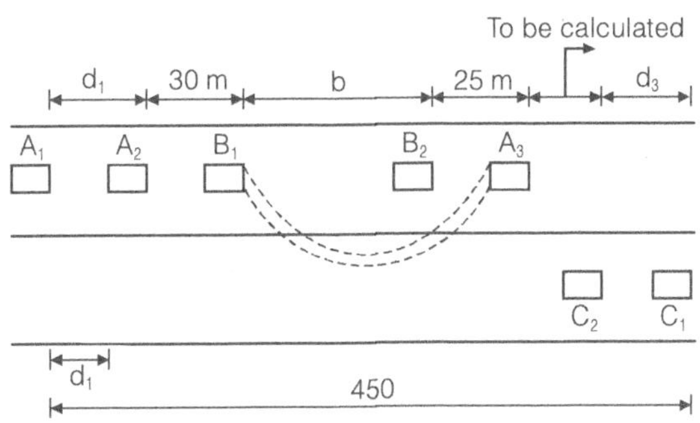

Both car A and truck B were travelling at speed of 40 kmph

$$V_b = 40 \times \frac{5}{18} = 11.11 \text{ m/sec}$$

Speed of car

$$C = \text{design speed of highway}$$
$$V = 80 \text{ kmph} = 22.22 \text{ m/sec}$$

Initial distance between A and C

= 450 m when car A is at A and car C is at C

Reaction time

$$t = 2 \text{ sec}$$

Distance covered by C during these 2 sec

$$= s \times t$$
$$= 22.22 \times 2 = 44.44 \text{ m}$$

Distance covered by car A in reaction time

$$d_1 = 11.11 \times 2 = 22.22$$

At the start of overtaking operations, distance between A and B = 30 m.

At the end of overtaking operation, distance between B and A = 25 m

Total distance covered by car A during overtaking

$$= 30 + b = 25$$

where b is the distance covered by truck B in overtaking time T

$$n = 11.11 \times T \quad \text{...(i)}$$

Distance covered by A during overtaking time 'T' is

$$VT + \frac{1}{2}aT^2 \quad \text{...(ii)}$$

Equating (i) and (ii), we have

$$30 + 11.11 \times T + 25$$

$$= 11.11 \times T + \frac{1}{2} \times 1.20T^2$$

or $T = 9.574$ sec

Distance covered during overtaking by A

$$= 30 + VT + 25$$

$$= 55 + 11.11 \times 9.574 = 106.27 \text{ m} + 55 = 161.27 \text{ m}$$

Similarly during overtaking time T, distance covered by car

$$C = V'T$$

$$= 22.22 \times 9.574 = 212.74 \text{ m}$$

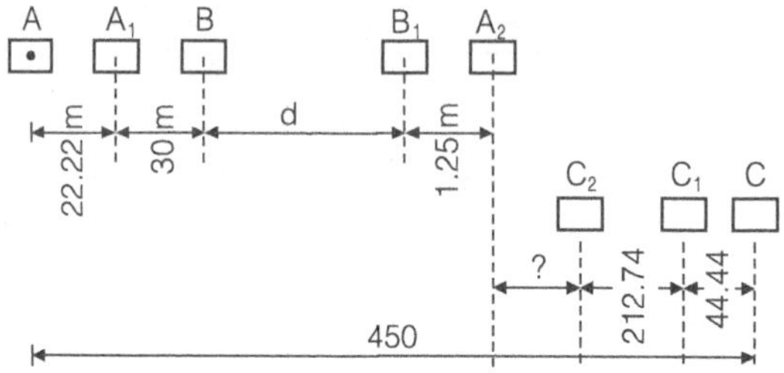

Final distance between A and C after completion of overtaking

$$= 450 - (212.74 + 44.44) - (22.22 + 30 + d + 25)$$

$$= 450 - 257.18 - 77.22 - 106.27 = 9.33 \text{ m}$$

2.17 **There is a horizontal highway curve of radius 400 m and length of 200 m on this highway. Compute the set-back distance required from the centre line on the inner side of curve so as to provide for**

(i) stopping sight distance of 100 m

(ii) safe overtaking sight distance of 250 m

The distance between the centre line of road and the inner lane is 1.9 m.

[20 marks]

Sol:

(i) The stopping sight distance (SSD) of 100 m is less than the circular curve length of 200 m i.e., $L >$ SSD.

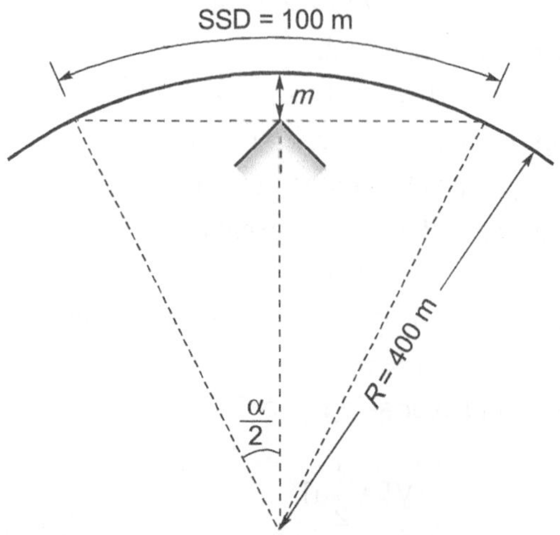

$$\therefore \quad \frac{\alpha}{2} = \frac{180S}{2\pi(R-D)} = \frac{180\times100}{2\pi(400-1.9)} = 7.196°$$

Set back distance,

$$m = R-(R-d)\cos\frac{\alpha}{2}$$

$$= 400-(400-1.9)\cos 7.196°$$

$$= 5.036 \text{ m}$$

(ii) The overtaking sight distance of 250 m is greater than circular curve length which is 200 m i.e., $L <$ OSD (S). Therefore the required set back distance is CF = (CG + GF).

$$S = 250 \text{ m}, L_c = 200 \text{ m}, R = 400 \text{ m}, d = 1.9 \text{ m}$$

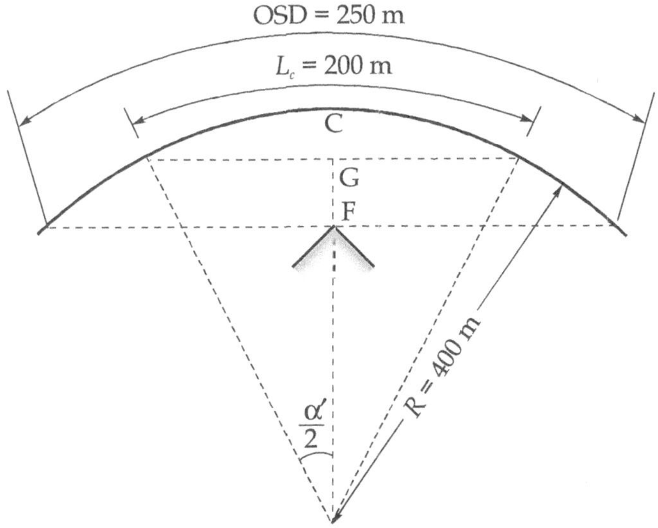

$$\frac{\alpha'}{2} = \frac{180L_c}{2\pi(R-d)} = \frac{180\times200}{2\pi(400-1.9)} = 14.39°$$

Set back distance,

$$m' = CF = CG + GF$$

$$= R-(R-d)\cos\frac{\alpha'}{2}+\frac{(S-L_c)}{2}\sin\frac{\alpha'}{2}$$

$$= 400-(400-1.9)\cos 14.39°+\frac{(250-200)}{2}\sin 14.39°$$

$$= 400 - 385.61 + 6.21 = 20.6 \text{ m}$$

Hence, minimum set-back distance required from the centre line of the roads on the inner side of pavement to provide an OSD of 250 m = 20.6 m.

2.18 Speed of overtaking vehicle = 80 kmph

Speed of overtaken vehicle = 50 kmph

Acceleration of vehicle = 1 m/s^2

Assume the traffic to be two-way and calculate

(i) Safe OSD

(ii) Minimum length of overtaking zone

(iii) Draw a proper sketch to show various positions of sign posts.

[15 Marks]

Sol:

$$\text{Design speed for vehicle 1} = 80 \text{ kmph}$$
$$v_1 = (0.277 \times 80) \text{ m/s} = 22.16 \text{ m/s}$$
$$v_2 = (50 \times 0.277) \text{ m/s} = 13.85 \text{ m/s}$$
$$\text{Acceleration, } a = 1 \text{ m/s}^2$$
$$\text{OSD} = D_1 + D_2 + D_3$$

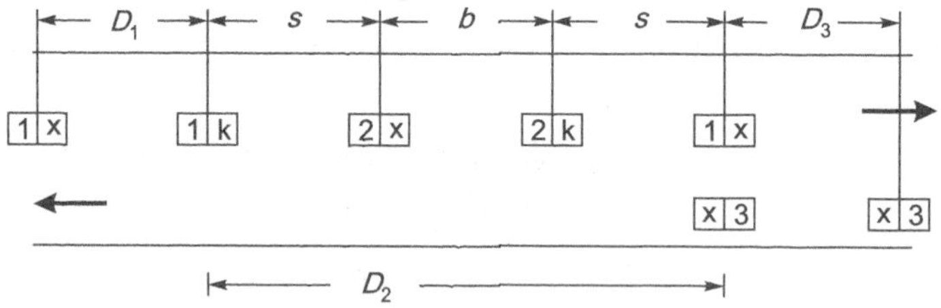

$$D_1 = v_2 t \qquad \text{(Take } t = 2 \text{ sec)}$$
$$= 13.85 \times 2 = 27.7 \text{ m}$$
$$D_2 = v_2 T + 2s$$
$$s = (0.7v_2 + 6) = (0.7 \times 13.85 + 6) = 15.7 \text{ m}$$
$$T = \sqrt{\frac{4s}{a}} = \sqrt{\frac{4 \times 15.7}{1}} = 7.92 \text{ sec}$$
$$D_2 = (13.85 \times 7.92) + (2 \times 15.7) = 141.1 \text{ m}$$
$$D_3 = V_3 T = V_1 T = 22.16 \times 7.92 = 175.5 \text{ m}$$

Note V_1 and V_3 are assumed as design speed of that road.

$$\therefore \quad \text{OSD} = D_1 + D_2 + D_3$$
$$= 27.7 + 141.1 + 175.5$$
$$= 344.3 \text{ m} \simeq 345 \text{ m}$$

(ii) Minimum length of overtaking zone = 3 × OSD = 1035 m

This minimum length is over two-way traffic.

Desirable length of overtaking zone = 5 × OSD = 1725 m

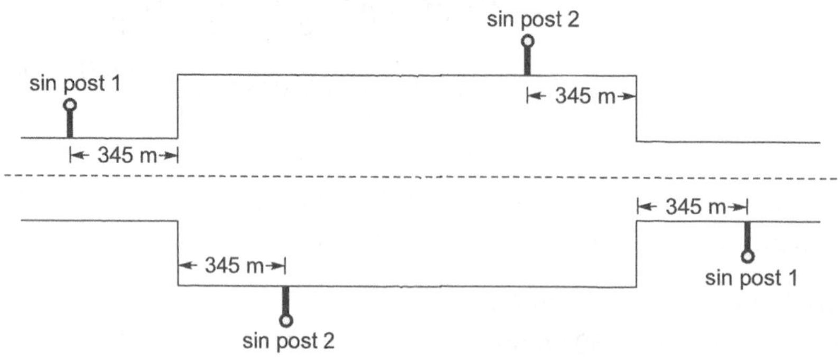

Sign post 1 rays = Over taking zone ahead

Sign post 2 rays = End of overtaking zone

2.19 Design the length of transition curve for national highway on hilly roads on Mawsynram, Shillong. No snowfall occurs at the place. The available horizontal curve is of radius 150 m.

[15 Marks]

Sol:

Design speed for NH in mountainous terrain = 50 kmph.
Camber for heavy rainfall and high bituminous surface = 2.0%
Rate of introduction of superelevation = 1 in 60 on hilly roads.
Pavement width, $W = 7.0$ m, radius of curve = 150 m

(a) Length of transition curve, by rate of change of centrifugal acceleration,

$$L_s = \frac{V^3}{CR}$$

$$C = \frac{80}{75+V} = \frac{80}{75+50} = 0.64 \qquad (\because \text{ range} = 0.5 \text{ to } 0.8)$$

$$L_8 = \frac{(0.277\times 50)^3}{0.64\times 150} = 27.67 \text{ m}$$

(b) Length of transition curve by rate of introduction of superelevation,

$$e = \frac{V^2}{225R} = \frac{50^2}{225\times 150} = 0.074 < 0.10 \qquad \text{Hence ok}$$

$$e_{allowable} = 10\% \qquad \because \text{ Hilly roads with no show}$$

Extra widening at curve, taking $l = 6$ m and 2-laned road.

$$W_e = \frac{nl^2}{2R} + \frac{V}{9.5\sqrt{R}} = \frac{2\times 6^2}{2\times 150} + \frac{50}{9.5\sqrt{150}} = 0.67 \text{ m}$$

Usually W_e is distributed on both sides of curve.
Provide this w_e on inside of the curve only since it is a hill road.

$$\text{Total width of pavement} = 7.0 + 0.67 = 7.67 \text{ m}$$

$$L_s = 7.67\times 0.074\times 60 = 34.05 \text{ m}$$

(c) Check by IRC method, $L_s = \frac{V^2}{R} = \frac{50^2}{150} = 16.67$ m

$\therefore$ L_s = Maximum of the three values above i.e. $L_s = 34.05$ m

2.20 **Design following geometric features for the curve by assuming suitable data. A national highway is passing through plain terrain, passing through a well built up area.**

Design :

(i) Ruling minimum radius
(ii) Superelevation
(iii) Extra widening
(iv) Length of transition curve
(v) SSD
(vi) ISD
(vii)Set back distance (assuming length of circular curve > SSD)

[20 Marks]

Sol:

Assuming design speed (V) = 100 kmph for plain terrain

(i) Ruling minimum radius

$$R_{\text{ruling}} = \frac{V^2}{127(e+f)} = \frac{100^2}{127(0.07+0.15)}$$

$$= 358 \text{ m} \simeq 360 \text{ m}$$

(ii) Superelevation

We know, $$e = \frac{V^2}{127R} = \frac{100^2}{225 \times 360} = 0.123$$

$$e_{\max} = 0.07$$

Also, $$\frac{V^2}{127R} = e + f \text{ or } f = \frac{V^2}{127R} - e$$

$$\frac{100^2}{225 \times 360} - 0.07 = 0.149 < 0.15$$ Hence OK

(iii) Extra widening

Assume 2-lane pavement and l = 6 m

$$W_e = \frac{nl^2}{2R} + \frac{V}{9.5\sqrt{R}}$$

$$= \frac{2 \times (6.1)^2}{2 \times 360} + \frac{100}{9.5\sqrt{360}} = 0.658 \text{ m}$$

Total width = 7.0 + 0.658 m = 7.658 m

(iv) Length of transition curve

Maximum of the 3 lengths calculation on following basis

- Rate of change of centrifugal acceleration,

$$C = \frac{80}{75+V} = \frac{80}{175} = 0.45$$

$\because$ C has to be within 0.50 to 0.80.

$\therefore$ $$L_s = \frac{(0.277V)^3}{CR} = \frac{(0.277 \times 100)^3}{0.5 \times 360} = 118.1 \text{ m}$$

- By rate of change of superelevation

$$L_s = \frac{eBN}{2}$$ [Assuming rotation about centre line]

$$e = 0.070$$

$$B = 7.658$$

$$N = 100$$ [$\because$ Built up area]

$$= \frac{0.07 \times 7.658 \times 100}{2} = 26.80 \text{ m}$$

- Minimum L_s by IRC formula

$$L_s = \frac{2.7V^2}{R} = \frac{2.7 \times 100^2}{360} = 75 \text{ m}$$

$\therefore$ L_s = 118.1 m (Maximum of all 3 values)

(v) SSD

$$SSD = 0.277Vt_R + \frac{(0.277V)^2}{2 \times g \times f}$$

$$= (0.277 \times 100) \times 2.5 + \frac{(0.277 \times 100)^2}{2 \times 9.81 \times 0.35}$$
$$= 180.99 \simeq 181 \text{ m}$$

(vi) ISD

$$\text{ISD} = 2 \times \text{SSD} = 2 \times 181 \text{ m} = 362 \text{ m}$$

(vii) Set back distance (assuming length of circular curve > SSD)

Set back distance on the figure = SR

$$\angle POR = \frac{\alpha}{2}$$

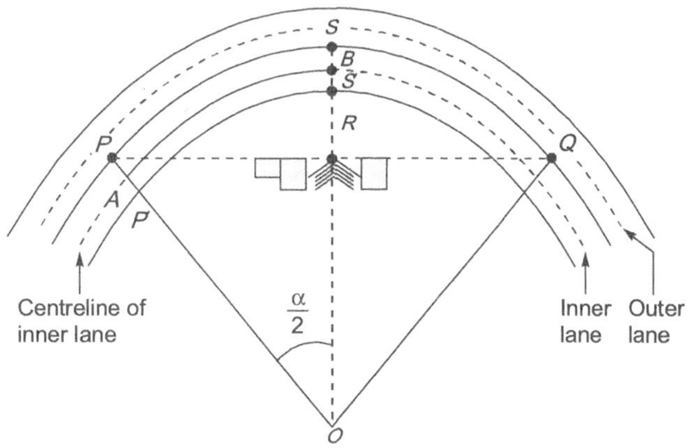

We know, $$\theta = \frac{\text{Length of arc}}{\text{Radius of arc}}$$

$\therefore$ $$\frac{\alpha}{2} = \left[\frac{AB}{OQ - \frac{\text{Lane width}}{2}}\right]$$

$$\frac{\alpha}{2} = \frac{(SSD/2)}{R - \frac{\text{Pavement width}}{4}} = \frac{(181/2)}{360 - \frac{7.658}{4}}$$
$$= 0.252733 \text{ radian}$$
$$= 14.48°$$

Setback distance (from centre of inner lane) = OS – SB – RO

$$\text{BR} = R - \frac{\text{Pavement width}}{4} - R\cos\frac{\alpha}{2}$$

$$\text{BR} = 360 - \frac{7.658}{4} - 360 \times \cos 14.48° = 9.521 \text{ m}$$

Atleast 9.521 m clearance is required an ISD = 362 m.

2.21 There is a horizontal curve on a highway with radius 300 m and length 180 m on the calculate the set back distance required from centre line of the inner lane so as to satisfy

(i) SSD = 100 m

(ii) OSD = 250 m

Take pavement width as 7.6 m.

[15 Marks]

Sol:

(i) SSD = 90 m, which is less than length of curve i.e. 180 m

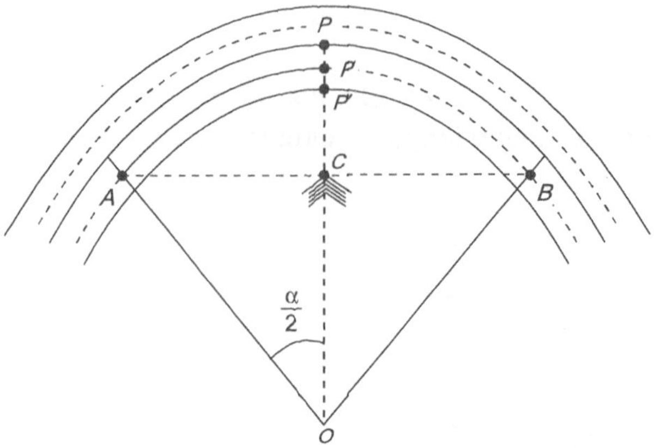

$$AP'B = SSD = 90 \text{ m}$$

$$\frac{\alpha}{2} = \frac{AP'}{R - \frac{W}{4}} = \frac{90/2}{300 - \frac{7.6}{4}}$$

$$= 0.151 \text{ radian} = 8.65°$$

Setback distance from centre of inner lane = P′C

where $P'C = OP - PP' - CO$

$$= R - \frac{W}{4} - \left(R - \frac{W}{4}\right)\cos\frac{\sigma}{2}$$

$$= 300 - \frac{7.6}{4} - \left(300 - \frac{7.6}{4}\right)\cos 8.65$$

(ii) OSD = 250 m

OSD > Length of curve i.e. 180 m

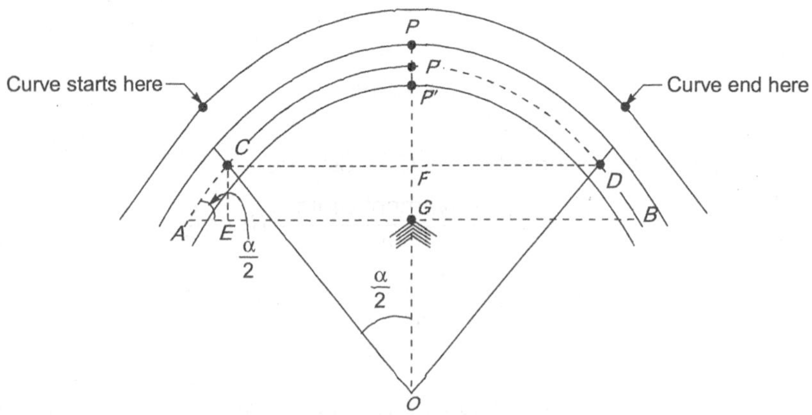

$$AP'B = \text{OSD required i.e. } 250 \text{ m}$$

$$CP'D = 180 \text{ m}$$

$$\therefore \quad AC = \frac{OSD - L_s}{2} = 35 \text{ m}$$

$$\angle CAE = \angle COP = \frac{\alpha}{2}$$

$$= \frac{(L/2)}{R - \frac{W}{4}} \times \frac{180}{\pi} = \frac{(180/2)}{300 - \frac{7.6}{4}} \times \frac{180}{\pi} = 17.30°$$

$\therefore$ Set back distance = (OP – PP′ – OG) = (P′F + FG)

(From centre of inner lane)

$$= \left(R - \frac{W}{4}\right)\left(1 - \cos\frac{\alpha}{2}\right) + AC \sin\frac{\alpha}{2}$$

$$= \left(300 - \frac{7.6}{4}\right)(1 - \cos 17.30°) + 35 \sin 17.30° = 23.90 \text{ m}$$

3. Traffic Engineering

Level-1

3.1 **A vehicle travelling at design spaced in rolling terrain on state highway was stopped within 2.65 seconds after brakes application. Determine the average skid resistance.**

[6 Marks]

Sol:

Design speed in rolling terrain on NH / SH = 80 kmph

$\therefore$ Initial speed, u = 80 kmph = 0.277 × 80 = 22.16 m/s

Stopping time, t = 2.65 sec

$$V = u + at$$

$V = 0$

$\therefore$ $0 = 22.16 + 2.64a$

$\therefore$ $a = -8.36 \text{ m/s}^2$

$$\text{Average skid resistance, } \mu = \frac{a}{g} = \frac{8.36}{9.81} = 0.85$$

3.2 **In reference to a busy road with mixed traffic condition, discuss the positives and negatives that a rotary will provide, if constructed.**

[12 Marks]

Sol:

A traffic rotary is an enlarged road intersection where all converging vehicles are forced to move round a large central island in one direction before they can weave out of traffic flow into their respective directions radiating from central island. The main objects of providing a rotary are to eliminate the necessity of stopping even for crossing streams of vehicles and to reduce the area of conflict. The crossing of vehicles is avoided by allowing all vehicles to merge into the streams around the rotary and then to diverge out to the desired radiating road.

Advantages of traffic rotary:

(a) Crossing manoeuvre is converted into weaving or merging and diverging operation so that there is no necessity of any of vehicles to stop and proceed within a traffic rotary. Thus journey is made more consistent and comfortable.

(b) All traffic including those turning right or going straight across the rotary have equal opportunity as those turning left.

(c) The variable cost of operation of automobile is less at a traffic rotary than at a signalized intersection where the vehicles have to stop and proceed.

(d) There is no necessity of traffic police or signal to control the traffic as the rotary could function by itself as a traffic controlled intersection and is the simplest of all controls.

(e) The possible number of accident and severity of accidents are quite low because of low relative speed. And also weaving, merging and diverging manoeuvres are easier and less dangerous operation than crossing or right turning check on speed of vehicles is automatically enforced by proper design.

(f) The vehicle capacity of the rotary intersection is the highest of all other intersection at grade which is about 3000 vehicles per hour.

(g) Rotaries can be constructed with advantage when the number of intersecting roads is between four and seven.

Limitation of traffic rotary:

(i) Rotary requires comparatively a large area of land and so where space is limited and costly as in built up areas, total cost may be very high.

(ii) Where pedestrian traffic is large as in urban areas, the rotary by itself cannot control the traffic and hence has to be supplemented by traffic police.

(iii) In place where there is mixed traffic and large number of cyclists and pedestrians the design of rotary becomes too elaborate and operation and control of traffic also become complex.

(iv) Where the angle of intersection of two roads is too acute or when there are more than seven intersecting roads, rotaries are unsuitable.

(v) When the distance between intersections on an important highway is less, rotaries become trouble-some.

(iv) When traffic volume is very low as in most of rural areas of India, construction of rotary cannot be justified.

3.3 **On a right-angled road intersection show the minor and major conflict points in a**
(i) Two-way traffic one both roads.
(ii) One-way traffic on both roads.
(iii) One-way traffic on one roads.

[8 Marks]

Sol:

(i)

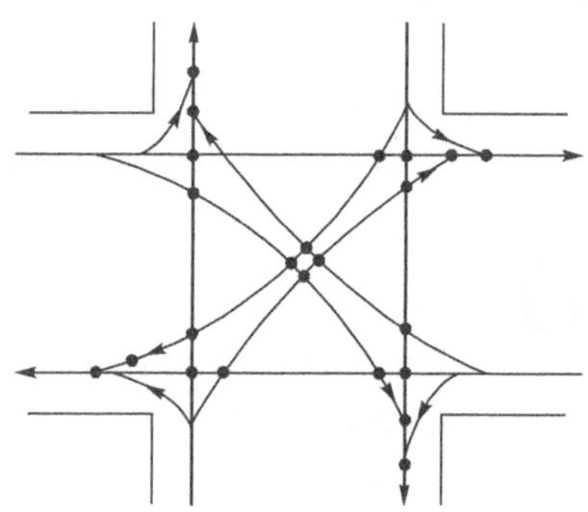

- Major conflicts = 16
- Minor conflicts = 8
- Total conflicts = 24

(ii)

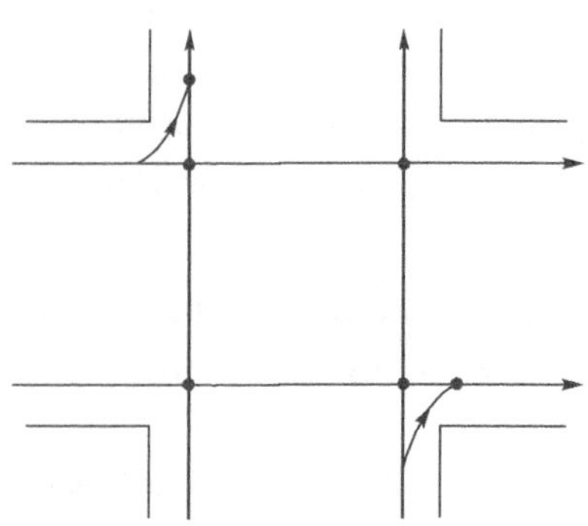

- Major conflicts = 4
- Minor conflicts = 2

- Total conflicts = 6

(iii)

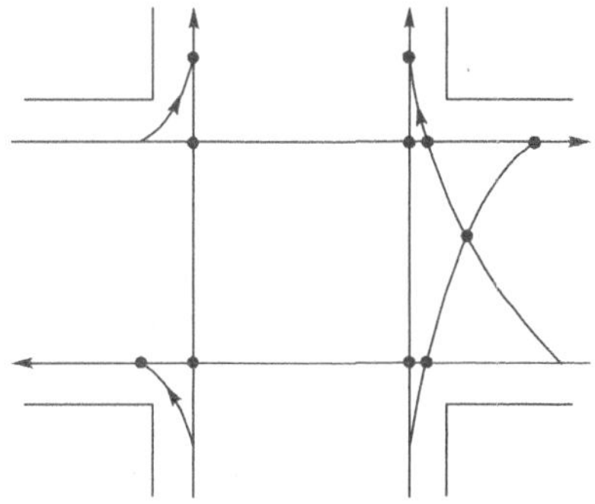

- Major conflicts = 7
- Minor conflicts = 4
- Total conflicts = 11

3.4 On 2 cross roads A and B, the average normal flow of traffic during design period are 500 and 300 pcu per hour. The saturation flow values on three roads are 1400 and 1100 pcu per hour respectively. The all red time required for pedestrian crossing is 10 sec. Design 2-phased traffic signal by Webster's method.

[8 Marks]

Sol:

$$y_a = \frac{q_a}{s_a} = \frac{500}{1400} = 0.357$$

$$y_b = \frac{q_b}{s_b} = \frac{300}{1100} = 0.273$$

$$y = y_a + y_b = 0.357 + 0.273 = 0.63$$

$$L = 2n + R = 2 \times 2 + 10 = 14 \text{ sec}$$

$$C_0 = \frac{1.5L+5}{1-y} = \frac{1.5\times 14+5}{1-0.63} = 70.27 \text{ sec} \simeq 71 \text{ sec}$$

$$G_a = \frac{y_a}{y}(C_0 - L) = \frac{0.35}{0.63}\times(70.27-14) = 31.89 \text{ sec} \simeq 32 \text{ sec}$$

$$G_b = \frac{y_b}{y}(C_0 - L) = \frac{0.273}{0.63}\times(70.27-14) = 24.38 \text{ sec} \simeq 25 \text{ sec}$$

All red time for pedestrian crossing = 10 sec

Provide amber time = 2 sec each for clearance

Total cycle time = 32 + 25 + 10 + 2 + 2 = 71 sec

3.5 Calculate the basic capacity for light weight commercial vehicle for a lane on a national highway in rolling terrain. Take length of light weight vehicle as 4.25 cm and commercial vehicle as 6.2 m. [Take f = 0.50]

[12 Marks]

Sol:

$$\text{Lane capacity} = \frac{1000V}{S} \text{ veh/hr}$$

where, $S = L + \text{SSD}$, L = Length of vehicle, SSD = Stopping sight distance

(i) Light weight vehicle,

$$S = 0.277Vt_r + \frac{(0.277V)^2}{2gf} + L$$

- V_{design} for NH on a rolling terrain = 80 kmph
- $t_r = 2.5$ sec

$$S = 0.277 \times 80 \times 2.5 + \frac{(0.277 \times 80)^2}{2 \times 9.81 \times 0.5} + 4.25 = 109.75 \text{ m}$$

$$\therefore \quad \text{Lane capacity} = \frac{1000 \times 80}{109.75} = 728.9 \text{ vph}$$

(ii) Commercial vehicle

$$S = 0.277Vt_r + \frac{(0.277V)^2}{2gf} + L$$

$$= 105.5 + 6.2 = 111.7 \text{ m}$$

$$\therefore \quad \text{Lane capacity} = \frac{1000 \times 80}{111.7} = 716.2 \text{ vph}$$

3.6 **A vehicle applies brakes and skids through a distance of 50 m before colliding with another parked vehicle. The weight of parked vehicle is 60% of the colliding vehicle. If distance travelled by both the vehicles after collision is 15 m before coming to rest, then determine the initial speed of the moving vehicle. [Take $f = 0.6$]**
Also write the assumptions used.

[10 marks]

Sol:

Assumptions:

1. Collision is perfectly plastic.
2. Brake efficiency = 100%

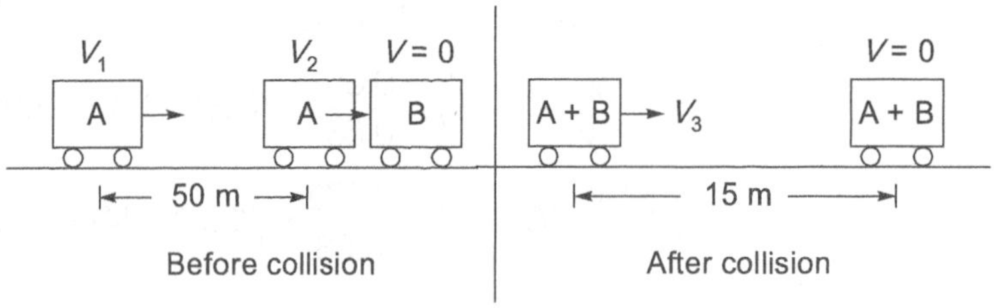

After collision for vehicle (A) + (B)

$$V_3^2 = 2gf \times S_2$$

$$V_3 = \sqrt{2 \times 9.81 \times 0.6 \times 15} = 13.288 \text{ m/s}$$

Applying momentum conservation equation,

$$m_A V_2 + m_B \times 0 = (m_A + m_B) V_3$$

$$\Rightarrow \quad V_2 = \frac{(m_A + 0.6m_A)}{m_A} \times 13.288$$

$$\Rightarrow \quad V_2 = 21.26 \text{ m/s}$$

Before collision for vehicle (A)

$$V_1^2 - V_2^2 = 2gfS_1$$

$$\Rightarrow \quad V_1 = \sqrt{(21.26)^2 + (2 \times 9.81 \times 0.6 \times 50)}$$

$$\Rightarrow \quad V_1 = 32.26 \text{ m/s}$$

3.7 Two vehicles A and B approach on intersection at right angle from East and North respectively and collide with each other. After the collision, vehicle A skids in a direction 40° South of East and vehicle B skids 50° West of South. Before the collision, A and B skid 40 m and 22 m respectively. After the collision, skid distances are 20 m and 40 respectively. If the weights of the vehicles are 3.8 and 5.0 tonnes calculate initial speeds of the vehicles. Take average skid resistance of the pavement as 0.61.

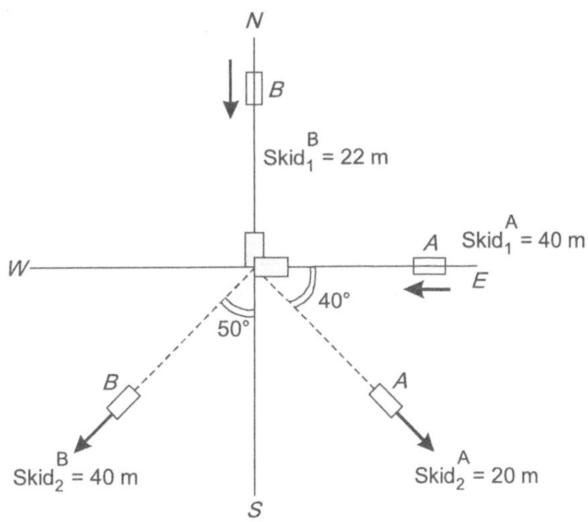

[12 Marks]

Sol:

- Speed of vehicle just after collision

$$V_{A_3} = \sqrt{254fS_{A_2}} = \sqrt{254 \times 0.61 \times 20} = 55.67 \text{ kmph}$$

$$V_{B_3} = \sqrt{254fS_{B_2}} = \sqrt{254 \times 0.61 \times 40} = 78.72 \text{ kmph}$$

- Speed of vehicles just before collision

$$V_{A_2} = \frac{W_B}{W_A} V_{B_3} \sin B - V_{A_3} \cos A$$

$$= \frac{5.0}{3.8} \times 78.72 \sin 50° - 55.67 \cos 40° = 36.7 \text{ kmph}$$

$$V_{B_2} = \frac{W_A}{W_B} V_{A_3} \sin A + V_{B_3} \cos B$$

$$= \frac{3.8}{5.0} \times 55.67 \sin 40° + 78.72 \cos 50° = 77.80 \text{ kmph}$$

- Speed of vehicles before application of brakes

$$V_{A_1} = \sqrt{254fS_{A_1} + V_{A_2}^2} = \sqrt{254 \times 0.61 \times 40 + 36.7^2} = 86.86 \text{ kmph}$$

$$V_{B_1} = \sqrt{254fS_{B_1} + V_{B_2}^2} = \sqrt{254 \times 0.61 \times 22 + 77.80^2} = 97.27 \text{ kmph}$$

3.8 The width of a carriage way approaching an intersection is given as 16 m. The width of rotary at entry and exit is 10.5 m. The traffic approaching the rotary is as shown. Find the capacity of the rotary using the given data.

[12 Marks]

Sol:

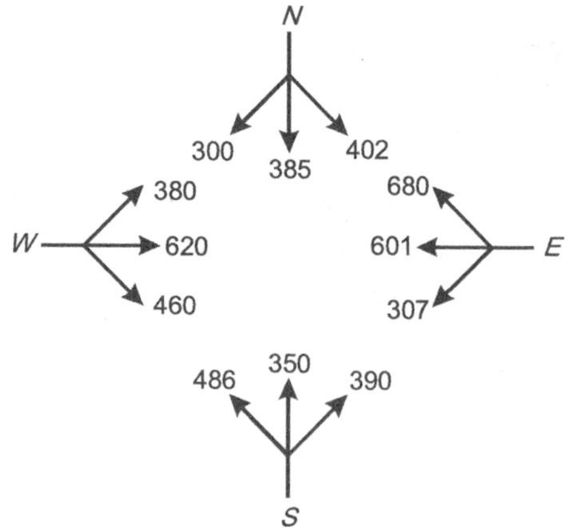

$$e_1 = e_2 = 10.5 \text{ m}$$

- $$\text{Weaving width} = \left(\frac{e_1 + e_2}{2}\right) + 3.5$$
$$= 10.5 + 3.5 = 14 \text{ m}$$
- $$\text{Weaving length} = 4w = 56 \text{ m}$$

(iii) Proportioning ratio, p

$$p_{N\text{-}E} = \frac{b+c}{a+b+c+d}$$

$a = 402$ [Moving between N and E]

b = 385 + 300 = 685 [Leaving N and crossing NE]

c = 620 + 390 = 1010 [Entering E and cross NE]

d = 460 [others crossing NE section]

$$\therefore \quad p_{NE} = \frac{1010+685}{1010+685+402+460} = 0.663$$

$$p_{ES} = \frac{b+c}{a+b+c+d}$$

$$= \frac{601+680+385+460}{307+601+680+385+460+300} = 0.778$$

$$p_{SW} = \frac{b+c}{a+b+c+d} = \frac{620+460+350+680}{380+620+460+350+680+390} = 0.733$$

$$P_{NW} = \frac{b+c}{a+b+c+d} = \frac{385+402+680+350}{680+350+390+380+620+460} = 0.631$$

$$\text{Hence, capacity of rotary in veh/hour} = \frac{280w\left[1+\frac{e}{w}\right]\left[1-\frac{p}{3}\right]}{\left[1+\frac{w}{l}\right]}$$

$$= \frac{280 \times 14 \times \left[1+\frac{10.5}{14}\right]\left[1-\frac{0.778}{3}\right]}{\left[1+\frac{14}{56}\right]}$$

$$= 4064.8 \text{ veh/hr} \simeq 4065 \text{ veh/hr}$$

Level-2

3.9 The 15 minute-traffic counts on cross roads 1 and 2 during peak hour are observed as 178 and 142 vehicles per lane respectively approaching the intersection in the direction of heavier traffic flow. If the amber times required are 3 and 2 seconds respectively for two lanes based on approach speeds, design the signal timings by trial cycle method. Assume an average time headway of 2.5 seconds during green phase.

[20 Marks]

Sol:

Trial (i)

$$\text{Assume a trial cycle } C_1 = 50 \text{ secs.}$$

$$\text{Number of cycles in 15 mins} = \frac{900}{50} = 18$$

Green time for road 1, allowing an average time headway of 2.5 secs. per vehicle

$$G_1 = \frac{178 \times 2.5}{18} = 24.7 \text{ secs}$$

$$\text{Green time for road 2, } G_2 = \frac{142 \times 2.5}{18} = 19.7 \text{ secs}$$

Amber times A_1 and A_2 are 3 and 2 secs. (given)

$$\text{Total cycle length} = 24.7 + 19.7 + 3.0 + 2.0 = 49.4 \text{ secs}$$

As this is lower than the assumed trial cycle of 50 secs, another lower cycle length may be tried.

Trial (ii)

$$\text{Assume a trial cycle } C_2 = 40 \text{ secs.}$$

$$\text{Number of cycles in 15 mins} = \frac{900}{40} = 22.5$$

$$\text{Green time for road 1, } G_1 = \frac{178 \times 2.5}{22.5} = 19.8 \text{ secs}$$

$$\text{Green time for road 2, } G_2 = \frac{142 \times 2.5}{22.5} = 15.8 \text{ secs}$$

$$\text{Total cycle length} = 19.8 + 15.8 + 3 + 2 = 40.6 \text{ secs}$$

Trial (iii)

$$\text{Assume trial cycle } C_3 = 45 \text{ secs.}$$

$$\text{Number of cycles in 15 min period} = \frac{900}{45} = 20$$

$$\text{Green time for road 1, } G_1 = \frac{178 \times 2.5}{20} = 22.25 \text{ secs}$$

$$\text{Green time for road 2, } G_2 = \frac{142 \times 2.5}{20} = 17.75 \text{ secs}$$

$$\text{Total cycle length} = 22.25 + 17.75 + 3 + 2 = 45.0 \text{ secs}$$

Therefore the trail cycle of 45 secs. may be adopted with the following signal phases:

$G_1 = 22.25$, $G_2 = 17.75$, $A_1 = 3.0$, $A_2 = 2.0$ and Cycle length = 45.0 secs.

Phase-I

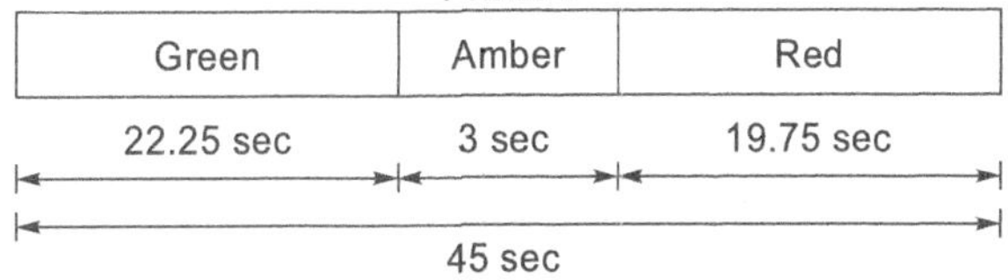

Phase-II

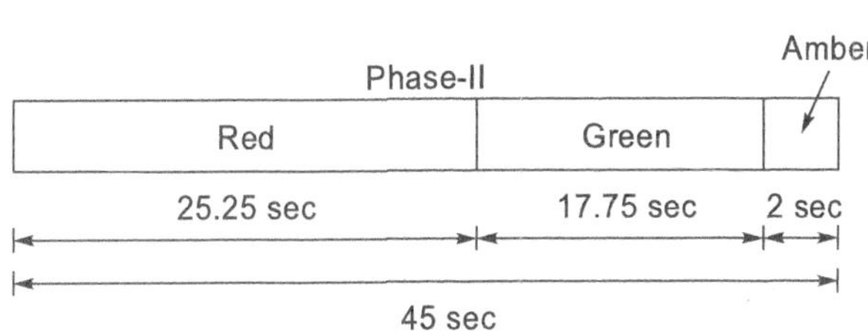

3.10 The following tables give the particulars collected for a section of road 0.7 km long during the source of a moving observer study:

North Bound							
Run No.	Journey Time (minutes)	Stopped Time (minutes)	Vehicles met with in the opposing direction			Vehicles in the same direction	
			Car	Bus	Truck	Over-taking Vehicles	Over-taken Vehicles
1.	1.01	0.04	11	0	5	1	0
2.	0.92	0.10	13	0	0	2	1
3.	0.77	0.08	19	2	11	1	1
4.	1.03	0.14	14	2	4	1	0

South Bound							
Run No.	Journey Time (minutes)	Stopped Time (minutes)	Vehicles met with in the opposing direction			Vehicles in the same direction	
			Car	Bus	Truck	Over-taking Vehicles	Over-taken Vehicles
1.	1.00	0.05	10	0	2	1	1
2.	0.87	0.07	2	0	3	1	0
3.	1.20	0.11	23	1	6	2	1
4.	1.18	0.12	7	0	1	2	0

Calculate the flow in PCU per hour in both directions of traffic assuming an equivalency factor of one for car, 3 for bus and 2 for trucks. Also calculate the journey speed and running speed in kmph.

[20 marks]

Sol:

North Bound							
Run No.	Journey Time (minutes)	Stopped Time (minutes)	Vehicles met with in the opposing direction			Vehicles in the same direction	
			Car	Bus	Truck	Over-taking Vehicles	Over-taken Vehicles
1.	1.01	0.04	11	0	5	1	0
2.	0.92	0.10	13	0	0	2	1
3.	0.77	0.08	19	2	11	1	1
4.	1.03	0.14	14	2	4	1	0
Total	3.73	0.36	57	4	20	5	2

$$\text{PCU} = 57 + 4 \times 3 + 20 \times 2 = 109$$

Average	0.9325	0.09	$\frac{109}{4} = 27.25$	1.25	0.5

South Bound							
Run No.	Journey Time (minutes)	Stopped Time (minutes)	Vehicles met with in the opposing direction			Vehicles in the same direction	
			Car	Bus	Truck	Over-taking Vehicles	Over-taken Vehicles
1.	1.00	0.05	10	0	2	1	1
2.	0.87	0.07	2	0	3	1	0
3.	1.20	0.11	23	1	6	2	1
4.	1.18	0.12	7	0	1	2	0
Total	4.25	0.35	42	1	12	6	2
			PCU = 42 + 1 × 3 + 12 × 2 = 69				
Average	1.0625	0.0875	$\frac{69}{4} = 17.25$			1.5	0.5

The conversion of the traffic into PCUs is done in the above tables.

Average volume, $$q = \frac{x+y}{(t_a+t_w)}$$

and using the suffix *n* for north bound and *s* for south bound traffic,

$$q_n = \frac{(x_s+y_n)}{(t_a+t_w)} \quad \text{...(i)}$$

$$= \frac{17.25+(1.25-0.5)}{1.0625+0.9325} = 9.023 \text{ PCUs/min} = 541.35 \text{ PCUs/hour}$$

$$q_s = \frac{x_n+y_s}{(t_n+t_s)} = \frac{27.25+(1.5-0.5)}{1.0625+0.9325}$$

$$= 14.16 \text{ PCUs/min} = 849.62 \text{ PCUs/hour}$$

$$\overline{t}_n = t_n - \frac{y_n}{q_n} = 0.9325 - \frac{(1.25-0.5)}{9.023} = 0.85 \text{ min} \simeq 51 \text{ sec}$$

$\therefore$ Mean journey speed in north bound direction $= \frac{0.7\times 60}{0.85} = 49.41$ kmph

$$\overline{t}_s = t_s - \frac{y_s}{q_s} = 1.0625 - \frac{1.5-0.5}{14.16} = 0.992 \text{ min} \simeq 59.5 \text{ sec}$$

$\therefore$ Mean journey speed in south bound direction $= \frac{0.70}{0.992}\times 60 = 42.34$ kmph

$\therefore$ Mean Running time in north bound direction

$$\overline{t}_s\text{, stopped time} = 0.85 - 0.09 = 0.76 \text{ min} = 45.6 \text{ sec}$$

$\therefore$ Running speed in north bound direction $= \frac{0.70}{0.76}\times 60 = 55.26$ kmph

Mean running time in south bound direction

$$\overline{t}_s\text{, stopped time} = 0.992 - 0.0875 = 0.9045 \text{ min} = 54.27 \text{ sec}$$

Running speed in south bound direction $= \frac{0.70}{0.9045}\times 60 = 46.43$ kmph

3.11 Spot speed studies carried out at certain stretch of a highway and the data is given below:

Speed range in kmph	0-10	10-20	20-30	30-40	40-50	50-60	60-70	70-80	80-90	90-100	100-110
Observed No. of vehicles	18	25	38	85	236	301	245	122	84	43	3

Determine:

(i) Design speed of the highway.

(ii) Upper and lower speed limits.

[15 Marks]

Sol:

Frequency distribution table is to be prepared. The data mentioned in it are self explanatory.

Speed range in kmph	Mid speed kmph	Frequency, f	% frequency	Cumulative frequency %
0-10	5	18	1.5	1.5
10-20	15	25	2.08	3.58
20-30	25	38	3.17	6.75
30-40	35	85	7.08	13.83
40-50	45	236	19.67	33.5
50-60	55	301	25.08	58.58
60-70	65	245	20.42	79
70-80	75	122	10.17	89.17
80-90	85	84	7.0	96.17
90-100	95	43	3.58	99.75
100-110	105	3	0.25	100.0
		Total : 1200	100.0	

(i) Design speed = 98th percentile speed

$$V_D = 85+\frac{95-85}{99.75-96.17}\times(98.0-96.17)$$

$$= 85+\frac{10}{3.58}\times 1.83 = 90.11 \text{ kmph} \simeq 90 \text{ kmph}$$

(ii) Upper speed limit = 85th percentile speed

$$V_u = 65+\frac{75-65}{89.17-79}\times(85-79) = 70.9 \text{ kmph} \simeq 70 \text{ kmph}$$

Lower speed limit = 15th percentile speed

$$V_L = 35+\frac{45-35}{33.5-13.83}\times(15-13.83) = 35.6 \text{ kmph} \simeq 35 \text{ kmph}$$

3.12 Speed and delay study data is mentioned below conducted by floating car method. The road stretch was 4.0 km long in New Delhi running *E-W*. Determine:

(i) Average vehicles volume.

(ii) Average journey time and speed.

(iii) Average stopped delay.

(iv) Average running time and speed.

Trip S. No.	Direction	Journey time Min : sec	Total Stopped Delay Min : sec	Number of vehicles Overtaking	Number of vehicles Overtaken	No. of vehicles from opposite direction
1	E - W	7 - 12	1 - 30	6	8	312
2	W - E	8 - 18	1 - 40	4	3	186
3	E - W	7 - 30	1 - 20	4	2	361
4	W - E	8 - 26	1 - 50	3	1	200
5	E - W	7 - 22	1 - 40	4	6	343
6	W - E	8 - 48	1 - 40	2	2	210
7	E - W	7 - 32	1 - 10	3	5	328
8	W - E	8 - 32	1 - 30	2	1	196

[20 Marks]

Sol:

Given: Road stretch = 4 km away

Average/ Mean values of speed and delay data:

Direction	Journey time Min : sec	Stopped Delay Min : sec	Number of vehicles		
			Overtaking	Overtaken	In opposite direction
E - W	7 - 12	1 - 30	6	8	312
	7 - 30	1 - 20	4	2	361
	7 - 22	1 - 40	4	6	343
	7 - 32	1 - 10	3	5	328
Total =	29 - 36	5 - 40	17	21	1344
Mean =	7 - 24	1 - 25	4.25	5.25	336
S - N	8 - 18	1 - 40	4	3	186
	8 - 26	1 - 50	3	1	200
	8 - 48	1 - 40	2	2	210
	8 - 32	1 - 30	2	1	196
Total =	34 - 4	6 - 40	11	7	792
Mean =	8 - 31	1 - 40	2.75	1.75	198

(i) East West Direction

n_y = Average number of vehicle (overtaking- overtaken)

= 4.25 − 5.25 = −1.0

n_a = Average number of vehicles during trips in opposite direction from west to east trips = 198

t_w = Average journey time when test vehicle travelling with the stream

= 7 min 24 sec = 7.4 sec

t_a = Average journey time when test vehicle is running against the stream

= 8 min 31 sec = 8.517 min

$$q = \text{Average volume} = \frac{n_a + n_y}{t_a + t_w} = \frac{198 + (-1)}{8.517 + 7.4} = 12.377 \text{ veh/min}$$

$$\bar{t} = \text{Average journey time} = \left(t_w - \frac{n_y}{q}\right)$$

$$= 7.4 - \frac{(-1)}{12.377} = 7.481 \text{ min}$$

$$\text{Average journey speed} = \frac{4}{7.481}\text{km/min} = \frac{4\times 60}{7.481}\text{ km/h} = 32.08 \text{ kmph}$$

$$\text{Average stopped delay} = 1 \text{ min } 25 \text{ sec} = 1.417 \text{ min}$$

$$\text{Average running time} = (\text{Average journey time} - \text{Average stopped delay})$$

$$= 7.481 - 1.417 = 6.064 \text{ min}$$

$$\text{Average running speed} = \frac{4.0\times 60}{6.064}\text{ km/h} = 39.58 \text{ kmph}$$

(ii) West East Direction

$$n_y = (2.75 - 1.75) = 1.0$$

$$n_a = 336$$

$$t_w = 8 \text{ min } 31 \text{ sec} = 8.517 \text{ sec}$$

$$t_a = 7 \text{ min } 24 \text{ sec} = 7.4 \text{ sec}$$

$$q = \frac{n_a + n_y}{t_a + t_w} = \frac{336 + 1.0}{7.4 + 8.517} = 21.17 \text{ veh/min}$$

$$\bar{t} = \left(t_w - \frac{n_y}{q}\right) = 8.517 - \frac{1}{21.17} = 8.47 \text{ min}$$

$$\text{Journey speed} = \frac{4.0\times 60}{8.47}\text{km/h} = 28.34 \text{ kmph}$$

$$\text{Average stopped delay} = 1 \text{ min } 40 \text{ sec} = 1.67 \text{ min}$$

$$\text{Average running time} = 8.47 - 1.67 = 6.8 \text{ min}$$

$$\text{Average running speed} = \frac{4.0\times 60}{6.8} = 35.29 \text{ kmph}$$

3.13 **Explain different road signs required for traffic regulations in India? Draw the figures of the signs.**
[15 Marks]

Sol:

The traffic signs should be backed by law in order to make them useful and effective.

Traffic signs have been divided into three categories according to Indian Motor Vehicles Act.

1. **Regulatory Signs:** Regulatory or mandatory signs are meant to inform the road users of certain laws, regulations and prohibitions, the violation of these signs is a legal offence.

   Some types of regulatory signs are classified as under:

   (a) Stop and Give-way signs
   (b) Prohibitory signs
   (c) No parking and No stopping signs
   (d) Speed limit and vehicle control signs
   (e) Restriction ends sign
   (f) Compulsory direction control and other signs

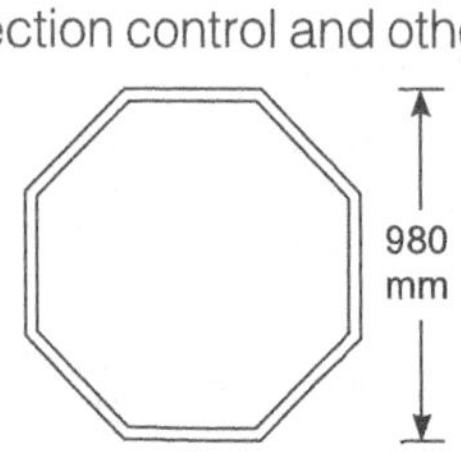

STOP

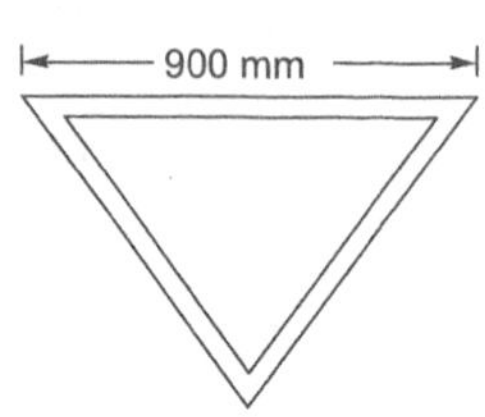

Give way

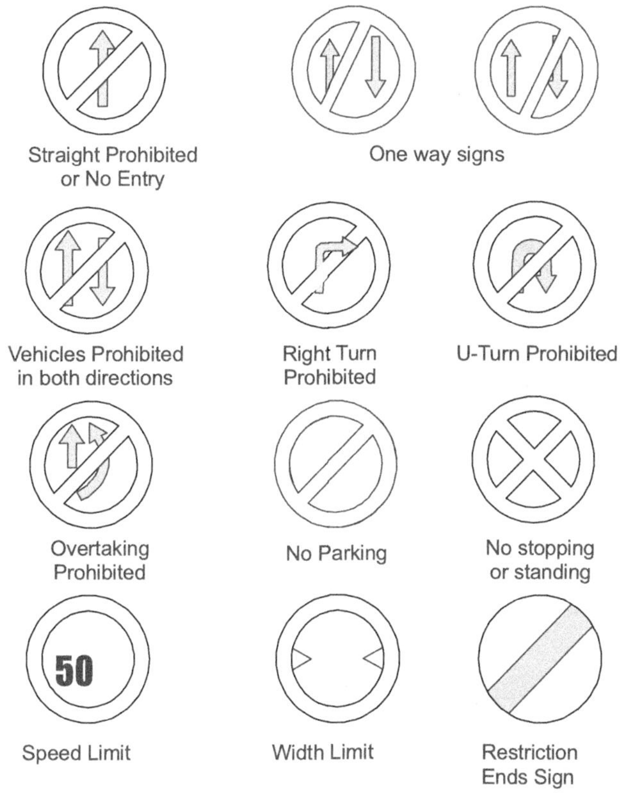

2. **Warning Signs:** Warning or cautionary signs are used to warn the road users of certain hazardous conditions that exist on or adjacent to the roadway. The warning signs are in the shape of equilateral triangle with its apex pointing upwards. These are to be located at sufficient distance in advance of the hazard warned against.

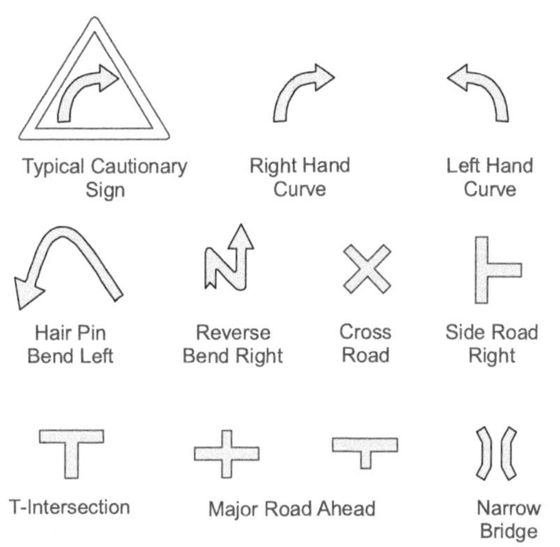

3. **Informatory signs:** These signs are used to guide the road users along routes, inform them of destination and distance and provide with information to make travel easier, safe and pleasant.

   The different types of informatory signs are under:

   (i) Direction and Place Identification Signs

   (ii) Facility Information Signs

   (iii) Parking Signs

   (iv) Flood Gauge

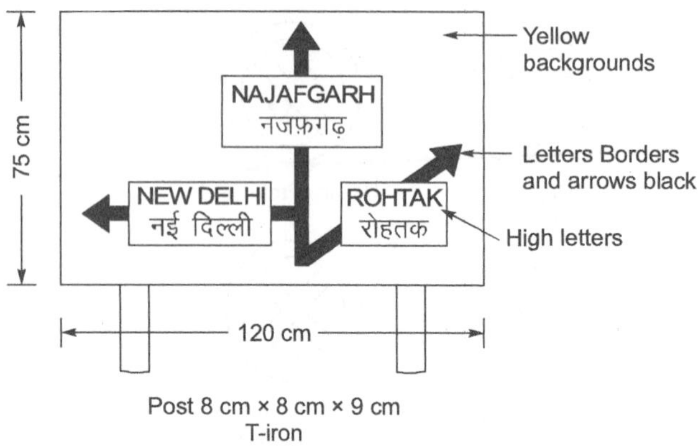

(a) Road Junction Approach

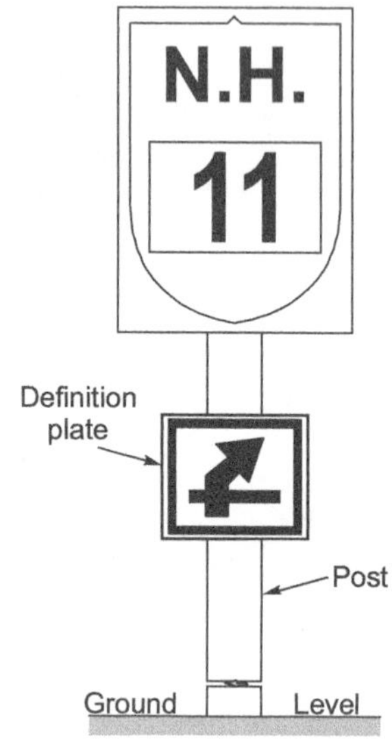

(b) Route Marker Sign

3.14 The width of a carriage way approaching an intersection is given as 15 m. The entry and exit width at the rotary is 10 m. The traffic approaching the intersection from the four sides is shown below. Calculate the capacity of rotary using the given data. [Assume suitable assumptions as per IRC consideration.]

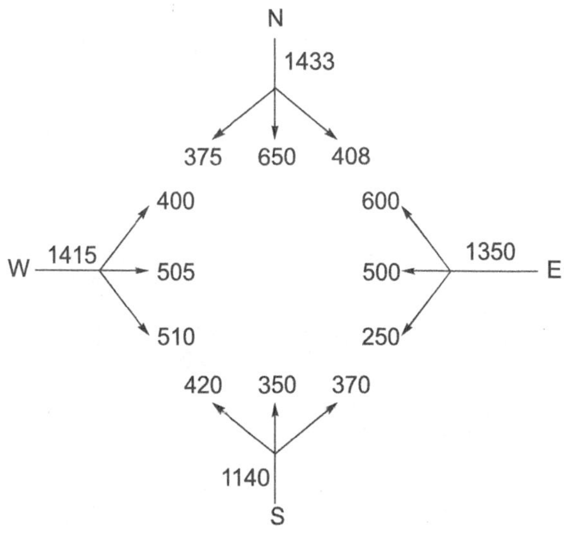

[20 marks]

Sol:

Weaving width, $$W = \left|\frac{e_1 + e_2}{2}\right| + 3.5 = \frac{10+10}{2} + 3.5 = 13.5 \text{ m}$$

$$\text{Weaving length} = 4W = 4 \times 13.5 = 54 \text{ m}$$

Now, from the diagram, we will calculate the proportion of weaving traffic to the non-weaving traffic in all the four approaches.

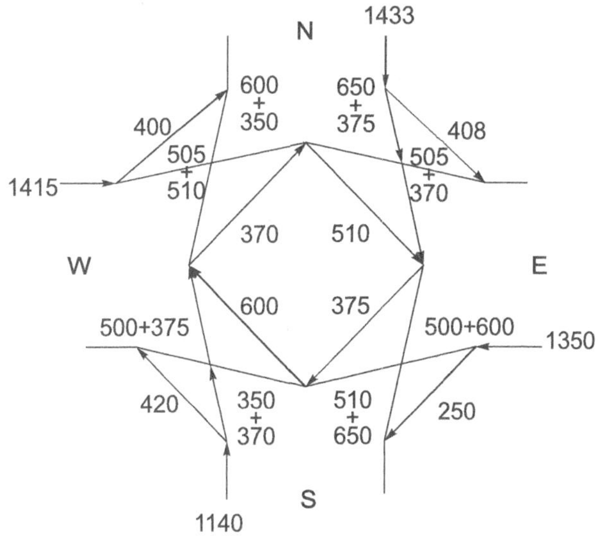

Traffic negotiating a rotary

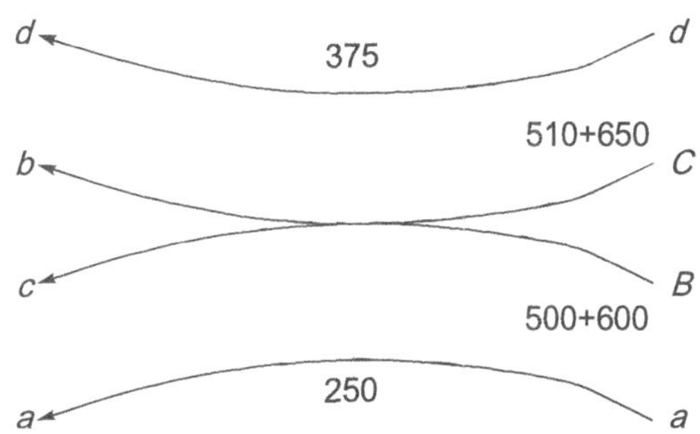

Traffic weaving in East-South direction

It is clear from the equation that highest proportion of weaving traffic to non-weaving traffic will give the minimum capacity.

Let the proportion of traffic in West-North direction be denoted by p_{WN}, in North-East direction as p_{NE}, in East-South direction as p_{ES} and finally in South-West direction as p_{SW}

Hence,

$$p_{ES} = \frac{510+650+500+600}{510+650+500+600+250+375} = \frac{2260}{2885} = 0.783$$

$$p_{WN} = \frac{505+510+350+600}{505+510+350+600+400+370} = \frac{1965}{2735} = 0.718$$

$$p_{NE} = \frac{650+375+505+370}{650+375+505+370+510+408} = \frac{1900}{2818} = 0.674$$

$$p_{SW} = \frac{350+370+500+375}{350+370+500+375+420+600} = \frac{1595}{2615} = 0.61$$

Hence proportion of weaving traffic to non-weaving traffic is highest in East South direction.

Weaving width, $W = \left(\frac{e_1 + e_2}{2}\right) + 3.5 = \left(\frac{10+10}{2}\right) + 3.5 = 13.5 \text{ m}$

Weaving length, $L = 4W \simeq 4 \times 13.5 = 54 \text{ m}$

Therefore, capacity of rotary will be the capacity of this weaving section.

So,
$$Q_{ES} = \frac{280W\left(1+\frac{e}{W}\right)\left(1-\frac{p_{max}}{3}\right)}{\left(1+\frac{W}{L}\right)} = \frac{280 \times 13.5\left(1+\frac{10}{13.5}\right)\left(1-\frac{0.783}{3}\right)}{\left(1+\frac{13.5}{54}\right)}$$
$$= 3890.1 \text{ veh/hr} \simeq 3890 \text{ veh/hr}$$

Alternatively:

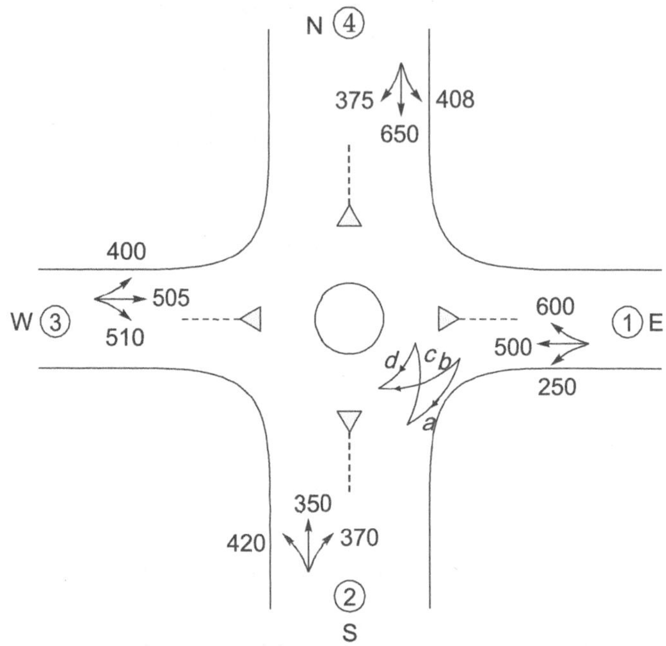

Practical capacity of rotary,
$$Q_p = \frac{280W\left(1+\frac{e}{W}\right)\left(1-\frac{p}{3}\right)}{\left(1+\frac{W}{L}\right)} \quad \text{...(i)}$$

Where, W = Width of weaving section

$$= \left(\frac{e_1 + e_2}{2}\right) + 3.5$$

e_1 = Entry width = 10 m

e_2 = Exit width = 10 m

$$W = \left(\frac{10+10}{2}\right) + 3.5 = 13.5 \text{ m}$$

L = Length of weaving section

= 4 × width of weaving section

$L = 4 \times 13.5 = 54$ m

p = Weaving ratio

$$= \frac{b+c}{a+b+c+d} = \frac{\text{weaving traffic}}{\text{total traffic}}$$

where, b and c are weaving traffic, a and d are non-weaving traffic

Calculation of weaving ratio:

For East-south:

$$a = V_{12} = 250$$
$$b = V_{13} + V_{14} = 500 + 600 = 1100$$
$$c = V_{32} + V_{42} = 510 + 650 = 1160$$
$$d = V_{43} = 375$$
$$p = \frac{1100+1160}{250+1100+1160+375} = 0.783$$

For South-west:

$$a = V_{23} = 420$$
$$b = V_{24} + V_{21} = 350 + 370 = 720$$
$$c = V_{13} + V_{43} = 500 + 375 = 875$$
$$d = V_{14} = 600$$
$$p = \frac{720+875}{420+720+875+600} = 0.61$$

For West-north:

$$a = V_{34} = 400$$
$$b = V_{32} + V_{31} = 505 + 510 = 1015$$
$$c = V_{24} + V_{14} = 350 + 600 = 950$$
$$d = V_{21} = 370$$
$$p = \frac{1015+450}{400+1015+950+370} = 0.718$$

For North-west:

$$a = V_{41} = 408$$
$$b = V_{42} + V_{43} = 650 + 375 = 1025$$
$$c = V_{31} + V_{21} = 505 + 370 = 875$$
$$d = V_{32} = 510$$
$$p = \frac{1025+875}{408+1025+875+510} = 0.674$$

It is clear from eq. (i) that the highest proportion of weaving traffic to non-weaving traffic will give minimum capacity

$$p = 0.783$$

Practical capacity of rotary

$$Q_p = \frac{280\times 13.5\left(1+\frac{10}{13.5}\right)\left(1-\frac{0.783}{3}\right)}{\left(1+\frac{13.5}{54}\right)}$$

$$= 3890.096 \text{ pcu/hr} \simeq 3890 \text{ pcu/hr}$$

3.15 A four-legged right angled intersection is to be signalised with a fixed time two phase signal. The design hourly flow and saturation flow are as under:

	North (N)	South (S)	East (E)	West (W)
Design hourly flow	900	500	800	700
Saturation flow	2500	2000	3200	3000

The lost time may be taken as 2 seconds per arm. Determine the optimum cycle time and apportion the green times in the two phases. Sketch the timing diagram for each phase.

[15 marks]

Sol:

$$\frac{f_N}{S_N} = \frac{900}{2500} = 0.36$$

$$\frac{f_S}{S_S} = \frac{500}{2000} = 0.25$$

$\therefore$ Maximum value of $\frac{f}{S}$ in the N-S direction, $X_{NS} = 0.36$

$$\frac{f_E}{S_E} = \frac{800}{3200} = 0.25$$

$$\frac{f_W}{S_W} = \frac{700}{3000} = 0.233$$

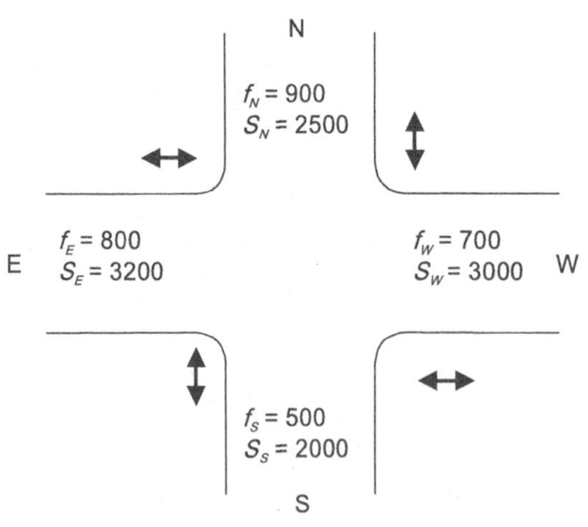

$\therefore$ Maximum value of $\frac{f}{S}$ in E-W direction, $X_{EW} = 0.25$

$$T = \text{Lost time due to starting and stopping}$$
$$= 2 + 2 + 2 + 2 = 8 \text{ sec}$$

Signal design by Webester's method is the optimum cycle length design.

Optimum cycle time, $C_0 = \frac{1.5T+5}{1-X}$

$$= \frac{1.5\times8+5}{1-(0.36+0.25)} = \frac{17}{1-0.61} = \frac{17}{0.39}$$

$$= 43.6 \text{ sec} \simeq 44 \text{ sec (say)}$$

$\therefore$ Effective green time per cycle $= C_0 - T$

$$= 44 - 8 = 36 \text{ seconds}$$

Effective green time per phase is given by

$$G_{NS} = \frac{X_{NS}}{X}(C_0 - T)$$

$$= \frac{0.36}{0.61}\times 36 = 21.25 \text{ sec} \simeq 21 \text{ sec (say)}$$

$$\therefore \quad G_{EW} = \frac{X_{EW}}{X}\times(C_0 - T) = \frac{0.25}{0.61}\times 36$$

$$= 14.75 \text{ sec} \simeq 15 \text{ sec (say)}$$

The cycle time diagrams are drawn below:

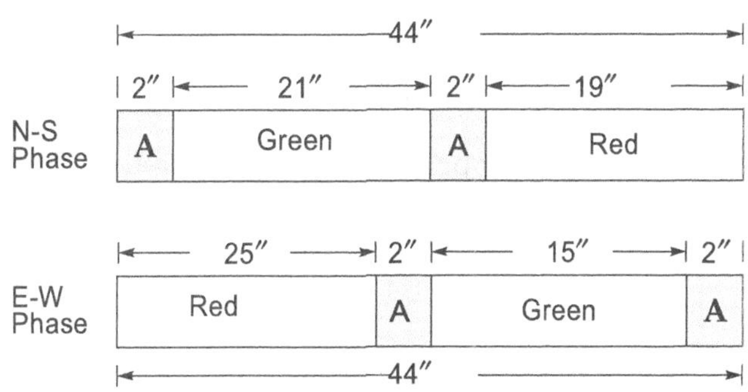

Cycle time diagram

4. Highway Materials and Pavement Design and Hill Roads

Level-1

4.1 The load penetration data from a California Bearing Ratio (CBR) test is provided in the following table. Indicate whether any correction is required for the calculated CBR value. Find the CBR value of the soil from the data provided.

Table: Load penetration data

Penetration(inmm)	0	0.5	1.0	1.5	2.0	2.5	3.0	4.0	5.0	7.5	10	125
Load in kgf(kg force)	0	4	13	29	40	50	58	70	78	93	103	112

Area of plunger is given as 19.6 cm^2. Pressure for standard crushed stones at 2.5 mm and 5.0 mm are 70 kgf/cm^2 and 105 kg/cm^2 respectively.

[10 Marks]

Sol:

The graphical plot is given here. From the graph we find correction is needed to calculate CBR value. That are that in graph

After correction, load at 2.5 mm penetration = 51 kg

5 mm penetration = 83 kg

After of plunger = 19.6 cm^2

$$\text{Pressure for 2.5 mm penetration} = \frac{51}{19.6} \text{kg/cm}^2$$

$$\text{Pressure for 5.0 mm penetration} = \frac{83}{19.6} \text{kg/cm}^2$$

(i) CBR value of soil at 2.5 mm

$$= \frac{\text{Pressure of plunger at 2.5 mm penetration of soil}}{\text{Pressure for standard crushed store at 2.5 mm}} \times 100$$

$$= \frac{51}{19.6} \times \frac{100}{70} = 3.71\%$$

$$\text{(ii)CBR of soil at 5.0 mm} = \frac{83}{19.6} \times \frac{100}{105} = 4\%$$

Hence adopt CBR value = 4%

Normally the CBR value at 2.5 mm penetration which is higher than that at 5.0 mm is reported as the CBR value of the material. However, if the CBR value obtained from the test at 5.0 mm penetration is higher than that at 2.5 mm, then the test is to be repeated for checking. If the check test again gives similar results, the higher value obtained at 5.0 mm penetration is reported as the CBR value.

4.2 **The specific gravities and weight proportions for aggregates and bitumen are as under for the preparation of Marshall moulds:**

	Aggregate-1	Aggregate-2	Aggregate-3	Aggregate-4	Bitumen
Weight(gm)	825	1200	325	150	100
Specific Gravity	2.63	2.51	2.46	2.43	1.045

The volume and weight of one Marshall mould was found to be 475 cc and 1100 gm. Assuming absorption of bitumen is aggregate is zero, find

(a) percentage air voids (V_v)

(b) percentage bitumen by volume (C_b)

(c) percentage and voids in mineral aggregates (VMA)

[10 Marks]

Sol:

(a) Percentage air void

$$V_v = \frac{G_t - G_m}{G_m} \times 100$$

Where,

G_m = bulk density or mass density of the specimen

G_t = theoretical specific gravity of the mixture

$$G_t = \frac{100}{\frac{W_1}{G_1} + \frac{W_2}{G_2} + \frac{W_3}{G_3} + \frac{W_4}{G_4}}$$

Where, W = % weight of aggregate

$$= \frac{100}{\frac{8.25}{2.63}+\frac{12}{2.51}+\frac{3.25}{2.46}+\frac{1.5}{2.43}}$$

$$= \frac{100}{3.14+4.78+1.32+0.62} = 10.14$$

$$G_m = \frac{1100}{475} = 2.315$$

$$V_v = \frac{1014-2.34}{2.34} = 3.33\%$$

(b)Percentage bitumen by vacuum,

$$V_b = G_m \frac{W_4}{G_4}$$

$$= 2.35 \times \frac{1.5}{2.43} = 1.45\%$$

(c) $\quad VMA = V_v + V_b$

$$= 3.33 + 1.43 = 4.78\%$$

4.3 Calculate the voids filled with bitumen (VFB) in % for a marshall sample with following data:

- Bulk specific gravity of mix = 2.292
- Bulk specific gravity of aggregate = 2.486
- % of bitumen by total weight = 4.72%
- Specific gravity of bitumen = 1.12
- Theoretical maximum S.G. of mix = 2.398

[8 Marks]

Sol:

Given: $G_m = 2.292$, Gb = 1.12, $W_b\% = 4.72\%$, $G_t = 2.398$

$$V_v = \frac{G_t - G_m}{G_t} \times 100 = \frac{2.398-2.292}{2.398} \times 100 = 4.42\%$$

$$V_b = \frac{W_b\%}{G_b} \times V_b = \frac{4.72}{1.12} \times 2.292 = 9.66\%$$

Voids in mineral aggregate, $VMA = V_V + V_b$

$$= 4.42 + 9.66 = 14.08\%$$

Voids filled with bitumen, $VFB = \frac{V_b\%}{VMA} \times 100 = \frac{9.66}{14.08} \times 100 = 68.71\%$

4.4 Marshall mix design was prepared as per the table below. The volume and weight of one specimen was 500 cc and 1200 g. Find V_v, V_b, VMA and VFB, assuming absorption of bitumen in aggregate as zero.

Item	Agg. I	Agg. II	Agg. III	Agg. IV	Bitumen
Weight	875	1275	345	160	120
Specific gravity	2.65	2.55	2.45	2.40	1.1

[8 Marks]

Sol:

$$\text{Theoretical specific gravity, } G_t = \frac{\Sigma W_i}{\Sigma \frac{W_i}{G_i}} = \frac{875+1275+345+160+120}{\frac{875}{2.65}+\frac{1275}{2.55}+\frac{345}{2.45}+\frac{160}{2.4}+\frac{120}{1.1}} = 2.419$$

$$\text{Mass specific gravity, } G_m = \frac{\text{Weight of marshall specimen}}{\text{Volume of marshall specimen}}$$

$$= \frac{1200}{500} = 2.40$$

$$\text{Percentage air voids, } V_v = \frac{G_t - G_m}{G_t} \times 100$$

$$= \frac{2.419 - 2.4}{2.419} \times 100 = 0.79\%$$

$$\text{Percentage volume of bitumen} = \frac{G_m}{G_b} \times W_b\% = \frac{2.40}{1.1} \times \frac{120}{2775} \times 100 = 9.435\%$$

$$\text{Voids in mineral aggregate} = V_v + V_b = 0.79 + 9.435 = 10.225\%$$

$$\text{VMA} = 0.225\%$$

$$\text{Voids filled with bitumen} = \text{VFB} = \frac{V_b\%}{VMA} \times 100$$

$$= \frac{9.435}{10.225} \times 100 = 92.27\%$$

4.5 Write a short note on drainage works on hill roads.

[12 Marks]

Sol:

Drainage can categorized in 2 parts:

(a) Surface drainage

(b) Sub-surface drainage

(a) Surface Drainage system on hill slopes. Road side drains are constructed and planned to drain the surface water from roadway surface only which causes severe damage to roadway. Hence water flowing from the hill side towards the road side drain should get intercepted at high level by catch water drains. The water from catch water drain is diverted to slopping drains and then taken across the hill roads through suitable cross drainage structures, such as culverts or scuppers. Surface drainage system comprises of following parts:

- Catch water drains
- Slopping drains
- Road side drains
- Gross drainage (as the site condition allows)

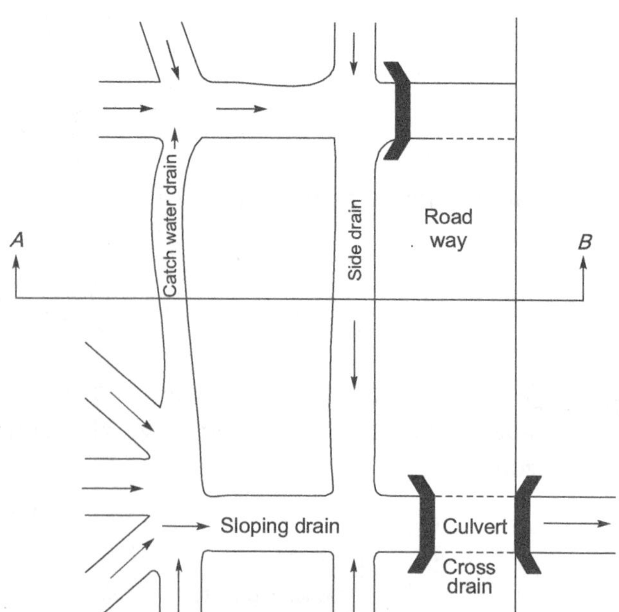

PLAN

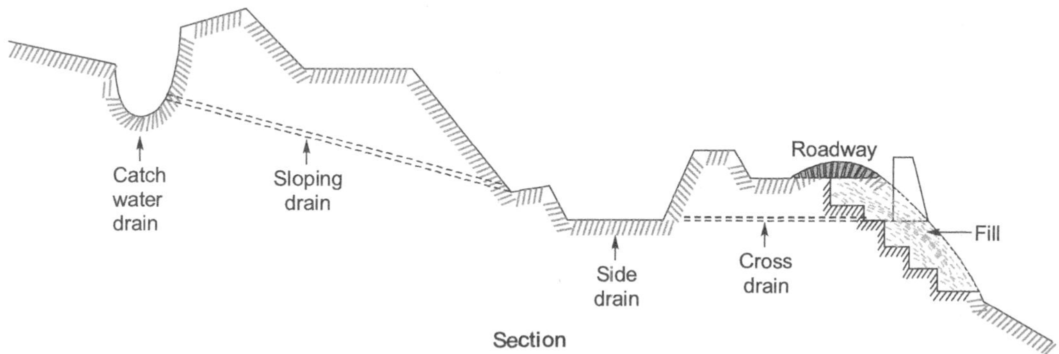

Section

(b) **Subsurface drainage** seepage flow of water on hilly roads is a great trouble. The water seeping in may also cut across the hill-side slope above, at or below the road level depending on

(i) Depth of hard stratum

(ii) Inclination of the stratum

(iii) Rainfall magnitude

(iv) Quantity of under ground flow

Seepage caused weakening of road bed and pavement. It can be countered by providing stone pitching, riprap or hand placed roaks, soil stabilization techniques and use of geosynthetics may also help.

4.6 **A cement concrete pavement has a thickness of 22 cm and has 2 lanes of total width 7.2 m, with longitudinal joint. Design the dimension and spacing of the bars, using the following data :**

1. **Allowable working stress in steel in tension = 1400 kg/cm^2.**
2. **Unit weight of concrete = 2500 kg/m^3.**
3. **Coefficient of friction = 1.5.**
4. **Allowable bond stress is concrete = 24.6 kg/cm^2.**

[10 marks]

Sol:

(i)

Area of longitudinal reinforcement per metre length of slab is given by

$$A_s = \frac{fbhw}{100s_s}$$

where, A_s = Area of longitudinal steel (cm^2)

f = Coefficient of friction = 1.5

b = Distance between joint and nearest free edge (m) = $\frac{7.2}{2} = 3.6$ m

h = Thickness of slab (in cm) = 22 cm

w = Unit weight of concrete (in kg/m^3) = 2500 kg/m^3

s_s = Allowable working stress in tension in steel (in kg/cm^2) = 1400 kg/cm^2

s_b = Allowable bond stress in concrete (in kg/cm^2) = 24.6 kg/cm^2

$$\therefore \quad A_s = \frac{1.5 \times 3.6 \times 22 \times 2500}{100 \times 1400} = 2.12 \text{ cm}^2$$

Assuming the diameter of the tie bar as 1 cm, number of tie bars required per metre will be

$$n = \frac{2.12}{\frac{\pi}{4} \times 1^2} = 2.699$$

$\therefore$ Spacing of the tie bars $= \dfrac{1000}{2.699} = 370.51 \text{ mm} \simeq 370 \text{ mm}$

Alternatively,

$$\text{Spacing of 1 cm diameter tie bars} = \frac{1000 \times \frac{\pi}{4} \times 1^2}{2.12} = 370.47 \simeq 370 \text{ mm} \quad \text{(say)}$$

Now, length of the tie bar, $L_t = \dfrac{ds_s}{2s_b}$

$$\Rightarrow \quad L_t = \frac{1 \times 1400}{2 \times 24.6} = 28.5 \text{ cm say 30 cm}$$

$\therefore$ Provide 1 cm diameter bars of length 30 cm at a spacing of 370 mm c/c.

4.7 Design the pavement section by triaxial section.
Given : Wheel load, $P = 4500$ kg, Radius of contact, $a = 15$ cm, Traffic coefficient, $x = 1.7$, Rainfall coefficient, $y = 0.85$, Design deflection, $\Delta = 0.40$ cm, E-value of subgrade soil, $E_s = 120$ kg/cm^2, E-value of base-course material, $E_b = 425$ kg/cm^2, E-value of 9 cm thick bituminous concrete surface course, $E_c = 1100$ kg/cm^2.

[12 Marks]

Sol:

Let the pavement to consist of single layer of base course material only. The pavement thickness is given by

$$T_P = \left[\sqrt{\left(\frac{3P \times y}{2\pi E_s \delta}\right)^2 - a^2}\right]\left(\frac{E_s}{E_b}\right)^{1/3}$$

$$= \left[\sqrt{\left(\frac{3 \times 4500 \times 1.7 \times 0.85}{2 \times 3.14 \times 120 \times 0.40}\right)^2 - 15^2}\right]\left(\frac{120}{425}\right)^{1/3} = 41.30 \text{ cm}$$

Let thickness t_b of base course be equivalent to 9 cm bituminous concrete surface of $E_c = 1100$ kg/cm^2

$$\therefore \quad \frac{t_b}{t_c} = \left(\frac{E_c}{E_b}\right)^{1/3}$$

$$t_b = t_c \cdot \left(\frac{E_c}{E_b}\right)^{1/3} = 9.0 \times \left(\frac{1100}{425}\right)^{1/3} = 12.36 \text{ cm}$$

Hence, base course thickness required = 41.30 – 12.36 = 28.94 cm

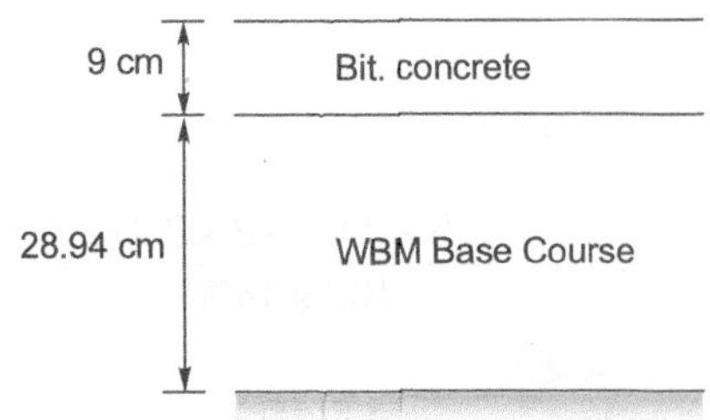

Level-2

4.8 CBR test was conducted on a soil subgrade to produce the following results:

Load	Penetration
89 kg	2.5 mm
115 kg	5.0 mm

Use the following materials and design the pavement
(i) Compacted soil CBR = 8%
(ii) Poorly graded gravel CBR = 14%
(iii) Well graded gravel
(iv) Bituminous surface of thickness 5 cm
Take wheel load of 5000 kg and tyre pressure as 7 kg/cm^2.

[12 Marks]

Sol:

P = 5000 kg, p = 7 kg/cm^2

Load (kg)	Penetration (mm)	Standard load (kg)	CBR %
89	2.5 mm	1370	6.50
115	5.0 mm	2055	5.60

CBR_{design} = Maximum value of the above two i.e. 6.50

Now, thickness of pavement above soil subgrade

$$= \sqrt{\frac{1.75P}{CBR} - \frac{P}{p\pi}} = \sqrt{\frac{1.75 \times 5000}{6.50} - \frac{5000}{7 \times 3.14}} = 33.45 \text{ cm}$$

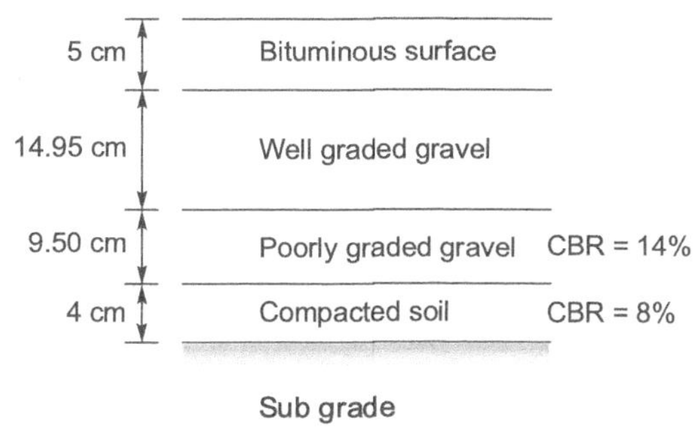

Thickness of pavement above compacted soil

$$= \sqrt{\frac{1.75 \times 5000}{8} - \frac{5000}{7 \times 3.14}} = 29.45 \text{ cm}$$

$\therefore$ Thickness of compacted soil = 33.45 – 29.45 = 4.0 cm

Thickness of pavement above poorly graded gravel $= \sqrt{\frac{1.75 \times 5000}{14} - \frac{5000}{7 \times 3.14}} = 19.95$ cm

$\therefore$ Thickness of poorly graded gravel = 29.45 – 19.95 = 9.50 cm

$\therefore$ Thickness of well graded gravel = 33.45 – 4.0 – 9.50 – 5 = 14.95 cm

4.9 Calculate the wheel load stresses using the data given below, at

(i) Interior

(ii) Edge

(iii) Corner regions of stress equations. Also calculate the location where the crack is likely to develop due to corner loading.

Wheel load, P = 5600 kg, Modulus of elasticity of cement concrete, $E = 3.1 \times 10^5$ kg/cm^2, Pavement thickness, h = 20 cm, Poisson ratio of concrete, μ = 0.15, Modulus of subgrade reaction, K = 5.5 kg/cm^3, Radius of contact area, a = 20 cm.

[20 Marks]

Sol:

Radius of relative stiffness (l) is given by

$$l = \left[\frac{Eh^3}{12k(1-\mu^2)}\right]^{1/4} = \left[\frac{3.1\times 10^5 \times 20^3}{12\times 5.5\times(1-0.15^2)}\right]^{1/4} = 78.74 \text{ cm}$$

Equivalent radius of resisting section is given by

$$b = \sqrt{1.6a^2 + h^2} - 0.675h,$$

when $\frac{a}{h} < 1.74$, $\frac{20}{20} = 1 \therefore < 1.74$

Hence,
$$b = \sqrt{1.6\times 20^2 + 20^2} - 0.675\times 20 = 18.75 \text{ cm}$$

(i) Stress at interior

$$S_i = \frac{0.316P}{h^2}\left[4\log_{10}\left(\frac{l}{b}\right) + 1.069\right]$$

$$= \frac{0.316\times 5600}{20^2}\left[4\log_{10}\left(\frac{78.74}{18.75}\right) + 1.069\right]$$

$$= 15.76 \text{ kg/cm}^2$$

(ii) Stress at edge

$$S_e = \frac{0.572}{h^2}\left[4\log_{10}\left(\frac{l}{b}\right) + 0.359\right]$$

$$= \frac{0.572}{20^2}\left[4\log_{10}\left(\frac{78.74}{18.75}\right) + 0.359\right] = 22.84 \text{ kg/cm}^2$$

(iii) Stress at corner

$$S_c = \frac{3P}{h^2}\left[1-\left(\frac{a\sqrt{2}}{l}\right)^{0.6}\right] = \frac{3\times 5600}{20^2}\left[1-\left(\frac{20\sqrt{2}}{78.74}\right)^{0.6}\right]$$

$$= 19.28 \text{ kg/cm}^2$$

Location where corner load crack develops:

The distance of crack from corner of the slab, x

$$x = 2.58\sqrt{l\cdot a} = 2.58\sqrt{78.74\times 20} = 102.38 \text{ cm}$$

4.10 **Design dowel bars at an expansion joint of concrete pavement of thickness 25 cm. Given the radius of relative stiffness of 100 cm, design wheel load 5000 kg, load capacity of dowel system is 40% of design wheel load. Joint width is 3 cm and permissible stress in shear, bending and the bearing stress in dowel bars are 900 kg/cm², 1600 kg/cm² and 120 kg/cm² respectively.**

[20 Marks]

Sol:

Length of dowel = L

Assume diameter of dowel bar, d = 25 mm i.e. 2.5 cm

Given : Design wheel load, P = 5000 kg, Radius of relative thickness, l = 100 cm, Thickness of pavement, h = 25 cm, Joint width, Δ = 3 cm, Permissible stress in shear, σ_s = 900 kg/cm^2, Permissible stress in bending, σ_b = 1600 kg/cm^2, Permissible stress in bearing, σ_{br} = 120 kg/cm^2, Load capacity of dowel system = 40% of P = 2000 kg.

In order to obtain balanced design for equal capacity in bending and bearing, the length of dowel bars is obtained by,

$$L_d = 5d\left[\frac{\sigma_b}{\sigma_{br}} \times \frac{L_d + 1.5\Delta}{L_d + 8.8\Delta}\right]^{1/2}$$

$$= 5 \times 2.5\sqrt{\frac{1600}{120} \times \frac{L_d + 1.5 \times 3}{L_d + 8.8 \times 3}}$$

$$= 12.5\sqrt{13.33 \times \frac{L_d + 4.5}{L_d + 26.4}}$$

Using, trial and error method

Taking, L_d = 40 cm

$$L_d = 12.5\sqrt{13.33 \times \frac{40 + 4.5}{40 + 26.4}} = 37.36 \text{ cm} < 40 \text{ cm}$$

Taking L_d = 35 cm

$$L_d = 12.5\sqrt{13.33 \times \frac{35 + 4.5}{35 + 26.4}} = 36.6 \text{ cm} > 35 \text{ cm}$$

Taking L_d = 36 cm

$$L_d = 12.5\sqrt{13.33 \times \frac{36 + 4.5}{36 + 26.4}} = 36.77 \text{ cm} > 36 \text{ cm}$$

∴ Total length of embedment, L_d = 36 cm

Minimum dowel bar length required = $(L_d + \Delta)$ = 39 cm

∴ Provide 2.5 cm diameter dowel bars of 40 cm length.

Load transfer capacity of single dowel bar, P'

- $P(\text{shear}) = 0.785 d^2 \sigma = 0.785 \times 2.5^2 \times 900 = 4415.6$ kg

  Actual value of $L_d = 40 - \Delta = 40 - 3 = 37$ cm

- $$P(\text{Bending}) = \frac{2d^3\sigma_b}{L_d + 8.8\Delta} = \frac{2 \times 2.5^3 \times 1600}{36 + 8.8 \times 3} = 801.28 \text{ kg}$$

- $$P(\text{Bearing}) = \frac{\sigma_{br} \times L_d^2 \times d}{12.5(L_d + 1.5\Delta)} = \frac{120 \times 36^2 \times 2.5}{12.5(36 + 1.5 \times 3)} = 768 \text{ kg}$$

P(Design) = Least of the above 3 values i.e. P(Design) = 768 kg

As given, load capacity of dowel group = $0.4P$ = 2000 kg

Required capacity factor for dowel group = $\frac{2000}{768} = 2.604$

Spacing of dowel bars

Effective distance upto which there is load transfer = 1.8 × radius of relative stiffness

$$= 1.8 \times l = 1.8 \times 100 = 180 \text{ cm}$$

Assume trial spacing of 28 cm between the dowel bars, the capacity factor available for the group

$$= 1 + \frac{180-30}{180} + \frac{180-60}{180} + \frac{180-90}{180} + \frac{180-120}{180} + \frac{180-150}{180} = 3.5$$

Since 3.5 > Capacity factor of 2.604.

∴ Let us assume spacing = 40 cm

$$\text{Capacity factor for group} = 1 + \frac{180-40}{180} + \frac{180-80}{180} = \frac{180-120}{180} + \frac{180-160}{180}$$

$$= 2.78 > 2.604, \text{ but close}$$

∴ Provide 2.5 cm diameter bars of total length of 40 cm at 40 cm spacing.

4.11 Estimate the wheel load stress at interior and edge of cement concrete pavement of 231 cm thickness, using stress coefficient.

Modulus of elasticity of concrete:2 × 10^7 kN/m^2

Poisson's ratio of concrete	:	**0.15**
Modulus of subgrade	:	**3.6 × 10^4 kN/m^3**
Wheel load	:	**40.8 kN**
Tyre pressure	:	**600 kN/m^2**

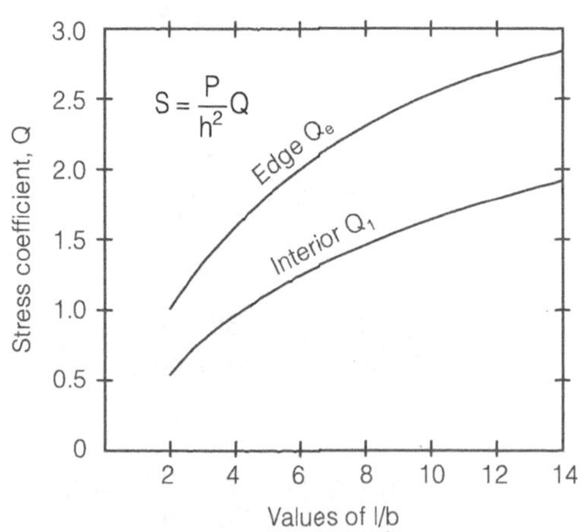

[15 Marks]

Sol:

$$S_i = Q_i \frac{P}{h^2}$$

As we know

$$l = \left[\frac{Eh^3}{12k(1-\mu^2)}\right]^{1/4}$$

$E = 2 \times 10^7$ kN/m^2

$h = 21$ cm $= 0.21$ m

$k = 3.6 \times 10^4$ kN/m^3

$\mu = 0.15$

$$\therefore \quad l = \left[\frac{2\times10^7\times(0.21)^3}{12\times3.6\times10^4\times(1-0.15^2)}\right]^{1/4}$$

$$= 0.81 \text{ m} = 81 \text{ cm}$$

Wheel load, P = 40.8 kN

Tyre pressure, p = 600 kN/m^2

Let radius of wheel load = a

$$\therefore \quad p = \frac{P}{\pi a^2}$$

$$\therefore \quad a = \sqrt{\frac{P}{\pi p}} = \sqrt{\frac{40.8}{\pi\times600}} = 0.147 \text{ m} = 14.7 \text{ cm}$$

Now,

$$1.724\,h = 1.724\times0.21$$

$$= 0.36 \text{ m} = 36 \text{ cm}$$

$$\Rightarrow \quad a < 1.724\,h$$

$$\therefore \quad b = \sqrt{1.6a^2+h^2} - 0.675h$$

$$= \sqrt{1.6\times(14.7)^2+21^2} - 0.675\times21 = 13.87 \text{ cm}$$

$$\therefore \quad \frac{l}{b} = \frac{81}{13.87} = 5.84$$

Now, from figure for $\frac{l}{b} = 5.84$

Interior stress coefficient

$$Q_i = 1.3$$

$$Q_e = 2$$

$\therefore$ Wheel load stress

$$S_i = \frac{P}{h^2}\times Q_i = \frac{40.8}{(0.21)^2}\times1.3$$

$$= 1202.72 \text{ kN/m}^2$$

$$S_e = \frac{40.8}{(0.21)^2}\times2$$

$$= 1850.3 \text{ kN/m}^2$$

4.12 The width of expansion joint gap is 2.5 cm in a cement concrete 20 cm thick pavement. If the laying temperature is 15°C and the maximum slab temperatures in the summer is 55°C, calculate

(a) the spacing between expansion joints, and

(b) the spacing between contraction joints

Coefficient of thermal expansion for concrete is 10×10^{-6} per degree centigrade. Ultimate stress in tension in cement concrete is 1.6×10^5 N/m^2. Ultimate tensile stress in steel is 1200×10^5 N/m^2. Factor of safety is to be taken as 2. Assume the pavement width to be 3.5 m. Unit weight of steel is 75000 N/m^3. Total reinforcement of 6 kg/m^2 is provided in the slab.

[12 Marks]

Sol:

Given data:

Width of expansion joint gap = 2.5 cm

Laying temperature t_1 = 15°C

Summer temperature t_2 = 55°C

Coefficient of thermal expansion

$$\alpha = 10 \times 10^{-6}/°C$$

∴ Spacing of expansion joint

$$L_e = \frac{\delta'}{100\alpha(t_2 - t_1)} = \frac{(2.5/2)}{100 \times 10 \times 10^{-6}(55-15)} = \frac{1.25}{10^{-3} \times 40} = 31.25 \text{ m}$$

Spacing of contraction joint

$$L_c = \frac{200 S_s A_s}{bhwf}$$

Here, pavement width

$$d = 3.5 \text{ m}$$

Reinforcement provided = 6 kg/m^2

Ultimate stress in concrete

$$S_c = 1.6 \times 10^5 \text{ N/m}^2$$

Ultimate stress in steel

$$= 1200 \times 10^5 \text{ N/m}^2$$

$$\text{FOS} = 2$$

$$\therefore \quad S_s = \frac{1200 \times 10^5}{2}$$

$$= 600 \times 10^5 \text{ N/m}^2$$

Unit weight of steel

$$= 7500 \text{ kg/m}^3$$

Assuming one meter length of slab, then the total cross-sectional area of steel A_s in one direction along slab is given by

$$b \times \frac{6}{2} = \frac{A_s \times 7500}{10^4}$$

or

$$A_s = \frac{3.5 \times 6 \times 10^4}{2 \times 7500} = 14 \text{ cm}^2$$

$$L_c = \frac{200 \times 600 \times 14}{35 \times 20 \times 2400 \times 1.5} = 6.66 \text{ mm} \approx 6.5 \text{ m c/c}$$

4.13 **A portion of a highway is to be constructed with 25 cm thick plain cement concrete slab. The design traffic intensity is estimated to be 3000 commercial vehicles per day. Using the data given below, check the adquacy of the slab thickness as per IRC.**

Dimensions of slab = 4.5 m × 3.5 m

Design wheel load, P = 5100 kg

Design tyre pressure, P = 7.2 kg/cm^2

Foundation strength, k = 6 kg/cm^3

Flexural strength of concrete = 40 kg/cm^2

Elastic modulus value of concrete,

E = 3.0 × 10^5 kg/cm^3

Poisson's ratio of concrete, μ = 0.15

Coefficient of thermal expansion of concrete, α = 10 × 10^{-6}/°C

Maximum value of temperature differential in the slab, Δt = 15°C
Radius of relative stiffness, l = 90.3 cm
Radius of tyre contact area, a = 15 cm
Radius of equivalent distribution of pressure, b = 14.5 cm
Use the following expressions whereever necessary:
Edge stress (due to load) in kg/cm²

$$= 0.529\frac{P}{h^2}(1+0.54\mu)\left(4\log_{10}\frac{l}{b}b-0.4048\right)$$

Corner stress (due to load) in kg/cm² $= \frac{3P}{h^2}\left[1-\left(\frac{a\sqrt{2}}{l}\right)^{1.2}\right]$

Edge stress (due to temperature differential) in kg/cm² = $E\alpha\frac{\Delta t}{2}C$

Bradbury coefficients (C) may be taken as 0.72.

[15 Marks]

Sol:

Edge stress (due to temperature difference) in kg/cm²

$$= E\alpha\frac{\Delta t}{2}C$$

Where, C = Bradbury Coefficient = 0.72

$$\Delta t = 15°C$$
$$\alpha = 10\times10^{-6}/°C$$
$$E = 3\times10^5 \text{ kg/cm}^2$$

Edge stress

$$= 3\times10^5\times10\times10^{-6}\times15\times\frac{0.72}{2}$$
$$= 16.2 \text{ kg/cm}^2 \text{ due to temp. difference}$$

Flexural strength of concrete

$$= 40 \text{ kg/cm}^2$$

(i) Edge stress (due to load)

$$= 0.529\frac{P}{h^2}(1+0.54\mu)\left(4\log_{10}\frac{l}{b}+\log_{10}b-0.4048\right)$$
$$= 0.529\times\frac{5100}{(25)^2}[1+0.54\times0.15]\left[4\log_{10}\frac{90.3}{14.5}+\log_{10}14.5-0.4048\right]$$
$$= 18.335 \text{ kg/cm}^2$$

The edge stress due to load and temperature difference

$$= 18.355 + 16.2 = 34.555 \text{ kg/cm}^2 < 40 \text{ kg/cm}^2$$

Therefore, thickness of slab is adequate to resist edge stress.

(ii) Corner stress due to load

$$= \frac{3P}{h^2}\left[1-\left(\frac{a\sqrt{2}}{l}\right)^{1.2}\right] = \frac{3\times5100}{(25)^2}\left[1-\left(\frac{15\times\sqrt{2}}{90.3}\right)^{1.2}\right]$$
$$= 20.176 \text{ kg/cm}^2 < 40 \text{ kg/cm}^2$$

Therefore, thickness of slab is adequate to resist corner stress.

5. Railway Engineering and Tunnel

Level-1

5.1 **Calculate the maximum permissible train load that can be pulled by locomotive having six pairs of driving wheels carrying an axle load of 20 tonnes each. Train has to run a speed of 90 kmph on a straight level broad gauge track. Also calculate the decrease in speed if the train climbs up a gradient of 1 in 180. If the train climbs the gradient with a 3° curve, then calculate the speed reduction?**

[8 Marks]

Sol:

Let hauling capacity of locomotive = $\frac{1}{6}\times$ load on driving wheels

$$= \frac{1}{6}\times 6\times 20 = 20 \text{ tonnes}$$

On a straight level track, train resistance is given by

0.0016 W + 0.00008 WV + 0.0000006 WV^2

W = Weight of train in tonnes including weight of locomotive

V = Speed of train in kmph

$$\therefore \quad 20 = 0.0016\text{ W} + 0.00008\text{ W}\times 90 + 0.0000006\, W\times 90^2$$

$$W = 1464.1 \text{ tonnes}$$

$$\therefore \quad \text{Permissible load} = 1464 \text{ tonnes}$$

Case 2

When train goes up the gradient of 1 in 180 i.e. $\frac{1}{180}\times 100 = 0.556\%$

$$\text{Total train resistance} = 0.0016W + 0.00008WV + 0.0000006WV^2 + W\times\frac{0.556}{100}$$

$$20 = (0.0016\times 1464) + (0.00008\times 1464\times V) + (0.0000006\times 1464V^2) + \left(\frac{0.556}{100}\times 1464\right) + \frac{0.04}{100}\times 1464\times 3$$

$$7.761 = 0.1171V + 0.000878V^2$$

$$\therefore \quad V = 48.58 \text{ kmph}$$

Hence reduction of (90 – 48.58) = 41.42 kmph occurs due to case 3.

5.2 **Design a diamond crossing between two BG tracks crossing each other at an angle of 1 in 12. [Consider, G = 1.676 m]**

[10 marks]

Sol:

Let α = Angle of crossing, N = Number of crossing, G = 1.676 m

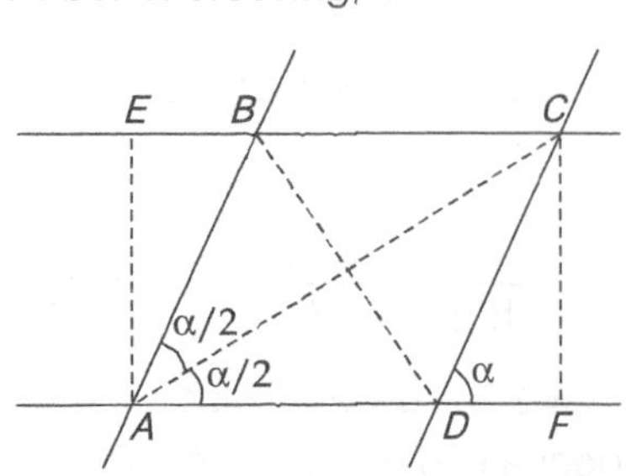

$$\because \quad \sin\alpha = \frac{FC}{DC}$$

$$\Rightarrow \quad \sin\alpha = \frac{G}{DC}$$

$$\Rightarrow \quad DC = G \operatorname{cosec}\alpha$$

$$AB = BC = CD = DA = G \operatorname{cosec}\alpha$$

Also,

$$N = \cot\alpha$$

$$\Rightarrow \quad \alpha = \cot^{-1}(12) = 4.764°$$

(i)

$$AB = BC = CD = DA = 1.676 \operatorname{cosec}(4.764°)$$
$$= 20.18 \text{ m}$$

(ii)

$$EB = DF = G\cot\alpha$$
$$= 1.676 \times 12 = 20.11 \text{ m}$$

(iii)

$$BD = G\sec\frac{\alpha}{2}$$
$$= \frac{1.676}{\cos\frac{\alpha}{2}} = \frac{1.676}{\cos\left(\frac{4.764°}{2}\right)} = 1.677 \text{ m}$$

(iv)

$$AC = G\operatorname{cosec}\frac{\alpha}{2} = \frac{1.676}{\sin\left(\frac{4.764°}{2}\right)} = 40.33 \text{ m}$$

5.3 **What should be the length of track (a) to overcome temperature stress, (b) to prevent creep for equilibrium?**

Given that: Cross-sectional area of rail (A) = 60 cm^2; Coefficient of thermal expansion (α) = 1.15 × 10^{-5}/°C; Modulus of elasticity of steel = 21.5 × 10^5 kg/cm^2; Rise in temperature (t) = 30°C; Resistance to track movement = 700 kg/km.

[10 Marks]

Sol:

(ii) Force required to prevent likely expension dut to temperature.

$$F = \alpha\, tAE$$
$$= 1.15 \times 10^{-5} \times 21.5 \times 10^5 \times 60 \times 30$$
$$= 44505 \text{ kg}$$

(a) Length of track required to overcome temperature stress

$$(L_t) = \frac{44505}{700} = 63.578 \text{ km}$$
$$\simeq 63.6 \text{ km}$$

(b) To prevent creep for equilibrium, the length of welded track,

$$= 2 \times L_t$$
$$= 2 \times 63.6 = 127.2 \text{ km}$$

5.4 Equilibrium cant is provided for speed of 80 kmph on a B.G. 2° Curve

(i) Calculate the allowable equilibrium cant.

(ii) Allowing a maximum cant deficiency, what would be the maximum permissible speed on the track?

[12 Marks]

Sol:

$$e_{\text{quilibrium}} = \frac{GV^2}{127R} = \frac{1.676 \times 80^2}{127 \times \dfrac{1720}{2}} = 0.0982 \text{ m} = 9.82 \text{ cm}$$

$$\text{Theoretical cant} = e_{eq} + \text{cant deficiency}$$
$$= 9.82 + 7.60 = 17.42 \text{ cm}$$

$$\therefore \quad \text{Equating } \frac{GV^2}{127R} = 17.42 \text{ cm}$$

$$\frac{167.6 \times V^2}{127 \times \dfrac{1720}{2}} = 17.42$$

$$\therefore \quad V = 106.5 \text{ kmph}$$

According to Railway Board's formula,

$$V_{max} = 4.35\sqrt{R-67} \text{ kmph}$$

$$= 4.35\sqrt{\frac{1720}{2} - 67} = 122.5 \text{ kmph}$$

Hence maximum allowable speed = 106.5 kmph

5.5 For the following given data, calculate the length of track to overcome temperature stress and also to prevent creep for equilibrium.

$A = 50$ cm², $\alpha = 1.5 \times 10^{-5}$/°C, $E = 22.5 \times 10^5$ kg/cm², $\Delta T = 40$°C

Take, resistance to track movement as 850 kg/cm²

[6 Marks]

Sol:

$$\text{Force to prevent expansion, } F = [\alpha \Delta TEA]$$
$$= [1.5 \times 10^{-5} \times 40 \times 22.5 \times 10^5 \times 50]$$
$$= 67500 \text{ kg}$$

- Length of track needed to overcome the stress due to temperature = $\dfrac{67500}{850} = 79341$ km
- To prevent creep for equilibrium, the length of welded track = $2 \times L = 2 \times 79.41 = 158.82$ km

5.6 What is composite steeper index? Explain in brief.

[6 Marks]

Sol:

C.S.I. or composite steeper index is an index to determine the suitability of a particular timber for use as a steeper. This index measures the mechanical strength of the timber, derived from its composite properties of strength and hardness. The C.S.I. is calculated by

$$CSI = \frac{S + 10H}{20}$$

S = Strength index, both the green and dry timber at 12% moisture content

H = Hardness index, both for green and dry timber at 12% moisture content

Minimum CSI values as per Indian Railways

Type of Sleeper	Minimum CSI
Track Sleeper	783
Crossing Sleeper	1352
Bridge Sleeper	1455

5.7 Determine all the necessary elements required to set out a 1 in 10 turnout which takes off from B.G. track with its curve starting from toe of the switch i.e. being tangential to the gauge face of the outer main rail and passing through T.N.C. Take heel divergence, d = 12.1 cm.

[8 Marks]

Sol:

Here, N = 10.0, G = 1.676 m, d = 12.1 cm = 0.121 m

I : Curve lead (CL) $= 2GN = 2 \times 1.676 \times 10 = 33.52$ m

II : $$\text{Radius, } R = R_0 - \frac{G}{2}$$

where, $$R_0 = 2GN^2 + 1.5G = 2 \times 1.676 \times 10^2 + 1.5 \times 1.676 = 253.91 \text{ m}$$

$$\therefore \quad R = 253.91 - \frac{1.676}{2} = 253.08 \text{ m}$$

III : $$\text{Switch lead } (SL) = \sqrt{2R_0 d} = \sqrt{2 \times 253.91 \times 121} = 7.84 \text{ m}$$

IV : Lead (L) = CL – SL = 33.52 – 7.84 = 25.68 m

Hence, note :

- Overall length of curve = 33.52 m
- Radius of outer curve = 253.91 m

Knowing the tangent point and above data curve can be set by perpendicular offsets from tangent point.

Level-2

5.8 A turnout is to be laid off a straight BG track with a 1 in 12 crossing. Determine the lead and radius of the turnout with the help of the following data: heel divergence (d) = 133 mm, crossing angle (α) = 4° 45′ 49″, switch angle (β) =1° 8′ 00″, straight length between the theoretical nose of crossing and the tangent point of crossing (h) = 1.418 m. Use the IRS method.

[12 marks]

Sol:

α = 4° 45′ 49″ $\qquad \beta$ = 1° 8′ 0″

G = 1.676 m $\qquad d$ = 0.133 m

N = 12 $\qquad h$ = 1.418 m

Crossing lead, $L = h\cos\alpha + (G - d - h\sin\alpha)\cot\dfrac{\alpha+\beta}{2}$

$= 1.418 \cos 4° 45' 49''$

$+ (1.676 - 0.133 - 1.418 \sin 4° 45' 49'') \times \cot 2° 56' 54''$

$= 1.418 \times 0.9965 + 1.425 \times 19.415$

$= 29.079$ m

Radius, $R = \dfrac{G - \alpha - h\sin\alpha}{2\sin\left(\dfrac{\alpha+\beta}{2}\right)\sin\left(\dfrac{\alpha-\beta}{2}\right)}$

$= \dfrac{1.676 - 0.133 - 1.418\sin 4°45'49''}{2(\sin 1°48'54.5'')(\sin 2°58'54'')} = 437.4$ m

5.9 **Calculate the steepest gradient permissible for the following data:**
Number of wagons in train = 25
Weight of each wagon = 15 tonnes
Rolling resistance of wagon =2.8 kg/tonnes
Speed of the train = 60 kmph
Weight of locomotive = 140 tonnes
Tractive effort of locomotive = 14 tonnes
Rolling resistance of locomotive = 4 kg/tonnes

[12 Marks]

Sol:

Total weight of train = Weight of (Wagons + Locomotive)
= 25 × 15 + 140 = 515 tonnes

Rolling resistance of each wagon = 2.8 × 15 = 42 kg

Rolling resistance of all wagons = 42 × 25 = 1050 kg = 1.05 tonnes

Rolling resistance of locomotive = 140 × 4 = 560 kg = 0.56 tonnes

Total resistance of locomotive and wagon = (1.05 + 0.56) = 1.61 tonnes

Resistance depending on speed = 0.00008 WV
= 0.00008 × 515 × 60
= 2.472 tonnes

Atmospheric resistance = $0.0000006\, WV^2$
$= 0.0000006 \times 515 \times 60^2$
= 1.112 tonnes

Train resistance on a straight track = Rolling resistance + Resistance due to speed + Atmospheric resistance + Resistance due to gradient

$$= 1.61 + 2.472 + 1.112 + \frac{1}{N} \times 515$$

Equating resistance to tractive effort of locomotive.

$$14 = 5.194 + \frac{1}{N} \times 515$$

$$N = 58.50 \simeq 60 \text{ approximately}$$

Hence the train can travel the maximum steepest gradient = 1 in 60

5.10 **A 7° curve diverges from 4° main curve in reverse direction in B.G. yard. If the speed on the branch line is limited to 30 kmph, determine the restricted speed on mainline.**

[12 Marks]

Sol:

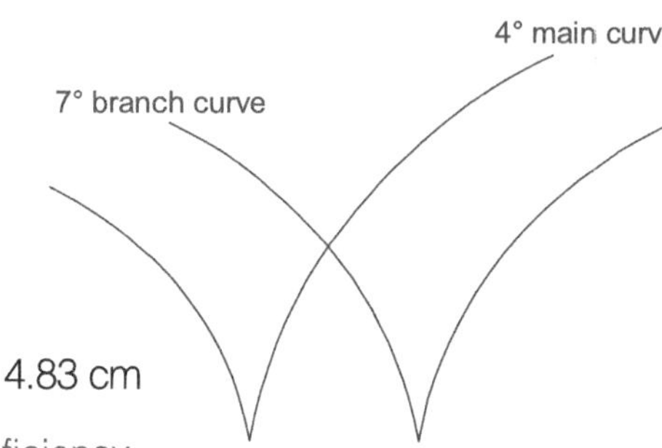

Superelevation on branch line, $e = \frac{GV^2}{127R}$

Here, G = 1.6 + 6 m, V = 40 kmph, $R = \frac{1720}{7} = 245.71$ m

Equilibrium superelevation on branch line

$$e = \frac{1.676 \times 30^2}{127 \times 245.71} \times 100 \text{ cm} = 4.83 \text{ cm}$$

Equilibrium superelevation = Negative cant. + Cant. Deficiency

$$4.83 = x + 7.60$$

$$\therefore \quad x = -2.77 \text{ cm}$$

Note : Negative cant is the maximum permissible superelevation on the main line i.e. 2.77 cm.

Theoretical equilibrium superelevation for main line = (2.77 + 7.6) cm = 10.37 cm

Hence, speed on the main track, V can be calculated from

$$e = \frac{GV^2}{127R}$$

$$0.1037 = \frac{1.676 \times V^2}{127 \times \frac{1720}{4}}$$

$$V = 58.13 \text{ kmph}$$

So, max (restricted) speed on main track = 58 kmph

5.11 **Calculate the length of transition curve for a B.G. track having a 3° curvature and a cant of 10 cm. The maximum allowable speed on curve is 90 kmph.**

[12 Marks]

Sol:

The length of T.C. is maximum of

(i) $7.20\, e_{cm}$

(ii) $0.073\, D\, V_{max}$

(iii) $0.073 e\, V_{max}$

(i) $$L = 7.20 e_{cm}$$

$$7.20 \times 10 = 72 \text{ m}$$

(ii) $$L = 0.073\, DV_{max}$$ where D = cant deficiency in cm

$$= 0.073 \times 7.6 \times 90$$

$$= 49.93 \text{ m}$$

(iii) $$L = 0.073 e_{cm} \times V_{max}$$

$$= 0.073 \times 10 \times 90$$

$$= 65.7 \text{ cm}$$

$\therefore$ Maximum value among (i), (ii) and (iii) = 72 m

Say $L = 75$ m

We know, $y = \frac{x^3}{6RL}$ is the cubic parabola equation adopted for transition curves setting usually.

$$R = \frac{1720}{D} = \frac{1720}{3} = 573.33 \text{ m}$$

$$\text{Offset at 10 m} = \frac{10^3}{6 \times 573.33 \times 75} = 3.875 \text{ mm}$$

$$\text{at 20 m} = \frac{20^3}{6 \times 573.33 \times 75} = 31.0 \text{ mm}$$

$$\text{at 30 m} = \frac{30^3}{6 \times 573.33 \times 75} = 104.65 \text{ mm}$$

$$\text{at 40 m} = \frac{40^3}{6 \times 573.33 \times 75} = 248.0 \text{ mm}$$

$$\text{at 50 m} = \frac{50^3}{6 \times 573.33 \times 75} = 484.5 \text{ mm}$$

$$\text{at 60 m} = \frac{60^3}{6 \times 573.33 \times 75} = 837.2 \text{ mm}$$

$$\text{at 70 m} = \frac{70^3}{6 \times 573.33 \times 75} = 1424.4 \text{ mm}$$

$$\text{at 75 m} = \frac{75^3}{6 \times 573.33 \times 75} = 1329.5 \text{ mm}$$

$$\text{Shift, } S = \frac{L^2}{24R} = \frac{75^2}{24 \times 573.33} = 408.7 \text{ mm}$$

5.12 Calculate the allowable speed on high speed broad gauge track, having following details:
Degree of curve = 1.5°
Superelevation = 90 mm
Length of transition curve = 150 m
Maximum speed likely to be approved = 150 kmph

[15 Marks]

Sol:

$$\text{Radius of curve} = \frac{1720}{D} = \frac{1720}{1.5} = 1146.7 \text{ m} \simeq 1150 \text{ m}$$

(i) Safe speed on the curve for higher speed track

$$V = 4.58\sqrt{R}$$

$$= 4.58\sqrt{1150} = 155.3 \text{ kmph} = 155 \text{ kmph}$$

(ii) From super elevation criterion

Actual superelevation, e = 9 cm

Maximum cant deficiency for H.S. track = 10 cm

∴ Theoretical superelevation, e_{th} = 19 cm

∴ For this e_{th} $$18 = \frac{GV^2}{127R}$$

$$\frac{167.6 \times V^2}{127 \times 1150} = 19$$

∴ $$V = 128.7 \text{ kmph} \simeq 128 \text{ kmph}$$

(iii) From length of transition curve

(a) $$L = \frac{eV_{max}}{198}, \ e \text{ in mm and } L \text{ in metres}$$

$$L = 150 \text{ m}, \ e = 9 \text{ cm} \simeq 90 \text{ mm}$$

$$\therefore \quad V_{max} = \frac{198 \times 150}{90} = 330 \text{ kmph}$$

(b) $$L = \frac{DV_{max}}{198}, \ D \text{ in mm and } L \text{ in metres}$$

$$150 = \frac{100 \times V_{max}}{198}$$

$$\therefore \quad V_{max} = 297 \text{ kmph}$$

Hence, maximum allowable speed is minimum of

- Safe speed on the curve = 155 kmph
- V_{max}, from superelevation condition = 128 kmph
- V_{max} from length of transition curve = 297 kmph
- Maximum speed that can be approved = 125 kmph

Hence, V_{max} = 125 kmph

5.13 Discuss various methods of tunnel construction in rocks.

[12 Marks]

Sol:

The following methods are commonly used for tunneling in rocks :

(1) **Full face method :** In this method, whole section of the tunnel is attacked at the same time. It is suitable for tunnels of small cross-section area upto 3 m diameter.

Advantages:

- Work is expedited since full section is tackled at once.
- Mucking track can be laid progressively along with the tunneling.
- Tunneling is continuous.

Disadvantages:

- Heavy mechanical equipments are used.
- Method is unsuitable for unstable rocks.
- Limited for short span only.

(2) **Heading and Bench Method :** In this method, top portion is driven in advance of the bottom portion. It is used when quantity of rocks is not satisfactory and tunnel section is very large.

Advantages:

- Simultaneous drilling and mucking is possible.
- Requires less explosive powder then full face method.

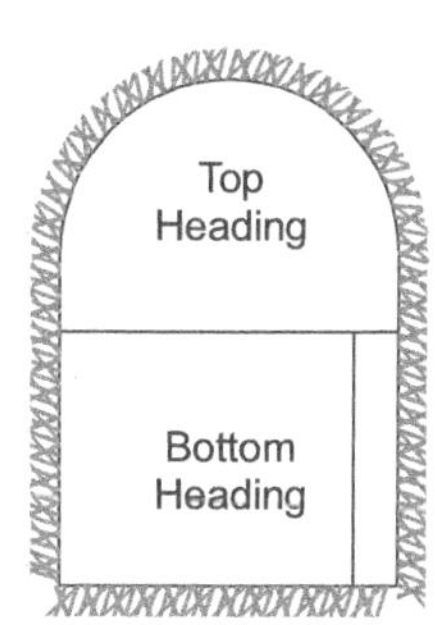

(3) **Drift method :** In this method, rock tunneling is carried out first in smaller section and then it is widened. Drift may be classified as centre, bottom or top drift depending upon its relative position with reference to main bore.

Advantages:

- It allows correction measures in case of presence of excessive water.
- Drift helps in providing ventilation.
- Explosive quantity is also less.
- Installation of timber supports becomes easy.

Disadvantages:

- Driving the main bore is delayed until the drift portion is completed.
- Cost of drilling and mucking in small drift is higher.

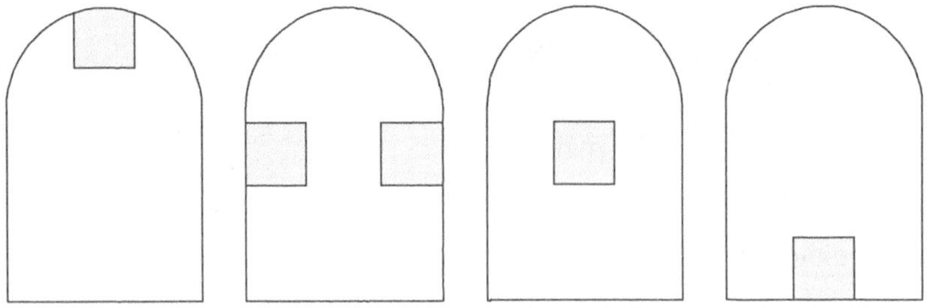

5.14 Discuss in brief the reasons and implications of creep in a rail track. Also discuss the creep prevention and remedial measures.

[20 Marks]

Sol:

Effects of creep in a rail track:

(a) Creep may leads to buckling of track in lateral directions.

(b) Square geometry arrangement of sleepers get disturbed leads to distorted gauge and alignment.

(c) Rail joints are opened out and bolt holes get elongated leads to premature fracture of fish plate and bolts.

(d) Point and crossing get disturbed.

(e) The joints get continuously jammed, and maintenance and replacement become difficult.

Creep: Creep is defined as longitudinal movement of the rail with respect to sleepers due to movement of traffic over a period of time.

Reason of creep:

(a) Wave motion theory due to wheel load a wave in rail is created in direction of traffic, resulting in creep.

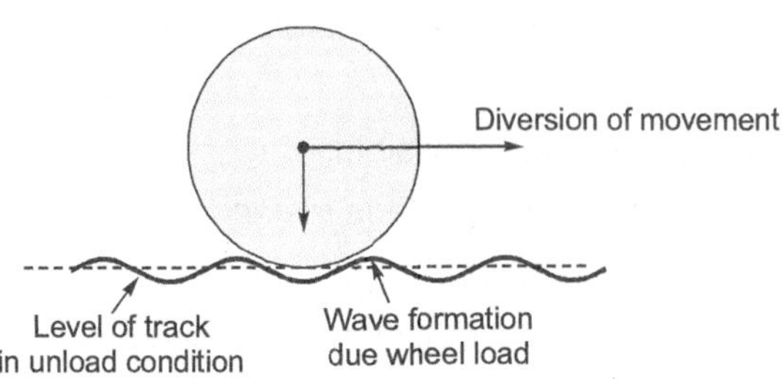

(b) **Percussion theory:** Due to horizontal component of (load + kinetic) force F, at expansion joint, creep is observed.

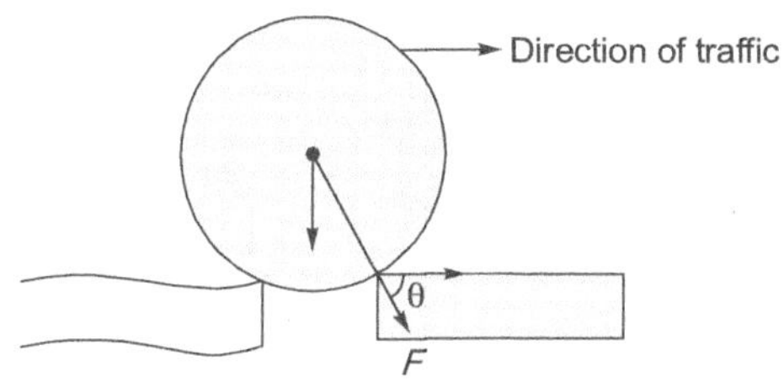

(c) **Dragging Theory:** Due to fraction force between wheel & rail a drag force generate opposite to movement of train, resulting in creep.

(d) Due to braking & acceleration (skidding + slipping).

Following are some of avoidable causes of creep:

(a) Rails are not property fastened with sleepers.

(b) Loose packing of ballast around sleepers.

(c) Improper expansion gaps.

(d) Too light rail for heavy traffic

(e) Sharp gradients and sharp curves.

(f) Uneven spacing of sleepers etc.

Preventive and remedial measures of creep:

(a) Rails should be firmly held to sleepers.

(b) Bearing load of fastenings should always be slightly more than the ballast resistance.

(c) Ballast should be properly packed specially around sleepers and shoulders.

(d) Use of creep anchors at adequate internal. (Fair 'T' and fir 'V' types).

(e) When creep became excessive (> 150 mm) causing maintenance problems. The same should be adjusted by pulling back manually or mechanically during pulling back operation survey of expansion joint, gaps and rail to rail joint with sleepers should be carried out properly.

Measurement of creep:

(a) Creep posts are erected every kilometer on either side of the track and position of joints is marked on one of the posts.

(b) Then measurement of creep is taken at an internal of about 3 month.

(c) Creep in excess of 150 mm should not be allowed on a straight and levelled track.

(d) In approaches of points and crossing. There should be no creep allowed.

6. Airport Planning and Design

Level-1

6.1 Refer the following data for proposed longitudinal section of runway.

End to end of runway	Gradient(%)
0.0 to 5.0 chains	1.2%
5.0 to 25.0 chains	–1.0%
25.0 to 40.0 chains	+0.7%
40.0 to 50.0 chains	+0.5%

One metric chain is of 20 m. Determine the effective gradient of runway.

[5 Marks]

Sol:

Let reference elevation of start point = 1000 m

Chainage	0	5	25	40	50
Elevation	100.0	101.2	97.2	99.3	100.3

$$\text{II}^{nd}\text{ elevation} = 100.0 + (5-0) \times 20 \text{ m} \times \frac{1.2}{100} = 101.2$$

$$\text{Maximum difference in elevation} = 101.2 - 97.2$$
$$= 4.0 \text{ m}$$
$$\text{Total running length, } L = 50 \times 20 = 1000 \text{ m}$$

$$\therefore \quad \text{Effective gradient} = \frac{4 \text{ m}}{1000 \text{ m}} \times 100 = 0.4\%$$

6.2 Determine the turning radius of a taxiway for Boeing jet, with following data:
Wheel base = 18 m
Tread of main landing gear = 6.5 m
Turning speed = 45 kmph
Coefficient of friction between tyre and pavement = 0.15

[10 Marks]

Sol:

- $$\text{Turning radius} = \frac{V^2}{125f} = \frac{45^2}{125 \times 0.15} = 108 \text{ m}$$
- From Horonjeff's equation, $$R = \frac{0.388W^2}{\left[\frac{T}{2} - S\right]}$$

where, $S = 6 + \frac{6.5}{2} = 9.25$ m, $W = 18$ m, $T = 22.5$ m

$$R = \frac{0.388 \times 18^2}{\frac{22.5}{2} - 9.25} = 62.86 \text{ m}$$

Turning radius is maximum of following three values.

(i) 108 m
(ii) 62.86 m
(iii) 120 m for subsonic
180 m for supersonic

$\therefore$ For Boeing supersonic jet $R_{min} = 180$ m

6.3 The runway gradation map indicates that these exists a rising gradient of 1.2% meeting a falling of 0.8%. There is again an upgrade of 0.80%. Design the runway profile as per FAA specifications.

[6 Marks]

Sol:

Description	Small Airport	Large Airport
Maximum grade change so that (a) and (b) should not exceed	2%	1.5%
Length of vertical curve (L_1 or L_2) for each one percent change	90 m	300 m
Distance between points of intersection of grade lines (d)	75 (a + b) m	300 (a + b) m

Here, $a = x - y = 1.2 - (-0.80) = 2.0\%,$

$b = y - z = -(0.80 + 0.80) = 1.60\%$

Since $a > 1.5\%$ for large airports. The design is carried out for small airports.

$$L_1 = 90 \times 2.0 = 180 \text{ m}$$

$$L_2 = 90 \times 1.6 = 144 \text{ m}$$

$$D = 75\,(a + b) = 75 \times 3.6 = 270 \text{ m}$$

6.4 **What is basic runway length? Calculate the capacity of a single gate which is available to all classes of aircraft as per details given in table below:**

Aircraft class	Mix %	Average occupancy time (min)
1	10	20
2	30	40
3	60	60

[10 marks]

Sol:

Basic runway length : It is the length of runway under the following assumed conditions at the airport:

- Airport altitude is at sea level.
- Temperature at the airport is standard (15°C).
- Runway is levelled in the longitudinal direction.
- No wind is blowing on runway.
- Aircraft is loaded to its full loading capacity.
- There is no wind blowing enroute to the destination.
- Enroute temperature is standard.

The basic runway length is determined from the performance characteristics of the aircraft using the airport. The following cases are usually considered:

- Normal landing case
- Normal take-off case
- Engine failure case

Since each gate is available for all aircrafts. The gate capacity for a single gate is given by

$$\frac{1}{\text{Weighted service time}} = \frac{1}{(0.1 \times 20) + (0.3 \times 40) + (0.6 \times 60)}$$

$$= 0.02 \text{ aircraft/min/gate}$$

Level-2

6.5 Length of a runway in standard conditions at an airport is 2500 m. The airport is at an elevation of 40 m above the MSL. The airport reference temperature (ART) is 94°C. The construction plan provides following data :

End to end of runway	Gradient(%)
0 to 300	+1.20
300 to 900	–0.60
900 to 1500	+0.50
1500 to 1800	+1.50
1800 to 2100	–0.75
2100 to 2700	–0.25
2700 to 3300	–0.40

Determine the length of runway. Apply correction for elevation and temperature as ICAO and for gradient as per FAA considerations.

[20 Marks]

Sol:

(i) Correction for elevation $= \frac{7}{100} \times 2500 \times \frac{400}{300} = 233.3$ m

Corrected length $= 2500 + 233.3 = 2733.3$ m

(ii) Correction for temperature

Standard atmospheric temperature at 400 m

$$\text{SAT} = 15 - 0.0065 \times 400 = 12.4°\text{C}$$

$$\therefore \quad \text{Rise} = 34 - 12.4 = 21.6°\text{C}$$

$$\text{Correction} = \frac{21.66}{100} \times 2733.3 = 590.4 \text{ m}$$

Corrected length $= 3323.7$ m

(iii) Check for total correction of elevation and temperature

$$\text{Total correction} = \frac{3323.7 - 2500}{2500} \times 100 = 32.94\%$$

Hence, less than 35%

$\therefore$ OK according to ICAO regulations.

(iv) Correction for gradient

Elevation of different points as per proposed gradients, taking reference elevation = 100 m

Distance	0	300	900	1500	1800	2100	2700	3300
Elevation	100	103.6	100	103	107.5	105.25	103.75	101.35

Length after correction $= 3325$ m

Maximum difference in elevation $= 107.5 - 100 = 7.5$ m

$$\therefore \quad \text{Effective gradient} = \frac{7.5}{3325} \times 100 = 0.226\%$$

$$\text{Correction for gradient} = \frac{20}{100} \times 3325 \times 0.226 = 150.3 \text{ m}$$

$$\therefore \quad \text{Revise length} = 3325 + 150.3 = 3475.3 \text{ m} \simeq 3475 \text{ m}$$

$$\therefore \quad \text{Design runway length} = 3475 \text{ m}$$

■■■■